Methods in Cell Biology

VOLUME 83

Cell Mechanics

Series Editors

Leslie Wilson

Department of Molecular, Cellular and Developmental Biology
University of California
Santa Barbara, California

Paul Matsudaira

Whitehead Institute for Biomedical Research
Department of Biology
Division of Biological Engineering
Massachusetts Institute of Technology
Cambridge, Massachusetts

Methods in Cell Biology

VOLUME 83

Cell Mechanics

Edited by

Yu-Li Wang

Department of Physiology
University of Massachusetts Medical School
Worcester, Massachusetts

Dennis E. Discher

Department of Chemical and Biomolecular Engineering
University of Pennsylvania
Philadelphia, Pennsylvania

AMSTERDAM • BOSTON • HEIDELBERG • LONDON
NEW YORK • OXFORD • PARIS • SAN DIEGO
SAN FRANCISCO • SINGAPORE • SYDNEY • TOKYO
Academic Press is an imprint of Elsevier

Cover Photo Credit: Schematic representation of intracellular forces
at the leading edge of migrating adherent cells. Image created by
James Lim, The Scripps Research Institute. From Chapter 9, Figure 1.

Academic Press is an imprint of Elsevier
525 B Street, Suite 1900, San Diego, California 92101-4495, USA
84 Theobald's Road, London WC1X 8RR, UK

For information on all Academic Press publications
visit our Web site at www.books.elsevier.com

ISBN: 978-0-12-370500-6

PRINTED IN THE UNITED STATES OF AMERICA
07   08   09   10   9   8   7   6   5   4   3   2   1

CONTENTS

PART II   Subcellular Mechanical Properties and Activities

PART III   Cellular and Embryonic Mechanical Properties and Activities

PART IV   Mechanical Stimuli to Cells

CONTRIBUTORS

Numbers in parentheses indicate the pages on which the authors' contributions begin.

Karen A. Beningo (29), Department of Biology, Wayne State University, Detroit, Michigan 48202

Timo Betz (495), Lehrstuhl für die Physik Weicher Materie, Fakultät für Physik und Geowissenschaften, Universität Leipzig, Linnéstr. 5, Leipzig D-04103, Germany

Nicolas Biais (473), Department of Biological Sciences, Columbia University, New York, New York 10027

Andrew D. Bicek (237), Department of Biomedical Engineering, University of Minnesota, Minneapolis, Minnesota 55455

James P. Butler (179), Physiology Program, Harvard School of Public Health, Boston, Massachusetts 02115

Christopher S. Chen (313), Departments of Bioengineering and Physiology, University of Pennsylvania, Philadelphia, Pennsylvania 19104

John C. Crocker (141), Department of Chemical and Biomolecular Engineering, Institute for Medicine and Engineering, University of Pennsylvania, Philadelphia, Pennsylvania 19104

Alfred J. Crosby (67), Polymer Science and Engineering Department, University of Massachusetts, Amherst, Massachusetts 01003

Kris Noel Dahl (269), Biomedical Engineering and Chemical Engineering, Carnegie Mellon University, Pittsburgh, Pennsylvania 15213

Brian R. Daniels (115), Department of Chemical and Biomolecular Engineering, The Johns Hopkins University, Baltimore, Maryland 21218

Gaudenz Danuser (199), The Scripps Research Institute, La Jolla, California 92037

Lance Davidson (425), Department of Bioengineering, University of Pittsburgh, Pittsburgh, Pennsylvania 15260

Dennis E. Discher (47, 269, 521), Department of Chemical and Biomolecular Engineering and Biophysical Engineering and Polymers Laboratory, School of Engineering and Applied Science, University of Pennsylvania, Philadelphia, Pennsylvania 19104

Susanne Ebert (397), Institut für Experimentelle Physik I, Universität Leipzig, Linnéstrasse 5, 04103 Leipzig, Germany

Allen Ehrlicher (495), Lehrstuhl für die Physik Weicher Materie, Fakultät für Physik und Geowissenschaften, Universität Leipzig, Linnéstr. 5, Leipzig D-04103, Germany

Adam J. Engler (47, 521), Department of Chemical and Biomolecular Engineering and Biophysical Engineering and Polymers Laboratory, School of Engineering and Applied Science, University of Pennsylvania, Philadelphia, Pennsylvania 19104

Evan Evans (373), Department of Biomedical Engineering, Boston University, Boston, Massachusetts 02215; Department of Physics and Astronomy, University of British Columbia, Vancouver, British Columbia, Canada V6T 1Z1; Department of Pathology and Laboratory Medicine, University of British Columbia, Vancouver, British Columbia, Canada V6T 2B5

Kristian Franze (495), Lehrstuhl für die Physik Weicher Materie, Fakultät für Physik und Geowissenschaften, Universität Leipzig, Linnéstr. 5, Leipzig D-04103, Germany

Margo T. Frey (47), Department of Physiology, University of Massachusetts Medical School, Worcester, Massachusetts 01605

Michael Gögler (495), Lehrstuhl für die Physik Weicher Materie, Fakultät für Physik und Geowissenschaften, Universität Leipzig, Linnéstr. 5, Leipzig D-04103, Germany

Andrés J. García (329), Woodruff School of Mechanical Engineering, Petit Institute for Bioengineering and Bioscience, Georgia Institute of Technology, Atlanta, Georgia 30332

Margaret Gardel (199), The Scripps Research Institute, La Jolla, California 92037

Benjamin Geiger (89), Department of Molecular Cell Biology, Weizmann Institute of Science, Rehovot 76100, Israel

Penelope C. Georges (1, 27), Department of Bioengineering, Institute for Medicine and Engineering, University of Pennsylvania, Philadelphia, Pennsylvania 19104

Jochen Guck (397), Institut für Experimentelle Physik I, Universität Leipzig, Linnéstrasse 5, 04103 Leipzig, Germany

Brenton D. Hoffman (141), Department of Chemical and Biomolecular Engineering, Institute for Medicine and Engineering, University of Pennsylvania, Philadelphia, Pennsylvania 19104

Shaohua Hu (179), Physiology Program, Harvard School of Public Health, Boston, Massachusetts 02115

Søren Hvidt (3), Department of Chemistry, Roskilde University, Roskilde DK-4000, Denmark

Donald. E. Ingber (443), Vascular Biology Program, Departments of Pathology and Surgery, Children's Hospital and Harvard Medical School, Boston, Massachusetts 02115

Paul A. Janmey (3, 29), Departments of Physiology and Physics and Department of Bioengineering, Institute for Medicine and Engineering, University of Pennsylvania, Philadelphia, Pennsylvania 19104

Lin Ji (199), The Scripps Research Institute, La Jolla, California 92037

Kandice R. Johnson (547), Institute for Medicine and Engineering and Department of Bioengineering, University of Pennsylvania, Philadelphia, Pennsylvania 19104

Josef Käs (495), Lehrstuhl für die Physik Weicher Materie, Fakultät für Physik und Geowissenschaften, Universität Leipzig, Linnéstr. 5, Leipzig D-04103, Germany

Roger D. Kamm (269), Departments of Biological Engineering and Mechanical Engineering, Massachusetts Institute of Technology, Cambridge, Massachusetts 02139

Casey E. Kandow (29), Department of Biology, Wayne State University, Detroit, Michigan 48202

Ray Keller (425), Department of Biology and Morphogenesis and Regenerative Medicine Institute, University of Virginia, Charlottesville, Virginia 22904

Koji Kinoshita (373), Department of Biomedical Engineering, Boston University, Boston, Massachusetts 02215

Daniel Koch (495), Lehrstuhl für die Physik Weicher Materie, Fakultät für Physik und Geowissenschaften, Universität Leipzig, Linnéstr. 5, Leipzig D-04103, Germany

Thomas P. Kole (115), Department of Chemical and Biomolecular Engineering, The Johns Hopkins University, Baltimore, Maryland 21218

Daniel M. Kroll (237), Department of Physics, North Dakota State University, Fargo, North Dakota 58105

Sanjay Kumar[*] (443), Vascular Biology Program, Departments of Pathology and Surgery, Children's Hospital and Harvard Medical School, Boston, Massachusetts 02115

Jan Lammerding (269), Department of Medicine, Cardiovascular Division, Brigham and Women's Hospital/Harvard Medical School, Boston, Massachusetts 02115

Jerry S. H. Lee (115), Department of Chemical and Biomolecular Engineering, The Johns Hopkins University, Baltimore, Maryland 21218

Juliet Lee (47, 297), Department of Molecular and Cell Biology, University of Connecticut, Storrs, Connecticut 06269

Jennifer L. Leight (547), Institute for Medicine and Engineering and Department of Bioengineering, University of Pennsylvania, Philadelphia, Pennsylvania 19104

Tanmay P. Lele[†] (443), Vascular Biology Program, Departments of Pathology and Surgery, Children's Hospital and Harvard Medical School, Boston, Massachusetts 02115

Bryan Lincoln (397), Institut für Experimentelle Physik I, Universität Leipzig, Linnéstrasse 5, 04103 Leipzig, Germany

Dinah Loerke (199), The Scripps Research Institute, La Jolla, California 92037

Yunbi Lu (495), Lehrstuhl für die Physik Weicher Materie, Fakultät für Physik und Geowissenschaften, Universität Leipzig, Linnéstr. 5, Leipzig D-04103, Germany

Benjamin D. Matthews (443), Vascular Biology Program, Departments of Pathology and Surgery, Children's Hospital and Harvard Medical School, Boston, Massachusetts 02115; Department of Pediatrics, Massachusetts General Hospital and Harvard Medical School, Boston, Massachusetts 02114

Kristin E. Michael (329), Woodruff School of Mechanical Engineering, Petit Institute of Bioengineering and Bioscience, Georgia Institute of Technology, Atlanta, Georgia 30332

Martin Montoya-Zavala (443), Vascular Biology Program, Departments of Pathology and Surgery, Children's Hospital and Harvard Medical School, Boston, Massachusetts 02115

David J. Odde (237), Department of Biomedical Engineering, University of Minnesota, Minneapolis, Minnesota 55455

Present addresses:
[*]Department of Bioengineering, University of California, Berkeley, Berkeley, California 94720.
[†]Department of Chemical Engineering, University of Florida, Gainesville, Florida 32611.

Darryl Overby[‡] (443), Vascular Biology Program, Departments of Pathology and Surgery, Children's Hospital and Harvard Medical School, Boston, Massachusetts 02115

Porntula Panorchan (115), Department of Chemical and Biomolecular Engineering, The Johns Hopkins University, Baltimore, Maryland 21218

Thomas Polte (443), Vascular Biology Program, Departments of Pathology and Surgery, Children's Hospital and Harvard Medical School, Boston, Massachusetts 02115

Manfred Radmacher (347), Institute of Biophysics, University of Bremen, Bremen 28334, Germany

Florian Rehfeldt (521), Biophysical Engineering and Polymers Laboratory, School of Engineering and Applied Science, University of Pennsylvania, Philadelphia, Pennsylvania 19104

Thomas P. Russell (67), Polymer Science and Engineering Department, University of Massachusetts, Amherst, Massachusetts 01003

Stefan Schinkinger (397), Institut für Experimentelle Physik I, Universität Leipzig, Linnéstrasse 5, 04103 Leipzig, Germany

Shamik Sen (521), Biophysical Engineering and Polymers Laboratory, School of Engineering and Applied Science, University of Pennsylvania, Philadelphia, Pennsylvania 19104

Julia E. Sero (443), Vascular Biology Program, Departments of Pathology and Surgery, Children's Hospital and Harvard Medical School, Boston, Massachusetts 02115

Michael Sheetz (473), Department of Biological Sciences, Columbia University, New York, New York 10027

Nathan J. Sniadecki (313), Department of Bioengineering, University of Pennsylvania, Philadelphia, Pennsylvania 19104

Joachim P. Spatz (89), Department of New Materials and Biosystems, Max Planck Institute for Metals Research, Stuttgart, Germany; and Department of Biophysical Chemistry, University of Heidelberg, Heidelberg, Germany

Björn Stuhrmann (495), Lehrstuhl für die Physik Weicher Materie, Fakultät für Physik und Geowissenschaften, Universität Leipzig, Linnéstr. 5, Leipzig D-04103, Germany

Erkan Tüzel (237), Supercomputing Institute and School of Physics and Astronomy, University of Minnesota, Minneapolis, Minnesota 55455

Monica Tanase (473), Department of Biological Sciences, Columbia University, New York, New York 10027

Irene Y. Tsai (67), Polymer Science and Engineering Department, University of Massachusetts, Amherst, Massachusetts 01003

Yiider Tseng (115), Department of Chemical and Biomolecular Engineering, The Johns Hopkins University, Baltimore, Maryland 21218; Department of Chemical Engineering, University of Florida, Gainesville, Florida 32611

Ning Wang (179, 443), Department of Mechanical Science and Engineering, University of Illinois at Urbana-Champaign, Urbana, Illinois 61801

Present address:
[‡]Department of Biomedical Engineering, Tulane University, New Orleans, Louisiana 70118.

Yu-Li Wang (47), Department of Physiology, University of Massachusetts Medical School, Worcester, Massachusetts 01605

Valerie M. Weaver (547), Departments of Surgery and Anatomy, Center for Bioengineering and Tissue Regeneration, University of California, San Francisco, California 94143; Department of Bioengineering, University of Pennsylvania, Philadelphia, Pennsylvania 19104

Denis Wirtz (115), Department of Chemical and Biomolecular Engineering, Department of Materials Science and Engineering, and Howard Hughes Medical Institute Graduate Training Program and Institute for NanoBioTechnology, The Johns Hopkins University, Baltimore, Maryland 21218

Falk Wottawah (397), Institut für Experimentelle Physik I, Universität Leipzig, Linnéstrasse 5, 04103 Leipzig, Germany

Shannon Xia (443), Vascular Biology Program, Departments of Pathology and Surgery, Children's Hospital and Harvard Medical School, Boston, Massachusetts 02115

PREFACE

Most eukaryotic cells possess a well-recognized sense of "taste or smell" mediated by diffusible factors, which bind to cellular receptors and initiate signaling cascades. However, cells also possess an oft-overlooked sense of "touch." This sense is rooted in cell mechanics and dependent on various cytoskeletal structures and processes that detect, modify, and respond to physical parameters of the environment. In particular, many types of cells are found to respond to mechanical force and substrate rigidity, and to transmit mechanical forces through the extracellular matrix or cell–cell adhesions to interact with other cells. Such combination of chemical and mechanical signals provides a wide range of means for cells to communicate with one another and with their extracellular environment, during processes such as embryogenesis and wound healing as well as pathological conditions such as cancerous invasion. Similar principles are also likely to be critical for success in regenerative medicine.

Advances in cell mechanics face many challenges. By its very nature, cell mechanics encompasses the fields of biology, chemistry, physics, engineering, and mathematics in which few scientists are comprehensively trained at an advanced level. In addition, the field, still in its infancy, lacks a defining text, an organized community, or established paradigms, while its experimental approaches are often adapted from multiple fields with substantial needs for customization and validation. However, these challenges also provide a fertile ground for creativity and discovery.

To foster the development in cell mechanics, this book aims not only to provide a collection of research methods, but, more fundamentally, to develop a common language among scientists who share the interest in cell mechanics but enter the field with diverse backgrounds. Indeed, before delving into technical details, it is much more important to introduce to biologists the basic concept of mechanics and rheology, and to physicists and engineers significant biological questions relevant to mechanics. Therefore, although the book represents a collective effort, all of the authors shared a similar vision and goal to explain in plain language the biological problems, the rationale for the approaches, in addition to the methods themselves. In addition, to balance practical utility against conceptual advances, the book has intentionally included both chapters that provide detailed recipes and those that emphasize basic principles. While the former is designed to facilitate the implementation of new laboratory procedures, the latter may find its main impact in promoting future interdisciplinary collaborations.

The book covers several key areas of cell mechanics. The first section introduces basic concepts and novel materials used for better understanding of cell mechanics.

The chapters highlight some of the driving forces that have extended cell biology much beyond what was feasible with conventional Petri dishes. The two subsequent sections deal with the measurements of mechanical characteristics and output, including rheological moduli, adhesion strength and coupling to cytoskeleton, as well as force generation and propagation in scales from subcellular structures to tissues and embryos. These chapters provide a comprehensive view of the broad range of cell biological questions that relate to mechanics, and an equally wide range of approaches from photolithography, atomic force microscopy, single particle tracking, optical trapping, micromanipulation to computation and modeling. The final section is focused on the complementary issue of applying mechanical stimuli to cells. The methods, including optical and magnetic tweezers, flexible materials, and 3D culture environments, are likely to draw increasing attention with the realization that mechanical and topographical stimuli may account for many important differences between cells cultured on Petri dishes and cells *in vivo*. Cell mechanics is thus presented here as a growing set of new or refined tools, protocols, and principles that are central to a deeper understanding of cell biology.

Yu-Li Wang
Denis E. Discher

PART I

Basic Concept and Preparation Culture Substrates for Cell Mechanical Studies

CHAPTER 1

Basic Rheology for Biologists

Paul A. Janmey,[*,†] Penelope C. Georges,[*] and Søren Hvidt[‡]

[*]Department of Bioengineering
Institute for Medicine and Engineering
University of Pennsylvania
Philadelphia, Pennsylvania 19104

[†]Departments of Physiology and Physics
Institute for Medicine and Engineering
University of Pennsylvania
Philadelphia, Pennsylvania 19104

[‡]Department of Chemistry
Roskilde University
Roskilde DK-4000, Denmark

Abstract

Many cellular processes lead to changes in elastic and viscous properties of cells. Rheology is the science that deals with deformation and flow of materials. Fundamental rheologic concepts are explained, and some of the main techniques are discussed. Nonperturbing oscillatory techniques are especially useful for monitoring structure formation including gelation, whereas other techniques such as steady shear flow and creep are useful for determining flow properties. Sample preparation is often a major obstacle, and advantages of different deformation geometries are discussed. Simple biological samples such as purified biopolymers can be investigated with a range of rheologic techniques, and factors affecting gelation of, for example, blood or cytoskeletal proteins can be studied in detail. More complex biological systems such as intact tissues can often only be studied with more qualitative techniques and results. With proper choice of experimental setup, rheologic techniques can give valuable information about cellular systems and dynamics on a timescale that is closely related to biological functions.

I. Introduction and Rationale

Rheology is the study of how materials deform when forces are applied to them. The word is derived from Heraclitus' expression "παντα ρει" translated as "everything flows" which was adapted to create the term rheology in 1929 when the American Society of Rheology was founded (Barnes *et al.*, 2001). Although a separate discipline of rheology is fairly new and often not familiar to biologists, key experimental results and concepts on which it is founded are among the most widely known discoveries in mechanics. The concept of viscoelasticity draws from theories describing ideal materials: that of Robert Hooke's description of ideal elastic behavior in "*True Theory of Elasticity*" (1678) and of Isaac Newton's definition of ideal liquid behavior in "*Principia*" (1687) (Barnes *et al.*, 2001). Hooke described elasticity as a state where the extension of a material produced is proportional to the load. A Hookean solid in a deformed state will remain deformed as long as the applied stress persists. In contrast, Newton's theory describing ideal liquids states that a flow will persist as long as a stress is applied and the stress will be proportional to the rate of flow.

The property that rheological studies are designed to quantify is conceptually simple, namely a value that predicts how a material will deform when a force of a certain magnitude is applied to it in a defined geometry for a given amount of time. Some of the complexity of the rheology field is due to the fact that quantitative measures of deformation are often not simple for samples with anything more than the most basic shapes, and the extent of deformation can depend strongly on how long and how quickly the force is applied. The molecular structure of a material can also change during the deformation, leaving the sample in a different state than it was before deformation, with different properties the next time it is deformed. Accounting for the different scenarios by which forces are applied and deformations are measured requires a certain amount of terminology, some of which has a more precise meaning as a rheological term than it does in common usage, and some of the jargon of rheology is likely to be unfamiliar to biologists. A basic set of terminology is listed in the glossary.

Two different ideal ways in which a material can deform are essentially related to the differences between liquids and solids. When a force is applied to an ideal solid, the material immediately deforms to a certain extent, and then stays put in that deformed state until the force is removed, when it returns to the shape it was before the deformation. This kind of perfectly recoverable deformation is called elastic, as the energy or work that was done to produce the deformation (the force integrated over the displacement) is stored within the deformed material and is entirely recovered when the force is removed. For such a simple elastic material in one dimension, most commonly exemplified by an ideal spring, a single number, the elastic constant—the ratio of force to displacement—suffices to predict how far the spring will stretch the next time a force of any given magnitude is imposed on it for any amount of time.

At the other extreme are ideal liquids. In this case, when a force is applied, the liquid will deform without limit for as long as the force is applied, and the liquid will remain in the deformed state when the force is removed. Although the extent of deformation is not limited, the rate at which the liquid will deform and flow is precisely determined by the magnitude of the force. The ratio of force to rate of deformation defines a viscosity that predicts how fast a liquid will flow whenever a force of any given magnitude is applied, and this rate will remain constant until the force is removed.

Real materials are neither ideal solids nor ideal liquids nor even ideal mixtures of the two. There are always effects due to molecular rearrangements and other factors that complicate deformation, transforming elastic and viscous constants to functions of time, and extent of deformation. Real materials, and especially biological materials, exhibit both elastic and viscous responses and are therefore called viscoelastic. They are also often highly anisotropic, showing different viscoelastic properties when deformed in one direction than when deformed in other directions. The goal of rheological experiments is to quantify the viscoelasticity of a material over as wide a range of time and deformation scales as possible, and ultimately to relate these viscoelastic properties to the molecular structure of the material.

Measuring the rheology of biological materials, cells, and subcellular organelles is currently motivated in part by recent demonstrations that cells respond to physical stimuli as strongly as they do to chemical agonists, and that the morphology and developmental program of cells and multicellular aggregates also depend very strongly on the viscoelasticity of the extracellular matrix on which or within which cells grow. To gain perspective on how cells respond to physical forces, it is essential to determine the mechanical properties of the cell itself as well as those of its environment and extracellular matrix. These allow one to evaluate how much deformation different magnitudes of force can cause, and to identify subcellular structures that will be significantly deformed that might thereby serve as sensors, transducers, or effectors of the forces. In addition, cells change their rheological properties in response to genetic changes or acute responses to signals, and differences in cell stiffness, a characteristic that can be explained in part by studying the material properties of intracellular constituents, show promise as a criterion to differentiate normal from transformed cells (Chapter 17 by Lincoln *et al.*, this volume). In the context of mechanosensing, defining the rheology of a cell and its individual components is as important as determining levels of intracellular messengers before and after stimulation (Bershadsky *et al.*, 2005).

II. Rheological Concepts

Two fundamentally different modes of deformation can result when a force acts on a system. The volume of the system can decrease or the shape of the system can change. The former deformation is measured by the compressibility of the system, and relates to the Poisson ratio, but it will not be discussed further in this chapter, since most biological tissues are hydrated and water is nearly incompressible under the forces generally exerted in biology (Poisson ratio equals 0.5). However, local volume changes as would occur when water flows across a cell membrane or within the interstices of a polymer matrix may play important roles in the microrheological methods described in other chapters of this volume. For most biological systems the macroscopic volume stays constant during deformation, but there will be a change in shape. For example, in a simple shear deformation, as illustrated in Fig. 1A, a force is applied along a surface plane. The system will deform but the volume stays constant. In the case of simple elongation or extensional deformation, the force acts in a direction normal to a plane (Fig. 1B), and increases in length couple to decreases in cross section. Simple shear or elongation yields the same information about the material properties as long as the volume does not change. For some materials, it is more convenient to test materials in other deformation geometries such as bending or twisting.

In order to determine the elasticity or viscosity of a biological or any other system, it is necessary to be able to quantify forces and deformations. For a given material, an applied force will result in different magnitudes of deformation depending on the size of the material. In order to determine material properties, which are independent of size and shape of the system, two key rheological concepts are used: stress and strain. Stress is defined as force per area and has the SI unit of Pa (N/m^2). In older literature

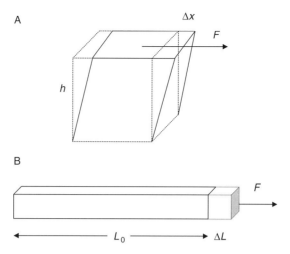

Fig. 1 Two common deformation geometries. In simple shear (A), a force F is applied along the surface of an undeformed box (fine lines) with height h. The top of the box is deformed Δx due to the force. In simple extension or elongation (B), the force is applied perpendicular to a surface. The undeformed length of a box, strip, or rod is L_0, and the extension due to the force is ΔL.

and in studies of vascular flow, stress is often reported in the cgs unit of dyne/cm^2, for which 1 Pa = 10 dyne/cm^2. The two forces shown in Fig. 1 result in a shear stress and in an elongational stress, respectively, and in both cases the stress is calculated as the force per area of the plane on which the force is acting. Strain is similarly defined as the relative deformation, which is the deformation divided by the height or length of the system and is a dimensionless quantity (length/length). The symbols σ and γ are often used for stress and strain, respectively, in a simple shear deformation. In simple shear $\gamma = \Delta x/h$, where Δx is the deformation and h is the height of the system (Fig. 1A). In simple elongation, the tensile strain $\varepsilon = \Delta L/L_0$, where L_0 is the unstrained length and ΔL is the extension resulting from the force, as shown in Fig. 1B. When the sample shape is more complicated than the simple forms in Fig. 1, calculation of strain is often difficult, requiring integration over complex volumes. Geometrical expressions called form factors allow calculation of dimensionless strains from measurable quantities such as the distance that a sample is stretched or the angle by which it is twisted in response to a given force. For some geometries, it becomes practically impossible to determine unambiguously the strain or the stress, and the rheological properties of a biological material are then reported in less informative empirical units such as force per displacement.

A. Elasticity

The mechanical responses of ideal elastic and viscous systems serve as important references when discussing more complicated biological systems, which often show both elastic and viscous characteristics. An ideal elastic system follows Hooke's law, which states that the force is proportional to the deformation, or in terms of stress and strain

$$\sigma = G\gamma \tag{1}$$

where G is the elastic shear modulus and $\gamma = \Delta x/h$ as defined earlier. The value of G is a measure of the rigidity of the system. Typical values of many biological gels and the cytoskeleton are of the order 10^3 Pa. A rubber band has a shear modulus of the order 10^6 Pa, and harder materials such as glass, wood, and steel have characteristic shear moduli of the order 10^{10} Pa. While such order of magnitude estimates of moduli are commonplace for materials as well as cells, more accurate and reproducible determinations require attention to sample preparation and history or aging. Nonetheless, knowing G for an ideal elastic material enables a prediction of the elastic deformation in any type of deformation.

Some systems are more conveniently studied in simple elongation than in shear deformation. The stress applied in simple elongation will be proportional to the extension, and the proportionality constant is called the Young's modulus or E modulus. For isotropic incompressible materials E equals $3G$, which illustrates that the same information may be obtained from either elongation or shear measurements.

B. Viscosity

In contrast to ideal elastic materials, for the ideal Newtonian liquid stress σ is independent of strain γ, but is proportional to the rate of strain, which has units of $\sec^{-1}$ and is related to the liquid flow rate or more precisely the time derivative of strain $d\gamma/dt$. The proportionality constant relating stress to shear rate is viscosity, η.

$$\sigma = \eta \frac{d\gamma}{dt} \tag{2}$$

Newtonian liquids are fully described by the value of the viscosity, which often depends on temperature and slightly on pressure but is independent of strain and strain rate (Fig. 2). The SI unit of viscosity is Pa sec, and characteristic values are 0.001 and 1 Pa sec for water and glycerol, respectively. In older literature, Poise (P) is often used as the cgs unit of viscosity. 10 P equals 1 Pa sec. Non-Newtonian liquids are liquids that cannot be characterized by a constant viscosity. Many biological systems show a constant viscosity at low shear rates but a decreasing viscosity above a characteristic shear rate (Fig. 2). Flow curves are plots of viscosity against shear rate and allow determinations of the zero shear rate viscosity, the critical shear rate (rate at which viscosity starts to decrease), and how rapidly the viscosity decreases at high rates.

When a constant stress is applied to an ideal elastic material, a constant strain is obtained [Eq. (1)], and a constant strain corresponds to zero shear rate (or zero flow). Ideal elastic materials are therefore characterized by an infinite viscosity [Eq. (2)]. As a result, viscosity measurements of elastic and many viscoelastic

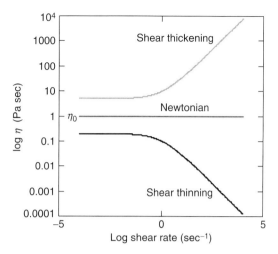

Fig. 2 Characteristic types of flow curves, showing the dependence of viscosity on shear rate. Newtonian liquids have a constant viscosity, whereas shear thinning and thickening solutions have a constant zero shear rate viscosity, η_0, only below a characteristic shear rate.

systems under steady shear are of limited use. Furthermore, in many cases, deformation caused by a steady shear rate will destroy fragile structures in the sample and provide little information about the unperturbed properties of the system. Empirical information about the amount of stress that a sample can withstand before it is damaged and flows, often called a yield stress, however, is generally useful for studies of bone fracture, blood vessel rupture, and other failure events.

C. Oscillatory Measurements

For reasons discussed above, rheological information for viscoelastic systems is often obtained by applying small amplitude oscillatory strains or stresses to the sample rather than steady flows. If an oscillatory strain deformation with an amplitude γ_0 and an angular frequency ω is applied, the stress will also oscillate in time t but will be phase shifted by δ with respect to the strain

$$\gamma(t) = \gamma_0 \sin \omega t \tag{3a}$$

$$\sigma(t) = \sigma_0 \sin (\omega t + \delta) \tag{3b}$$

where σ_0 is the stress amplitude and the angular frequency ω (in rad/sec) equals $2\pi v$, where v is the frequency in Hz (Fig. 3). The phase shift, δ, is always between $0°$ and $90°$. For an ideal elastic system the phase shift is $0°$ [Eqs. (1) and (3)], and for an ideal Newtonian liquid it is $90°$ [Eqs. (2) and (3)]. Materials with phase shifts between $0°$ and $90°$ are viscoelastic and the stress in Eq. (3b) is expressed as a sum of elastic and viscous contributions.

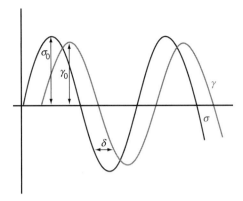

Fig. 3 Stress and strain against time in an oscillatory deformation. Stress and strain amplitudes are marked with σ_0 and γ_0, respectively. The stress and strain signals are phase shifted by an angle δ.

$$\sigma(t) = \gamma_0(G' \sin \omega t + G'' \cos \omega t) \tag{4}$$

In this equation, G' is referred to as the elastic storage shear modulus, and G'' is the loss shear modulus because it is related to the viscous properties and associated energy loss in the sample. For an ideal Newtonian liquid $G'' = \omega\eta$ and $G' = 0$. For an ideal elastic system $G' = G$ and $G'' = 0$. For viscoelastic materials G' and G'' depend on angular frequency, and oscillatory measurements as a function of time at a fixed frequency can be used to monitor structure formation in biological systems.

According to Eqs. (1) and (4), stress is proportional to the strain or strain amplitude. This is valid for all materials at small strains or amplitudes and is called the linear elastic or viscoelastic range. At larger strains or strain amplitudes, stress and strain will not be proportional and the material will be in the nonlinear range. Most materials will exhibit strain softening with a smaller G at large strains. However, some systems exhibit strain stiffening where G increases above a critical strain.

Steady shear and oscillatory measurements are probably the two most important types of rheological measurements. However, many biological systems experience a sustained force such as gravity or blood pressure. It is therefore useful to monitor how such systems deform under a constant load or stress. This type of measurement is called a creep experiment, and in such an experiment the strain $\gamma(t)$ is monitored as a function of time for a fixed stress σ_0 (Fig. 4). The compliance J is then obtained as

$$J(t) = \frac{\gamma(t)}{\sigma_0} \tag{5}$$

After a certain time period under a constant stress, the stress can be removed and the system's ability to recover toward the original undeformed state can be investigated in a creep recovery experiment.

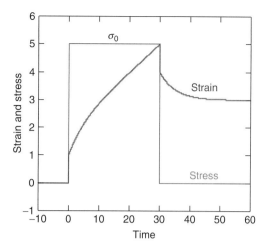

Fig. 4 An example of a creep recovery experiment. A constant stress, σ_0, is applied to the system at time 0. The stress (thin line) is removed again at time 30 and the rest of the experiment is the recovery measurement. The strain (thick line) shows an initial elastic response at time 0, followed by a more gradual increase in strain. When the stress is removed, a partial rapid recovery is seen, followed by a more gradual recovery of strain.

Creep experiments are generally performed on a controlled stress (CS)-type instrument, which applies a defined stress and measures the resulting change in strain. A closely related type of experiment, stress relaxation, can be performed on controlled rate (CR)-type instruments. In a stress–relaxation experiment, the sample is rapidly deformed to a fixed strain, γ_0, and the stress is monitored as a function of time (Fig. 5). The stress–relaxation modulus, $G(t)$, is then calculated as

$$G(t) = \frac{\sigma(t)}{\gamma_0} \qquad (6)$$

The relaxation modulus contains information about how rapidly structures can reorganize to relieve the stress in the system.

III. Rheological Instrumentation

A rheometer is any instrument that enables determination of rheological proper-ties. Very simple rheometers such as capillaries have been used for centuries to determine flow of liquids, and in the art of cooking people have used simple kitchen tools or teeth and fingers as a means to probe rheological properties of food products. Modern rheometers allow a more precise quantification of material properties in well-defined geometries, and they can be divided into two main types. In a CR-type instrument, a motor, often controlled by a computer, deforms

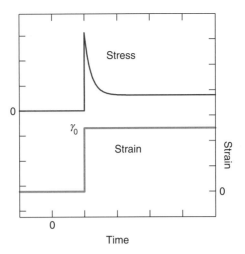

Fig. 5 Principle of a stress–relaxation experiment. A sample is quickly deformed to a constant strain γ_0 (lower curve) and the stress is measured as a function of time (top curve). The stress signal is characteristic for a viscoelastic solid with some stress relaxation and a finite equilibrium stress at long times.

the sample in a controlled manner and a force transducer monitors the force or stress resisting the deformation of the sample. The strain or strain rate is controlled and the forces or stresses are measured. In a CS-type instrument, a stress is applied to the sample and the resulting deformation is monitored. The flow of a liquid through a simple capillary viscometer is an example of a CS-type instrument where gravity determines the stress and the flow rate is determined from the flow time of a fixed volume of the flowing liquid.

Most modern rheometers are rotational type instruments. The sample to be measured is confined in a narrow gap between a stationary and a moving part of a measuring cell. Different measuring cells are used depending on the sample properties. The most common measuring cells are cone and plate (CP, Fig. 6A), parallel plate (PP, Fig. 6B), and concentric cylinders (couette). The couette geometry is often the choice for liquid samples. The PP cells are typically used for films or disk-shaped materials, but have the disadvantage that the sample confined between the plates is not deformed to the same degree, because the strain depends on the distance from the center of rotation. The maximum strain is obtained at the perimeter of the measuring plate, whereas the strain is zero at the rotational axis. The surface properties of the measuring cell are also important, since the sample has to stick to and follow the tool surface during measurements. Slippage between sample and cell surface, for example, as a result of syneresis, can be a problem for some biological systems.

Two signals from a rotational rheometer, the angular position and the torque, are used to compute strains and stresses. A position sensor registers the angular position of the moving cell part. The angle of rotation, ϕ, is proportional to the strain and the proportionality constant depends on gap size and geometry.

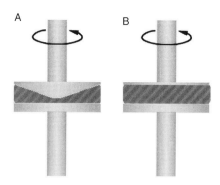

Fig. 6 Two common measurement geometries in rotational rheometers. The sample is confined between a cone and plate (A) or two parallel plates (B).

The stress is proportional to the torque (force times lever arm length) and the proportionality constant depends on the sample geometry and especially on the surface area of the measuring tool. Tools with a large area are used for soft materials and low-viscosity liquids, whereas smaller tools are used for harder materials. A summary of proportionality constants for various geometries can be found in (Ferry, 1980).

In oscillatory tests, the computer monitors both the position and torque signals as a function of time. The time-dependent strains and stresses are calculated and fitted to sine waves [Fig. 3; Eq. (3)]. The stress and strain amplitudes, as well as the phase shift, are obtained from the fits and from these G' and G'' are calculated. Modern rotational rheometers typically cover a frequency window from 0.001 to 100 Hz. The upper limit is often determined by mechanical resonances in the tool and detector. G' and G'' are only defined in the linear range where both strains and stresses are simple sinusoidal curves [Eq. (4)]. In order to ensure this, materials should always be tested as a function of stress or strain amplitude. In the linear range, G' and G'' are independent of these amplitudes.

IV. Experimental Design

Biological systems can be investigated by a number of different rheological techniques. The choice of technique depends on the type of information desired and the type of system. Some common techniques will be described with respect to typical uses and limitations. The techniques can be divided into steady and oscillatory measurements.

A. Stress–Strain Relation

Elastic systems can be studied in both simple shear and simple elongation by use of either CS or CR instruments. A stress–strain curve is a plot of steady stress against strain. The slope of this plot at low strains is the shear modulus (or the

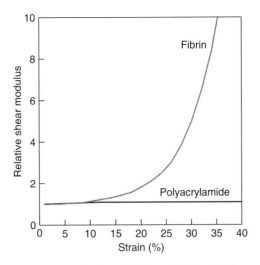

Fig. 7 Strain amplitude dependence of polyacrylamide and fibrin gels. The measured shear modulus divided by the modulus at small strain amplitudes is plotted against strain amplitude. G' of the polyacrylamide gel is seen to remain constant over the measured strain range, whereas fibrin gels are strain hardening.

Young's modulus in simple elongation). The part of the stress–strain curve, where stress is proportional to strain, is the linear range. At larger strains, stresses will not be proportional to strain and the measurement will be performed in the nonlinear range. Some systems are strain softening and some are strain stiffening as illustrated by data for a fibrin gel in Fig. 7.

Stress–strain measurements are most readily performed on systems that mechanically equilibrate in time rather than flow. Measurements can be done in both the linear and the nonlinear range. However, care should be taken that measurements be performed well below the yield stress, where structural integrity starts to be affected by the stress.

B. Stress Relaxation

Stress–relaxation measurements can be performed in both simple shear and simple elongation, and they are of special interest for viscoelastic systems. In a stress–relaxation experiment, the sample is rapidly deformed and the stress is monitored as a function of time, keeping the sample in the deformed state. The stress–relaxation curve is a plot of the shear modulus, $G(t)$, as a function of time. A typical example is shown in Fig. 8. For a viscoelastic material, $G(t)$ will decrease with time and the decrease will occur on a timescale which is determined by the relaxation time of the sample. At long (infinite) times, the relaxation modulus will

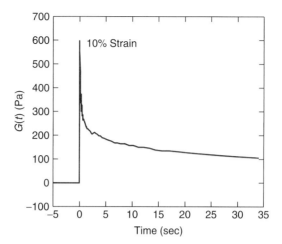

Fig. 8 Stress relaxation of adult rat brain exposed to 10% strain between parallel plates with a gap of ~3 mm. From a shear modulus of nearly 600 Pa at application of strain, the tissue will relax to a modulus of 125 Pa.

either relax to zero (characteristic of all liquids) or reach a constant value which will be the equilibrium modulus of the system characteristic of a viscoelastic solid.

Stress–relaxation experiments are especially useful for studies of systems with long relaxation times. Experiments should be performed in the linear range where the relaxation modulus is independent of the strain magnitude. Stress–relaxation measurements should be performed on CR instruments.

C. Creep and Creep Recovery

In a creep and creep recovery measurement, a constant stress is applied to the system for a period of time and then removed. An example of creep recovery of a blood plasma clot is shown in Fig. 9. The strain is monitored as a function of time. Results are often plotted as compliance, J, as a function of time, as shown in Eq. (5). Creep experiments are of interest primarily for viscoelastic materials. The elastic properties are seen in the rapid deformation and recovery when the stress is applied or removed. The slow subsequent deformation is characteristic of the viscoelastic time-dependent processes in materials. If this deformation (strain) increases linearly with time, one can obtain the viscosity at a low shear based on the inverse relationship between viscosity and the slope. The viscous or plastic deformation can also be calculated from the nonrecoverable compliance at long times.

Creep tests are also of interest for viscoelastic systems with long relaxation times because they measure flow at very low shear rate based. Care should be taken to ensure that the experiments are performed in the linear range where the compliance

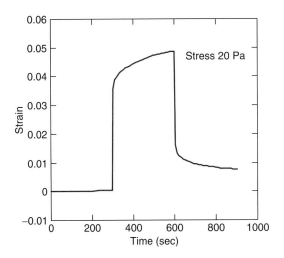

Fig. 9 Creep recovery experiment on a plasma clot. A stress of 20 Pa is applied to the clot between 300 and 600 sec. The Figure shows a rapid elastic deformation to a strain of 0.035, which corresponds to a shear modulus of 20 Pa/0.035 = 600 Pa. The clot nearly recovers completely at the end of the recovery part.

is independent of the stress magnitude. Creep measurements should be performed on CS-type instruments.

D. Frequency Sweep

Oscillatory measurements at low strain or stress amplitudes allow a determination of G' and G'' as a function of frequency. Such measurements are important because they can give information about both structure and dynamics. For viscoelastic liquids, G'' will dominate at low frequencies. At higher frequencies, relaxation of structures cannot take place within the oscillation cycle, resulting in an increase in G'. An example of a frequency sweep is given in Fig. 10, which shows the frequency dependencies of a hyaluronic acid solution. The Figure illustrates G'' dominating at low frequencies, where G'' is related to the zero shear rate viscosity and angular frequency through $G'' = \omega\eta$. At higher frequencies, G' dominates and a plateau is almost seen as expected for an ideal elastic material. The crossover angular frequency is approximately the inverse of the relaxation time, which for long biopolymers is strongly dependent on molecular mass and concentration.

Oscillatory tests can be performed on both CR and CS instruments. Care should be taken to perform measurements in the linear range where G' and G'' are independent of stress or strain amplitudes, and both stress and strain can be fit by sinusoidal functions as shown in Fig. 11A. Otherwise, as shown in Fig. 11B, the stress response to a sinusoidal strain is no longer purely sinusoidal, and G' and G'' are not well defined.

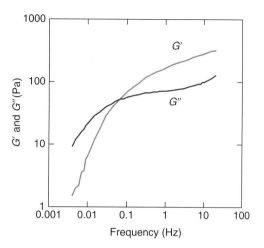

Fig. 10 Frequency sweep of a high molecular mass hyaluronic solution. G' and G'' are plotted against frequency. G'' dominates at frequencies below 0.1 Hz, whereas the elastic properties and G' dominate at high frequencies. The crossover frequency is determined by the longest relaxation time in the system.

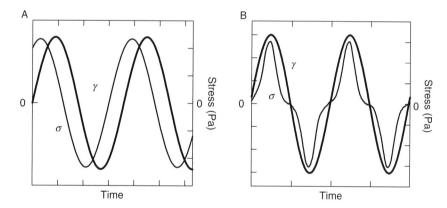

Fig. 11 Stress and strain in an oscillatory experiment on a hypothetical linear viscoelastic system (A) and on a strain-hardening system (B, plasma clot). The strain signal in B is a sine wave but the stress signal is not a sine wave. G' and G'' are therefore not well defined, but the very small phase shift demonstrates that the elastic properties dominate.

E. Time Sweep

A steady rate of shear should not be used to follow structural changes since structures are likely to rupture during steady shearing. Oscillatory measurements are an ideal method to overcome this problem for monitoring structural changes over time. In a time sweep, the system is subjected to small strain amplitude deformations at a single frequency. Time sweeps are useful to study, for example,

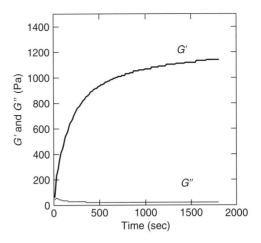

Fig. 12 Shear modulus measurement over 30 min of an 8 mg/ml fibrin gel exposed to 2% oscillatory shear strain at a frequency of 10 rad/sec. The slope of G' provides polymerization kinetics and the takeoff point indicates gelation time.

gelation or kinetics of enzymatic reactions. An example is given in Fig. 12, which shows a time sweep on a gelling system, a fibrin gel. The gelation time is sometimes defined as the time where G' and G'' are equal just before G' begins to rise rapidly.

Time sweeps should be performed at small strain amplitudes and preferentially on CR instruments, or on a CS instrument run in a controlled strain mode (Section VI). For experiments over a long time period, care should be taken to avoid evaporation or drying of the sample.

F. Strain or Stress Amplitude Sweeps

Oscillatory measurements at a fixed frequency with increasing strain or stress amplitudes are called strain sweeps. The linear range, where G' and G'' are constant, should always be observed at low amplitudes, but at larger strains G' and G'' will often depend on the strain amplitude. In many gel cases G' exceeds G'' at low strains, but G'' will dominate when the gel structure is broken down at higher stress amplitudes. The stress amplitude where nonlinear effects are observed has been used to determine the yield stress. In addition, if a strain sweep is repeated on the same sample in the rheometer, it is often possible to determine if the large strain has had an irreversible effect on structures (Kerst *et al.*, 1990). Some gel systems primarily consisting of long fairly stiff biopolymer structures show pronounced strain hardening (Storm *et al.*, 2005), that is G' increases with increasing strain amplitudes (Fig. 7).

The values of G' and G'' in the nonlinear range are not well defined, since the strain signal on a CS instrument and the stress on a CR instrument will not be true sine waves (Fig. 11B). The error made in estimating elastic moduli by fitting the data to sine functions, however, can be small, as demonstrated by direct comparison of oscillatory and steady shear measurements for a highly nonlinear material (Storm *et al.*, 2005).

G. Rate–Dependent Viscosity

Newtonian liquids are characterized by a viscosity that is independent of shear rate. Flow curves are plots of viscosity against shear rate and are of interest for non-Newtonian liquids. Any viscoelastic liquid should have a constant "zero shear rate viscosity," η_0, at sufficiently low shear rates. The example in Fig. 13A shows a typical flow curve for a biopolymer solution at fairly high concentrations. The η_0 value is very dependent on molecular mass and concentration. Above a critical shear rate, the viscosity decreases linearly with shear rate on the log–log plot. Such flow curves are very common for biopolymer solutions and the sample is said to behave as a power law fluid, since the viscosity is proportional to some power of the shear rate. The exponent should always be between 0 (Newtonian liquid) and a value not less than -1 (limit of indeterminate flow). The critical shear rate is related to the inverse of the longest relaxation time of the system.

Flow curves can also be used to determine yield stresses of elastic materials. If the shear rate is measured with increasing steady stresses, the elastic properties will dominate at small stresses, with a vanishing shear (flow) rate (Fig. 13B). At stresses above the yield stress, structures will be broken and the system will flow. The example in Fig. 13B is typical for many systems and is referred to as a

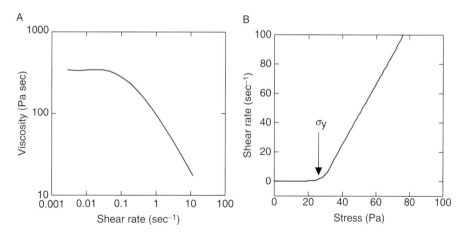

Fig. 13 Steady shear flow curves of hyaluronic solution (A) and a flocculated micellar system (B). The hyaluronic solution is a shear-thinning solution with a zero shear viscosity of about 300 Pa sec and power law behavior above shear rates of 0.3 sec^{-1}. The micellar system behaves like a Bingham liquid with an elastic response (i.e., shear rate ≈ 0) at stresses smaller than the yield stress σ_y. The linear relationship between shear rate and stress above the yield stress is characteristic for a Bingham liquid.

"Bingham liquid," with a Bingham viscosity which is the slope of the curve at higher stresses and a yield stress corresponding to the intercept of this curve at vanishing shear rates.

Flow curves should represent steady state values of the viscosity against shear rate. This means that viscosity should be independent of measurement time. It is important at each shear rate to allow sufficient time for the stress to reach a constant value before the stress is measured. Otherwise the calculated viscosity will not only depend on shear rate but also on time and history of the sample. Systems for which the viscosity depends on history are called thixotropic.

H. Flow Oscillation

During a steady shear flow, structures are often reversibly destroyed or aligned in the flow field. In some cases, it can be of interest to determine how rapidly structures are reformed. This can be investigated by a flow oscillation test, in which the system is first sheared at a given shear until steady shear flow is obtained, after which steady shearing is stopped and small amplitude oscillations as a function are immediately started. The resulting time dependence of especially G' contains information about the kinetics of recovery.

V. Sample Preparation

Shear, compression, and elongational deformations of soft biological materials require well-defined interactions of the material with the plates or holders of the rheometer, and these requirements often limit the types of measurements that can be made. Elongation strains require that the sample be held in place at two ends, usually by a clamp, while the sample is stretched. This manipulation generally works well for stiff elastic materials such as skin, tendon, or bone, but is usually impractical for fragile tissues like brain or liver, and impossible for viscoelastic liquids or gels that flow out of the holder. Compression studies are easier in the sense that clamping is unnecessary, but evaluation of the stress–strain curve requires accurate information about whether the surfaces resting on stationary plates are adherent or can slide as the sample is compressed, and whether liquid is squeezed out of the hydrated material during testing.

Most rheological measurements of biological gels, liquids, and soft tissues are done under shear deformation because shear strains preserve sample volume, and the materials can be more easily placed between the plates of the rheometer. Figure 6A illustrates a typical CP geometry in which liquids or gelling systems are held.

A. Solids

Solid tissues are generally excised and cut into disk-shaped samples with dimensions typically 5–25 mm in diameter and 0.1- to 5-mm thick. They are placed on a flat lower rheometer plate. A parallel top plate is lowered onto the sample until contact is made, as judged by a positive normal force measured by a force transducer. Generally, hydrated specimens bind well enough to the metal (usually stainless steel or titanium) rheometer plates so that slipping does not occur during the shear deformation required for the rheological measurements. Use of a serrated top and bottom metal plate can further prevent the problem of slipping. If required, adhesives can be used to glue the sample to the plate, as long as the adhesive causes no chemical changes in the sample and is much stiffer than the sample, such that it remains stationary as the sample deforms. Since the tissue is expected to be predominantly elastic, slippage can be detected by an irrecoverable deformation when the sample is subjected to creep and creep recovery, or as an anomalously high mechanical loss, characterized by a large phase angle between stress and strain in an oscillatory measurement. Slipping can also be eliminated by decreasing the gap between plates to compress the sample slightly. Typical compressions of 5% have been used in several studies of tissue rheology. There is a complication in compression, however, in that studies of brain and liver rheology show that the shear modulus is a strong function of compression (or squeezing), and therefore a series of shear modulus measurements should be made over a range of compressions and the modulus extrapolated to zero compression to reflect the shear elasticity in the uncompressed state (Fig. 14).

Since PPs are generally used for solid samples, shear moduli are only well defined for deformation in the linear range, since the shear strain in this geometry increases linearly with the distance from the center of the plate. Often, however, the physiologically relevant rheology occurs in the nonlinear range, and indeed the nonlinear elasticity, often in the form of strain stiffening, is an essential feature of tissues such

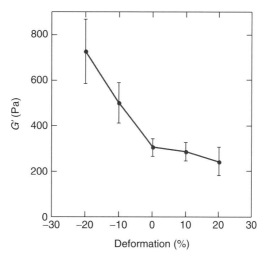

Fig. 14 Effects of compression or elongational deformation on shear modulus of intact rat brain tissue under 2% oscillatory shear strain. Compression of brain tissue (negative deformation) increases the shear modulus, and at 20% compression the shear modulus G' has increased by nearly a factor of 3.

as the arterial wall (Shadwick, 1999). The PP geometry will exaggerate strain stiffening since most of the stress is carried at the edge of the sample, and therefore a geometric correction needs to be made to relate these data to those made under uniform strains.

B. Liquids and Gelling Systems

Liquids and gels cannot be clamped and are therefore held in place between rheometer plates by surface tension and adhesion to the plate surface. This requirement limits the height of the sample to ~1 mm, depending on the tension at the sample interfaces. One important advantage of liquid systems is that they can be placed between a flat bottom plate and a truncated cone-shaped upper plate lowered to a prescribed separation so that the shear strain is uniform throughout the sample. This geometry is especially important for studying nonlinear materials such as strain-stiffening biopolymer gels.

The viscous properties of biological fluids such as blood and blood plasma have been extensively studied. Typical experiments measure viscosity as a function of shear rate. Shear-thinning properties of many fluids are important for their ability to transit through confined spaces in the vasculature, lymphatic system, and elsewhere. The range of biologically relevant flow rates is very large, from near zero in occluded blood vessels to over 1000 sec^{-1} in arteries.

Gelling systems require special consideration in sample preparation. Ideally, a gel is formed *in situ* between the rheometer plates, by initiation of the polymerization or cross-linking reaction right after the sample is placed in the rheometer so that the gel forms in an unstressed state before the rheological measurements begin. A common assay for gelling systems is a time-dependent oscillatory measurement of G' and G''. In a CR-type rheometer, a small amplitude strain is imposed during polymerization and the elastic and viscous components of the shear stress are measured. For a purely viscous solution placed before the gel forms, G' is near zero and G'' is relatively small. As the polymers form and become cross-linked, G' abruptly becomes finite, and then both G' and G'' will increase as more material is incorporated into the gel network. The time at which G' become nonzero is often denoted the gelation time or gel point. This method in principle is used clinically to define the clotting time of blood samples and is useful for monitoring the kinetics of polymerization reactions in many other systems.

VI. Special Considerations for Biological Samples

A. Biological Polymers

Biological samples are often more difficult to measure rheologically and their rheological characteristics are often more difficult to relate to specific molecular structures compared to synthetic polymer systems. In addition, biological samples are often fragile and nonlinear materials that can survive only a limited range of strains. Perhaps more importantly, they are also usually far from equilibrium, and the biologically relevant properties often depend on specific chemical states that can change during the course of an experiment. For example, one of the most commonly studied biopolymer gels—cross-linked F-actin—contains polymers that are continuously hydrolyzing ATP and exchanging subunits from the filament ends. The cross-links that bind filaments together have finite off rates that are often not known, and the geometry of the networks they form is usually kinetically determined, with slow rearrangements occurring as the rheological experiment proceeds. This slow time-dependent change in the chemical or morphological state of the material, even held at rest, makes studies such as frequency-dependent measurements difficult to interpret unless they are performed rapidly enough so that the chemical changes are negligible during the period of the measurement.

B. Intact Tissue

Intact tissues also undergo changes in rheology after isolation and storage. Biochemical changes during such reactions as ATP hydrolysis and proteolysis can be inhibited or allowed to reach a steady state, but other changes may be more subtle and potentially interesting. For example, most native tissues are in a state of tension due to the activity of motor proteins. Such internal stress within a biological material can greatly alter its rheology due to the nonlinear elasticity of biopolymer networks and to the specific nonrandom geometry of tissues. As a tissue ages and these internal tensions relax, they can alter tissue elasticity even without a gross change in tissue architecture.

The main complication in relating tissue rheology to molecular structure is perhaps the heterogeneity of biological materials. Tissues have complex architectures that are frequently connected to their *in vivo* function. They are often multilayered and composed of numerous cell types, each type with its own mechanical properties, contractility, strength, and orientation within the tissue. The situation might seem incomprehensibly complicated, but materials in common use today can also be complex and include carbon fiber composites (used in planes, cars, bikes) in which fibers are pretensed as they are layered into cured epoxy laminates.

C. Instrument Selection for Measuring Gelation Kinetics

Gelation kinetics generally require a CR instrument, rather than a CS instrument. Limiting the strain to a small, defined value ensures that the fragile polymer network that appears just at the point where filaments first make a continuous network throughout the sample—the so-called percolation threshold—is not destroyed by the measurement. When the network is very weak, the strain may be so small that the rheometer transducer cannot accurately measure the stress, and the data will be noisy. In contrast, in a CS rheometer, the value of stress chosen, even if very small, will usually be enough to induce continuous flow in the sample before it gels and becomes stiff enough to resist the stress. Therefore, the first measurements of a gelling system, before or near the gel point, can destroy the sample and cause flow alignment of filaments that have not yet formed a network, resulting in anomalously low values for G' at later time points. Many gels formed *in vitro* by purified cytoskeletal polymers such as actin and microtubules are very soft with G' on the order of 1 Pa and therefore very susceptible to damage in conventional CS rheometers.

VII. Conclusions

Forces and mechanical effects can direct cell function and tissue formation as specifically as chemical stimuli, and integrating physical studies into cell biology is increasingly used in bioengineering efforts and other biomedical studies. Since a cell or tissue's response to forces is defined by its viscoelastic parameters, quantitative measurements of cell and extracellular matrix rheology are necessary for a full explanation of how cells interact with their environment.

Rheology can be applied to a vast range of biological and medical problems (Gabelnick and Litt, 1973). The actin, intermediate filament, and microtubule networks of the cytoskeleton each have distinct rheological responses that affect multiple cell processes ranging from migration to division (Janmey, 1991). Soft tissues, as another example, maintain mechanical properties within a specific range that varies according to tissue function, development, or disease. Fibrotic liver is more than three times as stiff as healthy liver and this characteristic could possibly be used in surgery to assess which areas are extensively damaged and to be removed (Kusaka *et al.*, 2000). Stress–relaxation properties of tissues vary with their functions: the bladder exhibits a high degree of stress relaxation to minimize internal pressure despite large volume changes (Andersson *et al.*, 1989), while the aorta shows a lower degree of stress relaxation in response to volume change in order to maintain adequate blood pressure (Shadwick, 1999). Brain is a particularly soft mammalian tissue, with an elastic modulus $\sim$10 times lower than liver and nearly 50 times lower than muscle and may, therefore, be more susceptible to damage by shear (Donnelly and Medige, 1997; Liu and Bilston, 2002). Rheological oscillatory shear or stress–relaxation tests can be performed to attain short-term and long-term values of the stiffness of intact tissue or extracellular matrix, and the information used toward modeling relevant strains induced by cellular processes such as growth cone or lamellipodial advance. It is likely that these rheological properties of the tissue play an important role in maintaining specific cellular functions.

Macroscopic rheological methods as outlined in this chapter have been extensively used to provide empirical measurements of how biological materials deform when stressed, and some of these studies have suggested ways to relate the molecular structures within tissues to their rheology. However, macroscopic rheology has several important limitations that have motivated efforts to design and implement microrheological methods as are discussed in several chapters of this volume. The major limitations of macrorheology are associated with the size of materials needed, the magnitude of forces involved, and the heterogeneity of biological materials. Issues of sample size and forces that may disrupt the material have been addressed in Section VI and can in many instances be alleviated by appropriate sample preparation.

Heterogeneity of biological tissues presents a fundamental problem for macrorheology that in most cases makes a molecular or structural interpretation of the rheological behavior impossible. For example, it is difficult to relate the rheology of a tissue such as liver or brain to the rheology of the specific cell types or extracellular matrices that differentiate one tissue from another. Any tissue is a composite of cells linked to each other either directly by cell–cell junctions or indirectly through the extracellular matrices, and each cell undergoes biochemical reactions during the course of a measurement that can take between a few seconds and many minutes to perform. Therefore, the viscoelastic properties of the whole composite cannot be easily linked to the viscoelasticity of any particular element such as a cytoskeletal network, the membrane, or the extracellular network; however, mixture theories based on relative volume fractions have been successful in understanding materials with inclusions. Specific molecular manipulations such as the loss of a single actin cross-linking protein can be detected as a significant change in a macroscopic sample of cells each bearing this defect, but predicting the rheology of a tissue from knowledge of the concentration of various filaments and cross-linkers is presently possible only for simple systems of purified proteins.

As cell and tissue mechanics become more of an integral part of basic cell biologic studies, a comprehensive understanding of micro- and macrorheology may help develop a unified model for how specific structural elements are used to form the soft but durable and adaptable materials that make up most organisms. The results of these studies also have potential for developing materials and methods for wound healing, cell differentiation, artificial organ development, and many other applications in biomedical research.

Glossary

Anisotropic: variable properties with respect to direction

Compliance (J): the relative extent to which a body yields to deflection by force

Gel point: time at which shear modulus of a system becomes greater than zero

Elasticity: the property of a material to deform to a defined extent in response to a force and then return to its original state when the force is removed

Form factor: geometric expression that quantifies deformation through calculation of a dimensionless strain from an observable quantity such as distance moved or angle of rotation

Inertia: tendency of a body to resist acceleration

Isotropic: directionally invariant

Linear elasticity: Young's or shear modulus constant over range of strains

Loss modulus (G″): measure of energy lost during a strain cycle; often expressed as the imaginary part of the complex modulus: $G'' = (\sigma_0/\gamma_0)\sin(\delta)$

Newtonian viscosity: viscosity independent of shear strain rate; linear relationship between shear stress and shear strain rate

Nonlinear elasticity: Young's or shear modulus that changes with strain

Non-Newtonian viscosity: viscosity dependent on shear strain rate; nonlinear relationship between stress and strain rate (i.e., shear thickening or thinning)

Phase angle (δ): The angular shift between the sinusoidally varying stress and strain in an oscillatory measurement. The value of δ is $0°$ for a purely elastic solid and $90°$ for a purely viscous liquid

Shear Modulus (G): a constant describing a material's resistance to deformation in shear; $G = \sigma/\gamma$.

Shear strain (γ): unitless parameter quantifying the extent of deformation after application of shear stress. For a cube, shear strain is ratio of lateral displacement over sample height. For other shapes, the form factor relates measured displacement to unitless strain

Shear strain rate: rate of change of shear strain; $d\gamma/dt$

Shear stress (σ): force parallel to a material's axis per unit area; $\sigma F/A$

Stiffness: resistance of a body to deflection by force

Storage modulus (G′): measure of energy stored during a strain cycle; under sinusoidal conditions, the part of shear stress in phase with shear strain divided by shear strain; often expressed as the real part of the complex modulus: $G' = (\sigma_0/\gamma_0)\cos(\delta)$

Elongational strain (ε): fractional change in length or elongation; $\varepsilon = \delta/L$.

Stress (σ): force per unit area; $\sigma F/A$

Viscosity (η): measure of resistance of a fluid to shear stress; $\eta = \sigma/d\gamma/dt$

Young's modulus (E): a constant describing a material's resistance to deformation in extension; $E = \sigma/\varepsilon$

Yield stress (σy): maximum stress applicable to a system before rupture occurs

References

Andersson, S., Kronstrom, A., and Bjerle, P. (1989). Viscoelastic properties of the normal human bladder. *Scand. J. Urol. Nephrol.* **23**(2), 115–120.

Barnes, H. A., Hutton, J. F., and Walters, K. (2001). "An Introduction to Rheology." Elsevier Science Publication Co. Amsterdam.

Bershadsky, A. D., Ballestrem, C., Carramusa, L., Zilberman, Y., Gilquin, B., Khochbin, S., Alexandrova, A. Y., Verkhovsky, A. B., Shemesh, T., and Kozlov, M. M. (2005). Assembly and mechanosensory function of focal adhesions: Experiments and models. *Eur. J. Cell Biol.* **85**(3–4), 165–173.

Donnelly, B. R., and Medige, J. (1997). Shear properties of human brain tissue. *J. Biomech. Eng.* **119**(4), 423–432.

Ferry, J. D. (1980)."Viscoelastic Properties of Polymers," p. 641. Wiley, New York.

Gabelnick, H. L., and Litt, M. (1973). "Rheology of Biological Systems," p. 319. Thomas, Springfield, IL.

Janmey, P. A. (1991). Mechanical properties of cytoskeletal polymers. *Curr. Opin. Cell Biol.* **3**(1), 4–11.

Kerst, A., Chmielewski, C., Livesay, C., Buxbaum, R. E., and Heidemann, S. R. (1990). Liquid crystal domains and thixotropy of filamentous actin suspensions. *Proc. Natl. Acad. Sci. USA* **87**(11), 4241–4245.

Kusaka, K., Harihara, Y., Torzilli, G., Kubota, K., Takayama, T., Makuuchi, M., Mori, M., and Omata, S. (2000). Objective evaluation of liver consistency to estimate hepatic fibrosis and functional reserve for hepatectomy. *J. Am. Coll. Surg.* **191**(1), 47–53.

Liu, Z., and Bilston, L. E. (2002). Large deformation shear properties of liver tissue. *Biorheology* **39**(6), 735–742.

Shadwick, R. E. (1999). Mechanical design in arteries. *J. Exp. Biol.* **202**(Pt. 23), 3305–3313.

Storm, C., Pastore, J. J., MacKintosh, F. C., Lubensky, T. C., and Janmey, P. A. (2005). Nonlinear elasticity in biological gels. *Nature* **435**(7039), 191–194.

CHAPTER 2

Polyacrylamide Hydrogels for Cell Mechanics: Steps Toward Optimization and Alternative Uses

Casey E. Kandow,★ Penelope C. Georges,† Paul A. Janmey,† and Karen A. Beningo★

★Department of Biology
Wayne State University
Detroit, Michigan 48202

†Department of Bioengineering
Institute for Medicine and Engineering
University of Pennsylvania
Philadelphia, Pennsylvania 19104

METHODS IN CELL BIOLOGY, VOL. 83

0091-679X/07 $35.00
DOI: 10.1016/S0091-679X(07)83002-0

Abstract

Since their first introduction, polyacrylamide hydrogels have proven to be very useful for studies of mechanical interactions at the cell–substrate interface. In this chapter, we briefly review the basic concepts of this method and provide a series of modifications that have evolved since its inception. In addition, we have described several alternative uses of polyacrylamide hydrogels that have emerged for the study of cellular mechanics. Our intention is to provide users of this gel system with a number of improved and tested options as this method advances toward optimization.

I. Introduction

Mechanical interactions between cells and their substrates have become an area of intense study. Creative advances in the production of elastic substrates have fueled this interest, as research groups have sought an ideal experimental substrate for measuring traction forces and for probing cellular responses to changes in compliance (Balaban *et al.*, 2001; Beningo and Wang, 2002a; du Roure *et al.*, 2005; Pelham and Wang, 1997; Tan *et al.*, 2003; Chapter 6 by Panorchan *et al.*, Chapter 12 by Lee, and Chapter 22 by Engler *et al.*, this volume for applications). An ideal elastic substrate would be economical and versatile, easy to make and characterize, optically clear and inert, linearly elastic and tunable in stiffness, and would not require complex mathematical computations for traction forces. What have arisen over the last decade are a number of substrates that have some of these properties but not all.

Polyacrylamide-based hydrogels have been a popular choice for the study of cell–substrate mechanical interactions and are some of the easiest substrates to employ (Beningo *et al.*, 2002; Wang and Pelham, 1998). There are multiple mechanical, chemical, and optical advantages afforded by the use of these gels. The gels produce a linear deformation in response to a wide range of stress and show a rapid and complete recovery on removal of the stress. In addition, the rigidity of the gels can be easily manipulated by varying the concentration of the bis-acrylamide cross-linker, a feature that many other gel systems are incapable of providing. Polyacrylamide hydrogels are clear and nonfluorescent, which facilitate microscopic visualization of cellular processes. Furthermore, it is possible to covalently link proteins of interest to the otherwise nonadhesive substrate surface.

Despite these advantages, polyacrylamide hydrogels, as initially designed, fall short in meeting the ideal substrate described above due to limitations of the protein-coupling reaction, difficulties in measuring the compliance, incompatibility with full three-dimensional cell culture, and issues of optical interference when fluorescent microbeads are used for the detection of traction forces. The mathematical computation needed for determining traction forces is also quite complex (Dembo and Wang, 1999). Fortunately, a number of modified approaches that address these limitations have been described recently. It is the goal of this chapter to assemble some of these modifications into one unit and to compare and contrast them with the initial protocol as introduced by Pelham and Wang (Beningo *et al.*, 2002;

Wang and Pelham, 1998). The intention is to provide users of the polyacrylamide hydrogels with alternative options from the original protocol, and to highlight novel uses of these gels for the studies of cellular mechanics.

II. Principle of the Polyacrylamide Hydrogel

Polyacrylamide was used as a substrate for cultured cells as much as 30 years ago when hepatocytes were grown on nonadherent polyacrylamide gels (Schnaar *et al.*, 1978). In this early work, the polyacrylamide substrates were not used for mechanical studies but to generate a flat matrix for immobilizing sugars. In the late 1990s, Pelham and Wang (Pelham and Wang, 1997; Wang and Pelham, 1998) combined the use of polyacrylamide with improved reactions for protein coating and with the use of fluorescent microbeads as markers for determining the strain due to traction forces, as originally described for use with silicone rubber sheets (Oliver *et al.*, 1998). In addition, the initial design attached the polymerized substrates to a coverslip that had been activated by aminopropyltrimethoxysilane and gluteraldehyde, to limit swelling and to prevent the polyacrylamide from detaching. The unpolymerized gel solution included acrylamide, bis-acrylamide, and fluorescent microbeads.

A. Standard Method of Polymerization

The standard polymerization reaction of acrylamide is identical to that used for gel electrophoresis (Fig. 1). Briefly, acrylamide and its cross-linker, bis-acrylamide, are combined in solution with initiators (TEMED and ammonium persulfate) to trigger a free radical-dependent polymerization of double bonds (vinyl groups) in the otherwise stable acrylamide and bis-acrylamide monomers. As oxygen acts as a free radical trap and can slow or even prevent the polymerization reaction from proceeding, the solution is placed in a vacuum chamber prior to the addition of initiators to remove oxygen dissolved in the gel solutions (Menter, 2000). The polymerization reaction produces acrylamide chains of variable length which are statistically cross-linked by the bis-acrylamide (Charambach and Rodbard, 1971). To prepare a thin sheet of hydrogel, a small amount of the mixture was dropped onto a glutaraldehyde-activated coverslip and overlaid immediately with a small round coverslip, resulting in a thin gel once polymerized. In the initial design, the authors used a photoactivable heterobifunctional cross-linker (described below) to covalently link proteins to the gel surface. An outline of this method is illustrated in Fig. 2.

B. Light-Induced Initiation of Polymerization

The most common method for catalytic initiation of acrylamide polymerization involves the generation of free oxygen radicals by ammonium persulfate (Fig. 1). However, alternative methods of initiation are available as have been used previously in preparing electrophoretic gels. For example, Wong *et al.* (2003) have

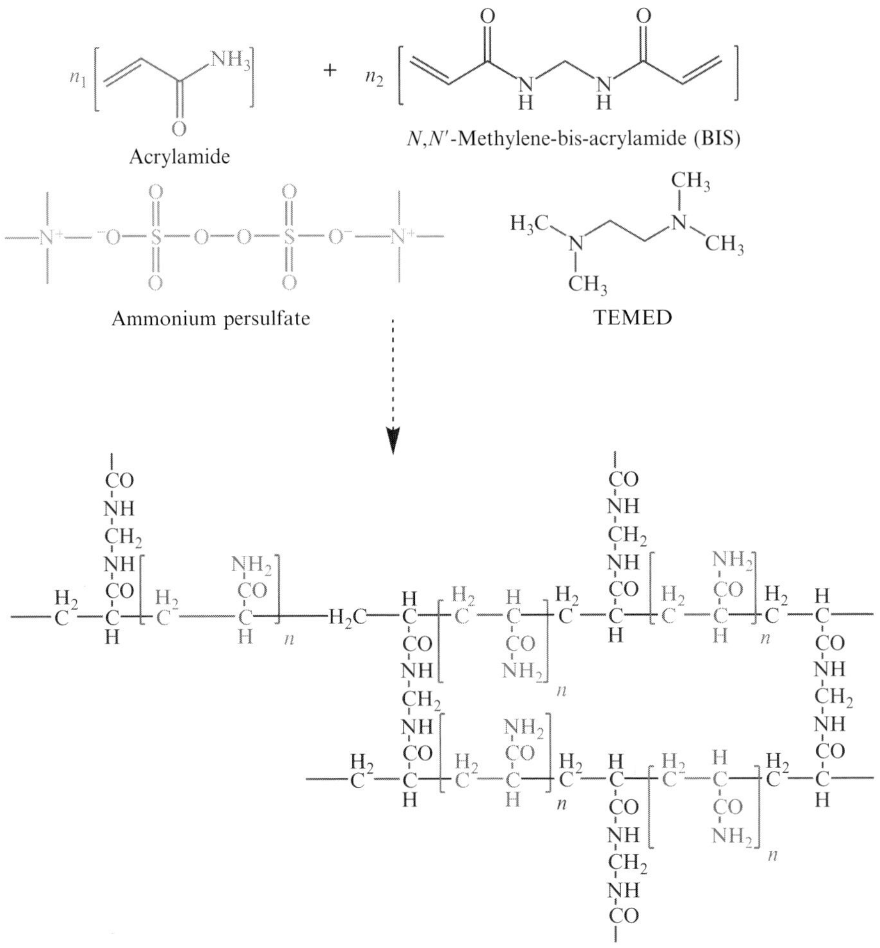

Fig. 1 The polymerization reaction of acrylamide. *N,N'*-Methylene-bis-acrylamide (BIS) acts as a cross-linker between acrylamide subunits. Catalytic initiation involves the production of free oxygen radicals by ammonium persulfate in the presence of TEMED. The final product is a meshwork in which the pore size can be controlled by the total concentration of acrylamide and BIS cross-linker. (See Plate 1 in the color insert section.)

described a particularly useful method of gradient polymerization involving photo-induced generation of free radicals. Unlike the traditional means of polymerization, the initiator ammonium persulfate is exchanged for 0.15 g/ml of the photoinitiator, Irgacure 2959 (1-[4-(2-hydroxyethoxy)-phenyl]-2-hydroxy-2-methyl-1-propan-1-one). Polymerization is initiated through exposure to ultraviolet (UV) light, allowing polymerization and cross-linking to be locally manipulated by varying UV light intensity or substrate exposure time. By placing a light diffusing or blocking mask over the gel, this method offers a unique approach

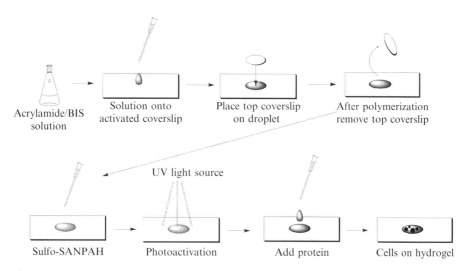

Fig. 2 The initial design of polyacrylamide hydrogels for cell culture. An activated acrylamide solution is dropped onto a coverslip with reactive surface that binds polyacrylamide. A second, nonreactive coverslip is then overlaid onto the acrylamide droplet. Following polymerization, the top coverslip is carefully removed. The hydrogel is then treated with the UV-activated heterobifunctional cross-linker sulfo-SANPAH for the covalent linkage of proteins. Cells are then seeded onto an acclimated substrate.

for creating a range of gradient or geometric patterns across the surface of a single substrate.

III. Conjugation of Proteins to Polyacrylamide

The chemical inertness of polyacrylamide is one of its greatest advantages and disadvantages. When first selected as the polymer of choice for PAGE, the lack of protein binding was an obvious benefit. The same property eliminates the issue of nonspecific cell adhesion when used as a culture substrate and allows for regulated control of adhesive interactions, a clear advantage when demonstrating the mechanical contribution of a given receptor or adhesive protein. Unfortunately, the degree of inertness is also a limitation, as the chemical nature of the polyacrylamide polymer does not allow for easy covalent attachment. In the initial design of the hydrogels, the photoactivatable heterobifunctional reagent sulfosuccinimidyl-6-(4′-azido-2′-nitrophenylamino) hexanoate, known as sulfo-SANPAH, was used. Sulfo-SANPAH contains a phenylazide group at one end that reacts nonspecifically with polyacrylamide on photoactivation, and a sulfosuccinimidyl group at the other end that reacts constitutively with primary amines (Fig. 3).

While the sulfo-SANPAH method for coupling has been used extensively and described in detail previously (Pelham and Wang, 1997), there are several limitations with this approach. Limited solubility and stability of the sulfo-SANPAH

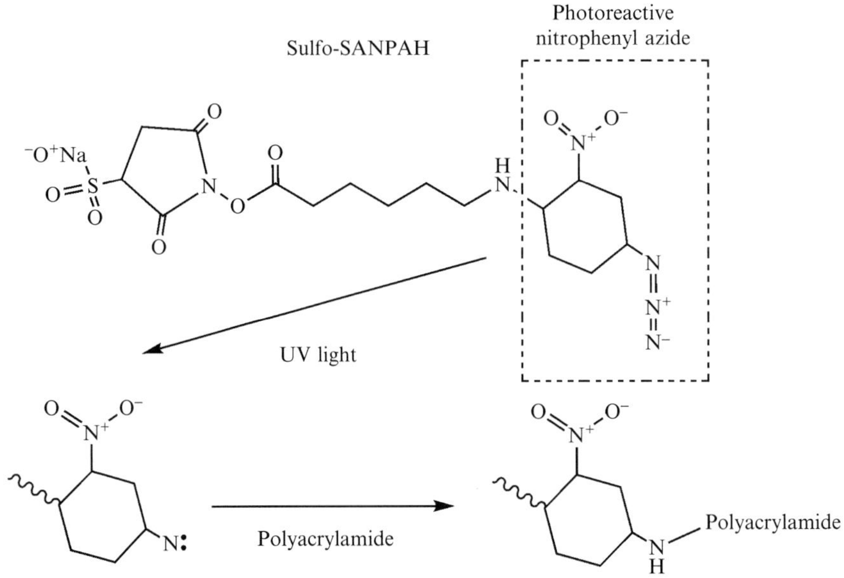

Fig. 3 The heterobifunctional photoreactive reagent sulfo-SANPAH. When the nitrophenyl azide reactive group is exposed to UV light (320–350 nm), it forms a nitrene group that can react nonspecifically with polyacrylamide. The sulfosuccinimydyl group at the other end reacts with primary amine groups of the protein.

can result in variations in the effectiveness of cross-linking. In addition, it is an expensive reagent with a limited shelf life. To overcome these shortcomings, a number of alternative approaches have arisen in the literature. In the sections below we will describe three alternative methods. Although, two of them involve novel reagents, have distinct advantages, and show potential as alternatives for sulfo-SANPAH.

A. Carbodiimide-Mediated Cross-Linking

Considered a classic method of protein coupling, the reagent 1-ethyl-3-(3-dimethylaminopropyl)carbodiimide-HCL [also known as *N*-(3-dimethylaminopropyl)-*N'*-ethylcarbodiimide hydrochloride, EDC or EDAC] has been used successfully for conjugating proteins to the polyacrylamide hydrogel (Beningo and Wang, 2002b). This method relies on the reaction of EDC with a free carboxyl group forming an amine-reactive intermediate, an *O*-acylisourea ester, for protein conjugation (Fig. 4). Since the polymer of acrylamide does not provide the necessary free carboxyl group, acrylic acid must be incorporated into the gel. The reaction scheme exploits the polymerization chemistry of acrylamide and bis-acrylamide. Acrylic acid, the deamidation product of acrylamide, copolymerizes with the acrylamide–bis-acrylamide mixture by the same free radical reaction

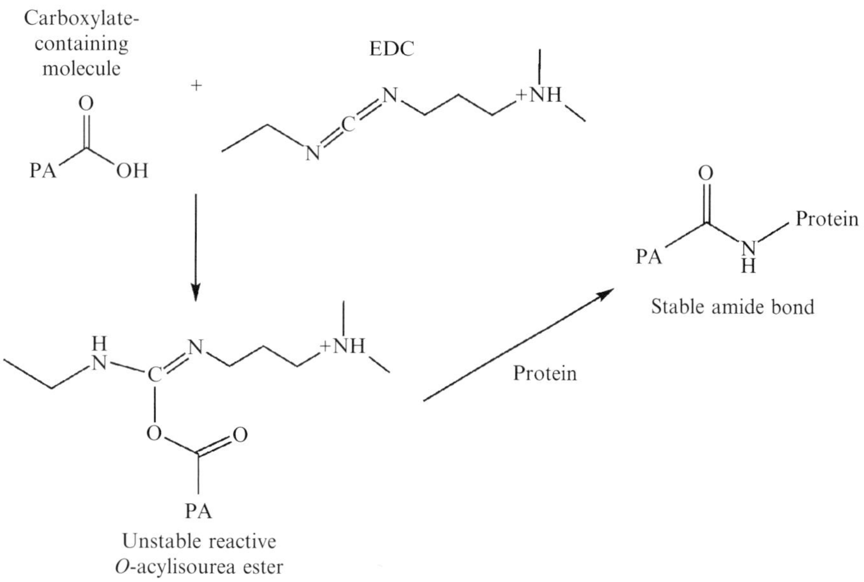

Fig. 4 The carbodiimide method of protein coupling. EDC reacts with acrylic acid incorporated into the polyacrylamide and activates the carboxyl group to form an unstable reactive *O*-acylisourea ester intermediate. This reaction allows it to be coupled to primary amines.

that drives the polymerization of polyacrylamide (Menter, 2000). A final concentration of 0.2% acrylic acid, which should be adjusted in proportion to the concentration of total acrylamide, has been used for a gel of 5% polyacrylamide.

The following protocol is adapted from a procedure described by Grabarek and Gergely (1990) and modified specifically for polyacrylamide substrates (Beningo and Wang, 2002b). After polymerization, the gels should be washed extensively in 0.1-M MES [2-(*N*-morpholino)ethanesulfonic acid], pH 6. EDC (Pierce Biotech, Rockford, Illinois) is dissolved at 26 mg/ml in 0.1-M MES and pooled onto the surface of the polyacrylamide substrate for 2 h at room temperature. Note that EDC breaks down easily on exposure to moisture, and must be stored in a desiccated, refrigerated environment and dissolved immediately before use. In addition, as the reactive intermediate is quite unstable, the reagent NHS (*N*-hydroxysuccinimide; Pierce, Illinois) is often included in the solution at a concentration of 0.6 mg/ml to increase the yield of protein conjugation. After incubation, the solution is rinsed well with 0.1-M MES to remove unreacted EDC. The protein of choice at 1 mg/ml in 0.1-M MES is pooled onto the substrate and incubated overnight at 4°C. Technically, the coupling reaction should occur rapidly as the intermediate is short-lived; however, in our experience overnight incubation yields the best result (K. A. B., unpublished observation). The substrates are rinsed with PBS and used immediately or stored in PBS at 4°C.

In comparison to sulfo-SANPAH, this method is somewhat more tedious, requiring the acclimation of the gel in MES and overnight incubation with the protein. However, it does not require a UV lamp and the reagents are relatively economical. Furthermore, it seems to work well on a number of proteins that do not couple effectively with sulfo-SANPAH, and in situations where the geometry of the polyacrylamide is not evenly accessible to the light source, for example with polyacrylamide beads.

B. Activation with Acrylic Acid *N*-Hydroxysuccinimide Ester

An alternate technique for conjugating proteins to polyacrylamide gel surfaces involves using the commercially available acrylic acid *N*-hydroxysuccinimide ester (NHS-acrylate, also known as *N*-hydroxysuccinimidyl acrylate; Collaborative Biomedical, Bedford, Massachusetts) (Fig. 5). The adapted method combines techniques from previous studies into a one-step polymerization reaction (Pelham and Wang, 1997; Schnaar *et al.*, 1978) to concentrate ligands at the surface of a gel by incorporating a reactive group during the acrylamide polymerization reaction rather than attaching a reactive group to a preformed gel. An aqueous layer containing acrylamide, bis-acrylamide, and reaction initiators was overlaid with an immiscible toluene solution containing succinimide acrylate (a protein-reactive ester of acrylic acid). This results in a reactive gel surface that once treated with an amine-containing ligand will produce a homogenous ligand surface (Fig. 6).

NHS-acrylate copolymerizes with acrylamide–bis-acrylamide as does acrylic acid described above. The functional NHS moiety, which is reactive with amine groups, will incorporate into the gel. Addition of any amine-containing ligand thereafter will displace the NHS and result in a covalent bond between the gel and the ligand. Ligand concentration can be controlled by varying the amount of NHS-acrylate in the organic layer of the polymerization (Schnaar *et al.*, 1978). Dissolving the NHS-acrylate in a solvent that is immiscible with water will create a nonaqueous NHS-containing layer at the surface of the aqueous polymerizing acrylamide solution. The NHS linker groups, therefore, will only incorporate at the gel surface.

The insoluble toluene layer overlaid on the polymerizing acrylamide layer also shields the process from air and improves the efficiency of the reaction by inhibiting the access of oxygen as described earlier. The advantage of incorporating NHS-acrylate only at the gel surface and not throughout the gel interior, as could be done with a water-soluble reactive acrylate derivative, is that when NHS-acrylate is hydrolyzed within the gel interior, the reaction product remaining linked to the gel becomes negatively charged at neutral pH. Fixed charges on a cross-linked polymer matrix cause large changes in gel swelling with minor changes in temperature, pH, and divalent cation concentration or after addition of nonaqueous solvents. Gel swelling will alter its stiffness and can cause cracks or breaks to appear in the gel as, for example, medium is changed or gels are moved from the benchtop to the incubator.

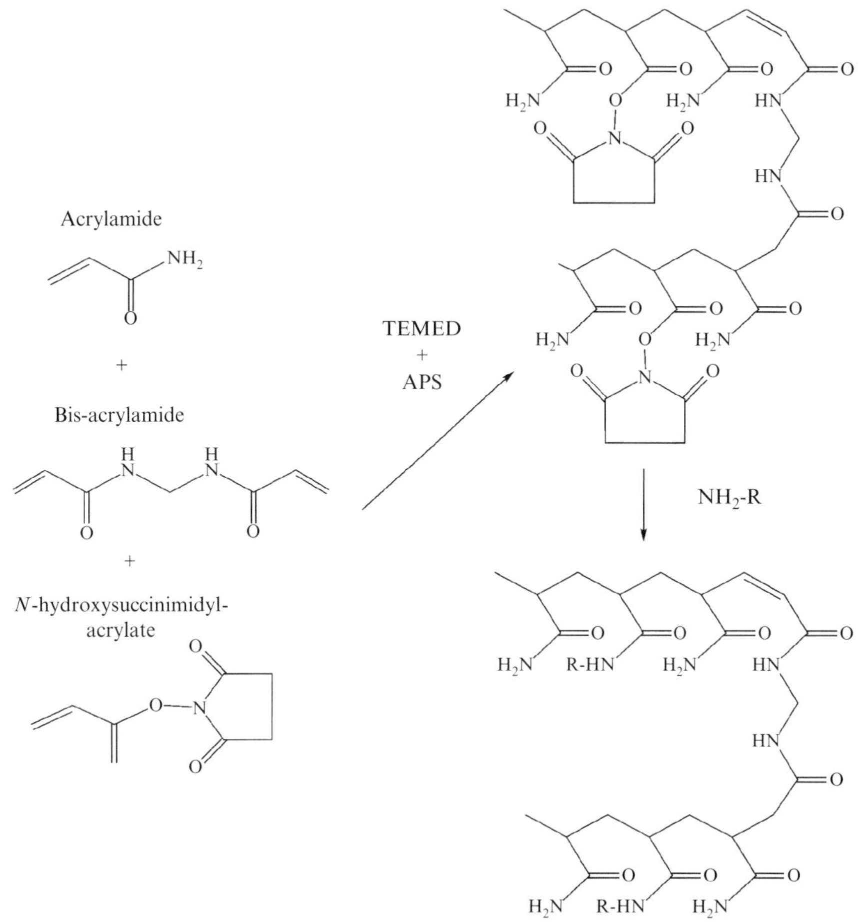

Fig. 5 Incorporation of NHS-acrylate into the hydrogel. Acrylamide monomers are made into free radicals by APS and TEMED, and polymerize into polyacrylamide chains, which are randomly cross-linked by bis-acrylamide that carries free radicals on both ends of their structures. NHS-acrylate also incorporates into the gel during the polymerization reaction. An amine-containing ligand (NH$_2$-R) covalently attaches via the NHS moiety.

Detailed explanation of polymerization with NHS-acrylate is as follows (Fig. 6); acrylamide and bis-acrylamide solutions are degassed in a vacuum chamber for 15 min. NHS-acrylate is dissolved in toluene (2%, w/v). To polymerize the solution, 1.5-μl TEMED and 5 μl of 10% ammonium persulfate are added to the degassed solution with the appropriate amount of 50-mM HEPES (pH 8.2) to yield a final volume of 500 μl. A fixed volume of 60 μl of the mixture is immediately pipetted onto the center of the 25-mm-diameter glass coverslip, which may be preactivated with glutaraldehyde as described earlier to provide stable gel attachment. A 10-μl drop of NHS-acrylate in toluene is pipetted onto the polyacrylamide solution.

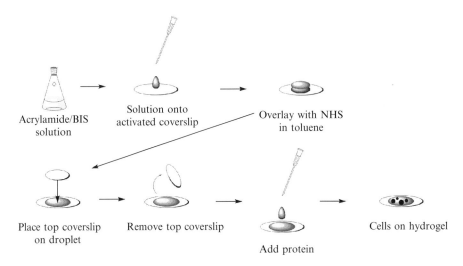

Fig. 6 Covalent coupling of proteins to the surface of polyacrylamide hydrogels with NHS-acrylate. Activated acrylamide solution is dropped onto a reactive coverslip. The top surface of the solution is covered with a nonaqueous solution of NHS-acrylate in toluene. The top coverslip is placed on the droplet to create a flat gel of defined width. After polymerization, the top coverslip is removed and the gel is incubated with the protein of choice.

An 18-mm-diameter coverslip is then carefully placed on top of the two copolymerizing layers. The polymerization is completed in minutes after which the top coverslip is slowly peeled off with a pair of fine tweezers. The bottom coverslip with the attached polyacrylamide gel is immersed in a six-well plate with 50-mM HEPES (pH 8.2). The gels are washed extensively in HEPES, and 100 μl of the adhesive protein of choice is pipetted onto the gel surface and allowed to incubate for at least 20 min at 37°C.

The advantage of this method is the one-step incorporation of the reactive species into the polymers itself and its confinement to the top surface of the gel. This enhances the effectiveness of the coupling and the speed at which the reactions can be carried out. In comparison to the sulfo-SANPAH and carbodiimide methods, this is a more efficient method offering effective, homogenous coupling that is less labor intensive.

C. Activation with the *N*-Succinimidyl Ester of Acrylamidohexanoic Acid (N6)

The third method of protein conjugation involves the agent *N*-succinimidyl ester of acrylaminohexanoic acid (N6), using methods previously described by Pless *et al.* (1983) and later modified for the polyacrylamide hydrogels (Reinhart-King *et al.*, 2005; Chapter 23 by Johnson *et al.*, this volume). As for NHS-acrylate, one end of the molecule mediates the incorporation into polyacrylamide whereas the other end is reactive toward primary amines. This procedure requires the synthesis of N6 in two stages (Pless *et al.*, 1983; Fig. 7). Basically, in alkaline aqueous solution, acryloyl chloride is added to 6-aminohexanoic acid resulting in 6-acrylamidohexanoic acid. The cross-linker EDC is used to catalyze the esterification with *N*-hydroxysuccinimide providing an *N*-succinimidyl ester. Treatment of *N*-succinimidyl ester with 6-aminohexanoic acid adds a second hexanoic moiety yielding 6-acrylamidohexylaminohexanoic acid. EDC is used a second time to catalyze esterification with *N*-hydroxysuccinimide resulting in the final product referred to as N6, 6-acrylaminohexylaminohexanoic acid *N*-succinimidyl ester.

For protein coupling to the hydrogels, the cross-linker N6 dissolved in ethanol is incorporated throughout the polyacrylamide hydrogels, by adding it at a concentration of 20 μM to the acrylamide and bis-acrylamide solution prior to initiating the polymerization reaction with TEMED and ammonium persulfate (Willcox *et al.*, 2005). The acrylamide copolymerizes with the N6, which presents an *N*-succinimidyl ester capable of linking the amine-containing protein. The protein of interest may then be pooled onto the surface and reacted. Alternatively, as described in further studies by the same group, 2-(2-pyridinyldithio) ethaneamine hydrochloride (PDEA) can be reacted with the *N*-succinimidyl ester of N6 to produce thiol-reactive sites (Willcox *et al.*, 2005). Peptides can then be synthesized that have terminal cysteines to be reacted with these sites on the gel. The reaction

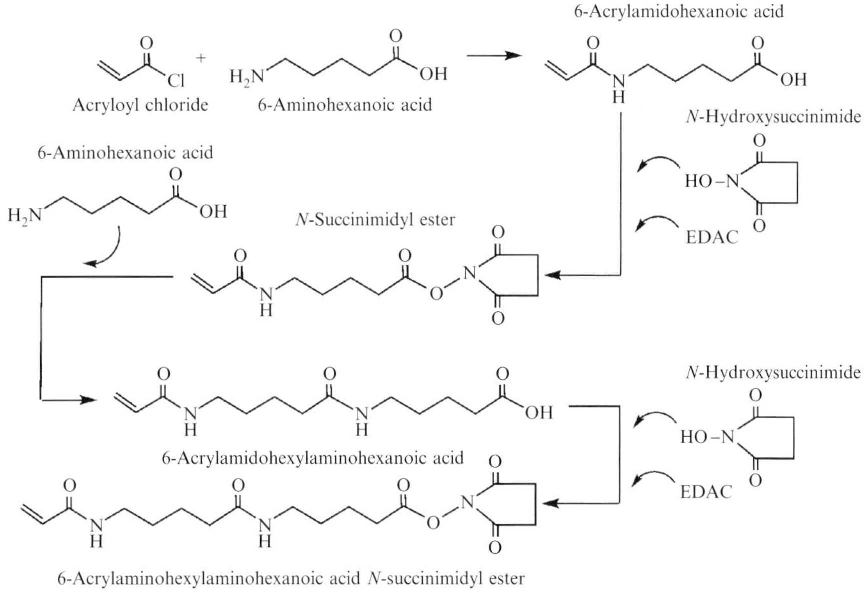

Fig. 7 Synthesis of the *N*-succinimidyl ester of acrylamidohexanoic acid (N6). N6 is prepared by a two-step process involving the reaction of acryloyl chloride with 6-aminohexanoic acid, and the reaction of the product with EDC and NHS. This process is repeated a second time resulting in the synthesis of N6. Figure adapted from Pless *et al.* (1983).

produces a chromophore by-product that can be quantified by spectrophotometry, thus providing information about the amount of ligand actually bound to the gel.

This technique provides several advantages over the original design. First, it incorporates the protein-coupling reagent throughout the gel, eliminating the issue of uniform distribution and allowing rapid reaction with amine groups. Second, it addresses a weakness in the initial design of the polyacrylamide hydrogels—an accurate method for quantifying the amount of protein actually coupled to the gel. However, one limitation to this method is the synthesis of N6, which may not be trivial for many biologically oriented laboratories. Furthermore, the stability of N6 as either a stock solution or incorporated into the gels is unclear. Nonetheless, this technique does afford an alternative method that would be useful for more accurate determination of conjugated protein concentrations.

D. Other Protein-Coupling Methods

In addition to the coupling methods described above, there are two other coupling methods that deserve consideration by the reader. First, the hydrazine method described by Damljanovic *et al.* (2005) employs the powerful reducing agent hydrazine hydrate to alter the amide groups of polyacrylamide yielding reactive hydrazide groups. Proteins that have been oxidized to contain aldehyde or ketone groups then react with the reduced polyacrylamide to form covalent bonds. While the additional chemical modifications may introduce problems for some proteins, the authors have demonstrated typical cell binding and increased protein conjugation efficiency over other coupling methods. Obvious drawbacks to this technique are the hazard of the hydrazine hydrate reagent and the additional time and labor needed in protein oxidation.

One of the easiest methods for rendering polyacylamide adhesive has been described by de Rooji *et al.* (2005). A positively charged derivative of acrylamide is incorporated into the normal acrylamide mixture at a desired ratio, followed by electrostatic adsorption of proteins of interest to the surface before cell culture. Unfortunately, at the time of publication, the modified acrylamide (triethylammonium-acrylamide) was no longer commercially available. However, gels incorporating negatively charged acrylic acid, as described earlier for EDC activation, may function in a similar manner without covalent conjugation for some cell types. While this general approach appears simple and promising, the efficiency for the adhesion of various cell types has yet to be characterized.

IV. Optimizing the Placement of Beads for Traction Force Microscopy

Polyacrylamide hydrogels were initially designed for the measurement of traction forces, which relies on the measurement of the displacements of fluorescent microbeads embedded in the hydrogel. Data collection of these displacements requires precise focusing on the beads within a single focal plane below the migrating cell. Since the beads are dispersed throughout the gel, it can be confusing to maintain the focal plane on the same set of beads during a time-lapse experiment, particularly since data collection typically involves focusing back and forth between beads and the cell. A second limitation is that the light scattering and out of focus signals from beads on other focal planes can sometimes interfere with the pattern recognition algorithm for the detection of displacements. By modifying published procedures, Bridgman *et al.* (2001) have developed an innovative solution to these problems.

To reduce the background fluorescence from beads on other focal planes, this group overlaid a thin layer of gel containing beads on a polyacrylamide substrate without beads (Bridgman *et al.*, 2001). Gels made in this fashion are thick enough to avoid significant mechanical constraint caused by adhesion of the hydrogel to the underlying coverslip, while confining the beads within a minimal thickness thereby tremendously reducing interference from out of focus signals. The bottom layer is first cast underneath a coverslip as a gel of 12-mm diameter and $\sim$75-μm thickness without fluorescent beads. The coverslip is then removed and the activated acrylamide solution, of the same composition but containing fluorescent beads, is pipetted on top of the first. A 22-mm-round coverslip is gently placed on the bead-containing solution and the second layer is allowed to polymerize. The larger diameter of the second coverslip allows excess solution to be collected around the first layer generating a very thin overlay. Removal of the top coverslip must be done slowly, with the gels fully immersed in buffer, to prevent distortion of the soft gel surface and to yield a thin but intact top layer of gel containing beads. Using a 0.75-NA 40$\times$ lens, all beads within the central ($\sim$8-mm diameter) area may be observed on a single focal plane. As for most gel cultures, the distance between cells from the coverslip precludes the use of most oil immersion lens, due to the limited working distance and/or spherical aberration. Therefore, a water immersion lens is required for high-magnification observations of hydrogel cultures that exceed 10 μm in thickness.

V. Manipulation of Gel Geometry

A. Preparation of Polyacrylamide Microbeads

Polyacrylamide is versatile because of its capacity to begin as a solution and to be molded before solidifying. Thus, in addition to being used as sheets for cell adhesion and migration, polyacrylamide can be molded into any number of shapes, including spherical beads (Beningo and Wang, 2002b). These beads can be used as phagocytic targets or microcarriers to support cell growth. Polyacrylamide microbeads have been used to demonstrate the mechanosensing aspect of phagocytosis (Beningo and Wang, 2002b).

Polyacrylamide microbeads were prepared in a microemulsion system, using bis(2-ethylhexyl) sulfosuccinate (AOT) (Fluka Chemicals, Buchs, Switzerland) in toluene as the emulsifier. This agent is capable of forming inverse micelles (i.e., micelles of aqueous solution in an organic solvent) of relatively uniform size. The size of the beads is determined by several factors, including the AOT concentration and the relative volume of the aqueous solution to the organic solvent (Kunioka and Ando, 1996).

Beads of 1–6 μm in diameter were prepared for phagocytosis studies. The solution of acrylamide is prepared as described earlier for the substrate; however, instead of fluorescent beads, a high molecular weight FITC dextran (464 kDa, Molecular Probes, California) is added at a concentration of 20 mg/ml. The dextran becomes trapped within the polyacrylamide beads and serves as a label. AOT is dissolved in toluene in a fume hood at a concentration of 10.2 mg in 1 ml. While stirring under a stream of nitrogen, the degassed acrylamide mixture is added to the AOT–toluene solution. After stirring under nitrogen for 30 min to allow polymerization, the beads are recovered by centrifugation at $23 \times g$ for 5 min in a microcentrifuge and washed repeatedly (three to five times) in methanol to remove the emulsifier. Following multiple washes in PBS, the beads may be stored at 4°C for up to 8 months. Conjugation of proteins to the beads is carried out with the EDC activation method as described in the previous section, which yields more consistent results than using sulfo-SANPAH. To perform Fc-receptor-mediated phagocytosis, bovine serum albumin (BSA) was conjugated to the beads overnight at room temperature, followed by incubation with affinity-purified rabbit anti-BSA IgG at 37°C for 1 h. Large particles and aggregates were removed by centrifugation at 20–$30 \times g$ in a microcentrifuge for 3 min. As with the polyacrylamide sheets, the beads swell and shrink in response to changes in osmolarity and temperature, thus it is important to acclimate them to the experimental condition before measuring the diameter and using them in a phagocytic assay. Protein-conjugated beads should be used within 2 days.

To create a microcarrier system for cell culture, larger beads can be made by optimizing the ratio of AOT to toluene and aqueous solution. The beads can then be conjugated with extracellular matrix proteins using the EDC reaction as described for substrate sheets. After extensive washing and acclimation in the appropriate medium and temperature, the suspension of microcarriers is transferred to a stir flask and used as in other microcarrier culture. While other microcarrier culture systems are commercially available, they can be expensive and may not be optimal for optical examinations due to their opaque quality or for protein recovery due to the charged nature. Polyacrylamide beads offer an attractive alternative to these systems.

B. Preparation of a Model Three-Dimensional Culture System

Cells grown in a two-dimensional culture display many characteristics not found in their native three-dimensional environment. Unfortunately, given the toxicity of acrylamide, it is neither possible to embed the cells within the matrix as with collagen nor reasonable to expect a cell to invade the covalently cross-linked hydrogels. In order to take advantage of the unique qualities of polyacrylamide, it would be ideal to create a model system that mimics at least some properties of a three-dimensional environment.

A simple approach involves sandwiching cells between two polyacrylamide hydrogels (Beningo and Wang, 2006; Beningo *et al.*, 2004; Fig. 8). This method does not fully embed the cells in the environment but it does engage at least part of the dorsal ECM receptors, enough to induce a morphology resembling that found *in vivo*. Furthermore, it allowed the manipulation of compliance and the measurement of traction forces.

Polyacrylamide substrates used in the double-hydrogel studies are prepared as described earlier for two-dimensional cultures. Cells are first seeded onto the designated bottom gel and allowed to adhere for a minimum of 2 h. The top coverslip, with the hydrogel attached, is trimmed with a diamond tip pen such that the shape of the glass matches the contour of the gel. This will allow the gel to fit on top of the bottom gel unimpeded by the sides of the culture chamber, thus maximizing the contact between the two substrates. Extraneous medium on the top

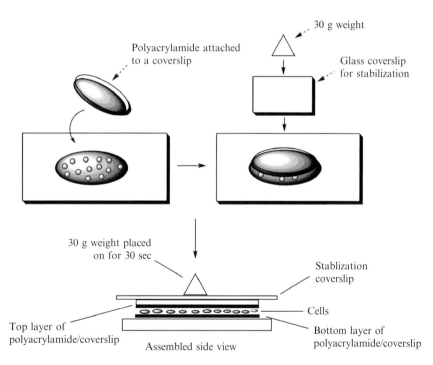

Fig. 8 The double hydrogel as a model system for three-dimensional culture. Cells are sandwiched between two polyacrylamide hydrogels to simulate a three-dimensional environment. The dorsal surface of the cell binds to the proteins conjugated to the top substrate; however, the cells are not entirely surrounded or embedded within the hydrogels. The engagement of dorsal receptors is sufficient to drive cultured fibroblasts into a more physiologically relevant, elongated morphology.

and bottom polyacrylamide gels is removed by aspiration with a Pasteur pipette. The top hydrogel is then gently laid over the bottom gel on which the cells are growing. A thick piece of glass (2 cm × 2 cm × 5 mm, 4.4 g) is then placed on top of the sandwich to push the top substrate down. However, in most cases this is insufficient to allow the adhesion of dorsal surface of the cell to the top substrate. A brass or stainless steel weight of 30 g is therefore applied on the weighing glass for 30 sec with a minimal amount of medium. The medium is then replenished on removal of the metal weight. Cells that make contact with the top substrate will adopt an *in vivo*-type morphology and reach a steady state as soon as 2 h after forming the sandwich. The responses require extracellular matrix coating on the top gel, indicating that it is not due only to the pressure although mechanical forces may contribute to the responses.

This semi-three-dimensional system affords all the benefits of the two-dimensional hydrogels, including variability of compliance, control of protein coating, optical quality, and the feasibility of measuring traction forces exerted on both the dorsal and ventral surfaces. The main limitation of this method is that the cells are not in a truly three-dimensional environment. In addition, it is evident that the dorsal surface of thin, peripheral regions of most cells remains separated from the top substrate. Due to the presence of debris, a fraction of cells may remain entirely in a two-dimensional environment, limiting the utility of this system for biochemical analyses. To identify cells that make contact with both top and bottom substrates, fluorescent beads (of a different color from those used on the bottom substrate) are added to the top gel, and cells that appear on the same focal plane as some beads in the top gel are chosen for observations. These beads also allow the detection of traction forces on the top substrate. Furthermore, it is possible to do immunofluorescence on cells within the double-substrate hydrogel by simply adding the fixative to the culture system after first rinsing with an appropriate buffer and removing the weighing glass. The addition of fixative will allow the top substrate to float off during fixation. Both the bottom and top substrates can be retained for further staining. Alternatively, GFP-labeled proteins of interest can be expressed in the cells prior to culturing in the double-substrate hydrogel.

The double-substrate hydrogel method could be extended to other gel systems. In addition, a modified version involves overlaying a two-dimensional polyacrylamide hydrogel with a layer of collagen matrix, agrose, alginate, or matrigel (Chapter 23 by Johnson *et al.*, this volume).

VI. Concluding Remarks

It was the objective of this chapter to collect a number of recently published improvements and alternative uses of the polyacrylamide hydrogel system in one place for evaluation by potential users. These improvements address the ultimate goal of using elastic substrates for the study of cellular mechanical interactions in a physiological relevant setting (Chapter 22 by Engler *et al.*, this volume for relation to tissue elasticity). The improvements in protein coupling (Balaban *et al.*, 2001; Beningo and Wang, 2002a,b, 2006; Beningo *et al.*, 2002, 2004; Bridgman *et al.*, 2001; Charambach and Rodbard, 1971; Damljanovic *et al.*, 2005; de Rooji *et al.*, 2005; Dembo and Wang, 1999; du Roure *et al.*, 2005; Grabarek and Gergely, 1990; Kunioka and Ando, 1996; Menter, 2000; Oliver *et al.*, 1998; Pelham and Wang, 1997; Pless *et al.*, 1983; Reinhart-King *et al.*, 2005; Schnaar *et al.*, 1978; Tan *et al.*, 2003; Wang and Pelham, 1998; Willcox *et al.*, 2005; Wong *et al.*, 2003), gel polymerization (Wong *et al.*, 2003), optics (Bridgman *et al.*, 2001), and alternative uses (Beningo and Wang, 2002b; Beningo and Wang, 2006) are likely to continue in the coming years. Further improvements in the spatial and temporal resolution of traction forces and in the development of substrates that allow real-time manipulations of mechanical properties will likely have a substantial impact on the understanding of complex mechanical interactions between cells and the environment.

References

Balaban, N. Q., Schwarz, U. S., Riveline, D., Goichberg, P., Tzur, G., Sabanay, I., Mahalu, D., Safran, S., Bershadsky, A., Addadi, L., and Geiger, B. (2001). Force and focal adhesion assembly: A close relationship studied using elastic micropatterned substrates. *Nat. Cell Biol.* **3**(5), 466–472.

Beningo, K. A., Dembo, M., and Wang, Y. L. (2004). Responses of fibroblasts to anchorage of dorsal extracellular matrix receptors. *Proc. Natl. Acad. Sci. USA* **101**, 18024–18029.

Beningo, K. A., Lo, C. M., and Wang, Y. L. (2002). Flexible polyacrylamide substrata for the analysis of mechanical interactions at cell-substratum adhesions. *Methods Cell Biol.* **69**, 325–339.

Beningo, K. A., and Wang, Y. L. (2002a). Flexible substrata for the detection of cellular traction forces. *Trends Cell Biol.* **12**(2), 79–84.

Beningo, K. A., and Wang, Y. L. (2002b). Fc-receptor mediated phagocytosis is regulated by mechanical properties of the target. *J. Cell Sci.* **115**, 849–856.

Beningo, K. A., and Wang, Y. L. (2006). Double-hydrogel substrate as a model system for 3D cell culture. *Methods Mol. Biol.* **370**, 203–211.

Bridgman, P. C., Dave, S., Asnes, C. F., Tullio, A. N., and Adelstein, R. S. (2001). Myosin IIB is required for growth cone motility. *J. Neurosci.* **21**, 6159–6169.

Charambach, A., and Rodbard, D. (1971). Polyacrylamide gel electrophoresis. *Science* **172**, 440–451.

Damljanovic, V., Lagerholm, B. C., and Jacobson, K. (2005). Bulk and micropatterned conjugation of extracellular matrix proteins to characterized polyacrylamide substrates for cell mechanotransduction assays. *Biotechniques* **39**, 847–851.

de Rooji, J., Kerstens, A., Danuser, G., Schwartz, M. A., and Waterman-Storer, C. M. (2005). Integrin-dependent actomyosin contraction regulates epithelial cell scattering. *J. Cell Biol.* **171**, 153–164.

Dembo, M., and Wang, Y. L. (1999). Stresses at the cell-to-substrate interface during locomotion of fibroblasts. *Biophys. J.* **76**, 2307–2316.

du Roure, O., Saez, A., Buguin, A., Austin, R. H., Chavrier, P., Silberzan, P., and Ladoux, B. (2005). Force mapping in epithelial cell migration. *Proc. Natl. Acad. Sci. USA* **102**, 2390–2395.

Grabarek, Z., and Gergely, J. (1990). Zero-length crosslinking procedure with the use of active esters. *Anal. Biochem.* **185**, 131–135.

Kunioka, Y., and Ando, T. (1996). Innocous labeling of the subfragment-2 region of skeletal muscle heavy meromyosin with a fluorescent polyacrylamide nanobead and visulization of heavy meromyosin molecules. *J. Biochem.* **119**, 1024–1032.

Menter, P. (2000). "Acrylamide Polymerization—A Practical Approach." Bio-Rad Laboratories, Hercules, CA.

Oliver, T., Jacobson, K., and Dembo, M. (1998). Design and use of substrata to measure traction forces exerted by cultured cells. *Methods Enzymol.* **298,** 497–521.

Pelham, R. J., and Wang, Y. L. (1997). Cell locomotion and focal adhesions are regulated by substrate flexibility. *Proc. Natl. Acad. Sci. USA* **94,** 13661–13665.

Pless, D. D., Lee, Y. C., Roseman, S., and Schnaar, R. L. (1983). Specific cell adhesion to immobilized glycoproteins demonstrated using new reagents for protein and glycoprotein immobilization. *J. Biol. Chem.* **258,** 2340–2349.

Reinhart-King, C. A., Dembo, M., and Hammer, D. A. (2005). The dynamics and mechanics of endothelial cell spreading. *Biophys. J.* **89,** 676–689.

Schnaar, R. L., Weigel, P. H., Kuhlenshmidt, M. S., Lee, Y. C., and Roseman, S. (1978). Adhesion of chicken hepatocytes to polyacrylamide gels derivatized with *N*-acetylglucosamine. *J. Biol. Chem.* **253,** 7940–7951.

Tan, J. L., Tien, J., Pirone, D. M., Gray, D. S., Bhadriraju, K., and Chen, C. S. (2003). Cells lying on a bed of microneedles: An approach to isolate mechanical force. *Proc. Natl. Acad. Sci. USA* **100,** 1484–1489.

Wang, Y. L., and Pelham, R. J., Jr. (1998). Preparation of a flexible, porous polyacrylamide substrate for mechanical studies of cultured cells. *Methods Enzymol.* **298,** 489–496.

Willcox, J. P., Reinhart-King, C. A., Lahr, S. J., DeGrado, W. F., and Hammer, D. A. (2005). Dynamic heterodimer-functionalized surfaces for endothelial cell adhesion. *Biomaterials* **26,** 4757–4766.

Wong, J. Y., Velasco, A., Rajagopalan, P., and Pham, Q. (2003). Directed movement of vascular smooth muscle cells on gradient-compliant hydrogels. *Langmuir* **19,** 1908–1913.

CHAPTER 3

Microscopic Methods for Measuring the Elasticity of Gel Substrates for Cell Culture: Microspheres, Microindenters, and Atomic Force Microscopy

Margo T. Frey,★ Adam Engler,† Dennis E. Discher,† Juliet Lee,‡ and Yu-Li Wang★

★Department of Physiology
University of Massachusetts Medical School
Worcester, Massachusetts 01605

†Department of Chemical and Biomolecular Engineering
University of Pennsylvania
Philadelphia, Pennsylvania 19104

‡Department of Molecular and Cell Biology
University of Connecticut
Storrs, Connecticut 06269

Abstract

In conjunction with surface chemistry, the mechanical properties of cell culture substrates provide important biological cues that affect cell behavior including growth, differentiation, spreading, and migration. The phenomenon has led to the increased use of biological and synthetic polymer-based flexible substrates in cell culture studies. However, widely used methods for measuring the Young's modulus have proven difficult in the characterization of these materials, as they tend to be relatively thin, soft, hydrated, and tethered to glass substrates. Here we describe three methods that have been applied successfully to probe the flexibility of soft culture substrates.

I. Introduction

A number of chapters in this volume have highlighted the important influence of soft substrates on cell adhesion, cell structure, and cell mechanics. These studies rely on reliable measurements of the Young's modulus (Chapter 1 by Janmey *et al.*, and Chapter 2 by Kandow *et al.*, this volume) of the culture substrate. However, classical methods for measuring material elasticity and other mechanical properties generally require macroscopic samples, often of a specific geometry, while gels intended for cell culture are generally formed as a thin layer adhered to the culture dish. Reliable measurements of such gels must be performed *in situ* because (1) gels can be so soft that macroscopic samples are difficult to handle; (2) gels are hydrated and sometimes temperature sensitive so smaller samples are more homogeneous and easier to control; (3) gels are often tethered to the underlying cover glass, which may affect the expansion or compression of gels; and (4) with cells adhering to the gel surfaces, measuring the elasticity of the gel surface instead of the bulk will provide values more relevant to cell behavior.

The purpose of this chapter is to describe three microscopic methods suitable for measuring the elasticity of gel substrates *in situ*. While all of them are based on Hertz contact mechanics (Hertz, 1882), namely the indentation in response to forces exerted with a probe of known geometry, they differ substantially in cost, simplicity, and resolution. Since indentations with metal microspheres under gravity and with atomic force microscopy (AFM; Chapter 15 by Radmacher, this volume) have been described in the literature, the description here will be limited to several key aspects critical to gel measurements. The method of microneedle indentation, which combines features of the two other methods, was developed recently as both a qualitative and a quantitative tool for probing substrates with modulatable mechanical properties. Since it has never been described, sufficient details will be provided for reproduction.

With all three methods, the measurements should be performed on fully acclimated substrates since hydrogel substrates are often sensitive to buffer conditions and temperature. In addition, the thickness of the testing sample should be close to that used for actual experiments. This is particularly important for polyacrylamide gels, since the surface rigidity decreases progressively with increasing thickness due to the covalent bonding of the gel to the coverslip to limit gel swelling and prevent detachment (Chapter 2 by Kandow *et al.*, this volume).

II. Probing with Microspheres Under Gravitational Forces

The microsphere indentation method is performed by measuring the indentation depth of a relatively heavy metal microsphere placed on top of the substrate (Lo *et al.*, 2000). It works with a high degree of reproducibility for both polyacrylamide and gelatin substrates with Young's moduli of $E \sim 2.2$–33 kPa. Hydrogel substrates are prepared with embedded fluorescent latex beads 0.1–0.2 μm in diameter. As the method relies strongly on the ability to focus accurately on the substrate surface, using these beads as markers, the beads may be confined to the top surface using the approach described in Chapter 2 by Kandow *et al.*, this volume. Since hydrogel substrates are often sensitive to buffer conditions and temperature, all the measurements should be performed with fully acclimated substrates (this also applies to the two other methods). With the substrate seated on the stage of an inverted microscope, a stainless steel microsphere of known diameter (0.3–0.62 mm) and density (e.g., 7.72 g/cm^3 for Pen grade 420SS stainless steel, Hoover Precision, East Granby, Connecticut) (Fig. 1) is gently placed near the field of observation with a pair of fine forceps. The measurement is made using an objective with a working distance larger than the thickness of the gel and with a numerical aperature as high as possible. Indentation is determined using the calibrated fine focus control of the microscope. For most microscopes the fine focus knob is marked in 1-μm increments, which can be confirmed and/or recalibrated by focusing through a second glass coverslip of known thickness with permanent ink marks on either side.

To avoid problems with backlash of the gears, the microscope should be focused first beneath and near the center of the metal microphere to mark the height of the indented surface, which is recognized as the region with maximal indentation. After the microscope is focused on the fluorescent beads near the gel surface of

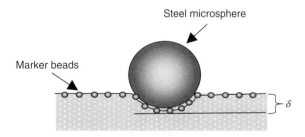

Fig. 1 Illustration of the indentation, δ, made by a steel microsphere placed on a gelatin substrate embedded with marker beads.

this region, the microsphere is removed with a magnet and the ensuing vertical movement of the gel surface measured by refocusing on the same group of fluorescent beads. Measurements are taken at five random locations on the substrate, and the average indentation is used to calculate the Young's modulus (E) given by the Hertz equation [Eq. (1)] (Hertz, 1882):

$$E = \frac{3(1 - v^2)f}{4r^{1/2}\delta^{3/2}} \tag{1}$$

where r is the bead radius, f is the force applied by the steel ball, δ is the indentation of the substratum, and v is the Poisson ratio of the hydrogel, assumed here to be 0.5 for gelatin (Li *et al.*, 1993) and 0.45 for polyacrylamide (Section IV.F). Note that f equals the gravitational forces minus the buoyancy, which equals the weight of the liquid displaced by the volume of the microsphere. By this method, the Young's modulus for 3% gelatin gels should be ~3 kPa.

Although the microsphere indentation method is simple and cost-effective, it tends to yield Young's moduli higher than those obtained with the other two methods. One possible source of error is that if δ is a significant fraction of the thickness of the gel, then the rigid glass surface below the gelatin or other gel substrate will significantly limit δ. Overestimation of E can also occur due to high systematic error at low indentation depths, for example, from a higher actual contact area than that expected from the Hertz equation (Yoffe, 1984; discussed later). To minimize these problems, the density and radius of microspheres should be chosen such that $\delta < 0.2t$ and $>0.3r$, where t is the thickness of the gel. In practice, the versatility of the microsphere indentation method is limited by the availability of microsphere to meet these criteria, particularly for stiff gels.

III. Atomic Force Microscopy

Several key aspects of the microsphere indentation technique are implemented and extended in a commercially available, high-resolution instrument known as the AFM. A number of designs are available from different AFM manufacturers, with prices starting at about half the cost of an inverted optical microscope. While imaging applications of AFM are relatively well known (Bennig *et al.*, 1986), the AFM in the "force mode" is highly suited to detailed mechanical assessments at microscale resolution (Weisenhorn *et al.*, 1989).

The key component of the AFM is a microfabricated flexible cantilever with a micrometer scale probe tip that is generally pyramidal in shape but can also be a sphere. These cantilevers are purchased individually or as wafers, with a spring constant that comes precalibrated to within about 50%; a more precise calibration is done on each cantilever at the time of use by standard methods supplied by the instrument manufacturer. The cantilever is displaced by a piezoelectric device to press the tip into an immobilized sample material, and the forced deflection of the

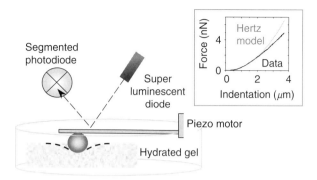

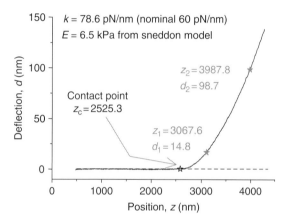

Fig. 2 Application of atomic force microscopy for probing the mechanical properties of gel substrates, by pressing into the surface and analyzing the resulting deflection. Light from a superintensified diode (or laser) is reflected off the end of the cantilever onto a segmented photodiode detector that magnifies small tip deflections into a detectable signal. When pressing the tip into the sample, indentation can be determined as the difference between tip deflection and the cantilever position and then plotted versus the force required to create the tip's deflection. The force–indentation plot can then be fit with a Hertz-type model (inset plot). The lower plot illustrates how the point of contact is determined from Eq. (4) in the text.

cantilever is measured by reflection of a laser off the backside of the cantilever and onto a position-sensitive photodetector (Fig. 2). Indentation of the material, δ, is determined by subtracting the deflection, d, of the cantilever from the distance driven by the piezoelectric device, z. The deflection is converted to force using the cantilever calibration curve, and force–indentation data can then be plotted as shown in the upper inset of Fig. 2.

A first consideration for an AFM cantilever is its spring constant, k. Spring constants have units of force per distance and range from relatively rigid ($k = 100$–1000 pN/nm) to soft (5–100 pN/nm). The latter yields a better signal-to-noise ratio when performing measurements on soft gel samples. Also of importance is the tip geometry, for example, pyramidal with different sharpness or spherical with different radii. Each probe type imparts a specific deformation field when indenting a material and as such, each has a mechanical model associated with it. For cantilevers with spherical tips such as those with $r = 2.5$-μm borosilicate beads mounted on 60-pN/nm cantilevers available commercially (e.g., Bioforce Nanoscience, Ames, Iowa), Eq. (1) given above for indentation is used, although the fit of f versus δ from AFM is conducted over many data points collected over just a few micrometers of indentation.

Pyramidal tips (of shape $\vee$) have different degrees of tip sharpness, and blunted tips are usually preferred for force mode (not so for imaging) as blunted tips tend to distribute the deformation field over a larger contact surface, minimizing the strain and potential damage to the sample. The relationship between the indentation force, f, and the deformation, δ, may be approximated with an axisymmetric cone model (Sneddon, 1965).

$$f = \left[\frac{2 \tan (\alpha)}{\pi} \right] \left[\frac{E}{1 - v^2} \right] \delta^2 \tag{2}$$

where α is the opening angle of the pyramid tip (i.e., how tapered the tip is; other symbols are the same as defined before). Blunted tips typically have an opening angle, α, of 35°, whereas sharpened tips have an angle of 18°. Note that the smaller contact area with the pyramidal tip relative to a spherical tip means that the indentation force scales more strongly with deformation, δ (as δ^2 rather than $\delta^{3/2}$). The difference gives spherical probes an advantage, as it translates to lower strains that allow proper corrections for soft, thin films below 20 μm (Dimitriadis *et al.*, 2002; Engler *et al.*, 2004a).

To measure E of hydrogels, samples are indented at a relatively modest rate of cantilever displacement 0.2–2.0 μm/sec (= dz/dt), which is generally sufficient to explore elastic rather than viscoelastic properties of substrates (Mahaffy *et al.*, 2000). The first step in determining the Young's modulus is to pinpoint when the tip first makes contact with the substrate, referred to as the contact point. While this is easy for a hard material, the transition for soft materials can be less obvious, but a formula from Domke and Radmacher (1998) can be used for such determination. It utilizes the deflection (d)–position (z) plot, by fitting the curve near two extremes of the region where the curve is to be used for rigidity analysis, and calculating the point of interception with the x-axis [Eq. (3)].

$$\text{Contact point} = \left(\frac{(d_2/d_1)^a [(z_2 - d_2) - (z_1 - d_1)]}{1 - (d_2/d_1)^a} \right) \tag{3}$$

The exponent $a = 1/2$ for a cone (approximating a pyramidal tip) and $a = 2/3$ for a sphere. With a properly chosen range of analysis—which is most accurate for a cantilever deflection range $d \sim 10$–100 nm as shown between the two gray stars in the lower plot of Fig. 2—the contact point calculated from Eq. (3) should accurately reflect where indentation starts. The difference between tip positions, starting from the contact point and moving down into the sample, generates the tip's actual deflection by the material, which is then converted into force with the force–deflection calibration curve and plotted against the indentation of the material. This force–indentation relationship may then be fit with the Hertz model for spherical tips or the Sneddon model for sharpened tips (Hertz, 1882; Sneddon, 1965), generally up to 2 μm of tip indentation, to determine E (Domke and Radmacher, 1998; Engler *et al.*, 2004a; Rotsch *et al.*, 1999).

A number of qualifications and conditions apply to such analyses. For example, for thin gels probed with spherical tips, a developed thin film approximation worked well (Dimitriadis *et al.*, 2002). Also, during measurement, attention should be paid to instrument parameters including indentation velocity, indentation distance, and data sampling rate. An example of such a relationship and fit is displayed as an inset in Fig. 2. For thick films, the resulting values of E typically do not depend on the range of indentation depth, but for thin films the measurements typically only fit the Hertz model at small indentations (Domke and Radmacher, 1998). Therefore, the range of analysis should be chosen according to the application.

The main drawback of AFM is the setup cost, and the limited longevity of the probe due to contaminations, which require replacement and calibration of new probes. Measurements of soft hydrogels can also be affected by adhesive interactions between the tip and the gel (Zhao *et al.*, 2003), which can bias the estimate of contact point. Additionally, unless the AFM is built on top of a high-quality light microscope, it will be necessary to move the sample between two different instruments to conduct both microscopic cellular studies and stiffness measurements of the gel. A third instrument may also be necessary to confirm gel thickness, required for a thin film correction of gels <20 μm in thickness (Richert *et al.*, 2004).

IV. Probing with Spherically Tipped Glass Microindenters

To address some of the limitations of microsphere indentation and AFM, a new approach was recently developed that uses a flexible glass microneedle with a spherical tip to probe the gel. The method is identical to AFM in principle, measuring the indentation of the gel in response to calibrated forces exerted at a spherical tip. However, the position of the probe and the indentation of the material are measured through an optical microscope with a calibrated focusing mechanism in conjunction with a micromanipulator. This allows both qualitative and quantitative measurements to be performed "*in situ*" on the microscope stage. The method may be viewed as an extension of an approach described by Lee *et al.* (1994), modified to incorporate Hertz contact physics for more reliable measurements. In addition, the method is closely related to the nanonewton force apparatus described in Chapter 18 of Davidson and Keller, this volume, and to the microindenter method described by Jacot *et al.* (2006).

The measurement is performed on a light microscope with a coupled imaging device. It is essential that the focusing mechanism move the objective lens, rather than the stage. The optimal objective lens should have a sufficient working distance to image the surface of the hydrogel, and as high a numerical aperture as possible to give a shallow depth of field. Optimally the microscope should be equipped with a motorized focusing mechanism and fluorescence optics. In addition, the method requires a micropipette puller, a microforge, and a micromanipulator with a fine, precise vertical control. Although specific instrument models are given in the description below, a range of designs by different manufacturers, in addition to simpler custom-built devices, may be acceptable. The results may be analyzed manually, or using programs written in Excel or MatLab (available on request).

A. Preparation and Calibration of the Spherically Tipped Microindenter

A borosilicate glass capillary tube, 1.2 mm OD × 0.9 mm ID (Frederick Haer & Co., Bowdoinham, Maine), is first pulled into a thin fiber with a micropipette puller (Vertical Pipette Puller, Model 720, David Kopf Instruments, Tujunga, California). The heating and pulling settings on the puller are adjusted to create a long taper, about 15–20 mm from the beginning of the taper to the tip. A microforge (MF-900, Narishige Co., Ltd., Tokyo, Japan) is then used to melt the tip into a semispherical shape ∼60–80 μm in diameter (Fig. 3A). Heat is also applied to the region ∼150 μm from the tip to create a bend of ∼45° (Fig. 3B). The spring constant of the microindenter may be varied over a wide range by experimenting with different glass materials, capillary sizes, and taper lengths.

B. Calibration of the Microscope and Micromanipulator

The magnification factor of the optical system should be calibrated with a micrometer standard to obtain the dimension imaged by each pixel, in micrometer per detector pixel. Vertical movement of the microscope focusing mechanism and the micromanipulator must be calibrated with a precision better than 0.5 μm. Calibration of the microscope-focusing knob may be performed by directly measuring a sample of known thickness, as described in the section of microbead

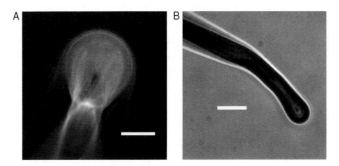

Fig. 3 Microindenter with a spherical tip, as viewed from the top (A, scale bar is 25 μm) or side (B, scale bar is 100 μm).

indentation. Once the focusing knob is calibrated, the movement of micromanipulator (a Leitz mechanical micromanipulator in this case) may be calibrated accordingly. This is performed by first focusing on a needle mounted on the micromanipulator (a Leitz mechanical micromanipulator in this case). The needle is then moved up by a known number of increments (e.g., 10 divisions on the micromanipulator knob or 45° turn), and the image of the needle is brought back into focus, keeping track of the position on the microscope focusing mechanism. These positions on the microscope are converted into distance according to the calibration of the focusing mechanism, and into the distance per micromanipulator increment.

C. Characterization and Calibration of the Microindenter

The first step in microindenter calibration is to obtain the radius r of the spherical tip, by taking an image and converting the radius in pixels into micrometers using the magnification factor obtained above. The next step is to collect a reference image of the microindenter, which is the phase-contrast image of the probe as the microscope is focused precisely on the bottom surface of the spherical tip.

The reference image is collected by pushing the microindenter against a glass surface sprinkled with 1-μm diameter fluorescent beads (FluoSpheres, Molecular Probes, Eugene, Oregon). The spherically tipped microindenter is mounted on the micromanipulator, at the tilting angle to be used for probing the gel and with the tip pointing down (see below), and is lowered slowly until the spherical tip just starts to thrust forward. This is the position where its bottom surface makes initial contact with the glass surface. Under the same magnification and optical conditions as will be used for probing the hydrogel (e.g., with an objective lens of 40×, 0.75 NA), bring the beads on the glass surface to sharp focus in fluorescence optics. Note that the use of fluorescence optics allows more precise focusing, although the beads can be readily visualized in phase-contrast optics. A phase-contrast image of the probe is then recorded. This image must be replaced again each time the microindenter is removed and remounted on the micromanipulator.

The spring constant of the microindenter is determined by measuring the bending after hanging a number of known weights near the tip. Basically, the microindenter is rotated such that the tip is pointing up, in order to hang a series of weights near the tip. The deflection (d) of the microindenter is measured using the focusing mechanism of the microscope (Fig. 4) and the spring constant, k, is determined from Hooke's law, $f = kd$ (Fig. 4C), where f is the gravitational force of the weight. The detailed procotol is given below.

1. Prepare a series of weights from thin electric wires of several different sizes, for example, by untwisting and separating individual conductors from thin telephone wires. Cut a piece 5–10 cm in length, measure the exact weight, and calculate the weight per mm. To match the stiffness of the microindenter, the linear density of the wire should generally be <0.1 mg/mm. Cut short pieces ~5 mm in length,

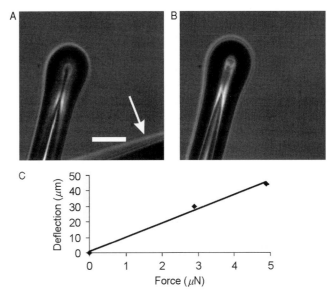

Fig. 4 Determination of the spring constant of the microindenter. The probe is deflected under applied weight load (A, arrow), and becomes out of focus on removal of the weight (B). Scale bar is 50 μm. The amount of deflection is determined from the change in focal plane and is plotted against the applied force (C).

and calculate the exact weight forces (f in nN $= 9.81 \times$ length in mm $\times$ density in mg/mm $\times$ 1000). Bend the wire into a V shape and bend one of the legs forward to facilitate handling.

2. Mount the microindenter on the micromanipulator at exactly the angle to be used for measuring the gel. As the sample typically sits on the bottom of a chamber, the angle of the microindenter will be limited by the accessibility to the sample. Therefore, even though this calibration does not require a sample chamber, it would be helpful to place one underneath the microindenter, both to monitor the approaching angle and to catch the calibrating weight that is blown off the microindenter during the calibration.

3. Rotate the microindenter and make sure that its bent tip is pointing *up*, which should be obvious by focusing the microscope up and down. In addition, no kink should be visible from its lateral profile under the microscope (compare Figs. 4A vs 3B).

4. Center and focus at the tip of the microindenter using low magnification, for example 10×.

5. Hang a wire weight on the microindenter with a pair of fine forceps, somewhere not too far from the tip. Gently tap the micromanipulator until the weight slips near the tip and gets caught at the bend. Make sure it does not touch the sample chamber. Monitor the last part of tapping under the microscope.

6. Switch to the magnification to be used for measuring the gel, or as high as the working distance of the lens allows (the tip of the microindenter is several milli-meters above the objective lens, due to the upward bend and to the dangling calibration weight). Bring the tip to focus and record an image (Fig. 4A).

7. Tap or blow the weight off the tip of the needle. The tip of the microindenter becomes out of focus (Fig. 4B).

8. Adjust the microscope focus to restore the sharp image of the tip of the microindenter (image to match Fig. 4A). This step is best performed by collect-ing a stack of optical sections with a motorized focusing mechanism, and searching for the image in the stack that best matches the weighted image. Using the calibrated focusing knob, determine the distance of vertical movement, d.

9. Repeat the measurement with different weights.

10. Plot d as a function of f. Stiffness, k, is determined from the slope of the f–d curve (Fig. 4C).

D. Measurement of the Indentations of Hydrogels in Response to Forces of the Microindenter

Qualitative monitoring of the change in material properties may be carried out simply by imaging the tip of a stationary microindenter, which becomes out of focus when the flexibility changes. This is an excellent tool for developing modulatable materials, as one can easily visualize the change in stiffness.

For quantitative measurement, E of the gel is calculated from its indentation as a function of forces exerted by the microindentor using the Hertz equation. To minimize error (discussed later) and to yield results relevant to actual experimental conditions, the sample thickness should be in the range of 100–150 μm. The following procedure is designed to facilitate consistent measurement of the posi-tion of the microindenter, the point of contact with the gel, and the resulting gel indentation. See Fig. 5 for the explanation of various symbols and the relationship between the probe and the sample.

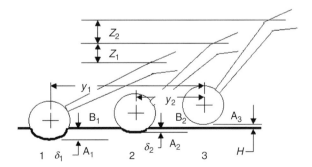

Fig. 5 Schematic of gel flexibility measurement using spherically tipped microindentation. See text for the definition of various positions marked by letters.

58

Margo T. **Frey** *et al.*

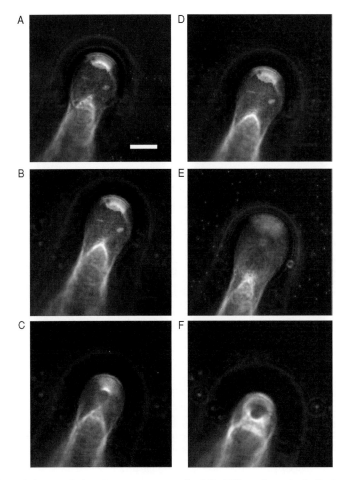

Fig. 6 Example images during the measurement of gel flexibility using spherically tipped microindentation. On initial indentation into a polyacrylamide gel, the image of the tip (B) is matched with the reference image (A), to make sure that the microscope is focusing on the bottom surface of the tip. The microindenter is then raised to a slightly higher position while remaining in contact with the sample (C). The new position of the tip and the gel surface are determined by serial optical sectioning and looking for an image that matches the reference image (D), and an image that shows sharply focused surface beads (E), respectively. The microscope focus is then returned to the point of initial indentation while the probe is raised off the sample surface (F). Serial optical sectioning is again performed to determine the position of the gel surface and the spherical tip off the surface. Scale bar is 25 μm.

1. Spinkle 1-μm fluorescent polystyrene beads (Molecular Probes) on the surface of the gel to be measured to mark the surface for focusing. A low density of beads is preferred ($\sim$2 beads/100 μm^2; Fig. 6E), as too high a density could interfere with the observation of the tip of the microindenter.

2. Set the microindenter to the angle it was calibrated. Rotate the microindenter such that the tip is pointing *down*. This is again confirmed by focusing the microscope up and down and by making sure that the lateral profile of the microindenter shows no kinks.

3. Estimate the thickness of the sample to make sure that it is at least 100 μm, which is required for reliable measurements of Young's modulus. This is done by using the calibrated focusing mechanism of the microscope, to focus from scratches or debris on the glass surface to the sprinkled fluorescent beads on the surface of the gel.

4. Use the micromanipulator to lower the microindenter until the tip starts to thrust forward, indicating that the tip is experiencing counter forces as it enters and exerts forces on the gel (Fig. 5). For optimal results (see below), the indentation, δ, should be $\sim 0.8r$, where r is the radius of the tip. The indentation may be estimated using the microscope focusing mechanism.

5. Focus the microscope until the phase-contrast image of the spherical tip (Fig. 6B) matches the reference image collected in Section IV.C (Fig. 6A). This focuses the microscope on the bottom surface of the spherical tip. This step may best be accomplished by taking serial image sections at <0.5-μm separation, searching for the best match within the image stack (note the two phase light dots near the tip), then moving the focal plane to the position where the best match with the reference image is found. The position is referred to as position A_1 (Fig. 5). Record a phase-contrast image of the spherical tip.

6. Raise the microindenter slightly, such that it is still indenting the sample but at a shallower depth, $>0.3r$ (Figs. 5 and 6C). This position is referred to as position A_2. Record the distance of vertical movement of the micromanipulator (z_1). Do not change the focus of the microscope.

7. Collect the first stack of optical sections, from position A_1 upward to somewhere above the surface of the gel (above the position in Fig. 6E), then return the microscope focus to the position A_1 using the focusing mechanism. If available, use a combination of transmitted and epi-illumination such that phase contrast and fluorescence are recorded as superimposed images.

8. Raise the tip of the microindenter such that it is above but not too far from the surface of the sample (Figs. 5 and 6F). This position is referred to as the off position or position A_3. Record the distance of vertical movement of the micromanipulator, z_2.

9. Collect a second stack of optical sections from position A_1 upward to somewhere above the spherical tip. If available, use a combination of transmitted and epi-illumination such that phase contrast and fluorescence are recorded as superimposed images.

10. In both stacks of optical sections, look for the section where the beads on the surface in the periphery of the image (i.e., away from the point of indentation) are in sharp focus (Fig. 6E). Calculate the corresponding position, which is referred to

as the surface position or positions B_1 and B_2, respectively (which theoretically should be at the same position, though B_2 is slightly more reliable due to the lack of indentation).

11. Search for the optical section in the first stack where the image of the tip matches the reference image (Fig. 6D). This marks the position of A_2. Distances B_1-A_1 and B_2-A_2 are the corresponding sample indentations, δ_1 and δ_2.

12. Search for the optical section in the second stack where the image of the tip matches the reference image. This marks the position of A_3, where the spherical tip is above the gel by a distance of H.

13. According to Fig. 5, the difference between $z_1 + z_2 - H$ and δ_1 is the net vertical deflection d_1 of the microindenter at position A_1, and the difference between $z_2 - H$ and δ_2 is the net vertical deflection d_2 of the microindenter at position A_2. However, direct measurement of H is prone to error due to surface interactions between the needle and the gel. A more reproducible approach is to plot the net deformation of the tip against the height of the micromanipulator $[z_1 + z_2, \sqrt{((A_3-A_1)^2 + y_1^2)}$, and $(z_2, \sqrt{(A_3-A_2)^2 + y_2^2})]$, where y is any lateral deflection of the microindenter tip (Fig. 7). If $H = 0$, the line connecting these two points should pass through the origin while maintaining the same slope. Therefore, its x-intercept is used as the estimate of H. It is important that the final position of the tip A_3 be very close to, but not touching, the surface. This position should yield a nearly linear relationship between net tip deformation and applied force, a condition which is required for the analysis.

14. Using the value of k obtained from the calibration curve obtained in Section IV.C, convert d_1 and d_2 to forces f_1 and f_2. This generates two pairs of force–indentation relationships, f_1 and f_2, and the corresponding indentations δ_1 and δ_2.

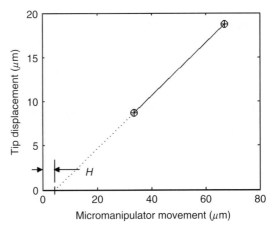

Fig. 7 Estimation of contact point from data taken on a polyacrylamide gel with 8% acrylamide and 0.08% bis.

15. Repeat the measurements at several different locations of the gel.

16. The microindenter may be stored and reused many times. Accumulated beads on the surface of the tip should be removed as much as possible by gentle sonication in a detergent solution, since they interfere with the phase-contrast imaging of the tip.

E. Data Analysis

Each set of f_1, δ_1 and f_2, δ_2 is first checked for consistency. The Young's modulus, E, is calculated for each pair of f, δ using the Hertz equation for a rigid sphere [Eq. (1); Fig. 8], assuming a Poisson's ratio of 0.45 for polyacrylamide (Engler *et al.*, 2004a). The resulting two Es should agree within 33% and those exceeding this criterion are removed. Inconsistency is typically due to one of the indentations being too deep or shallow (discussed later).

The use of the Hertz model is generally considered valid only for small indentations, due to behavior other than linear elasticity at greater indentation depths. However, macroscopic testing of polyacrylamide gels indicated that this limit may be overly conservative (Engler *et al.*, 2004a). AFM data with spherical tips also found good agreement to the Hertz model at indentations up to r (Engler *et al.*, 2004b). A similar conclusion was reached using conical tips, which involve higher strains that exceed this limit even at very low forces (Dimitriadis *et al.*, 2002; Engler *et al.*, 2004b). Most notably, Yoffe found that the Hertz model is valid to within 1%

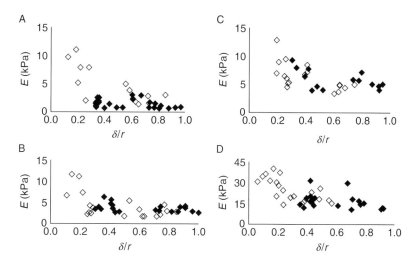

Fig. 8 Young's moduli of polyacrylamide samples calculated from discrete force–indentation data obtained with spherically tipped microindentation. The points that meet the selection criteria as described in the text are indicated by filled dots and those rejected by open dots. Data shows the upward trend of E at lower indentations for polyacrylamide gels made from 5% and 0.025% (A), 5% and 0.06% (B), 0.1% (C), and 8% and 0.08% (D) acrylamide and bis.

error for large indentations at least up to $a/r \sim 0.8$, where $a^2 = \delta r$ and a is the radius of the contact area, or $\delta \sim 0.64r$ for materials with a high Poisson's ratio (0.40; Yoffe, 1984). For the present measurements, Yoffe's correction for a material of $v = 0.40$ amounts to a difference of less than 2% in the value of E. These results suggest that indentations up to the radius of the microindenter may be applied in the present approach in conjunction with an uncorrected Hertz model, for elastic materials that are nearly incompressible (high v).

At small indentations, significant systematic error does occur. For example, experimental measurements indicated larger measured contact areas than those predicted by the model at low indentations (Fessler and Ollerton, 1957; Johnson, 1985), which leads to a value of E that is several times higher than those measured at higher indentations. There may be additional sources of error at small δ, for example, from surface interactions between the probe and the sample. An upward trend in E at low δ is indeed observed in both the present measurement (Fig. 8), and in several studies with AFM (Dimitriadis *et al.*, 2002; Richert *et al.*, 2004), where the values at low indentation depths were found to be two to three times higher than those at higher indentation depths. This systematic error may also explain the higher values of E obtained with the bead indentation measurement as discussed in Section II, particularly for stiff gels where the indentation may be very small compared with the radius of the bead.

On the basis of above considerations, the calculation of E for spherical tip microindentation was performed for data with $\delta_1 < r$ and $\delta_2 > 0.3r$. Moreover, the sample should be thick enough to avoid potential stress-stiffening artifacts from the underlying glass substrate. The values of E derived from qualified data sets turned out to be relatively consistent (Fig. 8), and were averaged to obtain the final value. The results with several polyacrylamide gels of different stiffness, as controlled by the concentrations of acrylamide/bis-acrylamide, are shown in Fig. 9. These measurements were conducted using three different probes with tip radii between 29.7 and 40.0 μm, spring constants between 0.099 and 0.143 nN/nm, and gels $\sim 150 \mu m$ in thickness. The values are in excellent agreement with those obtained with AFM as reported in the literature and measured independently (Guo *et al.*, 2006).

F. Discussion

As with any method, there are several sources of uncertainty. Values for Poisson's ratio of polyacrylamide gels commonly seen in the literature range between 0.3 and 0.5, and thus are responsible for the largest relative uncertainty. We chose to use a value of 0.45, as this value yields the best fit between E values obtained with macroscopic and microscopic testing (Engler *et al.*, 2004a).

A second significant source of error is associated with vertical measurements of the microindenter position based on visual inspection of images. However, although the field depth for a dry lens is typically several micrometers, the vertical position of the microindenter may be determined to an accuracy better than 1.0 μm,

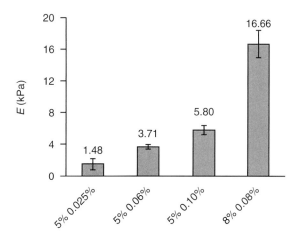

Fig. 9 Young's moduli for polyacrylamide gels of various stiffness obtained with spherically tipped microindentation, by averaging all the values that meet the selection criteria. Values are average ± SEM, $n = 10$, 7, 10, and 13.

using a stack of optical sections collected at ~ 0.25 μm per slice. The accuracy may be improved by replacing visual comparison with a cross-correlation-based algorithm for matching images. Other sources of error include the mechanical movements and calibration of the microscope-focusing mechanism and the micro-manipulator (dial reading), the weight of the calibration wire, and the deviation of the tip from a perfect spherical shape. The relative uncertainties for the measurement of microindenter spring constant, deflection, gel indentation, and radius are estimated to be around 13%, 11%, 6%, and 1%, respectively. The net relative uncertainty for E is $\sim 41\%$ (18% contributed by method), which is within 5% of our estimate of the uncertainty for AFM measurements (Guo *et al.*, 2006).

Finally, the method is based on the assumption that the spherical tip glides freely (slips) on the gel surface as it applies forces, as obstruction in axial movements affects the bending behavior of the probe. Judging from the vertical and lateral distribution of beads on the surface, this obstruction appears to occur with very soft gels (<1 kPa), where the surface undergoes complex shape changes as the tip digs into the gel. A similar problem may affect AFM. Another assumption of this method is that all measurements are static since the time required for measurements is such that stress relaxation, if the material is viscoelastic, has likely already occurred.

There are many advantages to the spherically tipped microindenter method. It is straightforward and the analysis is simple. The materials and equipment are inexpensive compared to AFM, and are often available for other purposes. Once calibrated and if handled carefully, the microsphere indenter can be used repeatedly, in contrast to the much shorter longevity of expensive AFM probes because of cantilever damage or corrosion of the reflective coating. Although the build up of beads on the surface after repeated use can impair visualization of the tip necessary for accurate measurements, the life of a microindenter can be extended greatly by cleaning after each use. This method also appears to be less sensitive to environmental factors than the AFM, as air currents and minimal amounts of vibration do not seem to significantly affect the outcome. This method can also test sticky samples that could not be tested with AFM due to its limited vertical scanner range necessary to pull the tip free from adhesive samples. One of the most significant advantages of this method is its ability to test substrates on a microscope, possibly in conjunction with the observations of cells. This is particularly useful for testing nonhomogenous substrates, and modulatable substrates that stiffen or soften in response to local manipulations. Even when used without quantification, it provides a convenient tool for assessing the mechanical properties of substrates during material development.

V. Conclusions

With the realization of the importance of mechanical signals, studies of cell biology are increasingly performed on polymer-based substrates with tunable mechanical properties in an effort to elucidate the mechanisms that regulate the complex cell behavior. The quality of these studies is directly affected by the proper characterization of the mechanical properties of these substrates.

While AFM represents a highly reliable and versatile approach, it is also the most costly and complicated compared to the two other methods. The optimization of AFM for high-resolution scanning of surfaces may in fact create unnecessary obstacles to the relatively simple task of stiffness measurements. This chapter shows that the measurements may be performed with much simpler and economical approaches. Indentation with microspheres is the easiest and cheapest method. Slightly more complicated but more reliable and versatile is the method of spherically tipped microindentation, which may be further improved by incorporating precise (e.g., piezoelectric) and automated positioning of the tip. Optimal choice of the method should take into consideration the requirements of the application, the acceptable uncertainty, as well as practical factors such as cost and equipment availability.

References

Bennig, G., Quate, C. F., and Gerber, C. (1986). Atomic force microscope. *Phys. Rev. Lett.* **56,** 930–933.
Dimitriadis, E. K., Horkay, F., Maresca, J., Kachar, B., and Chadwick, R. S. (2002). Determination of elastic moduli of thin layers of soft material using the atomic force microscope. *Biophys. J.* **82,** 2798–2810.
Domke, J., and Radmacher, M. (1998). Measuring the elastic properties of thin polymer films with the atomic force microscope. *Langmuir* **14,** 3320–3325.
Engler, A., Bacakova, L., Newman, C., Hategan, A., Griffin, M., and Discher, D. (2004a). Substrate compliance versus ligand density in cell on gel responses. *Biophys. J.* **86,** 617–628.
Engler, A., Richert, L., Wong, J. Y., Picart, C., and Discher, D. (2004b). Surface probe measurements of the elasticity of sectioned tissue, thin gels and polyelectrolyte multilayer films: Correlation between substrate stiffness and cell adhesion. *Surf. Sci.* **570,** 142–154.

Fessler, H., and Ollerton, E. (1957). Contact stresses in toroids under radial loads. *Br. J. Appl. Phys.* **8,** 387–393.

Guo, W. H., Frey, M. T., Burnham, N. A., and Wang, Y. L. (2006). Substrate rigidity regulates the formation and maintenance of tissues. *Biophys. J.* **90,** 2213–2220.

Hertz, H. J. (1882). Über die Berührung fester elastischer Körper. *J. Reine Angew. Mathematik* **92,** 156–171.

Jacot, J. G., Dianis, S., Schnall, J., and Wong, J. Y. (2006). A simple microindentation technique for mapping the microscale compliance of soft hydrated materials and tissues. *J. Biomed. Mater. Res. A* **79,** 485–494.

Johnson, K. L. (1985). "Contact Mechanics." Cambridge University Press, Cambridge.

Lee, J., Leonard, M., Oliver, T., Ishihara, A., and Jacobson, K. (1994). Traction forces generated by locomoting keratocytes. *J. Cell Biol.* **127,** 1957–1964.

Li, Y., Hu, Z., and Li, C. (1993). New method for measuring Poisson's ratio in polymer gels. *J. Appl. Polymer Sci.* **50,** 1107–1111.

Lo, C. M., Wang, H. B., Dembo, M., and Wang, Y. L. (2000). Cell movement is guided by the rigidity of the substrate. *Biophys. J.* **79,** 144–152.

Mahaffy, R. E., Shih, C. K., MacKintosh, F. C., and Kas, J. (2000). Scanning probe-based frequency-dependent microrheology of polymer gels and biological cells. *Phys. Rev. Lett.* **85,** 880–883.

Richert, L., Engler, A. J., Discher, D. E., and Picart, C. (2004). Elasticity of native and cross-linked polyelectrolyte multilayer films. *Biomacromolecules* **5,** 1908–1916.

Rotsch, C., Jacobson, K., and Radmacher, M. (1999). Dimensional and mechanical dynamics of active and stable edges in motile fibroblasts investigated by using atomic force microscopy. *Proc. Natl. Acad. Sci. USA* **96,** 921–926.

Sneddon, I. N. (1965). The relation between load and penetration in the axisymmetric Boussinesq problem for a punch of arbitrary profile. *Int. J. Eng. Sci.* **3,** 47–57.

Weisenhorn, A. L., Hansma, P. K., Albrecht, T. R., and Quate, C. F. (1989). Forces in atomic force microscopy in air and water. *Appl. Phys. Lett.* **54,** 2651–2653.

Yoffe, E. H. (1984). Modified Hertz theory for spherical indentation. *Philos. Mag. A* **50,** 813–828.

Zhao, Y. P., Shi, X., and Li, W. J. (2003). Effect of work of adhesion on nanoindentation. *Rev. Adv. Mater. Sci.* **5,** 348–353.

CHAPTER 4

Surface Patterning

Irene Y. Tsai, Alfred J. Crosby, and Thomas P. Russell

Polymer Science and Engineering Department
University of Massachusetts
Amherst, Massachusetts 01003

Abstract

Cell adhesion, migration and differentiation depend on a complex interaction between a cell and its microenvironment. A three-dimensional (3D) topographic substrate provides an invaluable tool to understand this interaction. Here, we present three distinct techniques to pattern a surface having 2-D and 3-D topographies to study cell behavior. The three methods are electrohydrodynamic instabilities of polymer films, photolithography and self-assembly of homopolymer blends and diblock copolymers. Depending on the technique used, the size scale of the surface pattern can be on the nanometer or micrometer level or both. These methods can easily be utilized in biological laboratories since they do not require the use of a cleanroom facility. We briefly discuss each technique and show its use in cell culture. The 3D topographic substrates are ideal system to understand cell adhesion, migration and differentiation that mimic cells in physiological conditions. The techniques described here have the potential to extend to other materials such as extracellular matrix proteins.

I. Introduction

The interaction between a cell and its surrounding matrix is important for cell adhesion, migration, and differentiation. Our current understanding of this interaction is largely derived from studies of cells on flat tissue culture dishes made either with synthetic polymers such as polystyrene (PS) or glass coverslips coated with extracellular matrix proteins, like collagen or fibronectin. However, compared to these flat surfaces, tissues present a three-dimensional (3D), more textured environment to cells. While some efforts have been made to examine cell adhesion (Chapter 23 by Johnson *et al.*, this volume; Weaver *et al.*, 1996), migration (Friedl and Brocker, 2000), and differentiation (Cukierman *et al.*, 2002) in various model 3D systems (Abbott, 2003), we found that surfaces with 3D topographic features are sufficient to induce some of the features seen in tissues. However, past efforts in such directions have exploited microfabrication methods that require clean rooms and costly equipment. In this chapter, we discuss several thin film techniques that are easy to implement in biological laboratories and that yield substrates with well-defined topographies. The methods described include the use of electrodynamic instabilities in polymer films, photolithography, and the self-assembly mixtures of homopolymers and diblock copolymers.

Several techniques have been developed to prepare substrates with well-defined topographies to mimic cell behavior in physiological conditions, but general use has been limited and the role of surface topography and surface chemistry remains poorly understood. Early studies have shown that cells preferentially adhere to the ridges of topographic features on the substrate, a phenomenon known as contact guidance (Dunn and Ebendal, 1978; Weiss, 1945). Studies on surfaces roughened by sandblasting, plasma spraying, or particle settling were performed to determine the effect of the lateral size scale of the surface features on cell spreading, cytoskeletal arrangements, or cell movements (Flemming *et al.*, 1999). These methods have provided some insight into the influence of surface roughness, but the inability to prepare surfaces with well-defined topographies precluded understanding the mechanism by which surface topography perturbs cellular behavior. The development of microfabrication techniques, like photolithography, to prepare well-defined surface features has stimulated a large number of experiments (Craighead *et al.*, 2001; Curtis and Wilkinson, 2001; Desai, 2000; Flemming *et al.*, 1999; Folch and Toner, 2000; Singhvi *et al.*, 1994; Wilkinson *et al.*, 2002). Tan *et al.* (2003) for example, generated hierarchically ordered structures, spanning length scales from the nanoscopic to microscopic, using hydroxyapatite, a mineral normally found in bones, teeth, and seashells, to investigate bone regeneration using human osteoblast-like cell lines (Tan and Saltzman, 2004).

Yet, there are few laboratories where the use of these methods is routine due to the inordinate cost of the clean room facilities necessary for processing.

Three versatile techniques are described here that can overcome this barrier. These include: (1) the application of an electric field to a thin film (electrodynamic instability), (2) a simplified photolithography process, and (3) the self-assembly of synthetic polymers in a thin film. In the first technique, a gradient in the electric field on the surface of a polymer is used to amplify instabilities on the surface of a polymer film. In the second, classic photolithography is modified for use in a biological setting that circumvents the need for clean room facilities. And finally, the self-assembly of block copolymers and their mixtures with homopolymers is described where one can easily make templates with feature sizes ranging from the nanoscopic to microscopic length scales, for patterning surfaces with topography and/or chemical functionality. For each case, a detailed description of the process is provided to facilitate the use of these techniques. The techniques can be used in combination with one another and/or with surface chemistries to modify the chemical nature of the surface. Although the methods described here are based on synthetic polymer systems, they can easily be extended to biologically relevant systems.

Common to each of the techniques described is spin coating where a polymer solution is placed on the surface of a substrate, like a silicon wafer (International Wafer Service, Santa Clara, California), and the substrate is rapidly spun at 1–4000 rpm in a centrifuge-like device (Headway Research, Inc., Garland, Texas). This process is identical to that used in spin art toys. During spinning, centrifugal force spreads the solution over the entire surface while the solvent is evaporating. The combination of these results in a polymer film that is uniformly spread over the surface with a roughness of $\sim$1 nm. The film thickness can be controlled by the spinning speed, concentration of polymer, and the vapor pressure or evaporation rate of the solvent. The thickness of the film can be measured optically with an ellipsometer, an interferometer, or a microscope, or mechanically with a profilometer or an atomic force microscope.

II. Patterning with Electrodynamic Instabilities

The generation of well-defined surface topographies on the submicron level by electrodynamic instabilities is simple and free of many harsh chemical processes. Consider the schematic diagram in Fig. 1A, we take a polymer film in a simple capacitive geometry where the thin film is spin coated onto a conducting or semiconducting surface. If a second electrode is placed over the surface of the film with an air gap, then when a voltage is applied to the two electrodes, the dielectric constant of the polymer film and the air between the film and the upper electrode define the electric fields. At the polymer film surface, there is a change in the density from that of the polymer film to air and, as such, there is a small gradient in the dielectric constant at the surface. This gradient in the dielectric constant translates into a gradient in the electric field, which, in turn, defines an

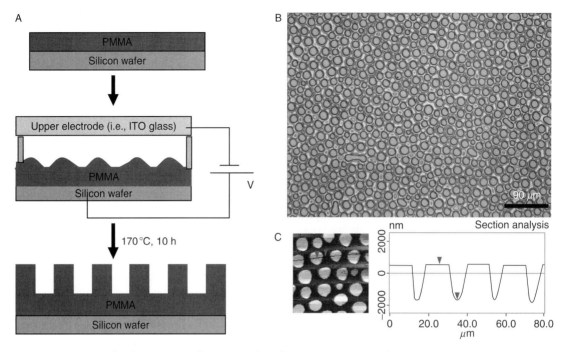

Fig. 1 (A) Schematic representation of the experimental setup for patterns formed by electrodynamic instability. (B) Optical image of poly(methyl methacrylate) posts with diameter of 15 μm, height of 2 μm, and center to center distance of 25 μm formed by electrodynamic. Instability using 800-nm-thick film and a voltage of 120 V. (C) Atomic force microscopic image of poly(methyl methacrylate) posts.

electrostatic pressure that tends to pull the polymer film toward the upper electrode. Any variations in the height or thickness of the film will cause a variation in this electrostatic pressure, which will tend to magnify any surface topography. However, amplifying topography comes at the expense of increasing the surface area, which costs interfacial energy and, as such, the Laplace pressure acts to smooth the surface. The balance of these two opposing pressures causes the amplification at a characteristic wavelength, which, with time, causes the formation of hexagonally packed columns or pillars of polymer that span between the two electrodes. Results from our laboratory quantitatively show that the diameter (typically ∼5 μm) and average separation distance between the columns (typically ∼20 μm) is completely defined by the film thickness and dielectric constant, the width of the air gap and the magnitude of the applied voltage (Schaffer *et al.*, 2000; Schafer *et al.*, 2001). One can also use an upper electrode that has a specific topography, for example rows of trenches that cause a lateral variation in the strength of the applied field. In typical applications, the initial polymer with surface topography is used as a mold so that the surface topography is replicated by casting the pattern onto polymer films. Features down to ∼140 nm (Schaffer *et al.*, 2000) have been replicated in this manner.

Lin *et al.* (2001, 2002) extended the work of Schaffer *et al.* (2001) to polymer–polymer bilayers. Here, the Laplace pressure is reduced, since the surface energy, typical $\sim$30 dyne/cm, is markedly reduced to $\sim$2 dyne/cm, which causes the wavelength of the surface topography to be reduced significantly (typically $\sim$1 μm). By further reducing the interfacial energy, for example with the use of a block copolymer that acts as a surfactant, the size scale and separation distance of the surface features can be further reduced.

A. Procedure for Patterning with Electrohydrodynamic Instabilities

Preparation of a polymer columnar pattern involves dissolving a polymer in a good solvent, spin coating the polymer solution to form a thin film, and applying an electric field across the polymer while it is in the liquid state (above its glass transition temperature). Once the desired pattern is formed, the polymer is rapidly cooled to freeze the structure. We have applied this process to a wide variety of polymers, although here we will focus on poly(methyl methacrylate) (PMMA, Polymer Source, Inc., Dorval, Quebec, California P7398-MMA), deposited and spin coated onto a freshly cleaned conducting surface (lower electrode). A 10% (w/v) toluene solution (Sigma-Aldrich Co., St. Louis, Missouri) of PMMA, with molecular weight of 29,000 g/mol and polydispersity index of 1.1, was spin coated onto a silicon substrate in a fume hood (silicon wafer with 2.5-Å silicon oxide, International Wafer Service), producing an 800-nm-thick PMMA film. Either a sheet of HF-etched glass coated with chromium or commercially available indium-tin-oxide (ITO, Delta Technologies Ltd., Minnesota) glass was brought to within 1 μm from the PMMA surface using spacers to keep the separation distance constant. The entire assemblage was heated to 170°C for 10 h while applying 120-V DC across the electrodes (power supply, Agilent, Model No. E3620A). The film was then rapidly cooled to room temperature, voltage was removed, and the upper electrode was lifted off. Figure 1B shows an optical micrograph (Olympus microscope, BX60) and Fig. 1C shows an atomic force microscopic (Digital Instruments, Dimension 3100) image of the PMMA surface. As can be seen, an array of 2-μm-diameter PMMA posts with a characteristic center-to-center distance of 15 μm was produced by this process.

While the surfaces produced by the electrohydrodynamic instabilities are useful in their own right, the patterned surface can, also, be used as a master template to form multiple samples of the same pattern. A prepolymer of poly(dimethylsiloxane) (PDMS), Sylgard 184, Dow-Corning, is mixed according to the manufacturer's instructions, poured over the PMMA film and allowed to cure in an oven at 65°C overnight to replicate the pattern in the PDMS film (Fig. 2A). The PDMS film can then be used as a secondary mold to form multiple samples by pressing the PDMS film against another polymer film above its glass transition temperature to allow capillary force to draw the polymer into the cavities in the PDMS. We demonstrate this with PS where the PDMS mold was placed in contact with a 1-μm-thick spin-coated PS film (molecular weight = 52,000, Polymer Source,

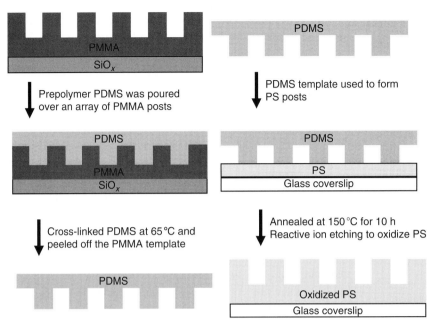

Fig. 2 The patterns formed by electrodynamic instability can be used as a master template for lithography to produce substrates of oxidized PS.

Inc.) for 1 day in a vacuum oven at 150°C. After peeling the PDMS from the PS, a PS film with posts that match those formed in the original PMMA film was formed (Fig. 2B). The surface of the PS can be oxidized using a commercially available reactive ion etcher (oxygen ion at 25 W for 30 sec, South Bay Technology, Inc., San Clemente, California RIE-200) to change the surface from being hydrophobic to hydrophilic in order to promote cell adhesion. Alternatively, the surface can be modified by direct chemical reactions (Mrksich, 2000, 2002; Oner and McCarthy, 2000). Proteins such as collagen or fibronectin can be used to coat the pattern or else an electric field can be applied directly to protein-based films.

B. Cell Migration on Topographic Surfaces

Fibroblasts on the hydrophilic PS posts, fabricated by electrodynamic instability, showed dramatic differences in morphology, cytoskeleton organization, and migration speed from those cultured on flat PS (Frey *et al.*, 2006). Fibroblasts on PS posts had a phenotype that was more elongated with many more extensions than cells on flat PS substrates. The actin bundles on the columnar-structured surfaces were distinctly different from those on flat substrates, showing a more diffuse distribution throughout the cell. Anti-vinculin immunostaining showed short but relatively stable patches of focal adhesions, where cells made contact with the columns structures. Quantitatively, cells were found to migrate more rapidly on the rough surfaces than on the flat surface. These characteristics fit the reported qualities of cells in 3D matrices (Friedl and Brocker, 2000), and thus, these topographic surfaces provide useful model systems to study cell migration or cell division in more physiologically relevant conditions.

III. Lithography Without a Clean Room

Although the self-assembly of 3D surface structures using techniques such as electrohydrodynamic instabilities are attractive for numerous reasons, the ability to produce exact length scales, spatial arrangements, and arbitrarily defined shapes of surface structures can be especially advantageous for cell-based investigations (Clark et al., 1992; Kaihara et al., 2000; Pirone and Chen, 2004; Whitesides et al., 2001). Direct assembly methods such as lithographic techniques allow for the fabrication of these structures. The two most popular and most accessible lithographic techniques are photolithography and imprint lithography (Bailey et al., 2000; Chou, 1997; Duffy et al., 1998; Geissler and Xia, 2004; Miyajima and Mehregany, 1995; Qin et al., 1998; Xia et al., 1996). Both techniques have had and will continue to have an important impact on the development of numerous technologies, including most significantly microelectronic devices.

In general, both photolithographic and imprint lithographic techniques rely on the transfer of patterns from a prefabricated mask or mold into a thin polymer coating on a rigid substrate. Depending on the specific application, this patterned polymer coating can serve as the final relief structure, or it can be used as an etch-resistant layer during the subsequent etching of the underlying substrate. In the latter process, the pattern from the prefabricated mold is displayed on the surface of the rigid substrate. If desired, this process of pattern transfer can be repeated with intermediate coating steps of metal, ceramic, or polymer layers to build an integrated 3D device. It is in this manner that integrated circuit devices, such as computer chips, have been fabricated for the past several decades.

Encompassed in this simple process are the primary advantages of lithographic techniques. First, the prefabricated molds or masks are created using well-controlled processes that permit the "drawing" of arbitrary shapes. Second, once the master mold or mask is fabricated, the pattern transfer process can be repeated indefinitely to produce identical surface structures in an economical and efficient manner. This quality is particularly useful for research on cell surface interactions where statistical analysis is critical for proper interpretation of observed behavior.

Conventionally, lithography is performed in a clean room environment where the presence of dust particles and other contaminants is minimized. This precaution is critical for the fabrication of surface structures with minimal generation of defects since dust or other contaminants can impede the proper transfer of a pattern. Although defects make an integrated device virtually useless, they can be adequately tolerated in experiments that are interested in statistically evaluating the influence of topographic structures on biological or material properties. In these experiments, researchers can simply neglect data gathered near the defect.

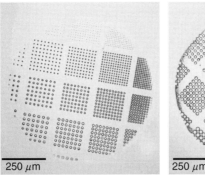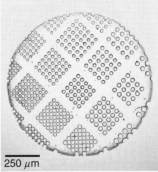

Fig. 3 Patterns fabricated from photolithography performed in general laboratory environment. Positive tone photoresist used for pattern on the left, and negative tone photoresist used for pattern on the right.

Therefore, for most biological and material characterization research on 3D surface structures, lithographic techniques can be performed successfully in a non-clean room environment. Additionally, as we discuss below, the use of inexpensive equipment and processes can make this approach for 3D structure fabrication quite feasible for most biological research laboratories. Figure 3 illustrates two patterns produced by photolithography in a non-clean room environment.

We discuss below the basic principles of photolithography, the experimental details of performing this technique in a modest, general laboratory setting, and a few examples of how these surface structures can provide insight into cell surface interactions. It is not the intention of the authors to comprehensively discuss the science and engineering details associated with this robust process. For these details, we refer the reader to several texts and review papers that have been written on these subjects. Rather, our goal is to provide a practical primer for a general scientist or engineer to initiate the use of this technique in their research laboratory. For related applications of photolithography or microcontact printing, see Chapter 19 by Polte *et al.* and Chapter 13 by Sniadecki and Chen, this volume.

A. Photolithography Basics

Photolithography involves the use of patterned light to create 3D surface structures in a polymer film. In this process, light is transmitted with uniform intensity through an aligned photomask, which blocks or significantly diminishes transmission in defined regions. The regions of transmitted light then interact with a film of polymers, referred to as photoresist, that is uniformly coated onto an underlying substrate. Depending on the chemistry of the photoresist, the exposed regions become either soluble or insoluble to a solvent used in a subsequent development stage. During the development of the polymer coating, the soluble regions are removed and the insoluble regions remain on the underlying substrate. These regions either can serve directly as the 3D surface structures or can protect the underlying substrate in these regions during a subsequent etching process. If etching is used, the photoresist-coated regions etch at a slower rate relative to the uncoated regions. Hence, the topographic pattern of the photoresist is transferred into the material of the underlying substrate with the depth of etch largely controlled by the relative time of the etching process. A schematic in Fig. 4 illustrates the key steps in this process.

In most photolithographic processes, the transmitted light is in the ultraviolet (UV) spectrum, but the necessary wavelength range is specified by the photochemistry of the polymer coating. In addition to ensuring the transmission of a specific wavelength for the initiation of the photochemical processes, the user should avoid potentially detrimental wavelengths such as infrared (IR) wavelengths. The intensity and uniformity of the light source is also critical to the success and efficiency of the photolithographic process. The intensity largely dictates the required time of exposure to activate the required photochemical processes in the photoresist. Increased time of exposure not only limits the efficiency of the photolithographic process but also can have detrimental effects on the resolution of the pattern transfer process. The uniformity of the light controls the aerial dimensions over which the pattern transfer process can reliably be conducted without significant defect generation.

The photomask is made from a material that is optimally transparent at the wavelengths that are required for conducting photochemistry in the polymer layer. Onto this material, a coating is deposited in regions where transmitted light is not permitted. These regions collectively define either the positive or negative image of the to-be-transferred pattern, depending on the chemistry of the photoresist layer. Typically, the photomask material is soda lime glass, fused silica, or quartz with deposited chrome serving as the transmission barrier layer. These masks are fabricated commercially, although several other materials and processes can be used to make economical photomasks for general structure fabrication. Specific examples can be found in the discussion of experimental procedures below.

The photoresist coating is perhaps the most important component, besides the photomask, in the photolithography process. The photoresist is a formulation of polymers, monomers, stabilizers, and photoinitiators that react in a prescribed manner in the presence of light of controlled wavelength. The product of the

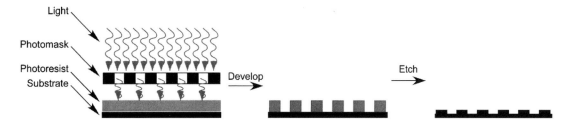

Fig. 4   Schematic of a basic photolithography process.

photoreaction defines the two primary classifications, or tones, of photoresists: positive and negative resists. A positive photoresist produces a positive rendition of the photomask pattern after exposure and subsequent development. In these materials, the exposure to light makes the material more soluble to a certain solvent compared to the unexposed regions. Therefore, after exposure of the positive photoresist, a developing solvent can be used to wash away the material regions that were exposed. Conversely, a negative photoresist produces a negative rendition of the photomask pattern. For a negative resist, the exposed regions become less soluble compared to the unexposed regions; therefore, these regions are not affected by the subsequent developing process. The ability to use both positive and negative resists for pattern fabrication is an advantage since a single photomask can be used to produce two alternative patterns simply by the choice of the appropriate photoresist. Although this flexibility is advantageous, the choice of photoresist tone is often accompanied with compromises associated with photoresist adhesion and pattern resolution limits.

The photoresist is deposited onto the substrate typically through a spin-coating process where the thickness is determined by concentration of the photoresist solution and the spin-coating speed. The photoresist thickness is critical in defining the depth of the 3D surface structures. In addition to specifying tone, the formulation of the photoresist is also optimized for different thickness regimes. This optimization should be considered in the selection of photoresists. Most commercial suppliers of photoresists will provide guidance in this selection process.

The substrate onto which the photoresist is coated is typically rigid with certain adhesion and optical reflection properties. Conventionally, silicon wafers are used as the preferred substrate, but this material is not the only choice. Practically, any material that does not change on exposure to light at the specified wavelength for the photoresist and is not soluble in the solvent and photoresist formulation can be used. In other words, glass, metals, or appropriately chosen polymers can also be used for the substrate. The primary advantages of silicon are the flatness, minimized roughness, and crystal structure of the silicon wafer. All of these parameters are critical for the controlled coating of the photoresist layer and the subsequent etching process after photoresist development. In addition to these parameters, appropriate primer coatings may be necessary to ensure adhesion of the photoresist to the substrate and to control the reflectivity of the substrate surface. The reflective properties of the substrate surface can negatively impact the resolution and fidelity of pattern transfer during the exposure stage. Similarly, the adhesion of the photoresist layer can have significant impact on the development and etching steps in the photolithographic processes.

The first step in producing photolithographic patterns is the design and purchase of an appropriate photomask. The design of a photomask pattern can be completed using software that produces files in an industry standard GDSII format (e.g., L-Edit®) or files in DXF format (e.g., AutoCAD®). If neither of these files can be produced using software such as CAD software, masks with simple pattern features, length scales, and densities may be designed using graphics software that represents graphical points in a vector format (e.g., Adobe® Illustrator), and converted to other formats by the manufacturer if necessary. After the design is complete, the mask needs to be fabricated. For photomask designs with feature sizes and spacings on the order of 50 μm or larger, it is often feasible and economical to print photomasks on transparency film using a commercial, high-resolution image setter. These services are available at several commercial printers (e.g., Pageworks, Cambridge, Massachusetts). Note that consumer ink-jet printers and laser printers can be used for printing simple photomasks, but pattern transfer will be poor due to the transmission rates of the transparency films and the inconsistent ink density in the printed features. If this method is used, a considerable amount of time will be spent optimizing the UV exposure conditions for appropriate pattern transfer. For photomasks with features smaller than 50 μm, a professional chrome-on-glass photomask should be produced. Several companies will produce these photomasks from submitted software design files (e.g., HTA Photomask, San Jose, California).

The next step in producing photolithographic patterns is purchasing appropriate photoresist and substrate materials. For substrates, silicon wafers are conventionally used, but glass slides that are flat, smooth, and cleaned rigorously can also be used for satisfactory pattern transfer for features larger than 1 μm. Silicon wafers should be at least single-sided polished and cleaned using standard wafer-cleaning procedures prior to coating. Several wafer supply companies exist. One such company is Wafer World, Inc. (West Palm Beach, Florida). There are also several manufacturers and suppliers that provide commercial photoresists. Examples of a negative tone resist and a positive tone resist are SU-8 (MicroChem Corp., Newton, Massachusetts) and SPR-220 (Rohm & Haas Electronic Materials Philadelphia, Pennsylvania), respectively. Both of these photoresists are supplied by MicroChem Corp. The appropriate developing solvents for these resists should be purchased based on the recommendation of the photoresist supplier. Note that the developing solvents are specifically designed for each photoresist formulation and significant quantities of developing solution will be consumed for each pattern transfer process. Note also that developing solvents are hazardous chemicals that require proper handling and disposal.

With the supplies purchased the following general procedure should be followed:

1. Clean substrate. Precleaned glass slides and coverslips may be cleaned by quickly passing them across a flame.

2. Coat substrate with adhesion promoting coating or antireflective coating as necessary (e.g., MicroChem Omnicoat, Newton, Massachusetts).

3. Spin coat substrate with photoresist to appropriate thickness. Approximate thickness versus spin-coating speed relationships are provided by the photoresist manufacturers.

4. Perform soft bake of photoresist coating at the temperature specified by the photoresist manufacturer. This process helps to eliminate the remaining solvent and densifies the photoresist film. A standard laboratory hot plate with a thick

aluminum block equilibrated to constant temperature can serve as an economical means of soft baking a photoresist film. The aluminum block will help maintain a uniform temperature across the wafer surface. Uniformity at the recommended temperatures for soft bake (and post bake) procedures is critical for best results in the pattern transfer process.

5. Place photoresist-coated substrate under the appropriate light source. For the suggested photoresists listed above, the formulations are optimized for exposure at wavelengths in the range of 350–400 nm. Economical UV sources can be purchased from many general laboratory supply vendors (Fisher Scientific, VWR, and so on) or UV optical sources (Jelight, Irvine, California). Professional grade UV sources and associated mask-aligning equipment can be purchased from companies such as OAI (San Jose, California).

6. Place photomask in contact with photoresist-coated substrate with patterned surface near the surface of photoresist coating. The distance between photomask and photoresist coating should be minimal to obtain optimal resolution of pattern transfer.

7. Expose photoresist for specified time limit. Most manufacturers will recommend approximate exposure conditions for their resists. If this information is not available or reliable, bracketing over a range of exposure times can be used to obtain optimal conditions for exposure. The correct exposure time is critical for best pattern transfer quality.

8. Postbake photoresist to complete the photochemical reactions (e.g., cross-linking in a negative tone resist). Again, standard laboratory hot plates can be used for this procedure, or optimized contact hot plates can be purchased for photoresist processing. The uniformity and appropriate magnitude of temperature, as provided by the manufacturer, are critical for achieving the best results.

9. Develop exposed photoresist coating in appropriate developer solvent in a fume hood. This process can be accomplished in several ways, but the most straightforward approach is to immerse the photoresist-coated substrate in a bath of the developer solvent and agitate the solution during the development process. Development time and the volume of the developer solution must be optimized for best pattern transfer results. Recommended times and volumes are often provided by the photoresist manufacturer.

10. If required, wet or dry etching should be used at this point to transfer the photoresist pattern into the underlying substrate. The details of this procedure and special considerations can be found in textbooks and review papers (Geissler and Xia, 2004; Madou, 2002).

Two final comments should be made in the context of producing 3D surface structures without clean room facilities. First, reduction of the pattern length scale can be accomplished in an economical and straightforward manner using an optical microscope with a fluorescent light source, as reported by Whitesides and coworkers (Love *et al.*, 2001). In this process, a transparent photomask is placed over the field aperture of epi-illuminator of an optical microscope and the photoresist-coated substrate is placed on the stage of the microscope. The microscope objective reduces the dimensions of the photomask accordingly when the objective's projected image is focused on the surface of the photoresist coating. Second, photoresist patterns and/or etched patterns in rigid substrates can serve as templates for imprint lithography. Imprint lithography, or microcontact printing, is a popular technique for the reproduction of chemical patterns and/or topographic patterns in a fast, reliable, and robust manner (Bailey *et al.*, 2000, 2002; Chou *et al.*, 1996; Guo, 2004; Krauss and Chou, 1997; Resnick *et al.*, 2003; Stewart *et al.*, 2005; Whitesides *et al.*, 2001; Xia *et al.*, 1996). Due to length constraints, the details of this process cannot be included here, but the interested reader is referred to several review papers on these economical and robust-patterning techniques. (Guo, 2004; Stewart *et al.*, 2005; Chapter 19 by Lele *et al.*, this volume).

The characterization of surface structures produced by lithographic techniques is similar to the procedures described previously for the characterization of 3D structures produced through the electrohydrodynamic instability methods. Prior to pattern transfer, the thickness and uniformity of the photoresist coating can be quantified using techniques such as ellipsometry and interferometry. Alternatively, profilometry can be used to quantify both of these properties for the preexposed photoresists.

To characterize the pattern transfer process, measurement techniques such as optical microscopy, scanning electron microscopy (SEM), atomic force microscopy (AFM), and profilometry can be used. Optical microscopy can be used to quickly assess the fidelity of the transferred patterns for features with dimensions greater than 1 μm. Optical microscopy essentially permits the quantitative assessment of resolution and alignment in two dimensions. SEM allows for similar measurements at smaller length scales. AFM, profilometry, and SEM can be used to quantify the height, edge shape, and edge roughness of the transferred patterns. These 3D parameters may have direct impact on the behavior of cells or deposited coatings.

Although numerous examples exist, one research investigation that clearly demonstrates the impact of lithographic techniques on our understanding of cellular behavior is the use of lithographically fabricated microneedles to measure cellular mechanics (du Roure *et al.*, 2005; Tan *et al.*, 2003; Chapter 13 by Sniadecki and Chen, this volume). Through a combination of conventional photolithography and imprint lithography, a surface of high-aspect ratio posts, or microneedles, was fabricated using a soft elastomer, cross-linked PDMS. Onto this surface of microneedles, smooth muscle cells were cultured and observed using a variety of microscope measurement methods. Not only were unique cellular morphologies observed due to the surface topography and spatial arrangement of the lithographic features, but image analysis permitted the measurement of the microneedle deformation imposed by the cellular traction forces. With the measured deformation profile and well-controlled geometry of the microneedles, the researchers used standard bending beam mechanics to directly quantify the magnitude of the cellular traction forces supported by the different cell lines. This measurement would not have been possible without the direct assembly of well-defined 3D surface structures through the processes of photolithography and imprint lithography.

IV. Patterning at the Micro- and Nanoscale with Polymer Mixtures and Block Copolymers

Many techniques are available to fabricate patterned surfaces, but these methods are limited by the size of the patterned area and/or minimum feature size of ~100 nm that can be routinely and reproducibly achieved. One strategy to overcome this limitation relies on the morphology generated by the microphase separation of diblock copolymers (explained below), the self-assembly of a mixture of homopolymer blend and diblock copolymer, and the macroscopic phase separation of homopolymer blends to fabricate surface patterns with feature sizes ranging from tens of nanometers to tens of micrometers over a large surface areas on a single substrate (Fredrickson and Bates, 1996; Hadjichristidis *et al.*, 2002; Hamley, 1999; Sperling, 2001; Chapter 5 by Spatz and Geiger, this volume). These surfaces may then be used either directly for cell culture or as molds for casting the pattern onto other polymer surfaces as described earlier. Due to the wide range of polymers and conditions that may be used for this purpose, the discussion will be focused on general principles rather than specific procedures.

Consider a mixture of two chemically different polymers. Phase separation in polymer mixtures occurs initially by a spinodal phase separation process where bicontinuous domains of the two polymers form with a characteristic wavelength on the micrometer size scale. With time, this morphology coarsens into macroscopic domains. Interfacial tension, of course, drives the domains to be as large as possible, while the slow diffusion of the polymer chains limits the maximum size of the domains achieved (Strobl, 1997; Young and Lovell, 1991). The features generated by this phase separation are used to produce 3D topographic substrates.

Homopolymers consist of a sequence of covalently bonded monomers of the same species into a long-chain molecule, like PS or PMMA. In general, homopolymers are immiscible with other homopolymers. As the total number of molecules, and thus the entropy, is already limited by the interconnectivity of the monomers, further reduction in the entropy as a result of small nonfavorable interactions will easily cause the polymers to phase separate on a macroscopic level. Block copolymers, on the other hand, consist of two or more chemically different polymers covalently bound together at one end (Fig. 5). For example, a diblock copolymer of polystyrene-*block*-poly(methyl methacrylate) (PS-*b*-MMA) is a chain composed of PS covalently bound to PMMA at one end. Block copolymers readily undergo spontaneous microphase separations in solution due to the distinct chemical nature of the blocks. By joining the chains together at one end, the size scale of the phase separation is limited to the size of the molecules, that is tens of nanometers. The relative lengths of the chains dictate the volume fraction of the components, which, in turn, determines the nature of the morphology on phase separation such as

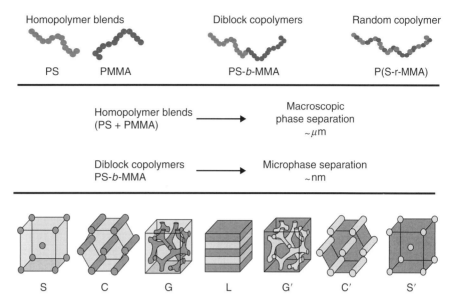

Fig. 5 Schematic representation of homopolymer blends, diblock copolymer, and random copolymer architectures. The phase diagram of diblock copolymer (Matsen and Bates, 1996).

body-centered cubic arrays of spheres, hexagonally packed cylinders, bicontinuous gyroids, or alternating lamellae (Fig. 5) (Bates and Fredrickson, 1990, 1999; Hamley, 2004; Hashimoto *et al.*, 1983). The architectures of block copolymers can be expanded further to multiblock copolymers, for example triblock or pentablock, star- or comb-shaped block copolymers, where the chemical composition of the blocks can be two, three, or more chemically different polymers. In addition, rather then linking complete polymer chains together, the monomers can be randomly placed along the chain forming a random copolymer (Fig. 5), for example P(S-r-MMA) consists of randomly dispersed monomers of styrene and methyl methacrylate.

A. Principles and Procedures for Controlling Pattern Formation with Polymer Mixtures and Block Copolymers

3D topographic substrates with micrometer-sized patterns can be prepared by coating a thin film of polymer mixtures with a well-defined thickness (ranging from tens of nanometers to micrometers in thickness) onto a substrate with a well-defined surface energy. Any deviation in controlling the film thickness or the surface of the substrate can dramatically change the final morphology of the phase-separated mixtures, and this consequently affects the 3D topographic substrates (Binder, 1998). For example, consider a thin film of a mixture of PS and PMMA, solution cast or spin coated onto a silicon wafer. In general, there is a thin (~2 nm) native oxide layers on the silicon, unless care is taken to strip this layer from the surface. PMMA, which is more polar than PS, will segregate to the polymer–silicon oxide interface (Reiter, 1992, 1993; Rockford *et al.*, 1999; Russell *et al.*, 1989), that is preferentially wetting or spreading on the substrate. On the other hand, the component with the lower surface energy, in this case PS, will preferentially segregate to the air surface. If, however, the substrate is coated with gold, then PS will preferentially wet the gold surface (Rockford *et al.*, 1999). Thus, interactions of the polymer mixture with the substrate can be controlled and, consequently, the morphology of the resultant mixture can be manipulated by chemically modifying the surface, for example, with functionalized chlorosilanes in the case of silicon oxide. The interactions may be fine-tuned by anchoring random copolymers with well-defined compositions to the surface (Hawker *et al.*, 1996; Huang *et al.*, 1998; Mansky *et al.*, 1997a). Such modified surfaces can then be used to control the lateral length scale of heterogeneities which, in turn, can be used to modify the topographic or chemical patterning of surfaces to investigate the effects on cell motility and adhesion (Tsai *et al.*, 2004b).

As mentioned previously, with block copolymers, due to the connectivity of the two chains, the phase separation is limited to the dimensions of the copolymer chain, ~5–20 nm in size and the relative volume fraction of the component blocks dictates the nature of the morphology. The versatility of block copolymers can be appreciated when one considers that the chemistry of the individual blocks can be tailored to perform a specific function, for example chemical reactivity, biological activity, conductivity, or degradability. While much is known about the bulk behavior of block copolymers, the number of applications that fully exploit their self-assembled morphologies has been limited. However, if the orientation and lateral ordering of the microdomains can be controlled in thin films and methods developed to bias the orientation and lateral ordering of the arrays of microdomains, then block copolymers have the potential to become a standard tool in the fabrication of nanostructured devices. We will discuss briefly various robust approaches for fabricating topographic structures at a scale of tens of nanometers.

The parameters that underpin the morphology in thin films of block copolymers include the segmental interactions between the components, the rigidity of each block, the surface energies of the components, and the interactions of the blocks with the underlying substrate. Similar to polymer mixtures described above, preferential interactions of one block with the substrate, or a lower surface energy of one component, will force a segregation of one block to either the surface or the substrate. However, due to the connectivity of the blocks, these interactions typically force an orientation of the microdomains parallel to the substrate. While this is desirable for some applications, the formation of 3D topographic pattern usually requires that the microdomains be oriented normal to the surface. The control of phase separation in thin films is further complicated by the anisotropic or asymmetric shape of the microdomains, as in the case with cylindrical or lamellar microdomains, since the interactions of the blocks with the substrate or air interface cause a preferential orientation of the microdomains. In addition to surface modification as described below, electric fields that operate on the differences in the dielectric constants of the microdomains, or solvents that affect interfacial interactions by dilution of each component, have been used with great success to control the orientation of the microdomains relative to the film surface (Kim *et al.*, 2004; Kimura *et al.*, 2003; Xu *et al.*, 2004). In the case of electric field, a voltage is applied across the film at liquid state or above the glass transition temperature; in the case of the solvent manipulation, the morphology is controlled by the direction of the solvent evaporation, that is, normal to the film surface. Figure 6 is an example of the nanostructures formed by the diblock copolymer P(S-*b*-MMA), measured by AFM.

Surfaces can be chemically modified to control the interfacial interactions of the blocks, in effect eliminating any preferential affinity either by anchoring a random copolymer or by removing the oxide layer with a buffered hydrogen fluoride (HF) (Kim and Russell, 2001; Kim *et al.*, 2001; Mansky *et al.*, 1997b). Balanced interfacial interactions, while necessary, are not sufficient, since either orientation of the block copolymer can satisfy this boundary constraint. Effective control may be achieved by slightly modifying one of the blocks by the random placement of a different chemical unit along the chain, to cause small changes in the surface energies, segmental interactions, and rigidity without sacrificing the overall characteristic of the block copolymer. This process is independent of the substrate and is not restricted to homogeneous surfaces, as shown by its successful applications to several different substrate such as gold, aluminum, silicon oxide, Mylar (polyethylene terephthalate), and Kapton (polyimide) surfaces (Ryu *et al.*, 2005).

One can further exploit the benefits of the self-assembly process of homopolymer blend and diblock copolymer to fabricate a gradient surface having nanometer to micrometer patterns (Tsai *et al.*, 2004a). This can be achieved by

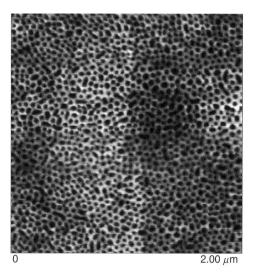

0   2.00 μm

Fig. 6 Atomic force microscopic image of a diblock copolymer of PS-*b*-PMMA template taken in the height mode. The dark areas are hole that have been made by removing the PMMA, using UV radiation and the brighter area is the matrix of PS that has been cross-linked by the UV radiation.

gradually varying the relative concentrations of homopolymers and block copolymers, and thus, the length scale of the domains from the tens of nanometers with pure diblock copolymers to hundreds of nanometers with a mixture of diblock copolymer and homopolymer blend, and, finally, to micrometers with homopolymer mixtures. Such gradient surfaces can be produced by drawing a solution across a surface, while the concentration of components in the solution is gradually varied at the same time. If the solvent evaporates rapidly, before mixing of the components in solution can occur, a film is left behind where the concentration of copolymer varies gradually across the surface. The procedure may involve placing two reservoirs on a surface, one containing a solution of the pure diblock copolymer (nano-sized morphology) while the other containing a solution of polymer mixtures (micro-sized morphology). As the two reservoirs spread over the surface, the solutions interdiffuse, producing a gradient in the concentration of the components. On solvent evaporation, a film is left where the concentration of copolymers varies linearly across the surface producing a gradient of morphology from nanometers to micrometers in scale.

In summary, we have discussed several routes to fabricate nano- to micropatterned surfaces using the self-assembly and phase separation of diblock copolymers with or without homopolymer blends. Although the principle may appear complex initially, it requires relatively inexpensive equipment (e.g., a spin coater, an oven, a UV lamp), except the usage of scanning force microscopy, SEM, or TEM, which can be borrowed from many material science centers, to characterize the structures. We have shown that NIH3T3 fibroblasts respond in dramatically different ways on surfaces with micrometer- and nanometer-sized heterogeneities (Tsai *et al.*, 2004b).

V. Summary

We have described several tools to study adhesion and migration of cells on 3D topographic substrates, which mimicks cells in physiological 3D environments. These techniques can be easily transferred to biological setting and applied to many types of cells such as epithelium, osteoblasts, and hepatocyte, during their differentiation and signal transduction. The method can be modified to pattern extracellular matrix proteins instead of synthetic polymers. In addition, such a substrates can be utilized to analyze mechanical interactions at cell–substrate adhesion and provide an easy approach to address cell behavior and function in 3D without drastic modifications to the current technology.

Acknowledgments

The authors are deeply indebted to the students and postdoctoral fellows in the laboratories of Russell, Wang, and Crosby for developing many of the methods described here. We acknowledge the support of the National Science Foundation Materials Research Science and Engineering Center on Polymers at the University of Massachusetts Amherst, and the US Department of Energy, Office of Basic Energy Science.

References

Abbott, A. (2003). Cell culture: Biology's new dimension. *Nature* **424**(6951), 870–872.

Bailey, T., Choi, B. J., Colburn, M., Meissl, M., Shaya, S., Ekerdt, J. G., Sreenivasan, S. V., and Willson, C. G. (2000). Step and flash imprint lithography: Template surface treatment and defect analysis. *J. Vac. Sci. Technol. B* **18**(6), 3572–3577.

Bailey, T. C., Johnson, S. C., Resnick, D. J., Sreenivasan, S. V., Ekerdt, J. G., and Willson, C. G. (2002). Step and flash imprint lithography: An efficient nanoscale printing technology. *J. Photopolym. Sci. Technol.* **15**(3), 481–486.

Bates, F. S., and Fredrickson, G. H. (1990). Block copolymer thermodynamics—theory and experiment. *Annu. Rev. Phys. Chem.* **41,** 525–557.

Bates, F. S., and Fredrickson, G. H. (1999). Block copolymers—designer soft materials. *Phys. Today* **52**(2), 32–38.

Binder, K. (1998). Spinodal decomposition in confined geometry. *J. Non-Equilib. Thermodyn.* **23**(1), 1–44.

Chou, S. (1997). Patterned magnetic nanostructures and quantized magnetic disks. *Proc. IEEE* **85**(4), 652–671.

Chou, S., Krauss, P., and Renstrom, P. (1996). Nanoimprint lithography. *J. Vac. Sci. Tech. B* **14**(6), 4129–4133.

Clark, P., Connolly, P., and Moores, G. R. (1992). Cell guidance by micropatterned adhesiveness *in vitro*. *J. Cell Sci.* **103,** 287–292.

Craighead, H. G., James, C. D., and Turner, A. M. P. (2001). Chemical and topographical patterning for directed cell attachment. *Curr. Opin. Solid State Mater. Sci.* **5**(2–3), 177–184.

Cukierman, E., Pankov, R., and Yamada, K. M. (2002). Cell interactions with three-dimensional matrices. *Curr. Opin. Cell Biol.* **14**(5), 633–639.

Curtis, A., and Wilkinson, C. (2001). Nantotechniques and approaches in biotechnology. *Trends Biotechnol.* **19**(3), 97–101.

Desai, T. A. (2000). Micro- and nanoscale structures for tissue engineering constructs. *Med. Eng. Phys.* **22**(9), 595–606.

du Roure, O., Saez, A., Buguin, A., Austin, R. H., Chavrier, P., Silberzan, P., and Ladoux, B. (2005). Force mapping in epithelial cell migration. *Proc. Natl. Acad. Sci. USA* **102**(7), 2390–2395.

Duffy, D. C., McDonald, J. C., Schueller, O. J. A., and Whitesides, G. M. (1998). Rapid prototyping of microfluidic systems in poly(dimethylsiloxane). *Anal. Chem.* **70**(23), 4974–4984.

Dunn, G. A., and Ebendal, T. (1978). Contact guidance on oriented collagen gels. *Exp. Cell Res.* **111**(2), 475–479.

Flemming, R. G., Murphy, C. J., Abrams, G. A., Goodman, S. L., and Nealey, P. F. (1999). Effects of synthetic micro- and nano-structured surfaces on cell behavior. *Biomaterials* **20**(6), 573–588.

Folch, A., and Toner, M. (2000). Microengineering of cellular interactions. *Annu. Rev. Biomed. Eng.* **2,** 227–256.

Fredrickson, G. H., and Bates, F. S. (1996). Dynamics of block copolymers: Theory and experiment. *Annu. Rev. Mater. Sci.* **26,** 501–550.

Frey, M., Tsai, I., Russell, T. P., Hanks, S., and Wang, Y. (2006). Cellular responses to substrate topography: Role of myosin II and focal adhesion kinase. *Biophys. J.* **90**(10), 3774–3782.

Friedl, P., and Brocker, E. B. (2000). The biology of cell locomotion within three-dimensional extracellular matrix. *Cell. Mol. Life Sci.* **57**(1), 41–64.

Geissler, M., and Xia, Y. N. (2004). Patterning: Principles and some new developments. *Adv. Mater.* **16**(15), 1249–1269.

Guo, L. J. (2004). Recent progress in nanoimprint technology and its applications. *J. Phys. D Appl. Phys.* **37**(11), R123–R141.

Hamley, I. W. (1999). "The Physics of Block Copolymers." Oxford University Press, USA.

Hamley, I. W. (2004). "Developments in Block Copolymer Science and Technology." John Wiley & Sons, Chichester, GB.

Hashimoto, T., Shibayma, M., Fujimura, M., and Kawai, H. (1983). "Block Copolymers: Science and Technology." Taylor & Francis, Miami.

Hawker, C. J., Elce, E., Dao, J. L., Volksen, W., Russell, T. P., and Barclay, G. G. (1996). Well-defined random copolymers by a 'living' free-radical polymerization process. *Macromolecules* **29**(7), 2686–2688.

Huang, E., Rockford, L., Russell, T. P., and Hawker, C. J. (1998). Nanodomain control in copolymer thin films. *Nature* **395**(6704), 757–758.

Kaihara, S., Borenstein, J., Koka, R., Lalan, S., Ochoa, E. R., Ravens, M., Pien, H., Cunningham, B., and Vacanti, J. P. (2000). Silicon micromachining to tissue engineer branched vascular channels for liver fabrication. *Tissue Eng.* **6**(2), 105–117.

Kim, H. C., Jia, X. Q., Stafford, C. M., Kim, D. H., McCarthy, T. J., Tuominen, M., Hawker, C. J., and Russell, T. P. (2001). A route to nanoscopic SiO_2 posts via block copolymer templates. *Adv. Mater.* **13**(11), 795.

Kim, H. C., and Russell, T. P. (2001). Ordering in thin films of asymmetric diblock copolymers. *J. Polym. Sci. Part B Polym. Phys.* **39**(6), 663–668.

Kim, S. H., Misner, M. J., Xu, T., Kimura, M., and Russell, T. P. (2004). Highly oriented and ordered arrays from block copolymers via solvent evaporation. *Adv. Mater.* **16**(3), 226–231.

Kimura, M., Misner, M. J., Xu, T., Kim, S. H., and Russell, T. P. (2003). Long-range ordering of diblock copolymers induced by droplet pinning. *Langmuir* **19**(23), 9910–9913.

Krauss, P., and Chou, S. (1997). Nano-compact disks with 400 Gbit/in^2 storage density fabricated using nanoimprint lithography and read with proximal probe. *Appl. Phys. Lett.* **71**(21), 3174–3176.

Lin, Z. Q., Kerle, T., Baker, S. M., Hoagland, D. A., Schäffer, E., Steiner, U., and Russell, T. P. (2001). Electric field induced instabilities at liquid/liquid interfaces. *J. Chem. Phys.* **114**(5), 2377–2381.

Lin, Z. Q., Kerle, T., Russell, T. P., Schäffer, E., and Steiner, U. (2002). Structure formation at the interface of liquid liquid bilayer in electric field. *Macromolecules* **35**(10), 3971–3976.

Love, J. C., Wolfe, D. B., Jacobs, H. O., and Whitesides, G. M. (2001). Microscope projection photo-lithography for rapid prototyping of masters with micron-scale features for use in soft lithography. *Langmuir* **17**(19), 6005–6012.

Madou, M. J. (2002). "Fundamentals of Microfabrication." CRC Press, Boca Raton, FL.

Mansky, P., Liu, Y., Huang, E., Russell, T. P., and Hawker, C. (1997a). Controlling polymer-surface interactions with random copolymer brushes. *Science* **275**(5305), 1458–1460.

Mansky, P., Russell, T. P., Hawker, C. J., Pitsikalis, M., and Mays, J. (1997b). Ordered diblock copolymer films on random copolymer brushes. *Macromolecules* **30**(22), 6810–6813.

Matsen, M., and Bates, F. (1996). Origins of complex self-assembly in block copolymers. *Macromolecules* **29**(23), 7641–7644.

Miyajima, H., and Mehregany, M. (1995). High-aspect-ratio photolithography for MEMS applications. *J. Microelectromech. Syst.* **4**(4), 220–229.

Mrksich, M. (2000). A surface chemistry approach to studying cell adhesion. *Chem. Soc. Rev.* **29**(4), 267–273.

Mrksich, M. (2002). What can surface chemistry do for cell biology? *Curr. Opin. Chem. Biol.* **6**(6), 794–797.

Hadjichristidis, N., Pispas, S., and Floudas, G. (2002). "Block Copolymers: Synthetic Strategies, Physical Properties, and Applications." Wiley-Interscience, New York, NY.

Oner, D., and McCarthy, T. (2000). Ultrahydrophobic surfaces. Effects of topography length scales on wettability. *Langmuir* **16**(20), 7777–7782.

Pirone, D. M., and Chen, C. S. (2004). Strategies for engineering the adhesive microenvironment. *J. Mammary Gland Biol. Neoplasia* **9**(4), 405–417.

Qin, D., Xia, Y. N., Rogers, J. A., Jackman, R. J., Zhao, X. M., and Whitesides, G. M. (1998). Microfabrication, microstructures and microsystems. *Microsyst. Technol. Chem. Life Sci.* **194**, 1–20.

Reiter, G. (1992). Dewetting of thin polymer-films. *Phys. Rev. Lett.* **68**(1), 75–78.

Reiter, G. (1993). Unstable thin polymer-films—rupture and dewetting processes. *Langmuir* **9**(5), 1344–1351.

Resnick, D. J., Dauksher, W. J., Mancini, D. P., Nordquist, K. J., Bailey, T. C., Johnson, S. C., Stacey, N. A., Ekerdt, J. G., Wilson, C. G., Sreenivasan, S. V., and Schumaker, N. E. (2003). Imprint lithography for integrated circuit fabrication. *J. Vac. Sci. Technol. B* **21**(6), 2624–2631.

Rockford, L., Liu, Y., Mansky, P., Russell, T. P., Yoon, M., and Mochrie, S. G. J. (1999). Polymers on nanoperiodic, heterogeneous surfaces. *Phys. Rev. Lett.* **82**(12), 2602–2605.

Russell, T. P., Coulon, G., Deline, V. R., and Miller, D. C. (1989). Characteristics of the surface-induced orientation for symmetric diblock Ps/Pmma copolymers. *Macromolecules* **22**(12), 4600–4606.

Ryu, D. Y., Shin, K., Drockenmuller, E., Hawker, C. J., and Russell, T. P. (2005). A generalized approach to the modification of solid surfaces. *Science* **308**(5719), 236–239.

Schaffer, E., Thurn-Albrecht, T., Russell, T. P., and Steiner, U. (2000). Electrically induced structure formation and pattern transfer. *Nature* **403**(6772), 874–877.

Schafer, E., Thurn-Albrecht, T., Russell, T. P., and Steiner, U. (2001). Electrohydrodynamic instabilities in polymer films. *Europhys. Lett.* **53**(4), 518–524.

Singhvi, R., Stephanopoulos, G., and Wang, D. I. C. (1994). Effects of substratum morphology on cell physiology—review. *Biotechnol. Bioeng.* **43**(8), 764–771.

Sperling, L. H. (2001). "Introduction to Physical Polymer Science." Wiley-Interscience, New York, NY.

Stewart, M., Sreenivasan, S., Resnick, D., and Wilson, C. (2005). Nanofabrication with step and flash imprint lithography. *J. Microlith. Microfab. Microsys.* **4**(1), 1537–1646.

Strobl, G. R. (1997). "The Physics of Polymers: Concepts for Understanding Their Structures and Behavior." Springer, Newark, DE.

Tan, J., and Saltzman, W. M. (2004). Biomaterials with hierarchically defined micro- and nanoscale structure. *Biomaterials* **25**(17), 3593–3601.

Tan, J. L., Tien, J., Pirone, D. M., Gray, D. S., Bhadriraju, K., and Chen, C. S. (2003). Cells lying on a bed of microneedles: An approach to isolate mechanical force. *Proc. Natl. Acad. Sci. USA* **100**(4), 1484–1489.

Tsai, I. Y., Kimura, M., and Russell, T. P. (2004a). Fabrication of a gradient heterogeneous surface using homopolymers and diblock copolymers. *Langmuir* **20**(14), 5952–5957.

Tsai, I. Y., Kimura, M., Stockton, R., Green, J. A., Puig, R., Jacobson, B., and Russell, T. P. (2004b). Fibroblast adhesion to micro- and nano-heterogeneous topography using diblock copolymers and homopolymers. *J. Biomed. Mater. Res. Part A* **71A**(3), 462–469.

Weaver, V. M., Fischer, A. H., Peterson, O. W., and Bissell, M. J. (1996). The importance of the microenvironment in breast cancer progression: Recapitulation of mammary tumorigenesis using a unique human mammary epithelial cell model and a three-dimensional culture assay. *Biochem. Cell Biol.* **74**(6), 833–851.

Weiss, P. (1945). Experiments on cell and axon orientation *in vitro*—the role of colloidal exudates in tissue organization. *J. Exp. Zool.* **100**(3), 353–386.

Whitesides, G. M., Ostuni, E., Takayama, S., Jiang, X., and Ingber, D. E. (2001). Soft lithography in biology and biochemistry. *Annu. Rev. Biomed. Eng.* **3**, 335–373.

Wilkinson, C. D. W., Riehle, M., Wood, M., Gallagher, J., and Curtis, A. S. G. (2002). The use of materials patterned on a nano- and micro-metric scale in cellular engineering. *Mater. Sci. Eng. C-Biomimetic Supramol. Syst.* **19**(1–2), 263–269.

Xia, Y. N., Zhao, X. M., and Whitesides, G. M. (1996). Pattern transfer: Self-assembled monolayers as ultrathin resists. *Microelectron. Eng.* **32**(1–4), 255–268.

Xu, T., Zhu, Y. Q., Gido, S. P., Hawker, C. J., and Russell, T. P. (2004). Electric field alignment of symmetric diblock copolymer thin films. *Macromolecules* **37**(7), 2625–2629.

Young, R. J., and Lovell, P. A. (1991). "Introduction to Polymers." CRC, Chapman & Hall, London.

CHAPTER 5

Molecular Engineering of Cellular Environments: Cell Adhesion to Nano-Digital Surfaces

Joachim P. Spatz[*] **and Benjamin Geiger**[†]

[*]Department of New Materials and Biosystems
Max Planck Institute for Metals Research
Stuttgart, Germany; and Department of Biophysical Chemistry
University of Heidelberg, Heidelberg, Germany

[†]Department of Molecular Cell Biology
Weizmann Institute of Science
Rehovot 76100, Israel

Abstract

Engineering of the cellular microenvironment has become a valuable means to guide cellular activities such as spreading, motility, differentiation, proliferation, or apoptosis. This chapter summarizes recent approaches to surface patterning such as topography and chemical patterning from the micrometer to the nanometer scale, and illustrates their application to cellular studies. Particular attention is devoted to nanolithography with self-assembled diblock copolymer micelles that are biofunctionalized with peptide ligands—a method that offers unsurpassed spatial resolution for the positioning of signaling molecules over extended surface areas. Such interfaces are defined here as "nano-digital surfaces," since they enable the counting of individual signaling complexes separated by a biologically inert background. The approach enables the testing of cellular responses to individual signaling molecules as well as their *spatial* ordering. Detailed consideration is also given to the fact that protein clusters such as those found at focal adhesion sites represent, to a large extent, hierarchically organized cooperativity among various proteins.

I. Introduction: Sensing Cellular Environments

Cell adhesion to the extracellular matrix (ECM) and to neighboring cells is a complex, tightly regulated process that plays a crucial role in fundamental cellular functions, including cell migration, proliferation, differentiation, and apoptosis (Blau and Baltimore, 1991; Ruoslahti and Obrink, 1996). By interacting directly with such external surfaces, cells gather information about the chemical and physical nature of the ECM, integrate and interpret it, and then generate an appropriate physiological response. The mechanism underlying the capacity of cells to perform such "intelligence missions" of data acquisition and processing is still poorly understood at the molecular level, although it is likely that the cell's adhesive machinery as well as associated cytoskeletal and signaling networks play a major role (Vogel and Sheetz, 2006).

What molecular features of the environment can cells sense? In principle, cells monitor differences between soluble and immobilized signaling molecules. Soluble factors are capable of activating specific signaling networks. Spatial resolution of different chemical concentrations is given by chemoattractant gradients. However, their spatial resolution is rather low, that is, in the range of a few micrometers when applying microfluidic devices (Lin and Butcher, 2006).

Examination of the structure of the ECM provides strong evidence that the presentation of signaling molecules to cells in a defined microscopic as well as nanoscopic geometry affects cell response with great consequences, provided that cells are in direct contact with the ECM. Figure 1A presents a transmission electron microscopy image of an epidermal cell, that is fibroblast, which is extensively embedded in a network of collagen fibers organized into bundles that run approximately at right angles to each other (Alberts *et al.*, 2001, p. 1098, Figs. 19–44; Ploetz *et al.*, 1991). Figure 1B presents a scanning force microscopy (SFM) image of a collagen fiber bundle as visualized in Fig. 1A. Since SFM provides nanoscopic resolution based on sensing topographic or mechanical differences on surfaces, fibers of collagen are displayed with a periodic corrugation of

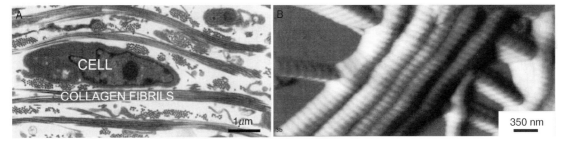

Fig. 1 (A) Fibroblast surrounded by collagen fibrils in the connective tissue (Alberts *et al.*, 2001 and modified from Ploetz *et al.*, 1991). (B) SFM of scleral collagen fibrils displaying ∼67-nm periodicity of collagen fibrils (from Meller *et al.*, 1997).

∼67 nm (Meller *et al.*, 1997). Such a periodicity is typically observed for collagen type one as also observed by scanning electron microscopy (SEM) or x-ray scattering (Fratzl *et al.*, 1998; Jiang *et al.*, 2004). From such structural observations, it becomes evident that the ECM presents a hierarchy of different length scales and viscoelasticities to cell surfaces. A key question in cell biology is how spatial positioning of signaling cues as well as variations in local rigidity of the ECM at different length scales affects cell responses.

Tissue cells respond to the presented environment with adhesion if appropriate conditions are given. Cells are able to sense differences in physical conditions such as different chemical or mechanical properties. Cells may distinguish between surfaces of different chemical properties due to the differential activation of distinct surface receptors on interactions with the particular repertoire of ligands on these surfaces. Such diverse interactions may then trigger a wide variety of cellular responses.

Rigidity of a cell's environment was identified as a marker that provides essential information to cells. Cells sense differences in rigidity of the adhesive environment which in turn strongly affects the fate of cells (Discher *et al.*, 2005; Georges and Janmey, 2005). A variety of receptors are involved in the cell's local interactions with the external surface (Fig. 2A). The best-characterized systems consist of micrometer-sized, multiprotein complexes called focal adhesions (FA) (Fig. 2A) and related structures, including focal complexes, which are small nascent adhesions formed under the leading lamellae (Ballestrem *et al.*, 2001; Zamir and Geiger, 2001a), and fibrillar adhesions formed with fibronectin networks and consisting of integrin receptors and a wide variety of cytoskeletal and signaling proteins (Fig. 1A) (Critchley, 2000; Geiger *et al.*, 2001; Giancotti and Ruoslahti, 1999; Hynes, 1987; Levenberg *et al.*, 1998; Miyamoto *et al.*, 1995; Zamir and Geiger, 2001a,b; for an overview, see Fig. 2B). The assembly of these molecules at nascent adhesion sites triggers the reorganization of the actin cytoskeleton, which in turn generates local forces by activating the motor activity of myosins (Balaban *et al.*, 2001).

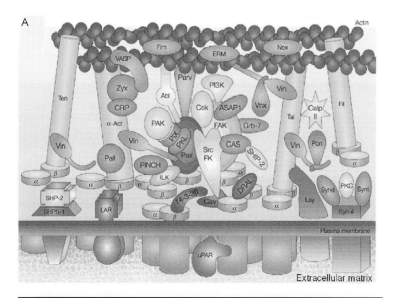

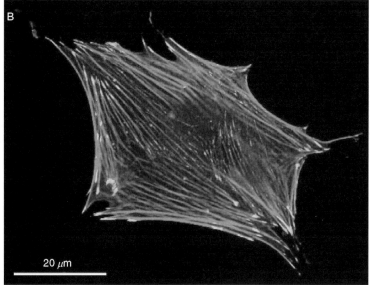

Fig. 2 (A) Schematic presentation (updated, 2001) of the complexity of the main molecular domains of cell–matrix adhesions (Geiger *et al.*, 2001; Zamir and Geiger, 2001a,b) The primary adhesion-mediating receptors in FA are heterodimeric (α and β) integrins, represented by the orange cylinders. For details concerning the various components, see (Geiger *et al.*, 2001; Zamir and Geiger, 2001a,b). (B) Fluorescent microscopy of a 3T3 fibroblast adhering to an RGD-functionalized glass plate. F-actin stress fibers are shown in red and clusters of the "plaque" protein, vinculin is shown in green (over-lapping areas are yellow). Vinculin marks the location of FA sites. (See Plate 2 in the color insert section.)

   The desire for surfaces to provide information on the nanometer scale is given by the fact that major players of FA are proteins with dimensions in the range of several nanometers such as integrin, talin, vinculin, paxillin, or α-actinin (see also Fig. 2A and B). In recent years, it has become more and more evident that these adhesion-associated proteins assemble into hierarchical and cooperative arrangements, although we do not understand the mechanism how synergetic interactions between adhesion-associated proteins regulate the assembly and signaling functions of FA. The assembly of integrins along the cell membrane is one of the first events observed during the formation of FA (Alberts *et al.*, 1994). Integrins are heterodimeric adhesion receptors which interact noncovalently with the RGD (arginine-glycine-aspartate) motif on ECM proteins, such as fibronectin and vitronectin. Ligand-binding affinity is influenced by conformational changes of the integrin molecule in response to both the extracellular environment and interactions with the cytoplasmic proteins of FA (Xiong *et al.*, 2002). Although the very first moments of adhesion must be essential for adhesion formation, only little is known how initial adhesive interaction activates single integrins and stimulates them to cluster formation (Cohen *et al.*, 2006; Li *et al.*, 2003). Notably, the projected area of an integrin hetrodimer is thought to be $\sim$25 nm^2. However, the density of integrin clusters can vary dynamically, even among cells attached to the same surface. For example, integrin density in focal complexes is usually approximately three times lower than that seen in FA (Ballestrem *et al.*, 2001). In this respect, integrin density and cluster structure are of most interest to be correlated with adhesion development and the consequent cellular response (Lasky, 1997). The size of integrins and other FA-associated proteins determines the length scale for the experimental manipulation of FA assembly, namely, in the order of 10 nm. A systems biology approach is needed to determine the effects of multiple perturbations—molecular-genetic manipulations, pharmacological treatments, and/or modifications of the adhesive surface—on FA structure and function. Currently, novel techniques are being developed to control structural surface arrangements by means of precise molecular nanopatterning.

   Accordingly, the design of an adhesive substrate with molecular precision must take into account this length scale of molecular dimensions as well as the packing density characteristics of molecules within FA sites. A straightforward approach to engineer such adhesive surfaces is to first synthesize fully nonadhesive surfaces where interaction of cells and proteins with surfaces is minimal. Specific adhesive epitopes can then be integrated into this nonadhesive background. This approach is quite powerful since it provides us with the ability to dissect the relationship among specific cellular responses to ligand presentation, the interaction of adhesion receptors with specific ligands, and the lateral positioning of single integrins along the substrate.

   Cell adhesion studies at nanoscale resolution may be divided into two general categories: (1) research addressing the responses of cells to variations in substrate topography (Dalby *et al.*, 2002a,b, 2003; Ebendal, 1976; Kemkemer *et al.*, 2004) and (2) the responses of cells to chemical variations along the substrate, which affect the specificity of the adhesive ligand and/or its distribution (Arnold *et al.*, 2004; Elbert and Hubbell, 2001; Maheshwari *et al.*, 2000).

Our aim, as implied in the title of this chapter, is to discuss novel approaches for the *molecular engineering* of adhesive surfaces—referred to as "nano-digital surfaces"—by means of chemical surface nanopatterning. Such artificial surfaces should be designed to allow individual control of their molecular composition and multiple physical features over a wide range of scales, from micro- down to nanometer levels without implying topographic features in surfaces to which cells do respond. Micro- and nanometer topographic corrugations in surfaces (Dalby *et al.*, 2002a,b, 2003; Ebendal, 1976; Kemkemer *et al.*, 2004) demonstrated that patterned surfaces may induce cell polarization and direct cell migration and that they may even regulate gene expression as well as cell signaling. Topographic features as small as 13 nm exert major effects on cell spreading, morphology, cytoskeletal organization, and even altering the cell's gene expression profile. Topographic features are avoided by, for example, plane film formation of polymers. Mixing signaling molecules with a biologically inert polymer such as polyethylene glycol (PEG) in combination with film formation (Elbert and Hubbell, 2001) as well as the use of oligo(ethylene oxide) functionalized self-assembled monolayers (SAMs) (Mrksich and Whitesides, 1997; Wang *et al.*, 1997) enabled the testing of cell responses to defined concentrations of signaling molecules.

Cell attachment to such passivated surfaces depends on many factors, such as the affinity and specificity of the corresponding cellular receptors for the ligand, the mechanical strength of ligand support and linkage, spacer length, overall ligand concentration, and spacing between ligand molecules (Roberts *et al.*, 1998). For example, the number of attached cells is clearly correlated to the average RGD surface density, as shown by the sigmoidal increase in attached cells with RGD concentration (Kantlehner *et al.*, 2000). This indicates that there is a minimal threshold ligand density for stable cell binding. In addition, it was found that, as a general rule, higher RGD surface density leads to an increase in cell spreading (up to a maximum), cell survival, and FA formation. It has also been established that the number of RGD molecules required for cell attachment is smaller than that needed for the induction of cell spreading and FA assembly. Furthermore, quantification of the average ligand density indicated that a threshold amount of as low as 6 (GRGDY) ligands/μm^2 is sufficient for maximal cell spreading, while ~60 (GRGDY) ligands/μm^2 are needed for FA and stress fiber formation (Massia and Hubbell, 1991). Subsequent work with monolayer-coated surfaces suggested that a much higher density of ligands ($>1000\,RGD/\mu m^2$) is required for maximum spreading of cells (Roberts *et al.*, 1998). Therefore, different biological responses exhibit different dependencies on ligand density.

Macromolecular designs, such as PEG stars, enable control over ligand homogeneity at the nanoscale. An average of 0, 1, 5, or 9 YGRGD ligands per star can be prepared, as demonstrated by Griffith and Lauffenburger in a series of publications (Elbert and Hubbell, 2001; Irvine *et al.*, 2001a,b; Koo *et al.*, 2002; Maheshwari *et al.*, 2000). Another advantage of the macromolecular approach is

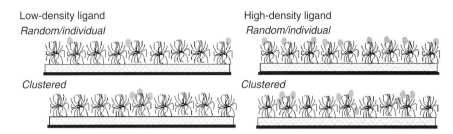

Fig. 3 Schematic illustration of star polymer as a tether to present ligands (shaded oval) in a manner in which the total average concentration (top vs bottom) and the spatial distribution, from homogeneous to highly clustered (left to right) can be independently varied (from Maheshwari *et al.*, 2000).

that the long and flexible chain arms may facilitate cell-binding activities by allowing local rearrangement of ligands at the cell membrane. However, such flexibility is unlikely to induce precise molecular ligand clustering. The interaction enables ligands to cluster either by pure chemical affinity or following cellular rearrangement (Koo *et al.*, 2002). Possible configurations for ligand patterns and local densities for two different macromolecular settings are shown in Fig. 3.

Precisely defined spatial distributions of ligands on an otherwise inert substrate could in principle shed light on many biological questions associated with matrix adhesion, formation, and cell signaling. A deeper understanding of how cell adhesion and signaling depend on the composition, size, and distribution of specific adhesion sites has remained elusive since most such studies have thus far been limited to micrometer or submicrometer areas (Sniadecki *et al.*, 2006). However, surfaces patterned with both adhesive and non-adhesive domains by means of microcontact printing have been prepared (Gates *et al.*, 2005; see Chapter 19), and successfully applied to geometrically control cell shape and viability on flat surfaces (Chen *et al.*, 1997). These experiments strongly indicate that cell shape and integrin distribution can control the survival or apoptosis of cells and can also "switch on" and "switch off" specific gene expression programs.

II. Nano–Digital Chemical Surfaces for Regulating Transmembrane–Receptor Clustering

A. Extended Nanopatterns and Biofunctionalization

Precise control of defined spacing between adhesive ligands on interfaces at a length scale of 10–200 nm remains a major challenge. However, it is this length scale on which protein clustering at FA occurs, thus surface patterning at this length scale is highly relevant to cell physiology. Most recently, nanopattern features have been found in adhesive fibers of collagen (Jiang *et al.*, 2004; Pompe *et al.*, 2005), which are major adhesive components of the natural ECM as already discussed above (see also Fig. 1).

Micelle diblock copolymer lithography technology represents a nanopatterning strategy that enables the modification of substrates at this length scale. This approach is based on the self-assembly of diblock copolymers of polystyrene-block-poly (2-vinylpyridine) (PS-b-P2VP) into reverse micelles in toluene (Glass *et al.*, 2003a,b; Spatz *et al.*, 1996a,b,c, 2000). Diblock copolymers consist of two blocks of polymers joined by a covalent bond. The core of the micelle consists of the P2VP block complexed with a metal precursor (HAuCl$_4$), which is added to the micellar solution during preparation. Dipping and retracting a substrate from such a solution results in uniform and extended monomicellar films on the substrate (Fig. 4A), and subsequent treatment of these films with oxygen or hydrogen gas plasma results in the deposition of highly regular Au nanoparticles. The gold "dots" form a nearly perfect hexagonal pattern on solid interfaces such as glass or silicon (Si) wafers (Fig. 4F). The size of the Au nanoparticles may be varied between 1 and 20 nm by adjusting the amount of HAuCl$_4$ added to the micellar solution. The spacing between Au nanoparticles may also be adjusted from 15 to 250 nm, by choosing the appropriate molecular weight of PS-b-P2VP and by changing the retraction speed. A preparation scheme is presented in Fig. 4A.

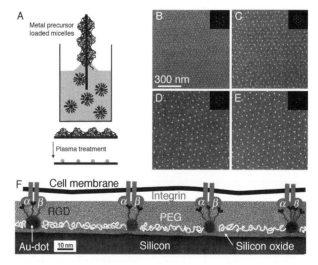

Fig. 4 Micellar block copolymer lithography and biofunctionalization. (A) Scheme of diblock copolymer micelle lithography. (B–E) Extended Au nanodot patterns are displayed by SEM. Uniform Au nanodots (bright spots) of (B) 3 nm by PS(190)-b-P[2VP(HAuCl$_4$)$_{0.5}$](190), (C) 5 nm by PS(500)-b-P[2VP (HAuCl$_4$)$_{0.5}$](270), (D) 6 nm by PS(990)-b-P[2VP(HAuCl$_4$)$_{0.5}$](385), and (E) 8 nm by PS(1350)-b-P[2VP (HAuCl$_4$)$_{0.5}$](400) deposited onto Si-wafers are shown. The number in brackets refers to the number of monomer units in each block which control the separation between Au dots. These varied between (B) 28, (C) 58, (D) 73, and (E) 85 nm. The Au dots form extended, nearly perfect hexagonally close-packed patterns as indicated by the Fourier transform images (inset) which show second order intensity spots. (F) Biofunctionalization of the Au nanodots pattern (Arnold *et al.*, 2004). Since the Au dot is sufficiently small, it is most likely that only one integrin transmembrane receptor directly interacts with one dot. The Au dots are presented as side view micrographs taken with a high-resolution transmission electron microscope (adapted from Arnold *et al.*, 2004).

SEM images of Si wafers coated with these Au nanoparticles show Au nano-particles as white spots arranged in quasi-hexagonal patterns (Fig. 4B–E). The patterns consist of Au nanoparticles 3, 5, 6, or 8 nm in diameter with spacings of 28, 58, 73, and 85 nm, respectively, between dots. A side view of the Au nanodots, taken by means of high-resolution transmission electron microscopy, is shown in Fig. 4F. These nanostructures serve as chemical templates for the spatial arrange-ment of RGD-based ligands, as shown schematically in Fig. 4F (Arnold *et al.*, 2004).

Passivation of the interface entails the binding of a polyethylene oxide (PEO) layer to the silicon oxide or glass substrate, covering the surface between the Au nanodots to avoid any interaction of the surface with proteins or the cell mem-brane. Subsequently, the Au dots are functionalized by the adhesive ligands that bind selectively to the Au nanoparticle. Cyclic RGD molecules (c(RGDfK)-thiol) recognized with high affinity by $\alpha v \beta 3$-integrin (Haubner *et al.*, 1997; Pfaff *et al.*, 1994) have been used successfully (Arnold *et al.*, 2004). In Fig. 4F, Au dots and integrins are drawn approximately to scale, indicating that the size of an Au nanodot provides a molecular anchor of a size that only a single integrin is likely to bind to. These nanodots thus constitute a very valuable tool to provide unique access to an important length scale for cell adhesion studies and allow to control the assembly of single integrins during the formation of FA clusters.

It is noteworthy that there are wide choices of ligands to link to the Au dots and different immobilization procedures are applicable for each ligand. Straightfor-ward coupling of ligands to Au nanoparticles is enabled by molecules or proteins with a thiol (−SH) group, for example, through a cysteine residue. Other possibi-lities include binding a carboxylic acid group (−COOH) to Au nanostructures which can then be activated by carbodiimides and covalently linked with amine groups. Since amine groups are commonly distributed in proteins, oriented immo-bilization is usually lacking. However, His-tagging, using either a synthetic or molecular genetic approach, is now coming into wide use, enabling oriented immobilization of nearly any ligand. Oriented binding of peptides or proteins to the Au nanoparticles has been achieved using $(His)_6$-tagging, which binds to NTA via nickel ion complexation. Lately, the stability of His-tag–NTA Ni complexes has been substantially improved by introducing multivalent com-plexation (Lata and Piehler, 2005). A biotin tag genetically engineered into a molecule can also be exploited as a linker for oriented binding of the molecule to surfaces, via a series of linkages mediated by streptavidin, for example (molecule-biotin)-streptavidin-(biotin-surface).

B. Micro-Nanopatterns for Spatially Controlled Molecular Clustering

In order to arrange locally a defined number of Au nanoparticles in a designated pattern, extended monomicellar layers are directly exposed to a focused ray of light in a conventional mask aligner or to an electron beam (e-beam) in an SEM. Either method will chemically modify the polymer located in exposed areas (Glass *et al.*, 2003a,b). This process is depicted in Fig. 5A. Locally exposed layers are washed in

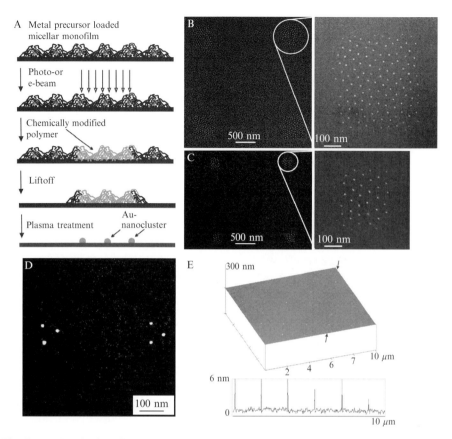

Fig. 5 (A) Application of monomicellar films as negative e-beam resist. The complementary length scales of a diblock copolymer micelle in which a single nanodot is perfectly positioned in the center and the resolution of photo- or e-beam lithography which reaches the diameter of a diblock copolymer micelle make this concept a suitable technology for the inexpensive generation of various micro-nanopatterns; for example, (B) squares made of ~100 × 6 nm Au particles, (C) squares made of ~40 × 6 nm Au particles, (D) 3 × 6 nm Au particles separated by ~400 nm, (E) single 6-nm Au particle separated by 2 μm in a square pattern (adapted from Glass *et al.*, 2003a,b).

dimethylformamide (DMF) or toluene in an ultrasound bath, and the remaining micelles are treated with gas plasma, resulting in the deposition of a pattern of nanoscopic Au particles in designated areas.

The SEM micrographs pictured in Fig. 5B–D show 6-nm Au particles localized within squares, each with ~100, 40, or 3 Au particles. Figure 5E shows a SEM image of single 6-nm Au particles separated by ~2 μm, arranged in a square pattern. Assuming that each Au nanoparticle represents a binding site for individual molecules, this method allows the size of ligand clusters to be controlled and quantified.

In the above procedure, the underlying surface had to be gas plasma resistant, which excludes the use of polymer surfaces as nanopattern supports. However, developments show that one can transfer such patterns to almost any type of soft surfaces (Fig. 6) (Graeter *et al.*, 2007), equipping one side of the gold dot surface area with a photosensitive linker molecule which then binds covalently to a polymer cast on top of the array. Pealing-off the polymer from the surfaces which originally carried the gold dots results in a complete transfer of the dots to the polymer matrix. This discovery is a major step forward in nanopattern substrate preparation since it enables the stiff glass or silicon oxide surfaces to be replaced with elastic or viscoelastic polymer surfaces. This new approach is regarded important for controlling surface properties that impact on cellular responses.

Figure 7A–C shows scanning electron micrographs of gold nanostructures 6 nm in size, represented as white spots on substrates made of either PS (A), PDMS (B), or PEGDA 20,000 hydrogels. Figure 7C represents a cryoscanning electron micrograph of a PEG hydrogel frozen in the water-swollen state at approximately $-120\,^{\circ}C$. Independently of the polymer, the nanostructures were successfully transferred one-to-one, with nanoscale precision, from the glass to the respective polymer substrate. These results demonstrate that transfer lithography is applicable to various polymers from rigid hydrophobic polystyrene to elastic silicone to soft hydrogels.

Another challenge for nanofabrication techniques is to prepare nanopatterns beyond planar surfaces. Curved surfaces are obviously meaningful for studies of cell behavior because most cells are three-dimensionally embedded in the ECM. This study confirms the feasibility of the block copolymer micelle nanolithography

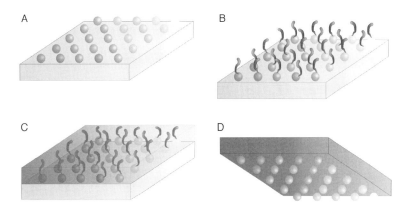

Fig. 6 Schematic presentation of the transfer nanolithography (Graeter *et al.*, 2007). (A) Gold nanostructures are deposited on glass supports by means of diblock copolymer micelle nanolithography, followed by chemical functionalization through linker molecules (B). (C) Coating of the glass support by a polymer and mechanical separation of the inorganic support and the polymer layer which transfers the inorganic structures from the glass to the polymer support (D). (See Plate 3 in the color insert section.)

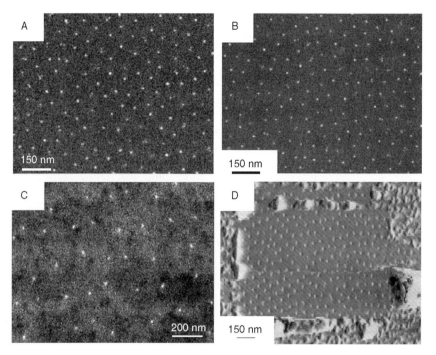

Fig. 7 Gold nanostructures on various polymer substrates as patterned by transfer lithography (Graeter *et al.*, 2007). SEM images of gold nanoparticles transferred from glass to PS (A) and PDMS (B). (C) A cryo-scanned electron micrograph of a gold nanopattern on a PEGDA 20,000 hydrogel. (D) Atomic force microscopy (AFM) micrograph of gold nanoparticles on a PEGDA 700 hydrogel immersed in water (Graeter *et al.*, 2007).

technique for nonplanar surfaces. As a demonstration, a tubelike hydrogel with a nanopatterned internal surface was prepared.

The diblock copolymer micelle nanolithographic technique was extended so as to decorate glass fibers with gold nanoparticles. Accordingly, glass fibers with diameters ranging from 60–500 μm were coated with nanostructures by dip-coating, as described for the planar substrates. The fibers were then treated with a hydrogen plasma to remove the micellar diblock copolymer from the gold particles as well as the glass fiber and to deposit the gold nanoparticles along the fiber. The remaining metal particles showed a typical hexagonal arrangement like that seen on the planar substrates (Fig. 8A). As with the transfer process of particles to a planar substrate, the gold particles were functionalized with a linker molecule. After embedding the fibers in PEGDA 700 and cross-linking PEGDA (Fig. 8B, step 1), the glass fiber was dissolved with hydrofluoric acid (Fig. 8B, step 2). This yielded channel structures internally decorated with gold nano-particles (Fig. 8D), as revealed by cryo-SEM of frozen and halved channels (Fig. 8C and D).

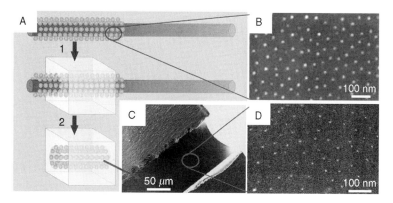

Fig. 8 Formation of nanostructured hydrogel microtubes (Graeter *et al.*, 2007). (A) Schematic drawing of the transfer lithography technique applied to curved surfaces, for example optical fibers. (B) SEM micrograph of a glass fiber decorated with gold nanoparticles by means of block copolymer micelle nanolithography. (C and D) Cryo-SEM images of a PEGDA 700 hydrogel channel decorated with gold nanoparticles. (See Plate 4 in the color insert section.)

C. Cellular Responses to Nano-Digital and Biofunctionalized Surfaces

Given the ability to functionalize adhesion surfaces with nanoscale resolution, one can test the diverse cellular responses to such patterns. In Fig. 9, MC3T3 osteoblasts were seeded on glass coverslips and examined by phase contrast microscopy. Only part of the glass substrate area was patterned with Au nanodots separated by different distances. The Au dots were functionalized by c(RGDfK)-thiols and the surrounding glass surface was passivated by PEG.

When plated on functionalized Au nanodot patterns separated by varying widths, MC3T3 osteoblasts display varied adhesive behaviors. It is evident that cells spread very well on the 58-nm pattern (Fig. 9A), comparable to their spreading on uniform RGD- or fibronectin-coated surfaces (data not shown). On the other hand, very limited cell spreading is observed on substrates with Au nanodots spaced 73 nm or more apart (Fig. 8B). Stationary cells appear rounded, while migrating cells are usually characterized by long extensions. Similar observations were made with other cell types, including REF52 fibroblasts, 3T3 fibroblasts, and B16 melanocytes.

Cell studies on nanostructured PEG hydrogels gold nanoparticles were transferred to PEG hydrogels and functionalized by cyclic RGDfK peptides. The gold nanostructured hydrogels were incubated in a 0.2-mmol solution of the cyclic RGDfK-thiol in water for 1 h. The hydrogels were then intensively washed with water to remove peptides which might have penetrated the gel. Subsequently, 3T3 fibroblasts were cultured for 24 h under standard cell culture conditions (in DMEM with 1% FBS) on PEGDA 700 hydrogels equipped with gold nanoparticles separated by distances of 40 nm. Figure 10A shows gold nanostructures that were not functionalized by cyclic RGDfK. Only a few cell aggregates are visible in

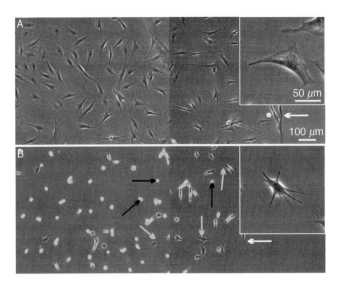

Fig. 9 Phase contrast microscopy of MC3T3-Osteoblasts attached to nanopatterned surfaces of different spacings: (A) ~58 and (B) ~73 nm. The right side was entirely passivated; thus, cell adhesion and attachment are only observed on the left side of the images. A line of cells marks the borderline of the nanopatterned area (white arrows). It is evident that cells spread very well on the 58 nm (Fig. 9A insert) pattern, comparable to their spreading on uniform RGD- or fibronectin-coated surfaces (not shown). On the other hand, very limited cell spreading is observed on substrates with 73-nm-spaced nanodots (Fig. 9B, insert). Stationary cells (black arrows) appear rounded, while migrating cells (gray arrows) are usually characterized by long extensions. Similar observations were made with other cell types, including REF52 fibroblasts, 3T3 fibroblasts, and B16 melanocytes (adapted from Arnold *et al.*, 2004).

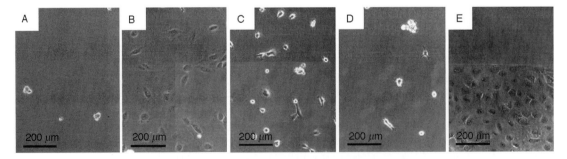

Fig. 10 Phase contrast optical micrographs of 3T3 fibroblasts on PEGDA 700 hydrogels. (A) Cells on a non-RGD-functionalized gold nanoparticle pattern. (B–D) Cells on cyclic RGDfK-functionalized gold particles; cyclic RGDfK patches are separated by varying distances (B) 40 nm, (C) 80 nm, and (D) 100 nm, after 24 h in culture. (E) Dense cell layer on a PEG support after 14 days in culture. The bottom part of the sample was patterned with cyclic RGDfK peptide-functionalized gold nanoparticles spaced 40 nm apart (adapted from Graeter *et al.*, 2007).

Fig. 10A, which indicates that cells neither spread nor survive. After functionalization of the gold nanostructures with cyclic RGDfK, cell attachment and spreading are clearly visible in Fig. 10B. Cells did not spread on the undecorated hydrogel, rather only on areas with gold particles, as illustrated by the partially gold-decorated substrate in Fig. 10E. With this substrate, the long-term biological activity of the gold-decorated hydrogel was investigated by culturing fibroblasts for more than 14 days on a PEGDA 700 hydrogel partially decorated with cyclic RGDfK-functionalized gold particles. A dense layer of cells was observed only in the region with a cyclic RGDfK-functionalized nanopattern. In agreement with previous studies of nanopatterned glass surfaces as shown in Fig. 9 (Arnold *et al.*, 2004), cell adhesion and spreading failed if cyclic RGDfK functionalized gold nanoparticles were separated by a critical distance near 80 nm (Fig. 10C). In the case of larger distances (Fig. 10D), cells could attach to the functionalized nano-dots but efficient cell adhesion and spreading did not happen. Our preliminary cell experiments illustrate that our nanolithography technique is biocompatible, and the results meanwhile imply effects of cell adhesion on the nanopatterned polymer surfaces.

In order to make the inside of the PEG hydrogel microchannels adhesive for cells, the tube was further biofunctionalized by a 0.2-mM water solution of cyclic RGDfK which was injected with a syringe into the channel; the microchannels were then incubated for 1 h. After the microchannels were intensively rinsed with water, HeLa cells were seeded into the channel and cultured under standard conditions for 24 h. Figure 11E and F depict cell spreading along the nanopatterned interior wall of the hydrogel tubes.

The molecular requirements for the formation of FA and the assembly of actin stress fibers in MC3T3 osteoblasts were further investigated by culturing these cells for 1 day, then fixing and staining them for the presence of vinculin and actin. Figure 12 shows fluorescence confocal micrographs of cells attached to c(RGDfK)-coated Au dots with spacings of (A) 58 nm and (B) 73 nm. It is apparent that an Au nanodot separation of 58nm induces the formation of large vinculin-rich FA and well-defined actin stress fibers, compared to the poorly organized vinculin and actin structures on nanodots separated by 73 nm (B) or more.

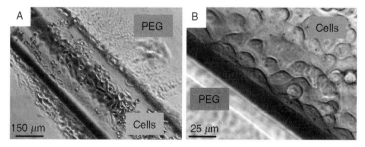

Fig. 11 (A and B) HeLa cells cultured in PEGDA 700 hydrogel tubes functionalized with RGDfK-functionalized gold nanoparticles (adapted from Graeter *et al.*, 2007).

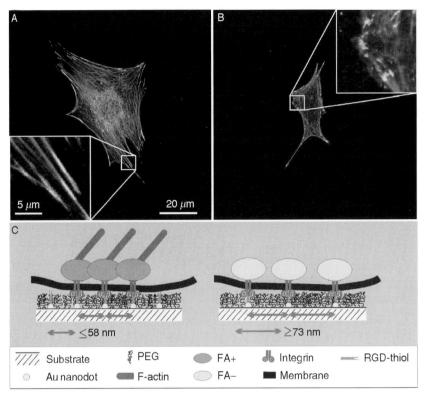

Fig. 12 Pair of confocal fluorescent micrographs of MC3T3 osteoblasts stained for vinculin (green) and actin (red) (Arnold *et al.*, 2004). Cells interacting with Au nanodot patterns with Au dot spacing of (A) 58 nm and (B) 73 nm. (C) Scheme depicting a hypothetical model that can explain the differential effects of the 58- and 73-nm-spaced, biofunctionalized nanopattern on integrin clustering and cytoskeletal organization. According to this model, a separation of Au/RGD dots by >73 nm causes limited cell attachment and spreading and actin stress fiber formation due to immobilization of integrin at intermolecular distances which are incompatible with transmembrane induction of FA formation. Intermolecular distances of <58 nm allow for such transmembrane interactions to take place, resulting in efficient FA formation (adapted from Arnold *et al.*, 2004). (See Plate 5 in the color insert section.)

D. Local Versus Global Effects of Ligand Density on Cell Adhesion

Surfaces bearing nanopatterns with different spacings also present different absolute numbers of nanodots per unit area to the attached cells. To determine whether the differential cellular responses to such surfaces might be attributable to the dot-to-dot distances (referred to as the local ligand density) or to the total number of nanodots present at the cell-surface interface (referred to as the global ligand density), the "micro"-nanostructured surfaces, prepared as described above (Glass *et al.*, 2003a,b), were used to deposit a fixed number of Au nanodots confined to a "micropatch" on the substrate. The surfaces were designed such that the global dot density was 90 dots/μm^2, thus significantly lower than either case of extended Au dot patterns, where the density was 280 dots/μm^2 for Au dots separated by 58 nm and 190 dots/μm^2 for Au dots separated by 73 nm. The local dot density present in 2×2 μm^2 patches of 58-nm-spaced dots was still 280 dots/μm^2 (Fig. 13A). A bright-field optical micrograph, taken 3 h after the plating of MC3T3 osteoblasts on the substrate, indicates that cells are confined to the structured area as the process of cell spreading advanced (Fig. 13B, inset). Twenty-four hours later, well-spread cells could be seen in this area (Fig. 13C), whereas cells located outside the frame spread poorly. A confocal fluorescent micrograph of a cell stained for vinculin and actin demonstrates the confinement of FA to the square micropattern and the association of the termini of actin stress fibers at these sites (Fig. 13D). The size distribution in FA lengths is remarkably narrow, indicating its confinement to the functionalized square.

These adhesion experiments indicate that local dot-to-dot separation (namely, local ligand density), rather than the global ligand number, is critical for inducing

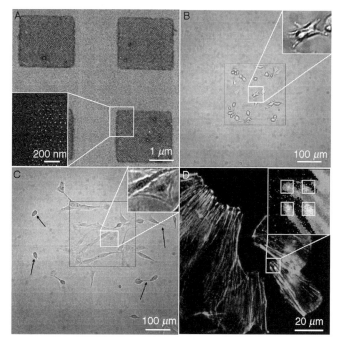

Fig. 13 MC3T3-osteoblast adhesion on "micro"-nanostructures functionalized by c(RGDfK)-thiols. (A) SEM image of "micro"-nanostructures: SEM micrograph of 5-nm Au dots separated by 58 nm in a hexagonally close-packed pattern, localized in 2×2 μm squares which are separated by 1.5 μm. Bright-field optical micrograph of adhesive MC3T3-osteoblasts growing on the pattern shown in (A), covering the area in the marked box. Cells were cultured for 3 h (B) or 24 h (C). (D) Fluorescence optical micrograph of MC3T3-osteoblast showing the location of FA by fluorescent staining for vinculin (green) and actin (red)(adapted from Arnold et al., 2004). (See Plate 6 in the color insert section.)

cell adhesion and FA assembly. Thus, for example, the global dot density of "micro"-nanostructured squares with 58-nm spacing is considerably lower than that of substrates uniformly patterned with dots at a spacing of 73 nm or more. Nevertheless, cells did form FA on the former surface and failed to do so on the latter surface. These findings are schematically summarized in Fig. 12C.

In future studies, the micro-nanopattern technique will allow for even larger variations in the organization of nano-adhesive sites. This will include alterations in ligand template pliability and presentations of small dot clusters—for example, pairs or triplets—which may shed light on the minimal molecular number that defines an effective integrin cluster for supporting cell attachment, spreading, or migration. It will also allow the exploration of pattern-specific features (i.e., molecularly defined adhesion "cues") that trigger cell adhesion-based signaling (Arnold *et al.*, 2004).

E. Cell Spreading and Migration on Different Nanopatterns

The difference in nanopattern spacing clearly affects the extent of cell spreading (measured as projected cell area). As shown in Fig. 14, the projected cell area is dramatically reduced on surfaces with an interparticle spacing $d > 73$ nm, supporting the notion that cell spreading is an active process triggered by specific (and density-dependent) integrin signals. Furthermore, dynamic analysis revealed the instability of FA in cells attached to the sparse RGD-nanopattern, associated with an increased migratory activity of cells (data not shown).

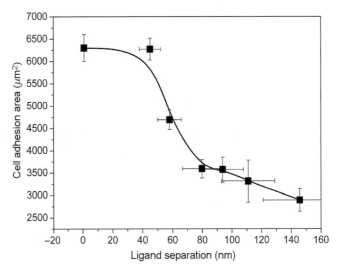

Fig. 14 Projected cell adhesion area per cell adhering to different ligand (dot) separation.

F. High-Resolution Visualization of Cells in Contact with Biofunctionalized Nanopatterns

Conventional optical microscopy is limited in spatial resolution to >200 nm, and thus cannot resolve individual adhesion points on nanopatterned surfaces. SEM was therefore utilized to visualize individual adhesions. Figure 15 shows a series of SEM micrographs, which illustrate cells attached to micro-nanopatterns (Fig. 15A and B) and to extended nanopatterns (Fig. 15C). The local interdot spacing was 58 nm in all samples. Nanoprotrusions attached to individual nano-dots could be seen (Fig. 15C). Examination of a large area along the surface indicated that the lamellipodium attaches to the micro-nanopattern "islands," rather than to the surrounding passivated area. A high-resolution micrograph illustrates the interaction of a cell protrusion with the nanopattern (Fig. 15C). The tiny nanoscopic membrane protrusions attach selectively to the RGD-functionalized Au nanoparticles and not to the PEG-passivated glass.

The latter observation is remarkable since it directly displays cellular adhesive nanostructures formed in response to molecularly defined interactions. The tiny

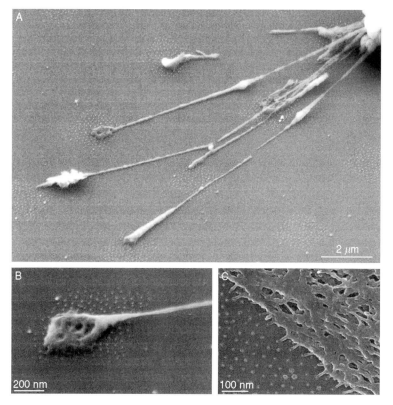

Fig. 15 SEM micrographs displaying 3T3 fibroblast adhering to micro-nanopattern (A and B) and to extended nanopattern (C). The dot spacing was 58 nm in all samples (adapted from Arnold *et al.*, 2004).

membrane protrusions may be as small as 20 nm in diameter and are most likely anchored to the adhesive nanoparticles (due to space restriction on each dot) via single-integrin molecules located at the tip of the membrane.

III. Outlook for the Future

Recent efforts to systematically control multiple features of adhesive substrates have greatly benefited from advanced research in materials science. The long-term objective of this effort is to design and fabricate so-called "cellular environments" with defined chemical and physical properties in order to exert desirable effects on cell structure, activity, and fate. Self-assembled micelle diblock copolymer nano-lithography and other associated biofunctionalization approaches described herein represent a new era in the development of tailored cellular environments, in which both the chemical and physical features can be exquisitely controlled with high, molecularly defined precision. For this reason, we refer to the design of such "biointerfaces" as "molecular engineering," and to the surfaces themselves as "nano-digital interfaces."

The most recent data show that surface variations at the nanoscale range, namely the typical size of single-protein complexes, can be sensed by cells and dramatically affect their behavior. This knowledge may be further expanded by combining such variations with manipulations of other properties of the adhesive surface such as topography, isotropy, rigidity, and texture. The replacement of rigid glass substrates as a support for nanoparticles by more flexible polymer surfaces is a major step forward in this regard (Graeter et al., 2007). The use of polymers as substrate not only allows for chemically nanopatterned surfaces of varying elasticity and viscoelasticity, even at microdimensions, but also presents a new challenge as one attempts to shape chemically nanopatterned surfaces in different ways (e.g., fabrication of nanopatterned supports in tube shapes for artificial vessels or stents or in other three-dimensional configurations). These novel substrates may ultimately modulate receptor presentation, occupation, and immobilization at the cell membrane, as well as alter the transcriptional program which, in turn, may affect multiple signaling events and cellular responses such as cell dynamics, differentiation, and fate.

Acknowledgments

This work has been made possible by a number of excellent students, postdocs, and colleagues from our groups, namely Dr. Ada Cavalcanti-Adam (MPI-MF & Uni Heidelberg), Dr. Marco Arnold (MPI-MF & Uni Heidelberg), Dr. Jacques Blümmel (MPI-MF & Uni Heidelberg), Dr. Roman Glass (MPI-MF & Uni Heidelberg), Nadine Perschmann (MPI-MF & Uni Heidelberg), Dr. Stefan Gräter (MPI-MF & Uni Heidelberg), Dr. Baruch Zimerman (Weizmann Institute), and Dr. Tova Volberg (Weizmann Institute). The RGD peptides were kindly provided by Professor Horst Kessler, TU Munich. The collaboration between B.G. and J.S. is supported by the Landesstiftung Baden-Württemberg, by the German-Israeli Foundation, and by the Max Planck Society. B.G. holds the Erwin Neter Professorial Chair in Cell and Tumor Biology.

References

Alberts, B., Bray, D., Lewis, J., Raff, M., Roberts, K., and Watson, J. D. (1994). "Molecular Biology of the Cell," 3rd edn. Garland Publishing, New York, USA.

Alberts, B., Johnson, A., Lewis, J., Raff, M., Roberts, K., and Walter, P. (2001). "Molecular Biology of the Cell," 4th edn. Garland Science, New York, USA.

Arnold, A., Cavalcanti-Adam, A., Glass, R., Blümmel, J., Eck, W., Kessler, H., and Spatz, J. P. (2004). Activation of integrin function by nanopatterned adhesive interfaces. *Chemphyschem.* **3**, 383–388.

Balaban, N. Q., Schwarz, U. S., Rivelin, D., Goichberg, P., Tzur, G., Sabanay, L., Mahalu, D., Safran, S., Bershadsky, A., Addadi, L., and Geiger, B. (2001). Force and focal adhesion assembly: A close relationship studied using elastic micropatterned substrates. *Nat. Cell Biol.* **3**, 466–472.

Ballestrem, C., Hinz, B., Imhof, B. A., and Wehrle-Haller, B. (2001). Marching at the front and dragging behind: Differential alphaVbeta3-integrin turnover regulates focal adhesion behavior. *J. Cell Biol.* **155**, 1319–1332.

Blau, H. M., and Baltimore, D. J. (1991). Differentiation requires continuous regulation. *J. Cell Biol.* **112**, 781–783.

Chen, C. S., Mrksich, M., Huang, S., Whitesides, G. M., and Ingber, D. E. (1997). Geometric control of cell life and death. *Science* **276**, 1425–1428.

Critchley, D. R. (2000). Focal adhesions—the cytoskeletal connection. *Curr. Opin. Cell Biol.* **12**, 133–139.

Cohen, M., Kam, Z., Addadi, L., and Geiger, B. (2006). Dynamic study of the transition from hyaluronan- to integrin-mediated adhesion in chondrocytes. *EMBO* **25**(2), 302–311.

Dalby, M. J., Yarwood, S. J., Riehle, M. O., Johnstone, H. J. H., Affrossman, S., and Curtis, A. S. G. (2002a). Increasing fibroblast response to materials using nanotopography: Morphological and genetic measurements of cell response to 13-nm-high polymer demixed islands. *Exp. Cell Res.* **276**, 1–9.

Dalby, M. J., Riehle, M. O., Johnstone, H. J. H., Affrossman, S., and Curtis, A. S. G. (2002b). *In vitro* reaction of endothelial cells to polymer demixed nanotopography. *Biomaterials* **23**, 2945–2954.

Dalby, M. J., Childs, S., Riehle, M. O., Johnstone, H. J. H., Affrossman, S., and Curtis, A. S. G. (2003). Fibroblast reaction to island topography: Changes in cytoskeleton and morphology with time. *Biomaterials* **24**, 927–935.

Discher, D. E., Janmey, P., and Wang, Y. L. (2005). Tissue cells feel and respond to the stiffness of their substrate. *Science* **310**, 1139–1143.

Ebendal, T. (1976). The relative roles of contact inhibition and contact guidance in orientation of axons extending on aligned collagen fibrils *in vitro. Exp. Cell Res.* **98**(1), 159–169.

Elbert, D. L., and Hubbell, J. A. (2001). Conjugate addition reactions combined with free-radical cross-linking for the design of materials for tissue engineering. *Biomacromolecules* **2**, 430–441.

Fratzl, P., Misof, K., Zizak, I., Rapp, G., Amenitsch, H., and Bernstorff, S. (1998). Fibrillar structure and mechanical properties of collagen. *J. Struct. Biol.* **122**, 119–122.

Gates, B. D., Xu, Q., Stewart, M., Ryan, D., Willson, C. G., and Whitesides, G. M. (2005). New approaches to nanofabrication: Molding, printing, and other techniques. *Chem. Rev.* **105**(4), 1171–1196.

Geiger, B., Bershadsky, A., Pankov, R., and Yamada, K. (2001). Transmembrane extracellular matrix cytoskeleton crosstalk. *Nat. Rev. Cell Biol.* **2**, 793–805.

Georges, P. C., and Janmey, P. A. (2005). Cell type-specific response to growth on soft materials. *J. Appl. Physiol.* **98**, 1547–1553.

Giancotti, F. G., and Ruoslahti, E. (1999). Integrin signaling. *Science* **285**, 1028–1032.

Glass, R., Arnold, M., Möller, M., and Spatz, J. P. (2003a). Micro-nanostructured interfaces fabricated by the use of inorganic block copolymer micellar monolayers as negative resist for electron beam lithography. *Adv. Funct. Mater.* **13**, 569–575.

Glass, R., Möller, M., and Spatz, J. P. (2003b). Block copolyer micelle nanolithography. *Nanotechnology* **14**, 1153–1160.

Graeter, S., Jinghuan, H., Perschmann, N., López, M., Kessler, H., Ding, J., and Spatz, J. P. (2007). Mimicking cellular environments by nónostructured soft interfaces. *Nano Lett.* **7**(5), 1413–1418.

Haubner, R., Finsinger, D., and Kessler, H. (1997). Stereoisomeric peptide libraries and peptidomimetics for designing selective inhibitors of the alpha(V)beta(3) integrin for a new cancer therapy. *Angew. Chem. Int. Ed.* **36**(13–14), 1375–1389.

Hynes, R. O. (1987). Integrins—a family of cell-surface receptors. *Cell* **48**, 549–554.

Irvine, D. J., Mayes, A. M., and Griffith, L. G. (2001a). Nanoscale clustering of RGD peptides at surfaces using comb polymers. 1. synthesis and characterization of comb thin films. *Biomacromolecules* **2**, 85–94.

Irvine, D. J., Ruzette, A. V. G., Mayes, A. M., and Griffith, L. G. (2001b). Nanoscale Clustering of RGD peptides at surfaces using comb polymers. 2. surface segregation of comb polymers in polylactide. *Biomacromolecules* **2**, 545–556.

Jiang, F. Z., Horber, H., Howard, J., and Müller, D. (2004). Assembly of collagen into microribbons: Effects of pH and electrolytes. *J. Struct. Biol.* **148**(3), 268–278.

Kantlehner, M., Schaffner, P., Finsinger, D. M. J., Meyer, J., Jonczyk, A., Diefenbach, B., Nies, B., Hölzemann, G., Goodman, S. L., and Kessler, H. (2000). Surface coating with cyclic RGD peptides stimulates osteoblast adhesion and proliferation as well as bone formation. *Chem. Bio. Chem.* **1**, 107–114.

Kemkemer, R., Schrank, S., Gruler, H., Kaufmann, D., and Spatz, J. P. (2004). Process of cell shape normalization of normal and haploinsufficient NF1-melanocytes through substrate interaction. *Chem. Phys. Chem.* **3**, 85–92.

Koo, L. Y., Irvine, D. J., Mayes, A. M., Lauffenburger, D. A., and Griffith, L. G. (2002). Co-regulation of cell adhesion by nanoscale RGD organization and mechanical stimulus. *J. Cell Sci.* **115**, 1423–1433.

Lasky, L. A. (1997). Cell adhesion: How integrins are activated. *Nature* **390**, 15–17.

Lata, S., and Piehler, J. (2005). Stable and functional immobilization of histidine-tagged proteins via multivalent chelator headgroups on a molecular poly(ethylene glycol) brush. *Anal. Chem.* **77**(4), 1096–1105.

Levenberg, S., Katz, B. Z., Yamada, K. M., and Geiger, B. (1998). Long-range and selective autoregulation of cell-cell or cell-matrix adhesions by cadherin or integrin ligands. *J. Cell Sci.* **111**, 347–357.

Li, R., Mitra, N., Gratkowski, H., Vilaire, G., Litvinov, R., Nagasami, C., Weisel, J. W., Lear, J. D., DeGrado, W. F., and Bennett, J. S. (2003). Activation of Integrin $\alpha IIb\beta 3$ by modulation of transmembrane helix associations. *Science* **300**, 795–798.

Lin, F., and Butcher, E. C. (2006). T cell chemotaxis in a simple microfluidic device. *Lab Chip* **6**(11), 1462–1469.

Maheshwari, G., Brown, G., Lauffenburger, D. A., Wells, A., and Griffith, L. G. (2000). Cell adhesion and motility depend on nanoscale RGD clustering. *J. Cell Sci.* **113**, 1677–1686.

Massia, S. P., and Hubbell, J. A. (1991). An RGD spacing of 440 nm is sufficient for integrin alpha V beta 3-mediated fibroblast spreading and 140 nm for focal contact and stress fiber formation. *J. Cell Biol.* **114**, 1089–1100.

Meller, D., Peters, K., and Meller, K. (1997). Human cornea and sclera studied by atomic force microscopy. *Cell Tissue Res.* **288**, 111–118.

Miyamoto, S., Akiyama, S. K., and Yamada, K. M. (1995). Synergistic roles for receptor occupancy and aggregation in integrin transmembrane function. *Science* **267**, 883–885.

Mrksich, M., and Whitesides, G. M. (1997). Using self-assembled monolayers that present oligo (ethylene glycol) groups to control the interactions of proteins with surfaces. *ACS Symp. Ser.* **680**, 361–373.

Pfaff, M., Tangemann, K., Muller, B., Gurrath, M., Muller, G., Kessler, H., Timpl, R., and Engel, J. (1994). Selective recognition of cyclic RGD peptides of NMR defined conformation by alpha IIb beta 3, alpha V beta 3 and alpha 5 beta 1 Integrins. *J. Biol. Chem.* **269**, 20233–20238.

Ploetz, C., Zycband, E. I., and Birk, D. E. (1991). Collagen fibril assembly and deposition in the developing dermis—segmental deposition in extracellular compartments. *J. Struct. Biol.* **106**, 73–81.

Pompe, T., Renner, L., and Werner, C. (2005). Nanoscale features of fibronectin fibrillogenesis depend on protein-substrate interaction and cytoskeleton structure. *Biophys. J.* **88**(1), 527–534.

Roberts, C., Chen, C. S., Mrksich, M., Martichonok, V., Ingber, D. E., and Whitesides, G. M. (1998). Using mixed self-assembled monolayers presenting RGD and (EG)(3)OH groups to characterize long-term attachment of bovine capillary endothelial cells to surfaces. *J. Am. Chem. Soc.* **120,** 6548–6555.

Ruoslahti, E., and Obrink, B. (1996). Common principles in cell adhesion. *Exp. Cell Res.* **227**, 1–11.

Sniadecki, N. J., Desai, R. A., Ruiz, S. A., and Chen, C. S. (2006). Nanotechnology for cell-substrate interactions. *Ann. Biomed. Eng.* **34**(1), 59–74.

Spatz, J. P., Mößmer, S., and Möller, M. (1996a). Mineralization of gold nanoparticles in a block copolymer microemulsion. *Chem. Eur. J.* **2,** 1552–1555.

Spatz, J. P., Roescher, A., and Möller, M. (1996b). Gold Nanoparticles in micellar Poly(styrene)-b-Poly(ethylene oxide) films–size and interparticle distance control in monoparticulate films. *Adv. Mater.* **8,** 337–340.

Spatz, J. P., Sheiko, S., and Möller, M. (1996c). Ion-stabilized block copolymer micelles: Film formation and intermicellar interaction. *Macromolecules* **29,** 3220–3226.

Spatz, J. P., Mößmer, S.,Hartmann, C., Möller, M., Herzog, T., Krieger, M., Boyen, H.-G., and Ziemann, P. (2000). Ordered deposition of inorganic clusters from micellar block copolymer films. *Langmuir* **16,** 407–415.

Vogel, V., and Sheetz, M. (2006). Local force and geometry sensing regulate cell functions. *Nat. Rev.* **7,** 265–275.

Wang, R. L. C., Kreuzer, H. J., and Grunze, M. (1997). Molecular conformation and solvation of oligo (ethylene glycol)-terminated self-assembled monolayers and their resistance to protein adsorption. *J. Phys. Chem.* **B101,** 9767–9773.

Xiong, J. P., Stehle, T., Zhang, R., Joachimiak, A., Frech, M., Goodman, S. L., and Arnaout, M. A. (2002). Crystal structure of the extracellular segment of integrin $\alpha V\beta 3$ in complex with an Arg-Gly-Asp ligand. *Science* **296**, 151–155.

Zamir, E., and Geiger, B. (2001a). Molecular complexity and dynamics of cell-matrix adhesions. *J. Cell Sci.* **114**, 3583–3590.

Zamir, E., and Geiger, B. (2001b). Components of cell-matrix adhesions. *J. Cell Sci.* **114**, 3577–3579.

PART II

Subcellular Mechanical Properties and Activities

CHAPTER 6

Probing Cellular Mechanical Responses to Stimuli Using Ballistic Intracellular Nanorheology

Porntula Panorchan,★ Jerry S. H. Lee,★ Brian R. Daniels,★ Thomas P. Kole,★ Yiider Tseng,★,† and Denis Wirtz★,‡,§

★Department of Chemical and Biomolecular Engineering
The Johns Hopkins University
Baltimore, Maryland 21218

†Department of Chemical Engineering
University of Florida, Gainesville, Florida 32611

‡Department of Materials Science and Engineering
The Johns Hopkins University
Baltimore, Maryland 21218

§Howard Hughes Medical Institute Graduate Training Program and
Institute for NanoBioTechnology
The Johns Hopkins University
Baltimore, Maryland 21218

Abstract

We describe a new method to measure the local and global micromechanical
properties of the cytoplasm of single living cells in their physiological milieu and
subjected to external stimuli. By tracking spontaneous, Brownian movements of
individual nanoparticles of diameter ≥ 100 nm distributed within the cell with
high spatial and temporal resolutions, the local viscoelastic properties of the
intracellular milieu can be measured in different locations within the cell. The
amplitude and the time-dependence of the mean-squared displacement of each
nanoparticle directly reflect the elasticity and the viscosity of the cytoplasm in the
vicinity of the nanoparticle. In our previous versions of particle tracking, we
delivered nanoparticles via microinjection, which limited the number of cells
amenable to measurement, rendering our technique incompatible with high-
throughput experiments. Here we introduce ballistic injection to effectively deliver
a large number of nanoparticles to a large number of cells simultaneously. When
coupled with multiple particle tracking, this new method—ballistic intracellular
nanorheology (BIN)—makes it now possible to probe the viscoelastic properties of
cells in high-throughput experiments, which require large quantities of injected

cells for seeding in various conditions. For instance, BIN allows us to probe an ensemble of cells embedded deeply inside a three-dimensional extracellular matrix or as a monolayer of cells subjected to shear flows.

I. Introduction

A. Why Cell Mechanics?

Mechanical responses of cells are essential for numerous cellular processes and human functions. The human body and ultimately its constitutive cells are constantly subjected to mechanical stresses. Mechanical stresses are created by blood flows on the surface of endothelial cells, the expansion and contraction produced by cardiac myocytes on surrounding tissues, and gravity or other external forces on chondrocytes—to name just a few examples. These forces are believed to have profound effects on cellular growth, differentiation, and migration. In addition, mechanical interactions between tumor cells and their environments have been linked to tumor cell invasion and metastasis (Chapter 23 by Johnson *et al.*, this volume).

An important goal of cell mechanics is to investigate mechanical responses of cells to chemical and physical signals. However, most current methods are unable to probe the mechanical properties of living cells under physiological conditions. For instance, one of the most important problems of cell mechanics, the study of cell migration in extracellular matrix, cannot be easily investigated with current cell mechanics techniques. Magnetic microspheres used in magnetic tweezers and magnetocytometry cannot be seeded on cells inside a matrix. Atomic force microscopy (AFM)- and micropipette-based methods require direct contact and are unable to probe cells deep within a matrix. Therefore, these methods are applicable predominantly for studying cell migration on nonphysiological, two-dimensional (2D) tissue culture dishes. Although our previous version of a particle-tracking method did not require direct contact with the cell during measurement, wider applications were hindered by the delivery of probe particles via microinjection (Tseng *et al.*, 2002b), which is tedious and requires special training. Due to the difficulties of microinjecting particles, hours of effort may yield fewer than 100 viable cells, while seeding of cells in 3D matrices typically requires an order of 10^5 cells. To circumvent the inefficiency of particle delivery, we introduce a method of ballistic intracellular nanorheology (BIN) for high-throughput application of particle tracking, by combining ballistic injection and multiple particle tracking. Ballistic injection allows for simultaneous injection of an entire culture of cells (of the order 10^6 cells). With BIN, we now have the ability to measure mechanical responses of cells migrating in 3D matrix and other complex environments and mechanical stress conditions such as cells subjected to shear flows.

B. Particle–Tracking Nanorheology: Measuring the Local Viscoelastic Properties of a Cell by Tracking the Brownian Motion of Individual Nanoparticles Embedded in the Cell

The fundamental principle of particle-tracking nanorheology (previously called microrheology) was introduced by Mason *et al.* (1997) and Gittes *et al.* (1997) (see also a complementary Chapter 7 by Crocker and Hoffman, this volume). Although nano-resolution tracking of the Brownian motions of particles on cell surfaces had previously been used to clarify mechanisms of active transport versus diffusion on migrating or stationary cells (Kucik *et al.*, 1989), little to no work had been done to extract viscoelastic properties from random motions of intracellular particles. Likewise, magnetically driven motions of magnetic particles inside cells have been used for decades to extract viscoelastic properties of cytoplasm (Valberg and Albertini, 1985), but Brownian motions had not been exploited. The first such applications of single particle (Lee and Discher, 2001; Yamada *et al.*, 2000) and multiple particle-tracking nanorheology to cells (Tseng *et al.*, 2002b) together with work since have provided important new insights into the complex passive and active mechanics of the cell and its cytoskeleton. The technique involves monitoring with high spatial and temporal resolutions, the random movements of *individual* inert nanoparticles embedded in the cytoplasm of live cells. The amplitude of such movements reflects both the elasticity and viscosity of the intracellular milieu.

The principle of microrheology is best illustrated by the two extreme examples of a nanoparticle suspended in a viscous liquid (e.g., water or glycerol) (Apgar *et al.*, 2000) and the same nanoparticle embedded in an elastic gel (e.g., a cross-linked polyacrylamide gel) (Flanagan *et al.*, 2002). A submicron particle suspended in a viscous liquid like water undergoes completely unrestricted motion and its mean-squared displacement (MSD) scales proportionally with time. Given enough time, a nanoparticle can reach any distance away from its initial position. The motion of the nanoparticles is driven by the constant bombardment of the water molecules surrounding the nanoparticle. Each time the nanoparticle makes a move in a given direction, it immediately loses all memory of where it just came from. The resulting motion of the nanoparticle is the familiar random walk (Berg, 1993; Haber *et al.*, 2000). Nanoparticles suspended in water move faster than nanoparticles suspended in glycerol, which is reflected by a larger slope of MSD against time since this slope is inversely proportional to the viscosity of the liquid in which the nanoparticle is suspended.

In contrast, an inert nanoparticle that is embedded in a stretchy or rubbery material such as a polyacrylamide gel will undergo highly restricted motion. Each time the nanoparticle makes a move in a given direction, it is instantaneously pushed by the surrounding elastic material back to where it just came from. The nanoparticle behaves like a particle tethered to a rubber band, with "perfect memory" of where it came from. In this case, the MSD of the nanoparticle is independent of time. In addition, the magnitude of MSD is inversely proportional to the material's elastic modulus (Chapter 1 by Janmey *et al.* and Chapter 2 by Kandow *et al.*, this volume).

The cytoplasm is both viscous and elastic. Therefore, a submicron nanoparticle embedded in the cytoplasm will display a MSD, $<\Delta r^2(\tau)>$, which is intermediate in its time dependence between the MSD in a viscous liquid, $<\Delta r^2(\tau)> \sim k\tau$, and the MSD of a nanoparticle in an elastic solid, $<\Delta r^2(\tau)> \sim C$, where k and C are constants. The two extreme cases, that of a viscous liquid and of a perfectly elastic material, teach us that the time dependence of the MSD and its actual value provides two crucial bits of information about the viscoelastic properties of the material surrounding the nanoparticle, that is whether it is only viscous or only elastic, or both elastic and viscous, and the values of the parameters describing the viscoelasticity of the material.

C. Local and Global Viscoelastic Properties of the Cell

Most methods for probing single-cell mechanics—micropipette suction, magnetic bead twisting, cell poking—cannot measure viscoelasticity of the cytoplasm at the nanoscale and are generally interpreted in terms of a single value of viscosity or elasticity—although exceptions include shear rate-dependent viscosities (Tsai *et al.*, 1993). In contrast, particle-tracking nanorheology is able to provide spatial maps of local viscoelastic properties of the cytoplasm. Using this method, we have shown that the effective viscoelastic property of the intracellular milieu can vary more than tenfold within a single cell (Tseng *et al.*, 2002b). For instance, the average viscosity of the nucleus is about 100 times that of the average viscosity of the cytoplasm (Tseng *et al.*, 2004b), and the average viscosity at the leading edge of a migrating fibroblast is about 10 times that at trailing edge (Kole *et al.*, 2005). Indeed, even a high-magnification view of the cytoplasm of a living, adherent cell should convince anyone that the physical properties look very different across the cell. Given these large spatial variations, global and local elasticity and viscosity are required to fully describe physical changes associated with cytoskeleton reorganization and organelle movements during processes such as cell migration (Heidemann and Wirtz, 2004).

D. Interstitial Versus Mesoscale Viscosity of the Cytoplasm

The viscosity of the intracellular space depends crucially on the length scale of interest: at small length scales, the viscosity is essentially that of water, at larger length scales, the apparent viscosity is much higher. The characteristic length scale that separates these two viscosities is the effective mesh size of the filamentous cytoskeleton network in the cytoplasm. This mesh size can be measured by probing the diffusion of fluorescently labeled dextran or DNA inside the cell. Dextran polymers of radius of gyration <50 nm diffuse within the cytoplasm (of fibroblasts) largely unhindered, while polymers >50 nm become immobilized (Luby-Phelps *et al.*, 1986; Lukacs *et al.*, 2000; Seksek *et al.*, 1997). The mesoscale viscosity is the viscosity that controls the rate of movements of mitochondria (Lacayo and Theriot, 2004), nucleus (Lee *et al.*, 2005), and phagosomes (Suh *et al.*, 2003), as well as viruses, bacteria, engineered drug, and delivery microcarriers (Suh *et al.*, 2003), because these entities have a size larger than the effective mesh size of the cytoskeleton. The mesoscale viscosity of the cytoplasm also controls the rates of cell spreading and migration (Gupton *et al.*, 2005). In what follows, we shall only discuss the rheology at the mesoscale, at length scales intermediate between the interstitial space ($\sim$50 nm) and the whole cell ($\sim$100 μm).

E. Viscoelastic Properties of the Cytoplasm Depend on the Timescale of Applied Forces

Viscoelastic properties of the cytoplasm generally depend not only on location but also on the rate and extent of mechanical stresses and strains applied to cells (Kole *et al.*, 2005; Tseng *et al.*, 2002b; Yamada *et al.*, 2000). A shear rate-dependent mechanical response is best illustrated by examining the rheology of a suspension of entangled actin filaments, a conventional *in vitro* system used by biophysicists to understand the rheological behavior of the cytoskeleton network from the bottom-up (Janmey, 1998; Janmey *et al.*, 1990, 1991; Ma *et al.*, 1999, 2001; Panorchan *et al.*, 2004a; Pollard *et al.*, 1992; Sato *et al.*, 1987; Tseng and Wirtz, 2001; Tseng *et al.*, 2001, 2004a; Wachsstock *et al.*, 1993, 1994; Xu *et al.*, 1998b,d, 2000). When this suspension is sheared slowly, actin filaments have the time to move with respect to each other and, therefore, to relax the mechanical stresses to which they are subjected (Wachsstock *et al.*, 1994). Therefore, at low rates of shear, the suspension under mechanical stress offers little elastic resistance and flows like a liquid. When the suspension is sheared rapidly, the entangled actin filaments do not have the time to move with respect to each other and to relax the stress. As a result, the suspension resists shear stresses elastically and does not flow. From a rheological standpoint, the suspension behaves like an elastic solid. This example illustrates the complex rheological response that a simple actin filament network *in vitro* can already display outside the intracellular milieu and without auxiliary proteins, behaving like a liquid at low rates of mechanical shear and like a solid at high rates of shear (Xu *et al.*, 2000).

Subcellular movements of organelles occur at very different timescales, from seconds for the reorganization of the cytoskeleton (Waterman-Storer and Salmon, 1999) to minutes for the directed migration of nucleus and the reorientation of centrosome (Lee *et al.*, 2005) to hours for large scale reorganization of the ER. Since these movements are directly affected by the viscoelastic properties of the intracellular milieu, it is essential for cell mechanics assays to measure viscoelastic parameters of the intracellular milieu over a wide range of timescales or, equivalently, over a wide range of frequencies or rates of shear (Heidemann and Wirtz, 2004).

F. The Viscoelastic Properties of the Cytoplasm Depend on the Amplitude of Applied Forces

Viscoelastic properties of the cytoplasm also depend on the amplitude of applied forces (Bausch *et al.*, 1999). Let us use a suspension of actin filaments cross-linked by α-actinin as a model (Xu *et al.*, 2000). At a fixed rate of shear and for increasing amplitude of shear, the stiffness or elasticity of an α-actinin-cross-linked actin filament network increases readily. This shear-induced stiffening (also called strain-hardening) is caused by the unfavorable deformation of the filaments, along their length between successive entanglements (Storm *et al.*, 2005; Xu *et al.*, 2000). In the absence of cross-linking proteins, the filaments slide and the uncross-linked actin filaments show only stiffening under very large deformations (Xu *et al.*, 2000). In the presence of cross-linking proteins, actin filaments cannot slide easily past one another and tend to bend under the applied mechanical stress. This causes the effective elasticity of the network to increase. Together, these examples show that a cross-linked actin filament network can display a rather complex rheological behavior, which depends on both the rate (the speed of application of the forces) and the amplitude of the mechanical stress to which it is subjected.

G. Measurements at the Cell Surface Versus in the Cytoplasm

Unlike most cell mechanics methods, particle-tracking nanorheology allows quantitative, absolute measurements of viscoelastic parameters (Table I). The viscoelastic properties of standard liquids and materials measured by particle-tracking nanorheology and traditional cone-and-plate rheometers compare favorably (Apgar *et al.*, 2000; Mason *et al.*, 1997; Xu *et al.*, 1998a,c). Therefore, the presence of the beads in materials like a polymer solution or the cytoplasm does not detectably change their mechanical properties. Moreover, the rheological parameters obtained by particle-tracking nanorheology are not relative, but absolute. This allows rheological parameters obtained by particle-tracking nanorheology to be compared among different types of cells (Kole *et al.*, 2004a, 2005; Lee *et al.*, 2006; Tseng *et al.*, 2002b, 2005) (Table I), compared to viscoelastic parameters observed in reconstituted cytoskeletal networks (Apgar *et al.*, 2000; Bousquet *et al.*, 2001; Coulombe *et al.*, 2000; Esue *et al.*, 2005a,b; Ma *et al.*, 1999, 2001; Mason *et al.*, 1997; Palmer *et al.*, 1998a,b, 1999; Panorchan *et al.*, 2004b; Tseng and Wirtz, 2001, 2004; Tseng *et al.*, 2001, 2002a, 2004a, 2005; Xu *et al.*, 1998a,c,d, 2000; Yamada *et al.*, 2002, 2003) (Table II), and compared with predictions of models (Atilgan *et al.*, 2005, 2006).

The size of the probe and associated device that make contact with the sample is much smaller for particle-tracking nanorheology than for most other techniques such as AFM or micropipette suction. This allows particle-tracking nanorheology to be applied to a wider range of samples and comparison to be made between standard fluids and cells. In addition, particle-tracking nanorheology does not require binding between the probe and the sample, unlike some techniques such as magnetic bead twisting or pulling. This avoids the ambiguities due to the binding interactions, although it also limits particle-tracking nanorheology from measuring these interactions.

Table I

Elasticity and Shear Viscosity of the Cytoplasm of Different Types of Cells Subjected to Various Biochemical and Biophysical Stimuli Measured by Particle-Tracking Nanorheology

	Average viscosity (P)	Average elasticity at 1 Hz (dyne/cm²)	References
Serum-starved Swiss3T3 fibroblast[a]	10 ± 3	50 ± 20	Kole et al., 2004a
Serum-starved Swiss 3T3 fibroblast treated with LPA[b]	95 ± 20	120 ± 30	Kole et al., 2004a
Serum-starved Swiss3T3 fibroblast subjected to shear flow[c]	300 ± 40	600 ± 50	Lee et al., 2006
Swiss 3T3 fibroblast at the edge of a wound[d]	45 ± 15	330 ± 30	Kole et al., 2005
Swiss 3T3 fibroblast treated with bradykinin[e]	22 ± 13	90 ± 20	Kole et al., 2005
Swiss 3T3 fibroblast treated with PDGF[f]	24 ± 8	190 ± 30	Kole et al., 2005
Mouse embryonic fibroblast (Lmna$^{+/+}$ MEF)[g]	18 ± 2	140 ± 30	JSHL, PP, DW, unpublished
MEF treated with latrunculin B[h]	NA	80 ± 4	JSHL, PP, DW, unpublished
MEF treated with nocodazole[i]	NA	50 ± 4	JSHL, PP, DW, unpublished
MEF deficient in lamin A/C (Lmna$^{-/-}$ MEF)[g]	8 ± 1	60 ± 4	JSHL, PP, DW, unpublished
HUVEC cell on a peptide 2D matrix[j]	17 ± 1	130 ± 10	PP, JSHL, DW, unpublished
HUVEC cell on a peptide 2D matrix treated with VEGF[k]	8 ± 1	100 ± 5	PP, JSHL, DW, unpublished
HUVEC inside a peptide 3D matrix[j]	14 ± 1	55 ± 4	Panorchan et al., 2006
HUVEC inside a peptide 3D matrix treated with VEGF[k]	18 ± 1	40 ± 3	Panorchan et al., 2006
Single-cell C. elegans embryo[l]	10 ± 1	Negligible	Daniels et al., 2006
Interphase nucleus of Swiss 3T3 fibroblast[m]	520 ± 50	180 ± 30	Tseng et al., 2004b

Particle-tracking nanorheology was used to study the mechanical properties of many varieties of cells under a wide range of conditions. Unless stated, all values of viscosity and elasticity in the table are for the cytoplasm. The elasticity was evaluated at a shear frequency of 1 sec^{-1} (1 sec^{-1} = 1 Hz) and the shear viscosity was estimated as the product of plateau modulus and the relaxation time. The plateau modulus is the value of the elastic modulus at intermediate frequencies where it reaches a quasi-plateau value. The relaxation time is the inverse of the frequency where elastic and viscous moduli are equal. All measurements are mean ± SEM. Unit conversions are 1 dyne/cm² = 0.1 Pa = 0.1 N/m² = 0.1 pN/μm². Pa, pascal; pN, piconewton.

[a]Cells placed on 50-μg/ml fibronectin deposited on glass were serum starved for 48 h before measurements.

[b]Serum-starved cells placed on 50-μg/ml fibronectin deposited on glass were treated with 1-μg/ml lysophosphatidic acid (LPA), which activates Rho-mediated actomyosin contractility. LPA was applied 15 min before measurements.

[c]Cells were grown on 20-μg/ml fibronectin for 24 h and exposed to shear flow (wall shear stress, 9.4 dynes/cm²) for 40 min before measurements.

[d]Cells in complete medium and grown on 50-μg/ml fibronectin to confluence were wounded to induce migration. Measurements were conducted 4 h after wounding.

[e]Cells in complete medium and grown on 50-μg/ml fibronectin were treated with 100-ng/ml bradykinin for 10 min before measurements.

[f]Cells in complete medium and grown on 50-μg/ml fibronectin were treated with 10-ng/ml PDGF for 10 min before measurements.

[g]Cells in complete medium were grown on glass.

[h]Lmna$^{+/+}$ MEFs in complete medium grown on glass were treated with 5-μg/ml actin-filament disassembly drug latrunculin B.

[i]Lmna$^{+/+}$ MEFs in complete medium and grown on glass were treated with 5-μg/ml microtubule disassembly drug nocodazole.

[j]Cells in complete medium were placed in a 0.5% puramatrix gel.

[k]Cells in complete medium were placed in a 0.5% puramatrix gel and treated with 4-ng/ml VEGF for 24 h prior to the measurements.

[l]Young C. elegans eggs were obtained by cutting gravid hermaphrodites from worms in egg salts. The nanoparticles were then microinjected into the syncytial gonads of gravid hermaphrodites.

[m]Cells in complete medium.

Table II

Viscosity and Elasticity of Common Liquids and of Cytoskeletal Filament Networks by a Cone-and-Plate Rheometer

	Average viscosity (P)	Average elasticity at 1 rad/sec (dyne/cm^2)	References
Water	0.01	0	
Blood	0.1	Negligible	
Glycerol	1	Negligible	
Corn syrup	20	Negligible	
Ketchup	500	Negligible	
Polyacrylamide gel		500	Flanagan et al., 2002
F-actin network		8 ± 3	Xu et al., 2000
F-actin + filamin		450 ± 60	Tseng et al., 2004a
F-actin + α-actinin		120 ± 20	Xu et al., 2000
F-actin + Arp2/3 complex/WASp		60 ± 15	Tseng and Wirtz, 2004
F-actin + fascin		80 ± 10	Tseng et al., 2001
F-actin + fimbrin		300 ± 30	Klein et al., 2004
Vimentin network		14 ± 2	Esue et al., 2006

Measurements are mean ± SEM. The elasticity was measured at a shear amplitude of 1% and a shear frequency of 1 rad/sec. Shear viscosity of the F-actin and vimentin networks was not measured because these filaments break under continuous shear.

The concentrations of actin and vimentin solutions are 24 μM. The concentrations of α-actinin, fascin, fimbrin, and filamin in solutions are 0.24 μM. The concentration of Arp2/3 complex is 0.12 μM and that of its activator WASp is 0.06 μM. The concentration of acrylamide and bis-acrylamide in the polyacrylamide gel is 0.04% and 0.05%, respectively.

H. Methods of Delivery of Nanoparticles to the Cytoplasm

For the simplest analysis of Brownian motion, there should be no binding interactions between the nanoparticles used for cell nanorheology measurements and subcellular structures or the organelle transport system (Tseng et al., 2002b). Such interactions would uncontrollably change the MSD profiles and affect the calculations of viscoelastic properties of the cytoplasm. Therefore, the nanoparticles cannot be transferred to the cytoplasm via endocytosis, by simply placing them on the surface of the cells (Tseng et al., 2002b). Nanoparticles delivered to the cytoplasm in this manner are enveloped in vesicles, which are shuttled in the cell by motor proteins such as dynein (Suh et al., 2003).

Therefore, it is essential to use a physical loading method for introducing the nanoparticles into the cytoplasm (Kole et al., 2004a, 2005; Tseng et al., 2002b). The most direct method is manual microinjection, which is a tedious procedure that necessitates special equipment and training and limits the number of cells amenable to measurements. There are also bulk loading methods such as scrape loading, electroporation (Gehl, 2003), peptide-mediated delivery (Plank *et al.*, 1998), and toxin-mediated membrane poration (Sandvig and van Deurs, 2002). However, these methods are generally limited to objects ≪100 nm in diameter, the size of the probe nanoparticles. They are also time-consuming, causing a large fraction of the nanoparticles to be engulfed by endocytosis.

We have employed a highly efficient method for transferring nanoparticles to cells. The method, referred to as ballistic injection, preserves the central advantage of microinjection while rendering the injection process high throughput. It was first introduced for the efficient delivery of DNA- or RNA-coated gold particles for transfection (Fire, 1986). However, the approach proves equally efficient for delivering the probes for particle-tracking nanorheology.

I. BIN: Proof of Principle

Here we illustrate the use of ballistic injection in a biological problem not easily addressed by other cell mechanics methods or by particle-tracking nanorheology coupled with traditional microinjection. Transferring cells from a 2D culture dish to a more physiological 3D matrix induces profound changes in cell morphology and function. In particular, endothelial cells placed on a thin matrix layer proliferate to confluence, while endothelial cells embedded in matrix can form extended tubular structures that resemble blood vessel capillaries *in vivo* (Folkman and Haudenschild, 1980). However, current rheological methods cannot measure the mechanical response of individual cells buried in a 3D matrix, since they are physically inaccessible to conventional cell mechanics probes, including AFM, glass microneedles, or suction micropipettes (Fig. 1). Moreover, intracellular microrheology in its current form cannot be easily utilized to probe cells in a 3D matrix because of the difficulties to find the small number of microinjected cells if they need to be detached and integrated within the 3D matrix.

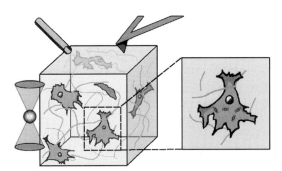

Fig. 1 Requirements of accessibility to cells for different single-cell mechanics methods. Cells embedded in an extended 3D matrix, while readily amenable to imaging, are not accessible to some cell mechanics approaches, including AFM and calibrated microneedles, and so on. The assay presented in this chapter is based on high-resolution tracking of nanoparticles ballistically bombarded into the cytoplasm; it is able to probe the micromechanical properties of single endothelial cells in a 3D matrix. Figure reprinted with permission of Panorchan *et al.* (2006) and the Biophysical Society.

Ballistic bombardment allows the injection of a large number of cells on a culture dish, which may subsequently be harvested and embedded in a 3D matrix (Fig. 2). In this method of injection, a pulse of high-pressure helium is fed into a gas chamber, which forces a macrocarrier disk coated with 100-nm fluorescent nanoparticles to crash into a stopping screen. The momentum of the macrocarrier is transferred to the nanoparticles, which penetrate into the target cells (Fig. 2). Problems could arise if nanoparticles do not penetrate directly into the cytoplasm on impact. They may then be engulfed by the cell through endocytosis, and undergo microtubule-mediated directed motion. We avoid this possible pitfall by thoroughly and repeatedly washing the cells with fresh culture medium right after ballistic bombardment. Compared to manual microinjection, ballistic injection increases ~1000-fold the number of cells available for data acquisition. This greatly facilitates microrheological measurements of cells inside an extended matrix.

To validate the results obtained with ballistic injection, we compared the viscosity and elasticity of Swiss 3T3 fibroblasts (Kole *et al.*, 2004a, 2005) with nanoparticles loaded by either microinjection or ballistic bombardment. These methods yielded similar mean cytoplasmic compliances (i.e., the deformability of the cytoplasm), albeit many more cells could be probed within the same period of time when using ballistic injection (Fig. 3).

J. Intracellular Micromechanics of Cells in 3D Matrix

Now we demonstrate how we use BIN to measure the rheology of cells embedded inside a matrix. Human umbilical venous endothelial cells (HUVECs) embedded at low density inside the 3D "puramatrix" peptide hydrogel (BD Biosciences, San Jose, California) displayed a dramatically different morphology from that of HUVECs plated at low density on a 100-μm layer of the same matrix (Bell *et al.*, 2001; Davis *et al.*, 2002). HUVECs plated on the surface of hydrogel matrix were elongated and displayed a wide lamella and extended stress fibers that spanned the length of the cell (not shown). HUVECs embedded within the hydrogel matrix showed extensive dendritic protrusions, a few disorganized actin filament bundles, and an actin-rich rim at the cell periphery (Fig. 4A).

To probe the intracellular mechanical behavior of single HUVECs embedded in a 3D extracellular matrix environment, the cells were plated on a culture dish, ballistically injected with 100-nm diameter fluorescent polystyrene nanoparticles, detached from their substrate (through trypsin treatment followed by extensive washing), and embedded in a 3D matrix. Following standard methods (Kole *et al.*, 2004b), the measurements of the MSDs of the nanoparticles were used for computing creep compliance (Fig. 4B) and viscoelastic moduli (Fig. 4C; Chapter 1 by Janmey *et al.* and Chapter 2 by Kandow *et al.*, this volume). The cytoplasm of HUVECs embedded in the matrix was significantly more compliant and, therefore, more deformable than the cytoplasm of HUVECs plated on the surface of the same matrix (data not shown). HUVECs in 3D were also significantly less viscous than those in 2D.

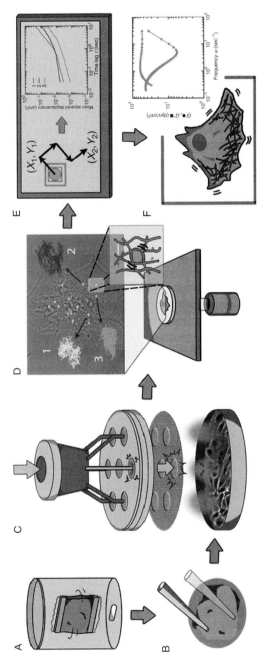

Fig. 2 Schematic of ballistic intracellular nanorheology (BIN). (A) To assist the coating process, nanoparticles are first prepared by dialysis against ethanol. (B) Dialyzed nanoparticles are then coated onto the macrocarrier. (C) Nanoparticles are delivered into the cytoplasm of adherent cells by ballistic injection. (D) Trajectories of embedded particles inside a cell are monitored by video-based particle tracking using high-magnification fluorescence microscopy. (E) The time-dependent x and y coordinates of the particles are tracked via measurement of intensity weighted centroid displacements and converted into mean-squared displacements. (F) Mean-squared displacements are then calculated and transformed into frequency-dependent viscoelastic moduli, which inform us about the local deformability and viscoelasticity of the cell. (See Plate 7 in the color insert section.)

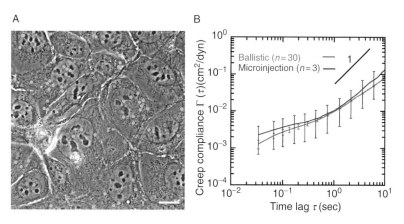

Fig. 3 Particle-tracking nanorheology of cells: manual injection versus ballistic injection of nanoparticles. (A) Phase contrast micrograph of Swiss 3T3 fibroblasts overlaid with a fluorescent micrograph of nanoparticles that were ballistically injected into their cytoplasm. (B) Comparison between the mean cellular compliance computed from the displacements of nanoparticles microinjected (*blue*) ($n = 3$) and bombarded (*red*) ($n = 30$) into the cytoplasm of Swiss 3T3 fibroblasts. Figures reprinted with permission of Panorchan *et al.* (2006) and the Biophysical Society. (See Plate 8 in the color insert section.)

II. Materials and Instrumentation

A. Preparation of Nanoparticles

One-hundred nanometer diameter fluorescent carboxylated nanoparticles (Cat. No. F8803) from Invitrogen (Carlsbad, California) are used for most applications of BIN. Nanoparticles are prepared by dialysis in 100% ethanol (Cat. No. 111000200; Pharmco, Brookfield, Connecticut) using 300,000 MWCO Spectra/Por cellulose ester dialysis membrane tubing (Cat. No. 131447; Spectrum Laboratories, Rancho Dominguez, California).

B. Cell Culture

HUVEC-C (Cat. No. CRL-1730), F-12K medium (Cat. No. 30-2004), and fetal bovine serum [used 10% (v/v) with culture media, Cat. No. 30-2020] are obtained from American Type Tissue Culture (ATCC, Manassas, Virginia). Cell culture medium is supplemented with 0.1-mg/ml heparin (Cat. No. H-3393; Sigma, St. Louis, Missouri), 0.05 mg/ml of endothelial cell growth supplement (ECGS; Cat. No. E2759; Sigma), and 1:100 dilution of Pen-Strep solution (Cat. No. SV30010; Hyclone, Logan, Utah). Hank's balanced salt solution (HBSS, Cat. No. 14170-112) and 0.25% trypsin—1-mM EDTA (Cat. No. 25200-056) are from Invitrogen. Poly-D-lysine coated 35-mm glass bottom dishes (Cat. No. P35GC-0-14-C) was from MatTek Corp. (Ashland, Massachusetts). Cultures are maintained at 37 °C in a humidified, 5% CO_2 environment.

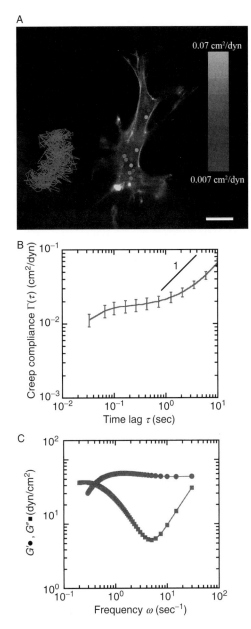

Fig. 4 Local micromechanics of single cells embedded in a 3D matrix. (A) Fluorescent micrograph of the actin filament architecture in a human umbilical venous endothelial cell (HUVEC) embedded in a 3D puramatrix gel, overlaid with a fluorescent micrograph of nanoparticles embedded in the cytoplasm. Large color-coded dots were drawn over each nanoparticle to aid presentation. Blue denotes the least deformable (stiffest) regions of the cell; red denotes the most deformable (softest) regions. Actin was visualized with green Alexa 488 phalloidin. Scale bar is 20 μm. Inset: typical trajectories of nanoparticles

C. Ballistic Injection of Nanoparticles

A biolistic particle delivery system (PDS-1000/He system, Cat. No. 165-2258) equipped with a hepta adaptor (Cat. No. 165-2225) from Bio-Rad (Hercules, California) is used to inject nanoparticles into live cells. Biolistic machine's accessories include 2200 psi rupture disks (Cat. No. 165-2334), macrocarriers (Cat. No. 165-2335), and stopping screens (Cat. No. 165-2336), all from Bio-Rad. Dry vacuum pump (model 2560, Cat. No. 2560C-02) is from Welch Rietschle Thomas (Stokie, Illinois). Rupture disks are pretreated with isopropanol (Cat. No. I9516; Sigma) prior to ballistic injection.

D. Encapsulation of Cells in a 3D Matrix

Puramatrix (Cat. No. 354250) from BD Biosciences (San Jose, California) is used as 3D matrix to study cell mechanics in 3D. Encapsulation is done in an 8-well, chamber glass slides (Cat. No. 155411; Lab-Tek). Sucrose (Cat. No. 4072-05) used during encapsulation is from JT Baker (Phillipsburg, New Jersey).

E. Imaging of Fluctuating Nanoparticles Embedded in the Cell

BIN experiments are conducted with a Nikon TE2000-E inverted microscope equipped for epifluorescence, using a Nikon PlanFluor $60\times$ oil immersion lens (NA 1.3). Movies of fluctuating fluorescent nanoparticles are captured onto the random access memory of a PC computer via a Cascade 1K camera (Roper Scientific, Tucson, Arizona) controlled by the Metavue software (Universal Imaging Corp., Sunnyvale, California), at a frame rate of 30 Hz. Time-dependent coordinates of the centroids of fluorescent nanoparticles are obtained using particle-tracking routines built into the Metamorph Imaging Suite (Universal Imaging Corp.).

III. Procedures

A. Preparation of Nanoparticles

Nanoparticles are coated onto macrocarriers prior to ballistic injection. To accomplish this, they are spread, then left to dry onto the macrocarrier. Since polystyrene nanoparticles are obtained as 2% (w/v) aqueous suspension, they are dialyzed against 100% ethanol prior to coating to accelerate the drying process.

in the cytoplasm of cells embedded in the matrix. (B) Mean cellular creep compliance (cytoplasmic deformability) and (C) mean frequency-dependent viscous and elastic moduli, $G'(\omega)$ (circles) and $G''(\omega)$ (squares), of HUVECs embedded in a 3D matrix. Figures reprinted with permission of Panorchan et al. (2006) and the Biophysical Society. (See Plate 9 in the color insert section.)

1. Preparation of Carboxylated Nanoparticles

    a. Steps

        i. Pipette 3 ml of stock nanoparticles into a 6-cm piece of dialysis tubing sealed at one end and prewet with ethanol.

        ii. Dialyze against 3 liters of ethanol with gentle stirring at 4 °C for 8 h. Repeat three times with fresh ethanol.

        iii. Aliquot the nanoparticle solution into tight, sterile 1.5-ml Eppendorf tubes, and store at 4 °C.

B. Ballistic Injection of Nanoparticles

A commercial biolistic particle delivery system (Bio-Rad) is used to deliver 100-nm diameter polystyrene fluorescent nanoparticles to culture cells. Helium gas at 2200 psi is used to force a macrocarrier disk coated with fluorescent nanoparticles to collide into a stopping screen. The forces of collision cause the nanoparticles to dissociate from the macrocarrier and bombard target cells.

Adherent cells are plated on 100-mm tissue culture dishes and grown to ~90% confluency. Prior to injection, the medium is aspirated and the dish is loaded onto the machine. Once ballistically injected, the cells are washed extensively and allowed to recover for 1 h, before replating onto 35-mm glass bottom dishes overnight for particle-tracking analysis.

1. Coating of Macrocarriers

    a. Materials

        i. Particle suspension prepared in step (Section III.A.1.*a*.iii)

        ii. Macrocarriers

    b. Steps

        i. Add a 17-μl drop of nanoparticle suspension onto a macrocarrier. Prepare seven macrocarriers per ballistic injection.

        ii. Holding a pipetter with a 20-μl pipette tip sideways, spread the drop of solution on the macrocarrier evenly with a brushing motion. Coat all seven macrocarriers. Wait 5 min and apply a second coating.

        iii. Let stand for 30 min to ensure complete dryness before ballistic injection.

2. Ballistic Injection of Fluorescent Nanoparticles

    a. Materials

        i. Dried nanoparticles on macrocarrier prepared in step (Section III.B.1.*b*.iii)

        ii. Isopropanol

        iii. HBSS

        iv. HUVEC's complete growth medium (F-12K medium, 10% FBS, 0.1-mg/ml heparin, 0.05-mg/ml ECGS, 1% Pen-Strep solution)

b. Steps

i. Autoclave hepta adaptor, stopping screens, and macrocarrier launch assembly.

ii. Soak the rupture disk in isopropanol for 2 sec and place a rupture disk in the round slot inside the hepta adaptor.

iii. Screw the hepta adaptor containing the rupture disk onto the helium gas acceleration tube inside the ballistic chamber. Insert torque wrench (provided with the machine) in the small upper ring of the hepta adapter assembly and tighten the hepta adaptor.

iv. Place the coated macrocarriers into seven-hole macrocarrier holder with the coated side down (particles toward the sample). Use the seating tool (provided with the machine) to fit macrocarriers firmly into the macrocarrier holder.

v. To eliminate aggregates of fluorescent nanoparticles on the macrocarrier (problematic for particle-tracking analysis), gently rub the macrocarrier with a clean finger or a small sterile scraper until no distinct pellets of nanoparticles is visible. Aggregates in cells can be distinguished by their anomalously high intensity relative to single beads.

vi. Enclose the macrocarrier holder with stopping screen and stopping screen holder. Place this assembly into the hepta macrocarrier shelf and slide the final assembly into the highest shelf slot position in the ballistic chamber. Align the seven pressure outlets of the hepta adapter with the centers of the seven holes in the macrocarrier holder.

vii. Aspirate medium from previously prepared 100-mm tissue culture dish. Place the tissue culture dish onto the target shelf, and slide it into the slot immediately below the hepta macrocarrier shelf.

viii. Lock the ballistic chamber and pull vacuum to ∼28-in. Hg. Hold the pressure and fire until the rupture disk breaks. Release pressure and retrieve tissue culture dish.

ix. Immediately wash cells three times with HBSS and replace medium with HUVEC complete growth medium. Allow cells to incubate for 1 h before replating.

C. Cell Seeding and Encapsulation in a 3D Matrix

Once ballistically injected, cells are now ready for culture. Cells can be either seeded on the surface or encapsulated within Puramatrix peptide hydrogels.

1. Seeding HUVECs on a 2D Surface of Matrix

a. Materials

i. Puramatrix peptide hydrogel (1% stock)

ii. HUVEC's complete growth medium (F-12K medium, 10% FBS, 0.1-mg/ml heparin, 0.05-mg/ml ECGS, 1% Pen-Strep solution)

b. Steps

    i. Sonicate hydrogel for 30 min in a bath sonicator.

    ii. Dilute the sonicated hydrogel 1:1 with sterile water. Add 100 μl of diluted hydrogel to an 8-well chamber slide (100 μl per well). Use the pipette tip to gently spread the hydrogel solution to cover the entire bottom of each well.

    iii. With a pipette tip touching the wall of the well, gently add 300 μl of culture medium to initiate assembly of the hydrogel.

    iv. Allow 1 h for gel assembly at room temperature. Change medium every 15 min within the hour.

    v. Typsinize and spin down cells at 2000 rpm for 5 min. Collect the pellet and resuspend at 1×10^5 cells/ml. Add 50 μl of the cell suspension to each well. Gently mix the solution with the pipette.

    vi. Allow 6–12 h incubation prior to BIN analysis.

2. Encapsulation of HUVECs Within a 3D Matrix

a. Materials

    i. Puramatrix peptide hydrogel (1%, stock)

    ii. 10% sucrose solution (use sterile water)

    iii. HUVEC's complete growth medium (F-12K medium, 10% FBS, 0.1-mg/ml heparin, 0.05-mg/ml ECGS, 1% Pen-Strep solution)

b. Steps

    i. Sonicate hydrogel for 30 min in a bath sonicator.

    ii. Typsinize and spin down cells at 2000 rpm for 5 min. Remove the supernatant and add 5 ml of 10% sucrose as wash solution. Resuspend cells and repellet, then remove the wash solution. Resuspend cells at 1×10^6 cells/ml with fresh 10% sucrose solution.

    iii. Mix the sonicated hydrogel with cell suspension at 1:1 ratio. Add 200 μl of the mixture to each well in an 8-well chamber slide. With a pipette tip touching the wall of the well, gently add 200 μl of culture medium to initiate assembly of the hydrogel. It is important to proceed as quickly as possible, as the hydrogel solution has a very low pH, which is harmful to the cells.

    iv. Allow 1 h for gel assembly at room temperature. Change medium every 10 min within the hour. Incubate overnight prior to BIN analysis.

D. BIN Analysis

BIN analysis involves capturing motions of embedded nanoparticles with time lapsed movies. These movies of trapped nanoparticles inside a cell are then analyzed by a custom particle-tracking routine incorporated into the Metamorph imaging suite as described (Tseng and Wirtz, 2001). Individual time-averaged MSDs, $\langle \Delta r^2(\tau) \rangle = \langle [x(t+\tau) - x(t)]^2 + [y(t+\tau) - y(t)]^2 \rangle$, where τ is the timescale, are calculated from the 2D trajectories of the centroids of the nanoparticles. With ballistic injection, a large number of cells are available for data acquisition (relative to conventional single cell studies). Typically, a sample size of 30 cells, each containing $\sim$10 nanoparticles, is used per condition. All measurements are performed in an incubator mounted on an inverted microscope maintained at 37 °C with 5% CO_2 and humidity. To correlate particle-tracking measurements with cellular structures, the cultures are fixed after measurements, for subsequent labeling with fluorescent antibodies (Fig. 4A).

1. Acquisition of Movies of Fluctuating Nanoparticles

a. Steps

i. Place previously prepared cell culture chamber on the stage of an epifluorescence microscope equipped with a live cell incubator maintained at 37 °C and 5% CO_2.

ii. Identify ballistically injected cells with a 60× Plan Fluor objective, using a combination of fluorescence and bright field illumination.

iii. Center on the cell of interest. Adjust the focus under fluorescence illumination to obtain the clearest image of nanoparticles and acquire 20–100 sec of streaming video with a Cascade 1K camera at a frame rate of 30 Hz. To achieve this frame rate, image acquisition region was limited to 500 × 300 pixels, with 3 × 3 binning. Adjustment may be necessary for different cameras. Acquire also a still fluorescence image of the nanoparticles after recording the movie.

iv. Switch to high-resolution bright field and acquire still image of the cell. Save the movie of fluorescence nanoparticles as a *stk* file and the still images as *tiff* files.

2. Analysis of Movies

a. Steps

i. Open an *stk* movie file of fluctuating nanoparticles in the Metamorph image analysis software. Calibrate pixel distances using a previously acquired image of a stage micrometer. Draw a region of interest, zoom 400%, and duplicate the entire movie sequence with zoom.

ii. Using the "track objects" command, create inner and outer regions around each nanoparticle in the first frame of the sequence. Do not create regions around aggregated nanoparticles, since these nanoparticles violate assumptions made in our constitutive viscoelastic equations and cannot be used for analysis.

iii. For each nanoparticle, adjust the inner region so that it extends just beyond the edge of the nanoparticle. Adjust the outer region so that it is large enough to encompass the nanoparticle in all the subsequent frames. Duplicate an image of the initial frame with labeled regions visible and save as a tiff file.

iv. Open "log to excel." Track the nanoparticles and log the ID number, frame number, and the time-dependent coordinates, $[x(t), y(t)]$, for each frame. Save this data as an excel spreadsheet.

3. Calculations of MSDs, Creep Compliance, and Viscoelastic Moduli

From the time-dependent coordinates, $[x(t), y(t)]$, the projections of the MSD for each nanoparticle in the x and y directions are calculated using the following formulas (Qian *et al.*, 1991):

$$\text{MSD}_x(\tau) = \frac{\sum_{i=1}^{N} (x(t_i + \tau) - x(t_i))^2}{N + 1}$$

$$\text{MSD}_y(\tau) = \frac{\sum_{i=1}^{N} (y(t_i + \tau) - y(t_i))^2}{N + 1} \tag{1}$$

Hence, the MSD of the nanoparticle, $\langle \Delta r^2(\tau) \rangle$, is simply

$$\text{MSD}(\tau) = \text{MSD}_x(\tau) + \text{MSD}_y(\tau) \tag{2}$$

The MSD of each probe nanosphere is directly related to the local creep compliance of the cytoplasm, $\Gamma(\tau)$, as (Xu *et al.*, 1998c)

$$\Gamma(\tau) = \frac{3\pi a}{2 k_B T} \langle \Delta r^2(\tau) \rangle \tag{3}$$

where k_B is Boltzmann's constant, T is the absolute temperature of the cell (in Kelvin), and a is the radius of the nanoparticle. Fluctuations due to active motors have also been considered (Lau *et al.*, 2003). The creep compliance (expressed in units of cm^2/dyn, the inverse of a pressure or modulus) describes the local deformation of the cytoplasm as a function of time created by the thermally excited displacements of the nanoparticles. The method to obtain frequency-dependent viscoelastic moduli has been described in details in Kole *et al.* (2004b). Briefly, the complex viscoelastic modulus, $G^*(\omega)$, is obtained using the following equation:

$$G^*(\omega) = \frac{k_B T}{\pi a i \omega \mathfrak{J}_u \{ \langle \Delta r^2(\tau) \rangle \}} \tag{4}$$

where $\omega = 1/\tau$ and $\mathfrak{J}_u\{ \langle \Delta r^2(t) \rangle \}$ is the Fourier transform of $\langle \Delta r^2(\tau) \rangle$. The elastic modulus is the real part of Eq. (4) and the viscous modulus is the imaginary part (Chapter 1 by Janmey *et al.* and Chapter 2 by Kandow *et al.*, this volume). While $G^*(\omega)$ may be calculated numerically, an analytical solution was obtained by Mason *et al.* (1997) by approximating the Fourier transformation of $\langle \Delta r^2(\tau) \rangle$ using a wedge assumption, which expands $\langle \Delta r^2(\tau) \rangle$ locally around the frequency of interest ω using a power law and retains the leading term (Mason *et al.*, 1997). The Fourier transform of $\langle \Delta r^2(\tau) \rangle$ then becomes:

$$i\omega \mathfrak{J}_u\{ \langle \Delta r^2(\tau) \rangle \} \approx \left\langle \Delta r^2 \left(\frac{1}{\omega} \right) \right\rangle \Gamma[1 + \alpha(\omega)] i^{-\alpha(\omega)} \tag{5}$$

where $\alpha(\omega) = d \ln\langle\Delta r^2(\tau)\rangle/d \ln\tau|_{\tau=1/\omega}$ is the local logarithmic slope of $\langle\Delta r^2(\tau)\rangle$ at the frequency of interest $\omega = 1/\tau$ and Γ is the gamma function. The frequency-dependent elastic and viscous moduli, G' and G'', can then be calculated algebraically using the following relationships:

$$G'(\omega) = |G^*(\omega)|\cos\left(\frac{\pi\alpha(\omega)}{2}\right) \tag{6}$$

$$G''(\omega) = |G^*(\omega)|\sin\left(\frac{\pi\alpha(\omega)}{2}\right) \tag{7}$$

where

$$|G^*(\omega)| = \frac{2k_B T}{3\pi a\langle\Delta r^2(1/\omega)\rangle\Gamma(1 + \alpha(\omega))} \tag{8}$$

IV. Pearls and Pitfalls

A. During data acquisition, the bright intensity of large aggregates in the sample will overwhelm the intensity of an individual particle. To circumvent this issue, there are several things to try. First, sonicate the particle suspension for 30 min prior to coating. Second, reduce the volume of particle suspension used for coating the macrocarrier. The particle suspension in ethanol usually becomes more concentrated after several cycles of usage/storage due to evaporation of ethanol, which may be compensated by using a smaller volume. Third, spread the particle suspension on the macrocarrier for longer to avoid visible collection of particles. Finally, while loading a coated macrocarrier onto the macrocarrier holder, be sure to thoroughly rub the macrocarrier with your finger or a scraper.

B. In case of low particle penetration efficiency, ensure while coating the macrocarrier, that the particles are suspended well in the solution. Once dialyzed in ethanol, particles tend to settle quickly to the bottom of the Eppendorf tube. Make sure to mix by pipetting up and down prior to coating every macrocarrier. In addition, ensure that the medium in the tissue culture is removed as much possible during aspiration. Wet layer on the tissue culture plate impedes particle penetration.

C. Maintain high confluence of culture prior to ballistic injection. Lower confluency lead to inefficient particle penetration and lower cell viability.

D. One of the most critical steps in ballistic injection is washing. Cells must be wash immediately and *vigorously* after ballistic injection to remove any lodged particles on the surface of the cells as well as free-floating particles. These particles may be endocytosed by the cells and will appear as extremely fast moving particles under BIN analysis since they move via active transport.

V. Concluding Remarks

A. Unique Advantages of BIN

BIN offers unique advantages compared to conventional particle-tracking nanorheology:

1. In a single ballistic injection, the number of injected cells amenable to measurements increases by a 1000-fold compare to the microinjection technique (Lee *et al.*, 2006; Panorchan *et al.*, 2006).

2. Microinjection of cells is highly inconsistent due to invasive nature of injection and the mechanical trauma to the cells that ensues. With ballistic injection, every cell is injected similarly thus decreasing cell-to-cell variations in the measurements sometimes found following microinjection.

3. With a large population of injected cells, BIN allows us to probe single-cell mechanics in complex geometries and in more physiological situations, including 3D model tissues (Panorchan *et al.*, 2006), cells subjected to shear flows (Lee *et al.*, 2006), and cancer cells at the rear and the edge of a wound (J. S. H. Lee, P. Panorchan, and D. Wirtz, unpublished data).

4. With a large sample size per condition (number of probed cells $\sim$30), our results become much more precise and significant. The small sample size (number of probed cells $\sim$5) of microinjection leads to cell-to-cell variations and potential random experimental errors. BIN provides a more precise and consistent values for global and local viscoelastic properties.

B. Advantages of Traditional Particle-Tracking Nanorheology Are Maintained by BIN

BIN also preserves all the advantages of particle-tracking nanorheology:

1. BIN can measure directly mechanical properties of the cytoplasm (Tseng *et al.*, 2004b). Most current single-cell mechanics methods rely on contact between the cell surface and a physical probe. Therefore, these methods cannot distinguish the contribution of the plasma membrane from those of the nucleus, cytoskeleton, and other organelles without making drastic assumptions. In contrast, BIN directly measures the mechanical properties of the cytoplasm.

2. BIN measures shear rate-dependent viscoelastic moduli. This is particularly crucial for the cytoskeleton, which behaves like a liquid at long timescales (or low rates of shear) and like an elastic solid at short timescales (or high rates of shear).

3. By tracking multiple nanoparticles simultaneously, BIN can measure simultaneously micromechanical responses to stimuli in various parts of the cell. By using video-based multiple particle tracking instead of laser deflection particle tracking (Mason *et al.*, 1997; Yamada *et al.*, 2000), hundreds of nanoparticles embedded in the body of cells can be tracked at the same time.

4. BIN rheological measurements are absolute and compare favorably with traditional rheometric measurements on standard fluids of known viscosity and elasticity (Apgar *et al.*, 2000; Mason *et al.*, 1997; Xu *et al.*, 1998a,c). This is not the case of some single-cell approaches that rely on the contact between the cell surface and the probe. It is now clear that the apparent viscoelastic moduli measured by magnetocytometry and AFM depend greatly on the type of ligands coated on the magnetic beads or AFM cantilever. Extracellular ligands—including fibronectin, RGD peptide, and ICAM1—coated on magnetic beads and AFM cantilevers lead to vastly different values of (apparent) cell stiffness. Therefore, the measurements of viscoelastic properties of standard materials using these methods cannot be compared to those obtained using a cone-and-plate rheometer.

5. BIN measures both elasticity and viscosity, while many other approaches cannot distinguish the elastic from the viscous responses of the cell directly.

Acknowledgments

This work was partially supported by an NIH grant (R01 GM075305-01) and a NASA grant (NAG9-1563). J.S.H. Lee was supported by a NASA graduate training grant (NNG04G054H). This work was also supported by a graduate training grant from the Howard Hughes Medical Institution.

References

Apgar, J., Tseng, Y., Fedorov, E., Herwig, M. B., Almo, S. C., and Wirtz, D. (2000). Multiple-particle tracking measurements of heterogeneities in solutions of actin filaments and actin bundles. *Biophys. J.* **79,** 1095–1106.

Atilgan, E., Wirtz, D., and Sun, S. X. (2005). Morphology of the lamellipodium and organization of actin filaments at the leading edge of crawling cells. *Biophys. J.* **89,** 3589–3602.

Atilgan, E., Wirtz, D., and Sun, S. X. (2006). Mechanics and dynamics of actin-driven thin membrane protrusions. *Biophys. J.* **90,** 65–76.

Bausch, A. R., Möller, W., and Sackmann, E. (1999). Measurement of local viscoelasticity and forces in living cells by magnetic tweezers. *Biophys. J.* **76,** 573–579.

Bell, S. E., Mavila, A., Salazar, R., Bayless, K. J., Kanagala, S., Maxwell, S. A., and Davis, G. E. (2001). Differential gene expression during capillary morphogenesis in 3D collagen matrices: Regulated expression of genes involved in basement membrane matrix assembly, cell cycle progression, cellular differentiation and G-protein signaling. *J. Cell Sci.* **114,** 2755–2773.

Berg, H. C. (1993). "Random Walks in Biology." Princeton University Press, Princeton, NJ.

Bousquet, O., Ma, L., Yamada, S., Gu, C., Idei, T., Takahashi, K., Wirtz, D., and Coulombe, P. A. (2001). The nonhelical tail domain of keratin 14 promotes filament bundling and enhances the mechanical properties of keratin intermediate filaments *in vitro. J. Cell Biol.* **155,** 747–754.

Coulombe, P. A., Bousquet, O., Ma, L., Yamada, S., and Wirtz, D. (2000). The 'ins' and 'outs' of intermediate filament organization. *Trends Cell Biol.* **10,** 420–428.

Daniels, B. R., Masi, B. C., and Wirtz, D. (2006). Probing single-cell micromechanics *in vivo*: The microrheology of *C. elegans* developing embryos. *Biophys. J.* **90,** 4712–4719.

Davis, G. E., Bayless, K. J., and Mavila, A. (2002). Molecular basis of endothelial cell morphogenesis in three-dimensional extracellular matrices. *Anat. Rec.* **268,** 252–275.

Esue, O., Carson, A. A., Tseng, Y., and Wirtz, D. (2006). A direct interaction between actin and vimentin filaments mediated by the tail domain of vimentin. *J. Biol. Chem.* **281,** 30393–30399.

Esue, O., Cordero, M., Wirtz, D., and Tseng, Y. (2005b). The assembly of MreB, a prokaryotic homolog of actin. *J. Biol. Chem.* **280**, 2628–2635.

Esue, O., Tseng, Y., and Wirtz, D. (2005a). Mechanical shear can accelerate the gelation of actin filament networks. *Phys. Rev. Lett.* **95**, 048301.

Fire, A. (1986). Integrative transformation of caenorhabditis elegans. *EMBO J.* **5**, 2673–2680.

Flanagan, L. A., Ju, Y. E., Marg, B., Osterfield, M., and Janmey, P. A. (2002). Neurite branching on deformable substrates. *Neuroreport* **13**, 2411–2415.

Folkman, J., and Haudenschild, C. (1980). Angiogenesis *in vitro*. *Nature* **288**, 551–556.

Gehl, J. (2003). Electroporation: Theory and methods, perspectives for drug delivery, gene therapy and research. *Acta Physiol. Scand.* **177**, 437–447.

Gittes, F., Schnurr, B., Olmsted, P. D., MacKintosh, F. C., and Schmidt, C. F. (1997). Microscopic viscoelasticity: Shear moduli of soft materials determined from thermal fluctuations. *Phys. Rev. Lett.* **79**, 3286–3289.

Gupton, S. L., Anderson, K. L., Kole, T. P., Fischer, R. S., Ponti, A., Hitchcock-DeGregori, S. E., Danuser, G., Fowler, V. M., Wirtz, D., Hanein, D., and Waterman-Storer, C. M. (2005). Cell migration without a lamellipodium: Translation of actin dynamics into cell movement mediated by tropomyosin. *J. Cell Biol.* **168**, 619–631.

Haber, C., Ruiz, S. A., and Wirtz, D. (2000). Shape anisotropy of a single random-walk polymer. *Proc. Natl. Acad. Sci. USA* **97**, 10792–10795.

Heidemann, S. R., and Wirtz, D. (2004). Towards a regional approach to cell mechanics. *Trends Cell Biol.* **14**, 160–166.

Janmey, P. A. (1998). The cytoskeleton and cell signaling: Component localization and mechanical coupling. *Physiol. Rev.* **78**, 763–781.

Janmey, P. A., Euteneuer, U., Traub, P., and Schliwa, M. (1991). Viscoelastic properties of vimentin compared with other filamentous biopolymer networks. *J. Cell Biol.* **113**, 155–160.

Janmey, P. A., Hvidt, S., Lamb, J., and Stossel, T. P. (1990). Resemblance of actin-binding protein/actin gels to covalently networks. *Nature* **345**, 89–92.

Klein, M. G., Shi, W., Ramagopal, U., Tseng, Y., Wirtz, D., Kovar, D. R., Staiger, C. J., and Almo, S. C. (2004). Structure of the actin crosslinking core of fimbrin. *Structure* **12**, 999–1013.

Kole, T. P., Tseng, Y., Huang, L., Katz, J. L., and Wirtz, D. (2004a). Rho kinase regulates the intracellular micromechanical response of adherent cells to rho activation. *Mol. Biol. Cell* **15**, 3475–3484.

Kole, T. P., Tseng, Y., Jiang, I., Katz, J. L., and Wirtz, D. (2005). Intracellular mechanics of migrating fibroblasts. *Mol. Biol. Cell* **16**, 328–338.

Kole, T. P., Tseng, Y., and Wirtz, D. (2004b). Intracellular microrheology as a tool for the measurement of the local mechanical properties of live cells. *Methods Cell Biol.* **78**, 45–64.

Kucik, D. F., Elson, E. L., and Sheetz, M. P. (1989). Forward transport of glycoproteins on leading lamellipodia in locomoting cells. *Nature* **340**, 315–317.

Lee, J. S., Chang, M. I., Tseng, Y., and Wirtz, D. (2005). Cdc42 mediates nucleus movement and MTOC polarization in Swiss 3T3 fibroblasts under mechanical shear stress. *Mol. Biol. Cell* **16**, 871–880.

Lee, J. C., and Discher, D. E. (2001). Deformation-enhanced fluctuations in the red cell skeleton with theoretical relations to elasticity, connectivity, and spectrin unfolding. *Biophys. J.* **81**, 3178–3192.

Lee, J. S. H., Panorchan, P., Hale, C. M., Khatau, S. B., Kole, T. P., Tseng, Y., and Wirtz, D. (2006). Ballistic intracellular nanorheology reveals ROCK-hard cytoplasmic stiffening response to fluid flow. *J. Cell Sci.* **119**, 1760–1768.

Lacayo, C. I., and Theriot, J. A. (2004). Listeria monocytogenes actin-based motility varies depending on subcellular location: A kinematic probe for cytoarchitecture. *Mol. Biol. Cell* **15**, 2164–2175.

Lau, A. W., Hoffman, B. D., Davies, A., Crocker, J. C., and Lubensky, T. C. (2003). Microrheology, stress fluctuations, and active behavior of living cells. *Phys. Rev. Lett.* **91**, 198101.

Luby-Phelps, K., Taylor, D. L., and Lanni, F. (1986). Probing the structure of cytoplasm. *J. Cell Biol.* **102**, 2015–2022.

Lukacs, G. L., Haggie, P., Seksek, O., Lechardeur, D., Freedman, N., and Verkman, A. S. (2000). Size-dependent DNA mobility in cytoplasm and nucleus. *J. Biol. Chem.* **275**, 1625–1629.

Mason, T. G., Ganesan, K., van Zanten, J. V., Wirtz, D., and Kuo, S. C. (1997). Particle-tracking microrheology of complex fluids. *Phys. Rev. Lett.* **79,** 3282–3285.

Ma, L., Xu, J., Coulombe, P. A., and Wirtz, D. (1999). Epidermal keratin suspensions have unique micromechanical properties. *J. Biol. Chem.* **274,** 19145–19151.

Ma, L., Yamada, S., Wirtz, D., and Coulombe, P. A. (2001). A hot-spot mutation alters the mechanical properties of keratin filament networks. *Nat. Cell Biol.* **3,** 503–506.

Palmer, A., Cha, B., and Wirtz, D. (1998a). Structure and dynamics of actin filament solutions in the presence of latrunculin A. *J. Polym. Sci. Phys. Ed.* **36,** 3007–3015.

Palmer, A., Xu, J., and Wirtz, D. (1998b). High-frequency rheology of crosslinked actin networks measured by diffusing wave spectroscopy. *Rheol. Acta* **37,** 97–108.

Palmer, A., Xu, J., Kuo, S. C., and Wirtz, D. (1999). Diffusing wave spectroscopy microrheology of actin filament networks. *Biophys. J.* **76,** 1063–1071.

Panorchan, P., Lee, J. S. H., Kole, T. P., Tseng, Y., and Wirtz, D. (2006). Microrheology and ROCK signaling of human endothelial cells embedded in a 3D matrix. *Biophys. J.* **91,** 3499–3507.

Panorchan, P., Schafer, B. W., Wirtz, D., and Tseng, Y. (2004a). Nuclear envelope breakdown requires overcoming the mechanical integrity of the nuclear lamina. *J. Biol. Chem.* **278,** 43462–43467.

Panorchan, P., Wirtz, D., and Tseng, Y. (2004b). Structure-function relationship of biological gels revealed by multiple particle tracking and differential interference contrast microscopy: The case of human lamin networks. *Phys. Rev. E. Stat. Nonlin. Soft Matter Phys.* **70,** 041906.

Plank, C., Zauner, W., and Wagner, E. (1998). Application of membrane-active peptides for drug and gene delivery across cellular membranes. *Adv. Drug Deliv. Rev.* **34,** 21–35.

Pollard, T. D., Goldberg, I., and Schwarz, W. H. (1992). Nucleotide exchange, structure, and mechanical properties of filaments assembled from ATP-actin and ADP-actin. *J. Biol. Chem.* **267,** 20339–20345.

Qian, H., Sheetz, M. P., and Elson, E. L. (1991). Single particle tracking. Analysis of diffusion and flow in two-dimensional systems. *Biophys. J.* **60,** 910–921.

Sandvig, K., and van Deurs, B. (2002). Membrane traffic exploited by protein toxins. *Annu. Rev. Cell Dev. Biol.* **18,** 1–24.

Sato, M., Schwarz, W. H., and Pollard, T. D. (1987). Dependence of the mechanical properties of actin/α-actinin gels on deformation rate. *Nature* **325,** 828–830.

Seksek, O., Biwersi, J., and Verkman, A. S. (1997). Translational diffusion of macromolecule-sized solutes in cytoplasm and nucleus. *J. Cell Biol.* **138,** 131–142.

Storm, C., Pastore, J. J., MacKintosh, F. C., Lubensky, T. C., and Janmey, P. A. (2005). Nonlinear elasticity in biological gels. *Nature* **435,** 191–194.

Suh, J., Wirtz, D., and Hanes, J. (2003). Efficient active transport of gene nanocarriers to the cell nucleus. *Proc. Natl. Acad. Sci. USA* **100,** 3878–3882.

Tsai, M. A., Frank, R. S., and Waugh, R. E. (1993). Passive mechanical behavior of human neutrophils: Power-law fluid. *Biophys. J.* **65,** 2078–2088.

Tseng, Y., An, K. M., Esue, O., and Wirtz, D. (2004a). The bimodal role of filamin in controlling the architecture and mechanics of F-actin networks. *J. Biol. Chem.* **279,** 1819–1826.

Tseng, Y., An, K. M., and Wirtz, D. (2002a). Microheterogeneity controls the rate of gelation of actin filament networks. *J. Biol. Chem.* **277,** 18143–18150.

Tseng, Y., Fedorov, E., McCaffery, J. M., Almo, S. C., and Wirtz, D. (2001). Micromechanics and microstructure of actin filament networks in the presence of the actin-bundling protein human fascin: A comparison with α-actinin. *J. Mol. Biol.* **310,** 351–366.

Tseng, Y., Kole, T. P., Lee, J. S., Fedorov, E., Almo, S. C., Schafer, B. W., and Wirtz, D. (2005). How actin crosslinking and bundling proteins cooperate to generate an enhanced cell mechanical response. *Biochem. Biophys. Res. Commun.* **334,** 183–192.

Tseng, Y., Kole, T. P., and Wirtz, D. (2002b). Micromechanical mapping of live cells by multiple-particle-tracking microrheology. *Biophys. J.* **83,** 3162–3176.

Tseng, Y., Lee, J. S., Kole, T. P., Jiang, I., and Wirtz, D. (2004b). Micro-organization and viscoelasticity of the interphase nucleus revealed by particle nanotracking. *J. Cell Sci.* **117,** 2159–2167.

Tseng, Y., and Wirtz, D. (2001). Mechanics and multiple-particle tracking microheterogeneity of alpha-actinin-cross-linked actin filament networks. *Biophys. J.* **81,** 1643–1656.

Tseng, Y., and Wirtz, D. (2004). Dendritic branching and homogenization of actin networks mediated by Arp2/3 complex. *Phys. Rev. Lett.* **93,** 258104.

Valberg, P. A., and Albertini, D. F. (1985). Cytoplasmic motions, rheology, and structure probed by a novel magnetic particle method. *J. Cell Biol.* **101,** 130–140.

Wachsstock, D., Schwarz, W. H., and Pollard, T. D. (1993). Affinity of a-actinin for actin determines the structure and mechanical properties of actin filament gels. *Biophys. J.* **65,** 205–214.

Wachsstock, D., Schwarz, W. H., and Pollard, T. D. (1994). Crosslinker dynamics determine the mechanical properties of actin gels. *Biophys. J.* **66,** 801–809.

Waterman-Storer, C. M., and Salmon, E. D. (1999). Positive feedback interactions between microtubule and actin dynamics during cell motility. *Curr. Opin. Cell Biol.* **11,** 61–67.

Xu, J., Palmer, A., and Wirtz, D. (1998a). Rheology and microrheology of semiflexible polymer solutions: Actin filament networks. *Macromolecules* **31,** 6486–6492.

Xu, J., Schwarz, W. H., Kas, J., Janmey, P. J., and Pollard, T. D. (1998b). Mechanical properties of actin filament networks depend on preparation, polymerization conditions, and storage of actin monomers. *Biophys. J.* **74,** 2731–2740.

Xu, J., Viasnoff, V., and Wirtz, D. (1998c). Compliance of actin filament networks measured by particle-tracking microrheology and diffusing wave spectroscopy. *Rheol. Acta* **37,** 387–398.

Xu, J., Wirtz, D., and Pollard, T. D. (1998d). Dynamic cross-linking by alpha-actinin determines the mechanical properties of actin filament networks. *J. Biol. Chem.* **273,** 9570–9576.

Xu, J., Tseng, Y., and Wirtz, D. (2000). Strain-hardening of actin filament networks—regulation by the dynamic crosslinking protein α-actinin. *J. Biol. Chem.* **275,** 35886–35892.

Yamada, S., Wirtz, D., and Coulombe, P. A. (2002). Pairwise assembly determines the intrinsic potential for self-organization and mechanical properties of keratin filaments. *Mol. Biol. Cell* **13,** 382–391.

Yamada, S., Wirtz, D., and Coulombe, P. A. (2003). The mechanical properties of simple epithelial keratins 8 and 18: Discriminating between interfacial and bulk elasticities. *J. Struct. Biol.* **143,** 45–55.

Yamada, S., Wirtz, D., and Kuo, S. C. (2000). Mechanics of living cells measured by laser tracking microrheology. *Biophys. J.* **78,** 1736–1747.

CHAPTER 7

Multiple-Particle Tracking and Two-Point Microrheology in Cells

John C. Crocker and Brenton D. Hoffman

Department of Chemical and Biomolecular Engineering
Institute for Medicine and Engineering
University of Pennsylvania
Philadelphia, Pennsylvania 19104

METHODS IN CELL BIOLOGY, VOL. 83

0091-679X/07 $35.00
DOI: 10.1016/S0091-679X(07)83007-X

Abstract

Mechanical stress and stiffness are increasingly recognized to play important roles in numerous cell biological processes, notably cell differentiation and tissue morphogenesis. Little definite is known, however, about how stress propagates through different cell structures or how it is converted to biochemical signals via mechanotransduction, due in large part to the difficulty of interpreting many cell mechanics experiments. A newly developed technique, two-point microrheology (TPM), can provide highly interpretable, quantitative measurements of cells' frequency-dependent shear moduli and spectra of their fluctuating intracellular stresses. TPM is a noninvasive method based on measuring the Brownian motion of large numbers of intracellular particles using multiple-particle tracking. While requiring only hardware available in many cell biology laboratories, a phase microscope and digital video camera, as a statistical technique, it also requires the automated analysis of many thousands of micrographs. Here we describe in detail the algorithms and software tools used for such large-scale multiple-particle tracking as well as common sources of error and the microscopy methods needed to minimize them. Moreover, we describe the physical principles behind TPM and other passive microrheological methods, their limitations, and typical results for cultured epithelial cells.

I. Introduction

Cell biologists have long studied the complicated biochemical and physiological responses of cells to mechanical stress or deformation (Orr *et al.*, 2006; Vogel and Sheetz, 2006). In addition to these responses, cells also show a purely mechanical, deformation response to applied stress, determined by their shear modulus (Chapter 1 by Janmey *et al.* and Chapter 2 by Kandow *et al.*, this volume). While mechanical and physiological responses can occur simultaneously and couple, complicating interpretation, it is usually assumed that the deformations occurring immediately after stress application or in response to small stresses are predominantly mechanical in origin.

Only in the last decade or so have techniques, termed *microrheology*, been developed that can characterize cells' dynamic shear modulus over a wide frequency range (Waigh, 2005; Weihs *et al.*, 2006). Initially, many researchers hoped that these new cell rheological measurements would display characteristic times corresponding to known molecular timescales, such as that of the myosin ATP hydrolysis cycle. Such contributions could then be dissected with genetic or pharmacological interventions to tie together cell behavior at the molecular and mesoscopic scales. Less ambitiously, it was hoped that the cell mechanical response would at least closely resemble the response of various purified biopolymer gel models (typically F-actin), whose rheology has been intensively studied since the 1980s (Kroy, 2006). In reality, however, the rheological responses of cells measured to date have not contained identifiable molecular timescales and have not been satisfactorily reproduced by any biopolymer gel model yet studied. Several experiments have found the (low frequency) shear modulus of cells to be well described by a weak power-law form, $G^*(\omega) \sim \omega^\beta$, with reported values of β varying over the range 0.1–0.3 (Alcaraz *et al.*, 2003; Desprat *et al.*, 2005; Fabry *et al.*, 2001; Hoffman *et al.*, 2006; Lenormand *et al.*, 2004; Yamada *et al.*, 2000). Such a power-law form has no characteristic times at all. In contrast, the low-frequency behavior of reconstituted gels is either purely elastic with $\beta = 0$ (Janmey *et al.*, 1990; Koenderink *et al.*, 2006) or nearly so with a very small β exponent (Gardel *et al.*, 2006; Xu *et al.*, 1998).

The challenges of interpreting and modeling cell rheological measurements have several causes (Weihs *et al.*, 2006). Primary among them is the obvious structural complexity within cells. Cells are composed of numerous chemically and spatially distinct subdomains that include the cell cortex, the nuclear envelope, lamellipodia and stress fibers, not to mention the microtubule and intermediate filaments networks, endoplasmic reticula, and other organelles that fill the cell interior. It seems likely that different cell microrheological methods will probe different mechanical subdomains (or different combinations of them), and that the target and response might differ among cell types or even among individual cells of the same type. Indeed, it is not clear *a priori* to what extent cells may be understood as a continuous viscoelastic solid, rather than a complex ensemble of discrete units. Finally, there is also the fact that microrheology comprises a new and emerging set of methodologies, with still unresolved technical issues regarding interpretation and measurement artifacts, even with comparatively simple synthetic or reconstituted biopolymer materials.

In this chapter, we will describe a cell microrheological method developed in our laboratory, two-point microrheology (TPM) (Crocker *et al.*, 2000; Lau *et al.*, 2003), which computes the rheology from measurements of the statistically cross-correlated Brownian motion of pairs of embedded, intracellular tracers (for complementary discussion of particle-based microrheology, see Chapter 6 by Panorchan *et al.*, this volume). Compared to other techniques, TPM has the advantage of being more interpretable: it is essentially immune to uncertainties related to cytoskeletal heterogeneity and the tracer/network connection, and provides additional positive controls regarding whether the cell even behaves as a viscoelastic continuum. TPM has the disadvantage that, like all methods based on

Brownian motion, it can be confounded by active intracellular processes, and is thus most reliable when applied to cells that have been depleted of ATP. Moreover, since TPM naturally probes a three-dimensional structure much larger than the tracers, it is best suited to measure the rheology of the thick central "body" of the cell, rather than thin structures such as the lamellipodium, cell cortex, or nuclear envelope.

The differences among the mechanical properties of different cell regions were highlighted in a study in our laboratory (Hoffman *et al.*, 2006), which compared the results of TPM with another technique, magnetic twisting cytometry (MTC; see Chapter 19 by Polte *et al.*, this volume), based on rocking external, integrin-attached microparticles using a magnetic field. While qualitatively similar at first glance (Fig. 1), the results of these two measurements on the same cell type are distinctly different on close inspection. Combined with literature and control measurements, these results indicate that the mechanical properties of cells' cortical and deeper intracellular regions are different, and thus presumably so are their predominant structural elements and organization. While this difference is in line with expectations from known cell physiology, it clearly represents a potential confounding factor for all cell rheological measurements; a given

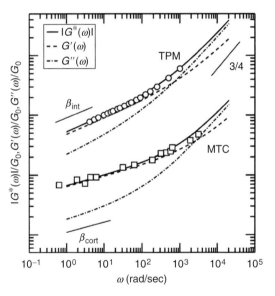

Fig. 1 Comparison of the cell rheological data from two different methods that probe either the deep interior of TC7 epithelial cells, TPM (circles), averaged over $N = 7$ cells, or the cortex, MTC (squares), $N = 8$ cells. Each rheological curve is a best fit to a sum of two power-law functions (Hoffman *et al.*, 2006). Curves indicate only the relative frequency dependence; the vertical positions of the two data sets are arbitrary. Absolute stiffness estimates are discussed in Section VII.B.

rheological method may potentially probe one or the other region, or a superposition of both. The unique interpretability of TPM was essential to sorting out this confounding factor.

Examining the cell rheological findings in Fig. 1, both the cortical and interior responses can be divided into two frequency regimes. At low frequencies, the structures are predominantly elastic ($G' > G''$), with the aforementioned power-law form reported by several groups, $G^*(\omega) \sim \omega^\beta$, where many literature values of β fall between the two values we find ($\beta_{int} = 0.26$, $\beta_{cort} = 0.16$). At high frequencies, both structures are more dissipative than elastic ($G'' > G'$), with a power-law exponent of about 3/4. Such high-frequency behavior has a simple physical explanation—a 3/4 exponent is characteristic of the high-frequency rheology of networks formed of semiflexible (i.e., filamentary) polymers (Deng *et al.*, 2006). While this clearly suggests that deformations at high frequencies are resisted by a filamentous network, it is an open question whether low-frequency deformations are resisted by the same biopolymer network or by a different structure altogether. Indeed, the physical origin and molecular determinants of cells' power-law rheology have not been compellingly identified; clearly, much research remains to be done. Given cells' structural complexity, it seems likely that no single-cell microrheological method can provide a complete or compelling description of cellular responses to mechanical stress. Rather, a sensible strategy may be to apply multiple microrheological techniques to single-cell types, with TPM contributing one particularly interpretable component, and to compare the results carefully for mutual consistency.

From a hardware point of view, TPM is accessible as it requires only hardware that is common in many cell biology laboratories: a high-magnification optical microscope with phase-contrast or differential interference-contrast optics and a high-intensity illuminator as well as a reasonably high-quality, low-noise camera that can collect images at or near video rate (tens of frames per second). While we use a specialized high-speed camera to collect the above reported TPM data, this is not required to measure rheology in the weak power-law regime. Unlike laser-deflection-based approaches, expertise with physical optics, analogue signal processing, or lasers is not required. As a statistical method, however, it does require the analysis of extremely large numbers of micrographs, at least 10,000 per single-cell measurement. With the use of automated image acquisition and analysis routines and increasingly fast microcomputers, this requirement presents little real impediment. Because of the small amplitude of the Brownian motion of intracellular particles, care must be taken to minimize the effects of microscope vibration and to maximize the precision of the particle-tracking process. Beyond describing the underlying algorithmic and mathematical procedures required for TPM, this chapter will also discuss how to achieve high-performance multiple-particle tracking using an imaging system. Such an instrument is potentially useful for a number of cell biology applications other than TPM, for example, for studies of endocytosis and intracellular trafficking.

II. Principles of Passive Tracer Microrheology

In general, there are two approaches to measuring the rheology of soft materials, active and passive. In the active approach, a known force (or deformation) is applied to the material, and the resulting deformation (or force) is measured. In the cell rheological context, this usually relies on atomic force microscope (AFM) (Alcaraz et al., 2003) or similar, force-calibrated microcantilever instruments (Desprat et al., 2005). The passive approach, which is our focus in this chapter, examines the Brownian motion of tracers embedded in the soft material. No force is applied at all, only the spontaneous motion of the tracers is observed and quantified, considerably simplifying instrumentation requirements. The "calibration" of the Brownian forces comes from simple physical principles: in general, the (squared) amplitude of the tracer motion is inversely proportional to the material's stiffness.

A. Conventional Passive Microrheology

The simplest and first described example of Brownian motion is a spherical tracer particle moving in a simple viscous fluid (like water or glycerol). Here, the relationship between mechanical properties and tracer motion is the familiar Stokes–Einstein relation:

$$\eta = \frac{k_B T}{\langle \Delta r(\tau)^2 \rangle \pi a} \tau \tag{1}$$

where η is the liquid's viscosity, a is the tracer radius, and the "driving force" is the energy of thermal fluctuations, $k_B T$, where k_B is Boltzman's constant and T the absolute temperature. The mean-squared displacement (MSD) of the particle's motion, $\langle \Delta r(\tau)^2 \rangle$, is simply the square of the net distance the tracer typically moves during a given time interval, τ, in this context called a lag time. We will discuss the computation of the MSD in a later section.

To model the Brownian motion of tracers embedded in viscoelastic materials, Eq. (1) needs to be modified. Several different, but mathematically consistent versions are in use; we employ the generalized Stokes–Einstein relation (GSER) (Mason and Weitz, 1995):

$$G^*(\omega) = \frac{k_B T}{i\omega \langle \Delta r(\omega)^2 \rangle \pi a} \tag{2}$$

where $\langle \Delta r(\omega)^2 \rangle$ is the unilateral Fourier transform, $f(\omega) = \int_0^\infty e^{-i\omega\tau} f(\tau) d\tau$ of the MSD and $i = \sqrt{-1}$, and the use of (Fourier) frequency, ω, rather than lag time facilitates comparison to conventional rheological models. Note that in order to determine the rheology at even a single frequency, we must formally evaluate the Fourier integral over all lag times from $\tau = 0$ to $\tau = \infty$, while it is obviously impossible to measure the MSD over that entire range. In practice, if we measure the MSD over a wide range of lag times, $\tau_1 < \tau < \tau_2$, then we can compute the rheology over a range of frequencies $\omega_2 = 1/\tau_2 < \omega < \omega_1 = 1/\tau_1$, with a little uncertainty at the frequency extrema. Numerically evaluating Fourier integrals with typical data can be quite challenging, but simple approximate methods have been developed, which we will describe in Section IV.

Equation (2) implies several relationships between the Brownian motion and the shear modulus. Roughly speaking, since the overall amplitude of the MSD and its Fourier transform are linearly proportional, the GSER implies an inverse relationship between mechanical properties and the tracer's MSD at a given frequency. In addition, since $\langle \Delta r(\omega)^2 \rangle$ is in general a complex function, so is $G^*(\omega)$. The real and imaginary parts of the shear modulus, $G'(\omega)$ and $G''(\omega)$, respectively called the storage modulus and loss modulus (Chapter 1 by Janmey et al. and Chapter 2 by Kandow et al., this volume), represent the solid-like and liquid-like behavior of the material. In simple viscous liquids, the above equation reduces to Eq. (1), with MSD $\sim \tau^1$ and $G^*(\omega) = i\omega\eta$. In simple elastic solids, the MSD is lag time independent, MSD $\sim \tau^0$. In a viscoelastic material, the MSD will have an intermediate form and will increase more slowly than linearly with lag time.

An important point is that the derivation of Eq. (2) assumed that the tracer is embedded in a completely homogeneous material, and has "no-slip" boundary conditions. In many complex synthetic materials, the rheology computed using the GSER has been found to be incorrect due to a failure of one or both of these assumptions (Crocker et al., 2000; Valentine et al., 2004). In porous materials, tracers tend to diffuse inside fluid-filled pores, and systematically report a softer and more fluid-like rheology. Tracers that associate strongly with the material, for example adhering strongly to segments of a porous material, tend to report rheology that is softer than the true bulk rheology but has the correct frequency dependence (Van Citters et al., 2006). Tracers in cells tend to show this latter behavior, implying that many microrheological techniques tend to yield reliable frequency dependences but may have systematically underestimated absolute stiffness values.

B. Expected Tracer Motion and Tracking Performance

The expected amplitude of the Brownian MSD for some simple cases can be readily estimated. Consider spherical particles 1 μm in diameter, which are easily visible under a high-magnification optical microscope. In water at room temperature, such particles will typically move 1 μm in each direction in a lag time of $\tau = 1$ sec. In an elastic material with a shear modulus of 1000 Pa = 1 kPa, a typical value reported for cell measurements (Fabry et al., 2001) and roughly that of very soft agar, the rattling motion of a 1-μm tracer has an expected amplitude of only 1 nm. Such a miniscule motion is quite invisible to all but the highest-performance particle-tracking instruments.

For comparison, consider the typical amplitudes of vibration and tracer position measurement error. For an optical microscope simply placed on an ordinary laboratory bench or table, typical sample vibration amplitudes are 10–100 nm. The same microscope on a well-engineered pneumatic vibration isolation platform or table will have vibration amplitude typically about 10 times smaller, but still comparable to or in excess of the expected Brownian motion signal, 1 nm, estimated above. Image-based particle-tracking routines that have not been carefully optimized typically locate micron-sized particles to 10- to 20-nm precision (Crocker *et al.*, 2000), and this can be improved to a few nanometers using the techniques described in later sections. While we find that intracellular tracers typically move somewhat more than the 1-nm estimate above, passive cell microrheology is nonetheless challenging for even a high-performance image-based particle-tracking system. This is the reason why many such experiments are performed with laser-deflection particle-tracking systems, which readily achieve subnanometer precision, or with longer lag times, where the motion is somewhat larger.

C. Two-Point Microrheology

We concluded above that interpreting cell rheological measurements is often difficult due to uncertainties related to tracer boundary conditions and tracer/network association. Fortunately, TPM removes most of the interpretation uncertainties of conventional microrheology. Unlike other microrheological methods, TPM can deliver a reliable, absolute measure of stiffness, rather than just its frequency dependence, even in heterogeneous materials and cases where the tracer boundary conditions are not known.

The basic principle of TPM is that all soft materials undergo a form of internal Brownian motion, like waves on the surface of a choppy ocean. Tracers are carried along by these random undulations of the medium, like corks bobbing on the ocean. The Brownian motion of two separate tracers will be statistically correlated because they will both be carried along by the Brownian motion of the segment of the material spanning between them. TPM computes the rheology from the amplitude of the tracers' resulting cross-correlated motion. Elementary calculations show that in three dimensions, the correlation between the particles motion is inversely proportional to their separation, R. This is a consequence of larger material segments having smaller Brownian motion amplitudes (just as larger particles diffuse more slowly than small ones). This $1/R$ dependence provides a useful positive control that the material is deforming like a continuous three-dimensional object, a prerequisite for our analytical framework (Crocker *et al.*, 2000).

The interpretive power of TPM comes from the fact that any local motion of the tracers relative to the material generally will not be correlated with one another (Levine and Lubensky, 2001). For example, in the troublesome case that tracers inhabit soft pores in the material, their diffusive rattling in their respective pores will be statistically independent, and will not contribute to the correlation measurement in two-particle microrheology. Details of how to compute the cross-correlation and rheology will be given later. While sample vibration does lead to correlated motion, it has a different character than that due to the material's

Brownian motion, and can simply be filtered out to a large extent. It turns out that tracer position error increases the noise of the two-point measurement, and must still be minimized. Provided it is not too much larger than the tracer motion, however, it does not systematically affect the inferred rheology, facilitating the use of image-based particle-tracking methods.

III. Multiple–Particle Tracking Algorithms

Multiple-particle tracking can be broken down into four processing stages: correcting imperfections in individual images, accurately locating particle positions, eliminating false or unwanted particles, and finally linking these positions in time to create a trajectory (Crocker and Grier, 1996). We perform these tasks with software we developed using the interactive data language (IDL; ITT Visual Information Solutions, Boulder, CO), a high-level programming language used extensively in astronomy and earth sciences, which we have made available as freeware. Each section below is broken into two parts: one describing the general approach and a second with specific instructions for using our IDL software.

A typical cell image is shown in Fig. 2. The image shows a TC7 green monkey kidney epithelial cell illuminated with a pulsed near-IR laser, visualized using shadow-cast DIC optics (bright-field Nomarski), and acquired using a non-interlaced

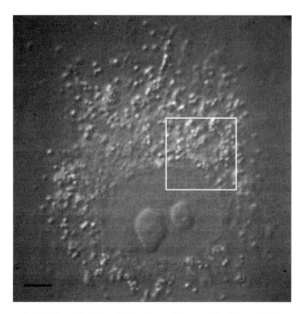

Fig. 2 An image of a TC7 epithelial cell illuminated by a pulsed near-IR laser, taken with high-performance shadow-cast DIC. White box shows a region of interest for which further analysis will be shown. Scale bar is 5 μm.

camera with 512 × 512 resolution at a magnification of 96 nm/pixel. In our example, we will track the ~400-nm diameter endogenous lipid granules that are common in these and many other cultured cells. Subsequently, we will highlight only the small region of interest marked in Fig. 2, even though the algorithms process the entire image. Other organelles or probes introduced experimentally, such as polystyrene particles (Chapter 6 by Panorchan *et al.*, this volume), may be used for a similar purpose.

A. Image Restoration

Even the apparently high-quality image in Fig. 2 contains imperfections that can frustrate particle-locating algorithms. These include both gradual variations in background brightness, "shading," and "snow," errors in individual pixel values. However, both effects can be sensibly removed by applying a spatial band-pass filter to the image, which reduces both low spatial frequency (shading) and high spatial frequency (snow) contributions, while leaving intact the intermediate frequencies corresponding to our tracers.

In IDL: *bpass.pro* is the program used to filter the images. It uses a "wavelet" technique corresponding to convolution with a "Mexican hat" kernel to remove the random digitization noise and background, respectively (Crocker and Grier, 1996). The program requires two parameters delimiting the spatial wavelengths in pixels. Set the first to 1 pixel and the second to the typical particle diameter, 5 pixels in the example. Filtered images will then retain features with a linear dimension between 1 and 5 pixels. The unfiltered and filtered images are shown in Fig. 3.

B. Locating Possible Particle Positions

The image now consists of bright spots on a dark background. As a first approximation for images collected at a high signal-to-noise ratio (SNR), the brightest pixel in each spot corresponds to its position rounded to the nearest pixel. The distribution of brightness in the rest of the spot can be used to further

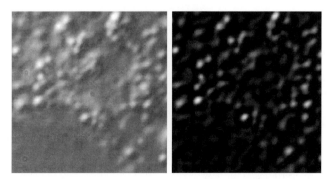

Fig. 3 Close-up of region of interest from Fig. 2 before (left) or after (right) spatial filtering. Notice subtraction of background intensity.

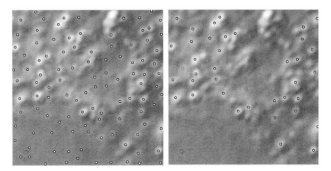

Fig. 4 Automatic detection of particles for tracking, showing region of interest with all local maxima in brightness marked with circles (left), and particles that will actually be tracked (right). Notice the removal of particles in the nucleus, particle aggregates, and dim out of focus particles.

refine the position to subpixel accuracy. To further refine the particle location, we use an algorithm that computes the brightness-weighted centroid within a circular mask that is slightly larger than the particle (Crocker and Grier, 1996). While many other methods exist, all perform similarly in the limit of large SNR (high-quality pictures) (Cheezum *et al.*, 2001). Details of this algorithm and the limits to its performance are the subject of Section V.

In IDL: *feature.pro* in our particle-tracking suite performs the centroiding operation. This program requires the user to specify the size of the mask to be used. To avoid finding multiple centers on the same particle and to minimize location error, it should be set to a value just larger than the average particle size. Since the sample has particles of different sizes, ranging from 5 to 9 pixels, we use a mask size of 11 pixels. All possible particle positions (local maxima of brightness) are shown in Fig. 4 (left). This algorithm also computes other properties of each particle image, including total image brightness, as well as average size (radius of gyration) and elongation of the particles. For convenience, this program and *bpass.pro* have been combined into one program, *pretrack.pro*, that can run on large, multiple image data sets once the correct *bpass.pro* and *feature.pro* parameters have been determined interactively on single images.

C. Eliminating Spurious or Unwanted Particle Trajectories

Because the previous stage of analysis identified all local maxima of brightness as potential particles, it readily identified a large number of maxima corresponding to very low-contrast intracellular structures or camera noise. Moreover, it also located particles in the nucleus, which presumably should not be pooled together with cytoplasmic lipid granules for analysis. In general, these unwanted tracer locations must be discriminated out based on some combinations of their spatial location and morphology, for example total brightness or apparent size.

In IDL: The discrimination process begins by making scatter plots of two of the desired properties (e.g., *x*- and *y*-coordinates or particle brightness and radius).

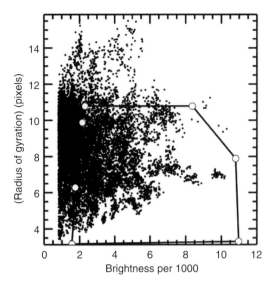

Fig. 5 An example polygonal cut in the brightness–radius plane. The left edge is set to isolate the 150–250 brightest tracers. The top edge is to eliminate the possibility of tracking multiple points on the same, oversized tracer, by limiting the feature size of the mask used during centroiding. The upper right edge is set to avoid tracking large clumps of aggregated particles.

Using *polycut.pro*, the user draws a polygon over the scatter plot using the mouse. The program then selectively retains or removes all of the objects within the polygon. In this example, first the particles in the nucleus are located and removed by drawing a polygon around the nucleus in the x–y scatter plot (polygon cut). If several discontinuous movies are being pooled together, we suggest a separate polygon cut for each, as the cell may drift between sets. Next, a polygonal cut in brightness-radius plane, Fig. 5, is used to further refine the particles by appearance. Particle positions due to (noise-induced) image artifacts are dim and small. Aggregates of endogenous particles are large and bright. Also points with sizes greater than the feature mask should not be used since they represent features found on large, low-contrast structures inside the cell such as the membrane. Typically, the user experiments with different polygon placements until only the desired population of features is selected for further processing. Provided the illuminator settings are not altered, the same polygon can be used for all the images of a single cell or even multiple cells collected in the same session. Note in Fig. 4 (right), there are no particles in the background or the nucleus and each particle contains only one center.

D. Linking Positions into Trajectories

Having determined a satisfactory set of particle positions, next we match locations in each image with corresponding locations in later images to produce trajectories. This involves determining which particle in a given image most likely corresponds to the particle in the next image. Since all tracer particles are fairly similar, proximity is the only real indication that particles in successive images correspond to the same physical object. Globally, the optimum identification of particle positions should minimize the total squared distance of travel (Crocker and Grier, 1996). To make such identification computationally feasible, a maximum possible particle displacement between images must be specified. This parameter should be sufficiently larger than the maximum distance of travel in a single-frame interval, to ensure that physical displacements are never rejected. The trajectories for our example are shown in Fig. 6. Notice that the particle motions resemble random walks more than continuous curves. For this reason, alternative tracking algorithms that identify particles by trying to extrapolate their motion or compare their velocities are not well suited to our data. How to extract rheological information from our particles' random trajectories will be discussed in Section IV.

It should be noted that particle tracking provides an additional layer of filtering for rejecting spurious, artifactual tracers. For instance, if a random fluctuation of the camera noise leads to a spurious feature being identified in a given frame (which survives discrimination), it is highly unlikely that a similar fluctuation will occur in several consecutive frames. As a result, while trajectories of physical particles often contain as many positions as movie frames, those of spurious noise particles are generally very short, and can be easily rejected based on the small number of continuous frames in which they are detected. For this reason, in the preceding discrimination stage it is usually better to adjust the settings to avoid occasional rejection of physical tracers, even if it allows a few spurious particles to be passed through.

In IDL: Particle locations from movies of many images are linked into trajectories using *track.pro*. The user needs to provide one parameter, the maximal distance that a particle may travel between consecutive frames. We find that 4 pixels is a suitable threshold for 50 frames per second data, but this is likely to vary with different cell types and magnification. In this application, setting the "goodenough" parameter to 10 rejects spurious tracer trajectories having fewer than 10 valid locations.

Figure 6B and C displays checks that we have found useful to verify the proper operation of the particle-locating and particle-tracking algorithms. Figure 6B shows a histogram of the particle displacements between consecutive frames ($\tau = 1/50$ sec), generated using *makepdf.pro*. Note that the histogram completely decays and is not truncated by the maximum displacement setting, 4 pixels, which is larger than the largest displacement. Moreover, note that the histogram has long tails. If particles were diffusing in water, this histogram would be completely Gaussian. Here, the long tails are an indication that different particles are trapped in heterogeneous local environments. Very long or nearly flat tails to large displacement are an indication of the tracking of spurious particle images. In Fig. 6C, a histogram of the fractional part of x-position of the particles, x mod 1, is shown; with proper subpixel accuracy in particle-tracking algorithms, there should be no favored subpixel value. That means a histogram of the x-position of all particle positions modulo 1 should be completely flat. Lack of flatness indicates poor imaging or mask settings and will be discussed in Section V.B.

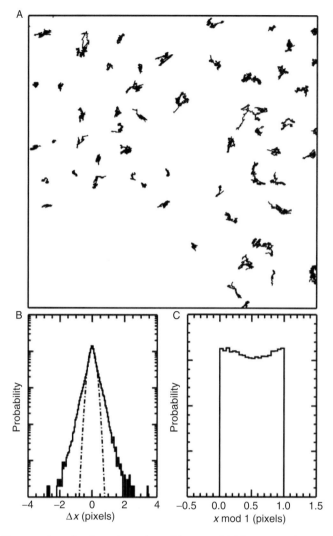

Fig. 6 (A) Tracer trajectories from the region of interest. (B) Histogram of tracer displacements between consecutive frames, which decays well before 4 pixels, the maximum displacement set in the particle-tracking software. (C) Histogram of the x-position of the particle modulo 1, demonstrating that there is negligible systematic error with these choices of tracking parameters.

E. Available Software Packages and Computing Resources

Source code for our IDL particle-tracking routines is available for free download at http://www.physics.emory.edu/~weeks/idl/ along with a short online tutorial. These particle-tracking routines have been translated into the increasingly popular MatLab environment. Both IDL and MatLab are available for a wide variety of computing platforms and many research institutions have a site license for at least one of them. Moreover, a "stand-alone" version (with a graphical user interface) has been compiled for the Windows, Macintosh, and Linux operating systems, and does not require purchasing the IDL software. Links to the MatLab and stand-alone versions may be found at the site above. Many commercially available particle-tracking software packages, such as Diatrack and Metamorph, should also be able to track intracellular tracers with comparable precision and ease of use. While these solutions can be quite costly, they have the advantage of offering user support, software customization to specific imaging needs, integrated image acquisition, and training courses. It should be remembered, however, that additional routines to compute rheology from the particle trajectories, described in the following section, are at this time only available in IDL form.

These software programs are somewhat computationally intensive, but can be run effectively on a high-performance personal computer. The use of large stacks of images can lead to large file sizes (~256 Mbyte). We use a dual 2.4-GHz Athlon processor Linux server, with 2 Gbyte of RAM and a 500-Gbyte RAID array hard drive. In one day, it can convert ~80,000 images, each containing several hundred tracers, into trajectories. A large amount of RAM storage allows for multiple users (ideally, at least 0.5 Gbyte of RAM per simultaneous user). Image- and particle-tracking data are analyzed, compressed, and then archived onto removable, external hard drives.

IV. Computing Rheology from Tracer Trajectories

The central idea behind passive microrheology is that the random, Brownian motion of small embedded tracers is determined by the stiffness of the material surrounding them; tracers in hard materials naturally move less than ones in soft materials. Here, we detail how the ensembles of random particle trajectories computed in the preceding section are converted to an MSD or its two-point equivalent, 2P-MSD. This is followed by a description of the algorithms that convert this data to the rheological parameters of the surroundings, along with a discussion of their physical and numerical limitations. As for earlier sections, a general description of the required procedures will be followed by specific instructions for using our IDL routines.

A. Computing Mean-Squared Displacements

In Fig. 7A, the x-component of a numerically generated Brownian particle is shown as a function of time. Note the random nature of this curve. Mathematically, the MSD is equal to the average, squared distance a particle travels in a given time interval, referred to as lag time, τ, expressed as:

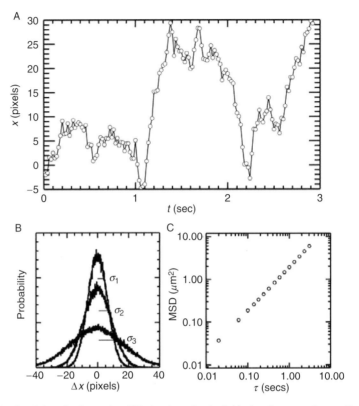

Fig. 7 Simulated data for Brownian diffusion in a simple fluid, showing a random walk trajectory (A), histograms of the tracer displacements at three lag times, with standard deviations labeled by horizontal lines (B), and the MSD calculated from a series of lag times (C).

$$\langle \Delta x^2(\tau) \rangle = \langle (x(t+\tau) - x(t))^2 \rangle \qquad (3)$$

where x indicates tracer position in one dimension, t is time, and $<\cdots>$ indicates time and ensemble (over all particles) average. However, it may be best understood graphically. Figure 7B shows a histogram of the distance of travel from 200 particles diffusing in a liquid for 0.02 and 0.4 sec, respectively. The MSD at given lag time is just the square of the standard deviation of particle positions after a given lag time. To convert a one-dimensional MSD, $\langle \Delta x^2(\tau) \rangle$, to a three-dimensional MSD, $\langle \Delta r^2(\tau) \rangle$, multiply its amplitude by 3. This assumes an isotropic material, which means the material has identical properties in the x-, y-, and z-directions.

In IDL: *msd.pro* calculates MSDs from the trajectory files. For the MSD to be in physical units, the pixel size and frame rate of the camera must be well known. The magnification can be readily calibrated using microscope stage micrometers (Edmund Optics, Barrington, NJ) or diffraction gratings.

B. Computing Two-Point Mean-Squared Displacements

The basic idea of TPM (Crocker *et al.*, 2000) is that inside a viscoelastic solid, a particle cannot move without moving its neighbors, like two people trying to jump on a trampoline without affecting each other. While in general the correlation of two vectorial displacements is a second-rank tensor, we typically correlate the motion of two particles along their line of centers, as this has the best signal to noise. This correlation may be represented mathematically as:

$$D_{rr}(R, \tau) = \langle \Delta r_1(\tau) \Delta r_2(\tau) \rangle \qquad (4)$$

where $\Delta r_1(\tau)$ and $\Delta r_2(\tau)$ are the stochastic motions of the first and second particle, respectively, along their mutual line of centers in lag time τ (Fig. 8). The brackets indicate an average over all pairs of particles in the data set with separation R. In practice pairs with similar R values are pooled together to create a function of both R and τ.

Since the TPM signal's origin is Brownian fluctuations of the material, the τ dependence should be related to the rheology in a manner resembling the conventional MSD. Indeed, this resemblance can be made exact (Crocker *et al.*, 2000), by rescaling the two-point correlation tensor D_{rr} with a geometric factor, $2R/a$:

$$\langle \Delta x^2(\tau) \rangle_2 = \frac{2R}{a} D_{rr} \qquad (5)$$

If the material is homogeneous and isotropic on length scales significantly smaller than the tracer, incompressible, and connected to the tracers by uniform no-slip boundary conditions over their entire surfaces, the two MSDs will be equal $\langle \Delta x^2(\tau) \rangle_2 = \langle \Delta x^2(\tau) \rangle$. If these boundary and homogeneity conditions are not satisfied, the two MSDs will be unequal. In this case, applying the two-point MSD in the GSER will still yield the "bulk" rheology of the material (on the long length scale "R"), while the conventional single-particle MSD will report a rheology that is a complicated superposition of the bulk rheology and the rheology of the material at the tracer boundary (Levine and Lubensky, 2001).

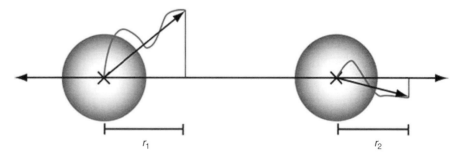

Fig. 8 Schematic of displacements used to compute the two-point MSD.

As mentioned earlier, two-point measurements will readily detect sample vibration and drift, as these effects lead to completely correlated tracer motion. When this artifactual motion is significant compared to the Brownian signal, we fit the $D_{rr}(R, \tau)$ function to a form $(A(\tau)/R) + B(\tau)$. By using the $A(\tau)$ component exclusively to compute the 2P-MSD and rheology, we can reliably remove artifacts due to sample vibration and drift.

In IDL: *msd2pnt.pro* is used to generate D_{rr} from trajectory data. Like *msd.pro*, it requires the pixel size and the frame rate of the camera to produce data in physical units. Also, it requires the user to input a range of separations over which to correlate particles. The images of closely spaced particles can overlap, adding a spurious cross-correlation to their motion and confounding the TPM measurement. We find that using a 2 μm minimum separation is sufficient to overcome this problem (Lau *et al.*, 2003). The upper limit is determined by the cell's finite thickness, which is about 4 μm for the cells we study. The fact that the cell is a thick slab rather than an infinite three-dimensional solid leads to deviations from $1/R$ decay. However, we find these deviations become negligible when R is less than about twice the cell thickness, so we typically use an upper limit of 8 μm and a smaller number for thinner cell types.

msdd.pro converts D_{rr} into $\langle \Delta x^2(\tau) \rangle_2$ and can further delimit the minimum and maximum separations used in the calculation. Since *msd2pnt.pro* can take hours to run, while *msdd.pro* takes seconds, it is more efficient to set R limits wide in *msd2pnt.pro* and then reduce them to different extents in *msdd.pro*, in order to test for the effects of changing the R limits on the final rheology. The "lfit" keyword in *msdd.pro* causes the routine to perform the vibration/drift correction described above.

C. Applying Automated Image Analysis for Statistics

The statistical error in particle MSDs is readily estimated. If we approximate the distribution of tracer displacements (as in Fig. 7B) as a Gaussian, the standard error for the variance is simply $2\langle x^2 \rangle / \sqrt{N_{eff}}$, where N_{eff} is the number of statistically uncorrelated measurements in the distribution. If an image series contains N_t tracers and spans a time interval T, then $N_{eff} \approx N_t T / \tau$. That is, if we image a single particle for 10 sec at 50 frames a second, we have roughly 500 independent samples of the displacement for a lag time of 1/50 sec, but only 10 independent samples for a lag time of 1 sec. This τ dependence causes the statistical errors to increase dramatically at longer lag times. As an example, if we were imaging a sample containing 100 tracers at 50 frames per second, and we wanted no more than 1% statistical error in the MSD over the lag times from 1/50 to 1 sec, then we need $N_{eff} = 10^4$ independent samples at $\tau = 1$ sec. Since we are pooling the results of 100 tracers, then we need $T = 100$ sec of data. Obviously, analyzing the corresponding 5000 images, and 500,000 tracer positions for this modest example requires efficient, automated image analysis. Algorithms that require the user to manually select particles to be tracked or to estimate their locations are not practical for this application.

In general, two-point correlation functions have much higher statistical noise, requiring the acquisition of significantly higher statistical power, higher tolerance of noisier data, or both. The origin of this is straightforward to understand. The value of D_{rr} is the mean of a distribution of numbers $\langle \Delta r_1 \Delta r_2 \rangle$, since both Δr_1 and Δr_2 are single-particle displacements, the widths of the distribution of $\langle \Delta r_1 \Delta r_2 \rangle$ is roughly $\langle \Delta x^2(\tau) \rangle$, the conventional MSD, in the limit of weak correlation. In general, the two-point correlated motion, D_{rr}, is much smaller than the single-particle MSD$\langle \Delta x^2(\tau) \rangle$. Indeed, under the most favorable case, the ratio of these two quantities according to Eq. (5) is $2R/a$, which typically has a value of 10–20. We then expect that reliable measurements of the two-point MSD would require averaging at least $(2R/a)^2$ or *several hundred times* [i.e., $(10-20)^2$] more $\Delta r_1 \Delta r_2$ measurements, relative to Δx^2 measurements to compute a conventional MSD, in order to reach a similar statistical noise. This simple estimate is consistent with our experience.

Does this mean that rather than the 5000 images in our example to measure the conventional MSD, we now need to collect 500,000 images? Fortunately, that is not the case. In a field of view containing 100 randomly located tracers, each tracer might have 10 or more neighbors within the proper distance range for computing two-point correlations. Thus, each image gives us not 100 samples of $\Delta r_1 \Delta r_2$ but more like several thousand. For this reason, the statistical noise of two-point measurements is highly sensitive to the number of tracers in the field of view. In general, if there are 100 or more tracers in a microscope field of view, then about 10 times as many images are required to accurately compute a two-point MSD than a conventional MSD. Alternatively, the statistical noise of the two-point measurement will be about $\sqrt{10}$, or just a few times higher than that of the conventional MSD computed from the same data. It should be noted, however, that more statistical power is required for materials where the conventional MSD is much larger than the two-point MSD, according to the square of MSD/2P-MSD ratio. In highly porous materials, the two-point signal can be so small compared to the "background noise" of uncorrelated tracer motion that it becomes hopelessly impractical to measure from a statistical point of view.

D. Converting MSDs to Rheology

Earlier we defined mathematical relationships that should allow analytical calculation of rheological properties from the MSD, a process called inversion typically performed in the Fourier space. However, notice that the Fourier transform integral spans all times from zero to infinity. This means we would need data sets spanning this same time interval in order to do the conversion analytically. While this cannot be achieved in practice, many numerical transform methods have been proposed to provide an approximation (Waigh, 2005), with many stressing the importance of collecting data at a high frequency over a very wide time range.

Here we describe a very simple method (Mason, 2000). At each lag time point, we estimate the logarithmic derivative (or slope on a log–log plot):

$$\alpha(\tau) = \frac{d \ln\langle \Delta r^2(\tau)\rangle}{d \ln \tau} \tag{6}$$

One way to do this is to first take the logarithm of both the MSD and τ values and then for each τ point, fit a line to a few points of the corresponding logarithms of MSD surrounding the chosen τ. The value of this fit at the chosen τ is therefore a smoothed approximation of the MSD, and its slope is the logarithmic derivative. We then use an approximate, algebraic form of the GSER:

$$|G^*(\omega)| \approx \frac{k_B T}{\pi a\langle \Delta r^2(\tau = 1/\omega)\rangle \Gamma[1 + \alpha(\tau = 1/\omega)]} \tag{7}$$

where Γ represents the Gamma function. In using this expression, one first computes a set of ω values that are reciprocals of the measured lag times. At each frequency point, the value of $|G^*(\omega)|$ is then computed using only the values of the MSD and its logarithmic derivative at $\tau = 1/\omega$. Formally, the value of $|G^*(\omega)|$ at each frequency would require the numerical Fourier integration of the MSD over all lag times in the interval (0 to infinity), but in practice that integral is dominated by the value of the integrand at $\omega\tau \approx 1$. Equation (7) is exact in the limit that the MSD has a purely power-law form. For other more general forms, Eq. (7) is an excellent approximation at lag times where the MSD is well approximated locally by a power-law, and is seldom more than 15% in error otherwise.

To make contact with more standard rheological representations, one can compute the following from $|G^*(\omega)|$:

$$\begin{aligned} \delta(\omega) &= \frac{\pi}{2}\frac{d \ln|G^*(\omega)|}{d \ln \omega} \\ G'(\omega) &= |G^*(\omega)| \cos(\delta(\omega)) \\ G''(\omega) &= |G^*(\omega)| \sin(\delta(\omega)) \end{aligned} \tag{8}$$

where $\delta(\omega)$ is the phase angle ($\delta = 0$ indicates solid-like behavior and $\delta = \pi/2$ liquid-like behavior), and G' and $G($ are the storage and loss moduli satisfying $G^* = G' + iG''$. In practice, we compute $\delta(\omega)$ using a procedure identical to that for computing $\alpha(\tau)$. Note that when the phase angle is near 0 or $\pi/2$, small errors in $\delta(\omega)$ [due either to statistical noise in the MSD or to systematic uncertainties in Eq. (7)] can get amplified tremendously in the smaller modulus, G'' or G', respectively. We have developed more accurate (but also more complicated) versions of Eqs. (7) and (8), which also rely on second logarithmic derivatives (Dasgupta et al., 2002). While these give more accurate results in general, for cell rheological data the difference is negligible.

Additionally, there are some artifacts that, no matter what algorithms are used, cannot be corrected for. Because the shear modulus formally depends on the MSD value at over a finite range of frequencies, shear moduli at the extrema of frequency range are subjected to additional "truncation" errors. The algorithm above implicitly assumes that the power-law behavior at the extrema of the data set extends indefinitely to higher or lower frequencies. Any curvature, even the slightest ripple, at the extremal lag times can cause a disproportionate change in the shear moduli. Consequently, unless there are physical motivations for believing this extrapolation is justified, confidence in this part of the data should be low. Strictly speaking, it may be best to compute the rheology from the entire lag time range and then ignore a decade (one order of magnitude) of frequency on each end of the curve.

MSDs themselves are often subject to artifacts at their frequency extrema as well, as described in Section V.E. Any systematic deviations of the MSD due to these effects will be further amplified in the shear moduli. Therefore, in some situations it might be wise to truncate the lag time range of the MSD prior to computing the rheology. In general, as the lag time range of the MSD being used is restricted, the ends of the curve will tend to "wiggle." This is a sign that the shear moduli being produced are subject to strong truncation errors.

One exercise we have found invaluable when utilizing this algorithm is to generate simulated MSD curves that resemble the actual data and calculate rheological properties from these curves. Changing the dynamic range of such simulated data allows the identification of truncation effects. Furthermore, adding Gaussian distributed noise to this data is helpful for determining the artifacts associated with it. A final warning is that a "reasonable" appearance of calculated rheology (e.g., resembling something from a rheology text book) does not mean the inversion is physical. The emergence of a noise floor on data from a particle diffusing in a liquid will lead to results that look almost exactly like the rheology of a Maxwell fluid, which is a common example in these texts.

In IDL: *micrheo.pro* is the program in our IDL suite that does these calculations. It has a built in smoothing to help reduce statistical noise. In addition, it uses second-order formula for smaller systematic errors (Dasgupta *et al.*, 2002) and provides warnings when the computed shear moduli are numerically unreliable. Its input is either an MSD or 2P-MSD.

V. Error Sources in Multiple-Particle Tracking

In this section, we describe several common sources of error in particle-tracking instruments. While some of these error sources can be mitigated by the use of high-quality equipment, most errors are due to irreducible physical limitations on imaging detector performance and illumination brightness. In practice, a good understanding of the origin of different errors, followed by careful adjustment of the imaging system to optimal settings, can lead to significant performance improvements. We will describe three classes of error: random error, systematic errors, and dynamic error.

A. Random Error (Camera Noise)

The more accurately a given particle can be located, the higher the quality of measurements of cellular rheology. In fact with optimal particles, illumination, imaging, and software, it is possible to measure the position of a 400-nm particle to within ~5 nm. While it may seem remarkable to be able to determine the particle's location this well with optical methods, the limiting precision of particle localization is quite different from optical resolution—the limit below which structural details in complicated specimens cannot be discerned (typically about a quarter of the wavelength of light used). Instead, the situation is analogous to a familiar problem in curve fitting, finding the position of a local maximum or peak in a curve. If we have a reasonable number of evaluations of a peaked function and the values are very precise, a least squares fitter can locate the peak center to an arbitrarily high precision (relative to the width of the peak). In this analogy, the peaked function is the light intensity distribution for a tracer, with a width set by the tracer size or optical resolution.

While some noise in a camera depends on the details of its construction and electronics, there are ultimate physical limits on the performance of all cameras due to the discrete nature of light itself. Imagine that we had a "perfect" camera that recorded the precise two-dimensional coordinates of all incoming photons arriving from a microscope. An "image" of a single small tracer might resemble a round cloud of points in a two-dimensional scatter plot (Fig. 9). From elementary statistics, we know that the standard error σ_x, σ_y when computing the mean x and y center positions of the cloud (corresponding to the particle location) is:

$$\sigma_x = \sigma_y = \frac{a}{\sqrt{N}} \tag{9}$$

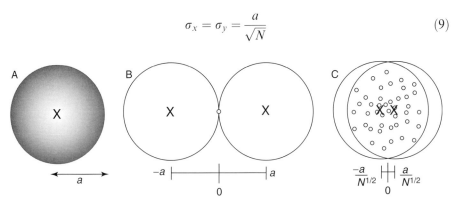

Fig. 9 Improvement in positioning precision by photon statistics. A particle forms an idealized, circular image of radius a whose center is marked by an × (A). If a single photon is detected, the position of the particle (set to be at zero) can be determined to ±a (B). If more photons are detected a cloud is detected (C). This limits the potential position (reduces measurement error) of the particle.

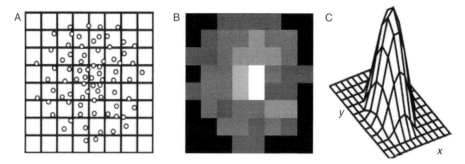

Fig. 10 Determination of particle centroid from a pixilated image. A hypothetical distribution of photons from a single particle on a grid (A) leads to the formation of a pixelated grayscale image (B) with a peaked brightness distribution (C). Particle-tracking algorithms find the center (peak) of this distribution very accurately.

where a is the apparent radius (technically the standard deviation) of the cloud, and N is the number of points/photons (assuming a Gaussian distribution).

In practice, well-made cameras approach this physical limit of performance when they are operating at illumination levels near the maximum allowed by detector saturation. Of course, rather than averaging the positions of individual photons, we determine the tracer's position by calculating the centroid of an intensity distribution, analogous to computing the "center of mass" for a continuous mass distribution (Fig. 10). Provided that we spread the tracer image over a sufficient number of pixels to reasonably represent the tracer's brightness distribution, the error in the centroid positions is precisely the same as that in the previous example.

In summary, whenever imaging a sample, there will be statistical fluctuations in the number of photons detected by the camera even under ideal conditions. These fluctuations are irreducible and lead to random errors in the reported brightness of individual pixels. In general, using a higher illumination will reduce this error, but that approach is ultimately limited by detector saturation.

B. Systematic Errors

In general, dividing the image of a small tracer into pixels does not necessarily introduce significant errors. However, improper image pixelation and masking can introduce systematic (as opposed to random) errors into the computed particle positions. One form of systematic error is "pixel biasing," the tendency of the algorithm to "round" the centroid position to the nearest pixel as mentioned in Section III.D. At its worst, a particle moving at a uniform speed across the field of view would appear to "hop" from one pixel to another (Fig. 11), introducing ± 0.5 pixels of position error.

This problem can largely be avoided by use of the proper magnification and image-processing settings. One source of pixel biasing is too low a magnification—ensure that single tracer images appear more than 3–4 pixels in diameter. This error

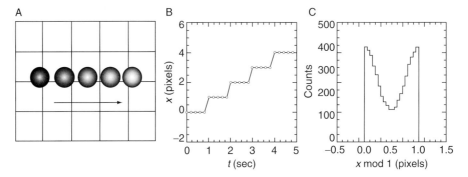

Fig. 11 An example of systematic error. The tracer, which forms an image smaller than the size of one pixel, moves slowly in the x-direction (represented as changing from dark to light) (A). The tracer's inferred position moves in 1-pixel increments (B), which leads to a minimum at half pixel in histograms of the fractional part of the x-coordinate, instead of being flat (C).

also commonly occurs during calculation of the centroid. When computing the centroid of a single tracer image, most algorithms mask off (force into zero brightness) pixels outside some radius to avoid contributions from other nearby particles. Error is introduced if the background of the image is not zero at the edge of the mask. Physically, common image defects, such as diffraction rings and out of focus background particles, or user errors, such as specifying too small of a mask size, can cause this problem. For images relatively free of obvious defects, it is usually possible to reduce this error to less than 0.1 pixels by using a suitably large mask size.

Another, rather subtle source of systematic error is due to the fact that the individual detectors corresponding to each camera pixel are not equally sensitive to light, with the typical variations between pixels being a few percent. This typically introduces a systematic error between the physical and measured position that is a few percent of a pixel, and which varies randomly with location in the field of view. This error can be largely eliminated by calibrating each pixel in the camera using a uniformly illuminated specimen (e.g., an empty bright field), and then numerically correcting all pixel intensity values by division against the calibration image prior to centroid finding. This algorithm, termed "flat-fielding," is routinely used by astronomers, but is seldom needed for particle-tracking applications.

For microrheology using large ensembles of particles, the contribution of all these systematic errors can often be ignored. For example, if a tracer moves a half a pixel, it may appear to move 0.4 or 0.6 pixels due to a 0.1 pixel systematic error. When many such measurements are squared and averaged together to compute an MSD, the deviations will largely cancel. However, this may not be the case in other particle-tracking situations with few particles, poor resolution, or short trajectory lengths in comparison to the lag times of interest. If one is computing a histogram of displacements, such as in Fig. 12, these systematic errors will lead to an obvious

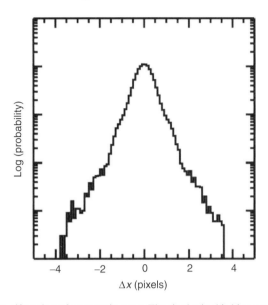

Fig. 12 Another manifestation of systematic error. The ripples in this histogram of tracer displacement indicate that tracers seem to move in increments of a pixel. This type of motion is nonphysical and is most likely an artifact from choosing poor tracking parameters.

"ripple" in the measurements, with displacements of integer multiples of one pixel length being more likely than half-integer multiples.

C. Dynamic Error

A subtle form of error that affects measurements of tracers' random motion is due to systematic underestimation of tracer motion when finite exposure times are used (Savin and Doyle, 2005). This error is absent when the duration of the exposure is infinitesimal compared to the time interval between successive movie frames. If the particle moves significantly during the camera exposure, the measured centroid reports the time-averaged position during the exposure (Fig. 13). Because random walks tend to loop back on themselves, this time-averaging effect artificially reduces the amplitude of the MSD. This selective reduction of short lag time MSD value will make diffusive Brownian motion look superdiffusive (the MSD will increase more rapidly than linearly with time) at short lag times.

The requirements to reduce dynamic error and optimize the MSD precision are at odds with each other. High precision demands as much light as possible be used to form the image, which is most easily achieved by increasing the exposure integration time, while dynamic error requires minimizing the exposure time. Obviously, the solution is to use the longest exposure possible that will not introduce dynamic error. We know of no rigorous mathematical criterion in determining the optimal exposure time for an arbitrary particle-tracking situation.

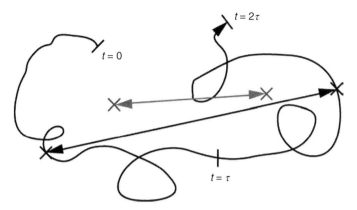

Fig. 13 An example of dynamic error in a random walk trajectory. Using a short shutter time, the position at a given instant is recorded at, for example, $t = 0.5$ and $1.5 \, \tau$ (black crosses). When a long shutter time, τ, is used, the time-averaged location of a tracer is found, for example t intervals from 0 to 1 τ and from 1 to 2 τ (gray crosses). The distance between time-averaged locations is systematically less than that for instantaneous locations.

In general, we find that if the amplitude of the mean-squared motion *during* the exposure interval is less than a quarter of the mean-squared motion *between* exposures, then the underestimation of MSD caused by dynamic error will be less than 10%. For tracers undergoing simple diffusive Brownian motion, this corresponds to an exposure time no longer than one quarter the time interval between frames, for example 1/200 sec for acquisition speeds of 50 frames per second. For the general case, the criterion is more difficult to apply, as it requires knowledge of the MSD at lag times shorter than those being measured. In that case, one is reduced to guessing the form of the short time MSD by extrapolating the available data or relying on physical models or literature results. Note that in the case where the MSD is nearly time independent (as in an elastic solid), the exposure time needs to be much smaller than a quarter of the frame interval to eliminate dynamic error.

D. Sample Drift, Computational Detrending, and Its Limitations

Drift in the sample stage contributes another source of uncertainty. Thermal expansion can cause the stage or sample chambers to slowly translate across the camera field of view during long experiments, often by several pixels. Computational correction of such global error is referred to as "detrending." One may correct each particle individually, by subtracting a portion of its net translocation, calculated by linear interpolation, from each point of the trajectory. However, this often introduces artifacts into the MSD at long times because it subtracts some of the tracer's actual motion. Alternatively, if a large number of tracers are available, one can compute the average motion of the tracers' center of mass and subtract it from each particle position. In general, this procedure reduces the amplitude of the

MSD by a factor $1/N$ (in addition to that caused by detrending), so is only recommended if $N \gg 10$.

The utility of all such "detrending" algorithms is significantly reduced by the systematic position errors described in the previous section. The nonlinear mapping of actual to measured position will cause even motionless tracers to move slightly relative to one another, over the timescale of detrending. In a similar vein, when studying very slowly moving tracers, it is often tempting to time average multiple measurements (perhaps with detrending) to yield spatial resolutions that are hypothetically much smaller than a nanometer. Systematic errors tend to limit the utility of such approaches because they are not reduced by averaging.

Finally, all of the measurements described here are in the two dimensions x and y. While the particles also fluctuate in z, which might be detected by fluctuations in particle intensity or diffraction pattern, the x and y positions of any spherical particle are still well determined as above. The 3-D separation of the particles will in general be slightly higher than the in-plane distance.

E. Effects of Measurement Errors on the MSD

Particle-tracking errors lead to several perturbations in measured MSDs (Martin *et al.*, 2002; Savin and Doyle, 2005). The most common error is due to the contribution from random position measurement error, as outlined in Section V.A, which contributes additively to the value of the actual, physical MSD:

$$\langle \Delta x^2 \rangle_{measured} = \langle \Delta x^2 \rangle_{physical} + 2\sigma_x^2 \qquad (10)$$

where σ_x is the standard deviation of the random position measurement error. At short lag times, where the MSD is small, the measured MSD will start to flatten and appear subdiffusive if the physical motion approaches the precision of the MSD. Conversely, if dynamic errors are significant, the MSD can appear super-diffusive at very short times.

Another common artifact is that MSDs will tend to flatten at long times. This is due to particle loss from the imaging plane. While we image in two dimensions, particles diffuse in three dimensions and at long times will diffuse out of the imaging plane. Since large particles move less than small particles, larger ones tend to stay in the focal plane longer. This can lead to downturns in the MSD, when the typical motion exceeds the square of the depth of focus. Our MSD calculation software tabulates the number of independent measurements, N_{eff}, used as a function of lag time. As discussed in Section IV.C, this quantity should decay as $1/\tau$; deviations from this indicate significant particle loss.

Reliably estimating the precision of measured MSDs can be quite difficult. A common method is for experimenters to estimate their total tracking precision by measuring the apparent MSD of immobilized tracers in "control" samples. This has obvious diagnostic value, and allows the effects of different microscope adjustments and ambient sample vibration to be rapidly determined. Still, the resulting error estimates must be treated with caution when applied to intracellular particles.

Obviously, ensembles of immobilized tracers will not display any dynamic error, and systematic error cannot be easily estimated from such samples. Moreover, the observed random position error will correspond to the intracellular case only if the optical parameters of that problem, such as tracer size and index of refraction, are faithfully reproduced, which is often difficult in practice.

VI. Instrument Requirements for High-Performance Tracking

Even though our software tools can often yield acceptable particle tracking under less than ideal imaging conditions, acquiring high-quality images facilitates image analysis and yields more reliable rheological data. The following sections discuss important practices of instrumentation that minimize errors of measurements.

A. Isolation from Vibration, Acoustic Noise, and Thermal Drift

Research laboratories often contain many sources of vibration (blowers for air handlers, fume hoods, cell culture hoods, as well as people walking and talking). All of these shake the microscope and the sample stage, often by a surprisingly large amount. In almost all research locations, a pneumatic isolation table is absolutely critical for dampening out these vibration sources. Furthermore, connections from the outside world (power cords, gas lines) can conduct vibrations to the floating tabletop; their number should be minimized. For example, rather than having several AC power cords hanging between the isolated table and the outside, these can be ganged into a power strip, and its single cord used to connect off the table. Moreover, many electronic devices contain small cooling fans and electrical transformers that can be significant sources of vibration. These should not be placed on the isolated table surface. For the same reason, we use an external lamp power supply in lieu of the supply built into the microscopic base (which has the additional advantage or reducing thermal expansion drift in the microscope body). Additionally, many inverted microscopes have a screw to lock the illumination stalk. We find tightening this down is helpful with reducing sample vibration.

Microscopes and sample chambers will expand and contract as they change temperature. If the scope is to be heated, let it reach the set point temperature for 10–20 min before starting experiments. Air currents from ventilation systems can cause substantial drift and periodic motion in a microscope over timescales as short as tens of seconds. Surrounding the microscopic with a curtain to prevent drafts from blowing directly on the microscope is very helpful and will largely eliminate these effects. Finally, stage drift can also be induced by the body heat and exhalations of an operator sitting too close to the microscope.

B. Microscopy—Generation of High-Contrast Tracer Images

Maximizing image quality and MSD precision requires juggling several imaging parameters. Specifically, we want to adjust the microscope to simultaneously maximize the contrast between the tracer image and its surrounding background and magnification, while using the maximal illumination intensity and minimal exposure time to avoid dynamic error.

In general, all other things being equal, the precision of MSD improves proportionally with the magnification and as the square root of the tracer brightness. Because it is difficult to predict the performance of a given microscopy condition beforehand, we take an empirical approach. Most image analysis software allows the user to acquire a single frame and to read out the brightness value of individual pixels selected with a cursor. For 8-bit cameras, this value is an integer between 0 and 255. We routinely use this function to determine the brightness difference between the center of a typical tracer and the background. Ideally, the brightness difference should be at least 100 of 255, while for some simple specimens and high-refractive index tracers it can be adjusted as high as 150–200. At the same time, it can be validated that image saturation, that is pixel values of 0 or 255, is not present.

Most image acquisition hardware and software allows the user to change image brightness and contrast, and to make an image collected at low illumination look like one collected at high illumination. However, it should be remembered that changing such "gain" will not affect the final tracking precision because both the signal and camera noise are amplified together. Even though the images may look the same to the eye, the image from the lower illumination setting will have a higher level of pixel noise and thus poorer resolution. For this reason, "gain" should be set at the minimum setting and any automatic controls in the camera (e.g., automatic gain control) should be turned off.

1. Contrast Generation Methods

Care should be taken when choosing and using optical contrast generation techniques to visualize low-refractive index tracers (such as the endogenous lipid granules we use here). Remember that our tracking software is not based on contrast as a ratio, but on the absolute difference in brightness. For example, extinction-mode DIC (also called dark-field Nomarski) creates a very dark background surrounding bright tracers; in other words, it produces a high brightness ratio. However, extinction mode also blocks much of the illumination, resulting in an image where the tracers have a low absolute brightness. As a result, extinction-mode DIC generally yields lower quality tracking than two other contrast enhancement techniques: shadow-cast DIC (bright-field Nomarski) or phase contrast, which generate a larger absolute difference in brightness for a given specimen illumination. Epifluorescence imaging of fluorescently tagged microspheres can also create images with high contrast ratio, although the overall brightness of the image is significantly lower than that provided by bright-field methods. Therefore, fluorescence imaging is not well suited for high-performance particle tracking.

With any contrast generation method, the microscope's performance depends critically on operator skill and proper alignment. For example, DIC will give images that "look right" but have degraded contrast if the wrong DIC prism is used, the polarizers are misaligned or there are small bubbles in the immersion oil.

Similarly, phase contrast will have degraded performance if the phase ring is not properly centered. Given all that can go wrong, quickly checking the actual brightness difference achieved (and verifying that it is comparable with earlier experiments) is usually a worthwhile practice prior to data collection.

2. Numerical Aperture of Condenser

For conventional bright-field microscopy, the numerical aperture (NA) of the illumination (i.e., the aperture diaphragm setting in the condenser) is critical to maximizing tracking precision. For example, with low-NA illumination, the tracer images have a very high contrast, large depth of focus, and prominent diffraction rings. Because much of the light is stopped by the aperture diaphragm, the absolute brightness of the image is low. Conversely, at high NA, the absolute image brightness is maximized, but the contrast between the tracers and the background can be quite small. The optimum setting is clearly in between the two extremes: where tracer contrast is good, and total image brightness is not too attenuated. For our microscope operating in conventional bright-field or DIC microscopy with an oil immersion condenser, a condenser aperture setting of near NA $= 0.4$ yields the best brightness difference and particle-tracking performance.

3. Magnification

The absolute minimum magnification is set by the size of the particles that are being tracked. Particles images must be at least 3–4 pixels in diameter. However, higher magnifications could be used to increase spatial precision, so long as total illumination can also be increased to keep the detector near saturation levels. With the higher magnification, the intensity of the particle can be spread over more pixels, allowing more accurate centroiding. However, it is important to note that, without increasing the illumination intensity, the signal at each pixel decreases in proportion to the square of magnification. Therefore, simply increasing the magnification without simultaneously increasing the illumination will not increase performance.

C. Using Low-Noise, Non-interlaced Camera

The amount of random pixel noise present in a camera can be readily measured. One acquires a series of "blank" images where the illumination is uniform across the field of view and near the saturation value. Because of camera noise, the brightness value of a given pixel will vary from image to image. One can then take the differences of intensities at a given pixel between pairs of images, and compute the standard deviation of the resulting differences. Division of this standard deviation by $\sqrt{2}$ yields the root-mean-squared (RMS) pixel noise. For a quality camera, the RMS noise amplitude should be 0.5–1% of the pixel brightness, which is calculated by subtracting out the average reading in dark images. The pixel detectors on many cameras hold about 40,000 photoelectrons at saturation (called the "well capacity"); the corresponding statistical noise is $1/\sqrt{40000} = 0.5\%$. Some inexpensive consumer cameras can have noise figures significantly higher than this, and they are not recommended for particle tracking. Specialized camera with larger well capacities might have significantly better performance, but will likely require 12-bit digitization to take advantage of the lower noise. 12-bit digitization is typically used to improve the dynamic range of intensity readout, rather than to reduce camera readout noise, and thus has a limited impact on particle-tracking applications. It should also be noted that lossy compression algorithms (such as JPEG or MPEG) tend to contribute low-amplitude defects to image data that can degrade particle-tracking performance. If high precision is needed, uncompressed images or lossless compression such as LZW should be used exclusively.

Interlacing is a standard leftover from the early days of video technology. To eliminate image flicker (which requires 50–60 images per second) while reducing transmission bandwidth, the display redraws half of the image at a time, alternating between the even and odd rows. This eliminates visual flicker while only transmitting 30 frames per second (for NTSC video in North America and Japan) or 25 frames per second (for PAL or SECAM video elsewhere). While the odd and even rows can be separately analyzed to perform particle tracking at equivalent rates of 50 or 60 frames per second, interlacing degrades tracking performance because only half as many pixels are used when computing the centroid. When spatial precision is critical, a noninterlaced (also called a progressive scan) camera should be employed.

D. Using High-Intensity, Filtered Illuminator

1. Intensity and Controlling Dynamic Error

As described earlier, the precision of particle-tracking techniques is ultimately limited by the intensity of the illumination. In fact the ideal illumination would be a very short, very bright pulse of light, like a camera flash, which simultaneously reduces both random and dynamic error. A practical approach to illuminating samples for particle tracking is to use the microscope illuminator at its maximum setting and then vary the exposure time in the camera or the pulse length of light to obtain the required SNR for image processing. In general longer exposure time will decrease random position error, while increasing dynamic error. While random error has a fixed amplitude, dynamic error is proportional to the tracer's MSD. Thus, if the MSD at the shortest lag time is much larger than the precision of the determined MSD, one can safely reduce the exposure time to minimize the dynamic error. Conversely, if the MSD at the shortest lag time is comparable or smaller than its precision, then a longer exposure is probably warranted.

Even at the maximum setting on a 100-W halogen illuminator, exposure times of several milliseconds are typically required to produce optimal, near-saturation bright-field images at a magnification of 100 nm/pixel. Dynamic error is thus likely even at frame rates in excess of 50 frames per second. When we require faster acquisition rates to perform high-frequency microrheology, we employ a custom-built, pulsed near-IR laser illuminator that can fully expose the detector in tens of microseconds, allowing the use of a high-speed camera operating at several thousand frames per second without dynamic error. Pulsing the laser also minimizes the possibility of heating and photodamage effects.

2. Filtering

One downside to the use of high dose of photons is the tendency to heat or photodamage living cells. In general, halogen lamps emit copious long-wave UV light, and their intensity actually peaks well into the near-IR. Many living cells under the continuous illumination of unfiltered, maximum setting (100 W) halogen illumination die in less than an hour, while cells just outside the illuminated field appear unaffected. To minimize radiation damage, we place two optical filters in the illumination light path (purchased from Edmund Optics, Barrington, New Jersey): one (Cat. No. F54–049) that blocks UV wavelengths shorter than 400 nm and another (Cat. No. F46–386) that blocks wavelengths from 700 to ~1800 nm. With both filters in place, cells in and out of the illumination field show indistinguishable long-term viability and proliferation on a heated 37 °C microscope stage.

E. Using Synthetic Tracers to Increase Visibility

One way to enhance the contrast of tracers is to use synthetic microparticles of a high-refractive index n, for example polystyrene ($n = 1.6$). Many such probes have been used in cellular rheological experiments, being introduced by phagocytosis (Caspi et al., 2000; Hoffman et al., 2006) or microinjection (Tseng et al., 2002). Moreover, such particles could potentially be designed to target a specific structure (actin filaments or microtubules) or a specific protein (different integrins or other surface receptors) (Puig-de-Morales et al., 2004). In in vitro studies, the surface chemistry of the bead has large effects on the resulting rheological measurements (Valentine et al., 2004). The effect of tracer surface chemistry on the inferred "one-point" microrheology of cells is still an issue without a clear consensus. On the other hand, tracer chemistry is expected to have a much more limited effect on TPM measurements.

VII. Example: Cultured Epithelial Cells

A. Particle-Tracking Results for TC7 Epithelial Cells

The results of particle tracking in TC7 cells are shown in Fig. 14. The top panel shows the MSD and 2P-MSD for the endogenous tracers found in living cells, computed using 18,000 images comparable to those shown in Section III. We find

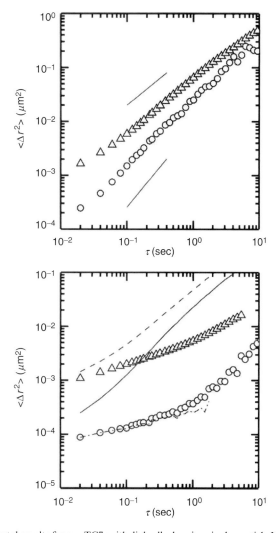

Fig. 14 Experimental results from a TC7 epithelial cell, showing single-particle MSD (top, triangles) and 2P-MSD (top, circles). Lines are eye guides with a slope 1 (top) and 1.5 (bottom), respectively. The curves change substantially for ATP-depleted TC7 cells (bottom), averaged over $N = 7$ cells, each with 18,000 images. MSDs are systematically larger than 2P-MSDs. Curves from untreated cells are reproduced in the bottom panel for comparison.

that the tracers' conventional MSD looks almost purely diffusive. If we assumed the motion to be Brownian (rather than due to ATP-dependent processes), we would conclude the cell interior is a simple fluid with a viscosity roughly 10 times that of water. This explains why the random motion of organelles in cells so strongly resembles the Brownian motion of tracers in water—both are diffusive motion with comparable amplitudes. The results of the two-point calculation give a 2P-MSD with a superdiffusive lag time dependence, 2P-MSD $\sim \tau^{1.5}$, which is faster than a linear dependence on lag time expected for Brownian motion and clearly indicative of active intracellular processes (Lau *et al.*, 2003). The functional form of D_{rr} consistently displays a $1/R$ dependence, providing a positive control that cells can be treated as three-dimensional viscoelastic continua. However, the fact that single-particle MSD and 2P-MSD differ in amplitude implies that the no-slip, homogeneous boundary condition assumptions needed for the conventional GSER are not valid (Section IV.B). The difference in functional form either indicates the material in the tracers' local environment has different rheological characteristics from the bulk material or differential sensitivity to the (yet unknown) non-Brownian driving forces. All in all, it is difficult to definitively conclude much else from such live cell data alone.

The bottom panel of Fig. 14 shows data for cells that have been depleted of intracellular ATP using a combination of sodium azide and deoxyglucose. ATP depletion inhibits motor-dependent processes but may also cause structural changes, for example, by inhibiting a wide range of kinase-dependent processes. The amplitudes of both the MSD and the 2P-MSD have dropped dramatically, as expected if the active processes have been slowed or eliminated. Both curves now display a simple power-law frequency dependence on lag time: MSD $\sim$ 2P-MSD $\sim \tau^{0.3}$. Both curves still turn upward at lag times greater than 1 sec. Measurements with varied dosages of ATP depletion agents suggest that this upturn is due to residual ATP activity, while the MSDs at shorter times are ATP independent and Brownian in origin (Hoffman *et al.*, 2006). The fact that two types of MSD have the same lag time dependence suggests that they are now probing the same viscoelastic structure. The observation that they differ in amplitude by an order of magnitude suggests that individual tracers do not have the simple, homogeneous no-slip boundary conditions commonly assumed. One simple explanation would be if the tracers are strongly adhered or attached to a viscoelastic network that is porous on their length scale (Van Citters *et al.*, 2006), which seems plausible given our knowledge of cytoskeletal ultrastructure. Finally, note that the 2P-MSD has noticeably higher (but still acceptable) statistical noise versus the conventional MSD, as expected from Section VI.C.

B. TPM of TC7 Epithelial Cells

The storage and loss modulus computed from our example data, specifically the 2P-MSD of ATP-depleted cells, is shown in Fig. 15. The response has a simple power-law form, $G'(\omega) \sim \omega^{0.3}$, over the range of accessible frequencies, similar to our published results (Hoffman *et al.*, 2006). Notice that the statistical noise (the point to point amplitude variations) in the computed rheology appears much lower than in the corresponding 2P-MSD. This is because our algorithm averages the MSD over the time domain to draw a "smooth curve" through the MSD prior to estimating the logarithmic slope as described in Section IV.D. This smoothing does

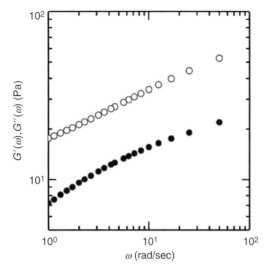

Fig. 15 Inferred two-point microrheology of ATP-depleted TC7 cells computed using the data in Fig. 13. The response is dominated by the storage modulus, G' (open circles). The curvature of the smaller loss modulus (closed circles) is artifactual.

not eliminate the noise, but replaces it with a slow "ripple" in the computed shear moduli, which is expected to follow a pure power-law in the frequency window, corresponding to two parallel lines for G' and G'' (compare with Fig. 1). The artifactual curvature shown in Fig. 15 is typical of results computed from 2P-MSD with the noise shown in Fig. 14. The curvature is always much more pronounced in the smaller of the two shear moduli, G'' in this example. Nevertheless, this rheological measurement would provide reliable power-law fits to infer both the amplitude of the shear modulus (i.e., stiffness) and power-law exponent β.

As we noted in Section II.C, TPM can provide reliable stiffness information even in heterogeneous media, or when the connections between the tracers and the network do not satisfy the requirements of the GSER, Eq. (2). Averaging the results of TPM measurements on many TC7 cells yields a mean stiffness value of 40 Pa at $\omega = 10$ rad/sec ($\sim$1.6 Hz). Cell to cell variations are considerable with a log-standard deviation of 1.6, meaning that $\sim$70% of the values fall within 1/1.6 and 1.6 times the mean value, or in the range 25–60 Pa. It should be remembered that TPM reports a whole-cell average; the magnitude of the stiffness variations across a cell has not yet been determined using TPM. Moreover, probes for TPM may not uniformly sample all regions of the cell. This mean stiffness is more than an order of magnitude lower than the value ($\sim$1 kPa) inferred by several other cell rheological techniques, which we presume are probing the cell cortex. From a cell-trafficking point of view, it is much easier to imagine organelles being trafficked through a soft 40-Pa intracellular space than a stiff 1-kPa network.

C. Computing Stress Fluctuation Spectra

In living cells where the rheology is known, it is also possible to use TPM to quantify the amplitude and frequency dependence of the cells' non-Brownian stress fluctuations (Lau *et al.*, 2003), which are presumably due to forces generated by molecular motors or filament treadmilling. The basic idea is similar to that of the linear response of a spring. If we apply a force F to a spring having a spring constant k, the deflection x satisfies: $F = -kx$. If we know any two of the three variables, we can compute the third. Similarly, if we know the stiffness of a material (its shear modulus) and how much it is deforming (by measuring the 2P-MSD of embedded tracers), we can infer the magnitude of the force or stress driving the deformation (all at a given frequency, ω). In a viscoelastic material, the formula is:

$$\mathfrak{J}(\omega) \approx 3\pi a |G^*(\omega)|^2 \langle \Delta x^2(\omega) \rangle_2 \qquad (11)$$

where $\langle \Delta x^2(\omega) \rangle_2$ is the unilateral Fourier transform of the 2P-MSD, as used in the GSER, Eq. (2), and $\mathfrak{J}(\omega)$ is the power spectrum of the fluctuating stresses causing the material to deform. The power spectrum of a fluctuating function is the squared amplitude of its Fourier transform, that is, the relative contributions of stresses at low versus high frequencies. In a simple case where all three functions are power-law functions of time/frequency, for example $\langle \Delta x^2(\tau) \rangle_2 \sim \tau^\alpha$, $G^*(\omega) \sim \omega^\beta$, and $\mathfrak{J}(\omega) \sim \omega^{-\gamma}$, then this formula takes the form of a simple relation: $\alpha + 1 = 2\beta + \gamma$. In an earlier study (Lau *et al.*, 2003), we applied this method to murine J774A.1 macrophage-like and F9 carcinoma cells, and found that their stress fluctuation spectrum had a nearly power-law form, with an exponent $\gamma \approx 2$. If this equality were exact, this implies the relation $\alpha = 1 + 2\beta$. For the example data in TC7 cells, a value of $\beta_{\text{int}} \approx 0.26$ would imply $\alpha \approx 1.52$, consistent with the observed power-law slope of the 2P-MSD in living TC7 cells (Fig. 14(top)).

A power-law stress fluctuation spectrum of the form $\Delta(\omega) \sim \omega^{-2}$ has a simple physical explanation. Such a spectrum occurs if the fluctuating stresses are due to discrete, rapid "step" changes. For example, such steps could be the result of processive molecular motors gradually building stresses on a cytoskeletal element, which is then abruptly released when the motor detaches (Mizuno *et al.*, 2007). Alternatively, in a prestressed cytoskeleton, if filaments or cross-link proteins occasionally rupture or release, this would also manifest as a $\Delta(\omega) \sim \omega^{-2}$ spectrum. All that is required is that the rate of the step change is faster relative to our fastest observed timescale (20 msec in the example data), and that the repetition time is longer than the longest lag time observed (a few seconds in the example). In contrast, if we imagine the network receiving short-duration force impulses or "kicks," the expected spectrum would be $\Delta(\omega) \sim \omega^0$, which is clearly excluded by the data. While a cell's shear modulus does not contain any readily identifiable molecular timescales, it remains an open question whether molecular timescales can be found in the stress fluctuation spectrum. This question might be addressed by extending the frequency range of the measurements.

========== ## VIII. Conclusions and Future Directions

TPM measurements on cultured cells provide a uniquely interpretable, quantitative contribution to the field of cell mechanics. They indicate that cells contain two mechanically distinct intracellular regions. The deeper intracellular interior probed by TPM is found to have a power-law frequency dependence, whose origin remains unknown. This region is remarkably soft, with a modulus that is only a few tens of pascals. In living cells, this intracellular structure exhibits ATP-dependent stress fluctuations with a "step-like" rather than a "kick-like" character. This appears compatible with intracellular trafficking, slow cytoskeletal remodeling, or both. While these results provide novel insight into the molecular-scale mechanics of cells, they lack molecular specificity. Future studies are needed to "dissect" the molecular determinants of both the rheology and non-Brownian fluctuations, for example, through the use of specific cytoskeletal disruptors, motor inhibitors, and genetic manipulations, as well as externally applied stresses (Lee and Discher, 2001). Studies from our laboratory (Van Citters et al., 2006) show, remarkably, that neither the intracellular rheology nor the stress fluctuations measured by TPM depend on either F-actin or myosin. Such results remind us that the field of cell mechanics is just beginning to understand the generation, propagation, and relaxation of intracellular stress.

References

Alcaraz, J., Buscemi, L., Grabulosa, M., Trepat, X., Fabry, B., Farre, R., and Navajas, D. (2003). Microrheology of human lung epithelial cells measured by atomic force microscopy. *Biophys. J.* **84,** 2071–2079.

Caspi, A., Granek, R., and Elbaum, M. (2000). Enhanced diffusion in active intracellular transport. *Phys. Rev. Lett.* **85,** 5655–5658.

Cheezum, M. K., Walker, W. F., and Guilford, W. H. (2001). Quantitative comparison of algorithms for tracking single fluorescent particles. *Biophys. J.* **81,** 2378–2388.

Crocker, J. C., and Grier, D. G. (1996). Methods of digital video microscopy for colloidal studies. *J. Colloid Interface Sci.* **179,** 298–310.

Crocker, J. C., Valentine, M. T., Weeks, E. R., Gisler, T., Kaplan, P. D., Yodh, A. G., and Weitz, D. A. (2000). Two-point microrheology of inhomogeneous soft materials. *Phys. Rev. Lett.* **85,** 888–891.

Dasgupta, B. R., Tee, S. Y., Crocker, J. C., Frisken, B. J., and Weitz, D. A. (2002). Microrheology of polyethylene oxide using diffusing wave spectroscopy and single scattering. *Phys. Rev. E* **65,** 051505.

Deng, L. H., Trepat, X., Butler, J. P., Millet, E., Morgan, K. G., Weitz, D. A., and Fredberg, J. J. (2006). Fast and slow dynamics of the cytoskeleton. *Nat. Mater.* **5,** 636–640.

Desprat, N., Richert, A., Simeon, J., and Asnacios, A. (2005). Creep function of a single living cell. *Biophys. J.* **88,** 2224–2233.

Fabry, B., Maksym, G., Butler, J., Glogauer, M., Navajas, D., and Fredberg, J. (2001). Scaling the microrheology of living cells. *Phys. Rev. Lett.* **87**(14), 148102.

Gardel, M. L., Nakamura, F., Hartwig, J., Crocker, J. C., Stossel, T. P., and Weitz, D. A. (2006). Stress-dependent elasticity of composite actin networks as a model for cell behavior. *Phys. Rev. Lett.* **96,** 088102.

Hoffman, B., Massiera, G., Van Citters, K., and Crocker, J. (2006). The consensus mechanics of cultured mammalian cells. *Proc. Natl. Acad. Sci. USA* **103,** 10259–10264.

Janmey, P. A., Hvidt, S., Lamb, J., and Stossel, T. P. (1990). Resemblance of actin-binding protein actin gels to covalently cross-linked networks. *Nature* **345**, 89–92.

Koenderink, G. H., Atakhorrami, M., MacKintosh, F. C., and Schmidt, C. F. (2006). High-frequency stress relaxation in semiflexible polymer solutions and networks. *Phys. Rev. Lett.* **96**, 138307-(1–4).

Kroy, K. (2006). Elasticity, dynamics and relaxation in biopolymer networks. *Curr. Opin. Colloid Interface Sci.* **11**, 56–64.

Lau, A., Hoffman, B., Davies, A., Crocker, J., and Lubensky, T. (2003). Microrheology, stress fluctuations, and active behavior of living cells. *Phys. Rev. Lett.* **91**, 198101.

Lee, J. C.-M., and Discher, D. E. (2001). Deformation-enhanced fluctuations in the red cell skeleton with theoretical relations to elasticity, connectivity, and spectrin unfolding. *Biophys. J.* **81**, 3178–3192.

Lenormand, G., Millet, E., Fabry, B., Butler, J. P., and Fredberg, J. J. (2004). Linearity and time-scale invariance of the creep function in living cells. *J. R. Soc. Interface* **1**, 91–97.

Levine, A. J., and Lubensky, T. C. (2001). Response function of a sphere in a viscoelastic two-fluid medium. *Phys. Rev. E.* **63**, 041510/1–041510/12.

Martin, D. S., Forstner, M. B., and Kas, J. A. (2002). Apparent subdiffusion inherent to single particle tracking. *Biophys. J.* **83**, 2109–2117.

Mason, T. G. (2000). Estimating the viscoelastic moduli of complex fluids using the generalized stokes-einstein equation. *Rheologica Acta* **39**, 371–378.

Mason, T. G., and Weitz, D. A. (1995). Optical measurements of frequency-dependent linear viscoelastic moduli of complex fluids. *Phys. Rev. Lett.* **74**, 1250–1253.

Mizuno, D., Tardin, C., Schmidt, C. F., and MacKintosh, F. C. (2007). Nonequilibrium mechanics of active cytoskeletal networks. *Science* **315**, 370–373.

Orr, A. W., Helmke, B. P., Blackman, B. R., and Schwartz, M. A. (2006). Mechanisms of mechanotransduction. *Dev. Cell.* **10**, 11–20.

Puig-de-Morales, M., Millet, E., Fabry, B., Navajas, D., Wang, N., Butler, J. P., and Fredberg, J. J. (2004). Cytoskeletal mechanics in adherent human airway smooth muscle cells: Probe specificity and scaling of protein-protein dynamics. *Am. J. Physiol. Cell Physiol.* **287**, C643–C654.

Savin, T., and Doyle, P. S. (2005). Static and dynamic errors in particle tracking microrheology. *Biophys. J.* **88**, 623–638.

Tseng, Y., Kole, T. P., and Wirtz, D. (2002). Micromechanical mapping of live cells by multiple-particle-tracking microrheology. *Biophys. J.* **83**, 3162–3176.

Valentine, M. T., Perlman, Z. E., Gardel, M. L., Shin, J. H., Matsudaira, P., Mitchison, T. J., and Weitz, D. A. (2004). Colloid surface chemistry critically affects multiple particle tracking measurements of biomaterials. *Biophys. J.* **86**, 4004–4014.

Van Citters, K. M., Hoffman, B. D., Massiera, G., and Crocker, J. C. (2006). The role of F-actin and myosin in epithelial cell rheology. *Biophys. J.* **91**, 3946–3956.

Vogel, V., and Sheetz, M. (2006). Local force and geometry sensing regulate cell functions. *Nat. Rev. Mol. Cell Biol.* **7**, 265–275.

Waigh, T. (2005). Microrheology of complex fluids. *Rep. Prog. Phys.* **68**, 685–742.

Weihs, D., Mason, T. G., and Teitell, M. A. (2006). Bio-microrheology: A frontier in microrheology. *Biophys. J.* **91**(11), 4296–4305.

Xu, J. Y., Wirtz, D., and Pollard, T. D. (1998). Dynamic cross-linking by alpha-actinin determines the mechanical properties of actin filament networks. *J. Biol. Chem.* **273**, 9570–9576.

Yamada, S., Wirtz, D., and Kuo, S. C. (2000). Mechanics of living cells measured by laser tracking microrheology. *Biophys. J.* **78**, 1736–1747.

CHAPTER 8

Imaging Stress Propagation in the Cytoplasm of a Living Cell

Ning Wang,* Shaohua Hu,† and James P. Butler†

*Department of Mechanical Science and Engineering
University of Illinois at Urbana-Champaign
Urbana, Illinois 61801

†Physiology Program
Harvard School of Public Health
Boston, Massachusetts 02115

Abstract

A fundamental issue in mechanotransduction is to determine pathways of stress propagation in the cytoplasm. We describe a recently developed synchronous detection approach that can be used to map nanoscale distortions of cytoskeletal elements and nuclear structures in living individual cells using green fluorescent protein technology and 3D magnetic twisting cytometry. This approach could be combined with single-cell biochemical and biological assays to help elucidate mechanisms of mechanotransduction.

I. Introduction

By what mechanisms do living cells sense and respond to mechanical stresses applied at the cell surface? Although established evidence has shown that mechanical forces are critical for many cell functions, including growth, proliferation, protein synthesis, and gene expression (Ingber, 2006), the specific mechanisms of mechanical force transmission and transduction remain largely unknown. One major gap in the field is the elusiveness of force transmission pathways inside the cytoplasm and their importance in mechanotransduction. For example, if one needs to know whether an intracellular structure or molecule (including intranuclear structures) responds directly to surface applied forces, one must know the mechanical pathways of force transmission to that specific structure or molecule. It is generally assumed that if a cytoplasmic molecule or a nuclear molecule can be *directly* influenced by a surface force or deformation, the influence could occur in at least three different, but not mutually exclusive, ways: (1) direct deformation or displacement of this molecule or structure, which leads to alteration of its activity via conformational change of the molecule (e.g., domain unfolding) or via changes in the proximity of this molecule to another molecule; (2) alteration of the local force balance, and thus the on- and off-rate of this molecule binding to another molecule; and (3) alteration of the transloca-tion speed, direction, or site of the molecule along the cytoskeleton, the DNA, or other structures. An essential step to better understand the mechanisms of mechano-transduction in the cytoplasm and nucleus is to map external force-induced displacement and stress distribution in the cytoplasm.

In this chapter, we will describe a simple but sensitive method of mapping estimates of strains and stresses inside the cytoplasm or nucleus of a living cell. The cell is labeled with a fluorescent tag and subjected to periodic mechanical stress using a number of approaches as described in other chapters of this volume. Displacements and deformation of the cell that occur synchronously with input loads are detected with fluorescence imaging (Hu *et al.*, 2003, 2004, 2005). This simple approach can be used to directly quantify external stress-induced nanoscale displacements and deformation in the cytoskeleton, even in the presence of dynamic and spontaneous movements of cytoskeletal structures or molecules. In principle, this approach can be used to quantify external force-induced displacements and to obtain deformation maps with high spatial (~5 nm) and temporal

($\sim$100 msec) resolution, for structures or organelles in the cytoplasm that are physically associated with the cytoskeleton, structures in the vicinity of cytoskeletal stress-bearing elements, or structures in the nucleus that are associated with the nuclear matrix (Penman, 1995).

II. Detecting External Stress-Induced Displacements in the Cytoplasm

A. Green Fluorescent Protein Transfection and Cell Culture

There are numerous studies on delivering or expressing fluorescently tagged cytoplasmic or nuclear molecules in living cells. These molecules may be introduced into the cytoplasm in a number of ways, such as lipofectamine, electroporation, or peptide-mediated membrane penetration, but the key is to choose a method that targets a specific cytoplasmic molecule or structure, that is nontoxic, and that causes minimal perturbation to the cell. The fluorescent dye should also be relatively resistant to photobleaching. Green fluorescent protein (GFP) conjugation is now a commonly used and reliable approach, although the number of cells successfully transfected with a desired expression plasmid is sometimes low when transfected with common reagents such as lipofectamine. In our experiments, to increase the efficiency of expression, adenovirus vectors have been used to transiently infect cells with GFP-caldesmon, GFP-tubulin, YFP (yellow fluorescent protein)-actin, or YFP-cytochrome c oxidases, which are proteins expressed exclusively at the inner membrane of the mitochondria. Cells of passages 3–8 are plated sparsely on type-1 collagen-coated six-well dishes in serum-free media overnight, and allowed to reach 70–80% confluency before adenovirus containing the DNA sequence of a fluorescently conjugated protein is added at 150 μl per well (viral concentration is $\sim$10^8 pfu/ml; Numaguchi et al., 2003) for a 2-day period of infection. By this approach, 50–80% of the cells can be infected. If one is interested in cell cycle-dependent stress propagation, serum or growth factors may also be added to G_0 or G_1 cells during and following the infection. One may also use tetracycline (Tet)-inducer (adenovirus-mediated Tet-off) system to achieve efficient, synchronous, and tunable expression of GFPs (Numaguchi et al., 2003).

After the incorporation/expression of a fluorescent protein, cells are replated at low density onto a culture dish coated with extracellular matrix proteins or with molecules such as polylysine that promote nonspecific cell attachment. For the purposes of studying how a local stress applied to one cell can be transmitted to another cell, the plating density can be increased so that there are cell–cell physical contacts. Micropatterning technologies can also be used to constrain a single cell or a group of cells to a defined shape and size. If one would like to measure interfacial tractions between the cell and the substrate simultaneously, or to study how extracellular matrix rigidity might influence force propagation, flexible substrates, such as collagen gels, polyacrylamide gels, or other transparent, nonfluorescent, biocompatible substrates (Chapter 2 by Kandow et al., this volume) can be used. Three-dimensional (3D) substrates can also be used in order to simulate more closely *in vivo* physiological conditions.

B. Application of Periodic Mechanical Stress

Once the fluorescently labeled cells are plated and appear viable, they are stressed mechanically by various means in a periodic fashion. Attention should be paid to the magnitude, timescale, and physiological relevance of the stress (including the frequency of oscillatory forces). In addition, the characteristics of the probe–cell surface interactions, as determined by the chemical and physical nature of the probe, are critical. The probe may be coated with ligands for integrin receptors or other adhesion molecules. Commonly used approaches that allow periodic deformation of a living adherent cell include atomic force microscopy (AFM) (Radmacher *et al.*, 1996; Shroff *et al.*, 1995; Chapter 15 by Radmacher, this volume), optical tweezers (Dai and Sheetz, 1995), magnetic twisting cytometry (MTC; Chapter 19 by Lele *et al.*, this volume) using uniform magnetic fields (Wang *et al.*, 1993) with optical detection (Fabry *et al.*, 2001), or magnetic pulling devices using magnetic field gradients (Bausch *et al.*, 1998; Chapter 19 by Lele *et al.* and Chapter 20 by Tanase *et al.*, this volume). One can also use microfluidic channels to deliver local fluid shear stresses to a small portion of the cell surface such as the apical surface (Helmke *et al.*, 2001). Alternatively, a micropipette can be used to aspirate a small portion of the cell surface and motor-driven microprobes may be used to directly deform the surface of the cell (Petersen *et al.*, 1982; Chapter 16 by Evans and Kinoshita, this volume). A local force probe, such as a micrometer-sized or submicrometer-sized magnetic bead, can also be introduced into the cytoplasm by endocytosis to apply a load intracellularly (Crick and Hughs, 1950; Valberg and Butler, 1987). In addition, an antibody-coated (sub)micrometer-sized magnetic bead can be injected into the cytoplasm to probe a specific molecule/structure.

To define the force propagation pathway and to allow proper interpretation of the measurements, one needs to understand local interactions between the probe and its environment (e.g., the cytoskeleton). In addition, since many cells are elongated, it is important to specify the direction of the force or torque application relative to the direction of the cell axis and also mean orientation of local cytoskeletal filaments in order to accurately characterize and control for the influence of the applied stresses on cytoplasmic stress propagation (see 3D-MTC technology below). Finally, it is important that the location of the probe on the cell surface be specified. Since the stiffness of a living cell is different at different sites (Petersen *et al.*, 1982; Yanai *et al.*, 1999), cytoplasmic strain and stress patterns are also heterogeneous (Hu *et al.*, 2003).

C. Image Acquisition

Images are collected by a charged-coupled device (CCD) camera connected to a Windows XP computer. Figure 1 shows a magnetic current with a sinusoidal waveform and the time points for image collection over a complete cycle of force

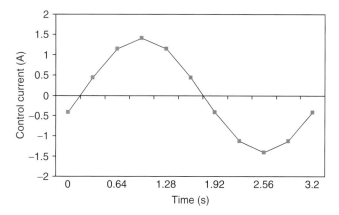

Fig. 1 A current waveform for controlling the magnets and the time points to collect microscope images (pink squares). Only one cycle is plotted (loading frequency = 0.3 Hz).

loading. To lower the noise level of the images, three to eight images at corresponding time points in each consecutive cycle are averaged to generate one image in the final image sequence. In general, this strategy can improve the signal-to-noise ratio (SNR) by a factor of 2–3. For N time points per cycle, $(N + 1)$ averaged images are needed, which are labeled as 0, 1, ..., N. Periodicity forces the last point in the cycle (the $N + 1$) to be (by definition) the same as the first point. In order to trace the entire sinusoidal waveform at a proper temporal resolution, N should be greater than 6. We generally set N to be 10.

D. Image Partitioning

To identify local displacements within an image, we first divide the image into arrays of small, half-overlapping subimages as shown in Fig. 2. We assume that the original image has a size of $a \times b$ pixels and that each subimage is square, of size $(2c + 1) \times (2c + 1)$ pixels, where c is the distance in pixels between the centers of two adjacent arrays in the horizontal or vertical direction. The total number of rows and columns of subimages are given by

$$\text{Row} = \text{Int}\left[\frac{a - 2 - 2 - (c + 1)}{c}\right] \tag{1}$$

and

$$\text{Col} = \text{Int}\left[\frac{b - 2 - 2 - (c + 1)}{c}\right] \tag{2}$$

respectively. The Int function is the greatest integer less than or equal to its argument. The presence of two 2s in Eqs. (1) and (2) accounts for a 2-pixel margin

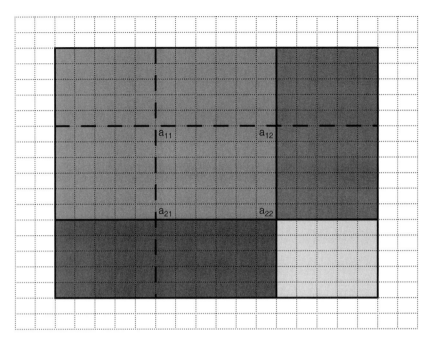

Fig. 2 Image partitioning. A 20 × 20 pixel image is divided into four 11 × 11 pixel subimages indicated by different colors; a_{11} is the center of pink subimage; a_{12} is the center of green subimage; a_{21} is the center of blue subimage; and a_{22} is the center of gray subimage. Only the pink subimage can be seen in whole; other subimages are partially seen due to overlaps. (See Plate 10 in the color insert section.)

bounding the detected image. In general, the smaller the c, the finer the grid size, and thus the more displacement data can be detected. However, there is a tradeoff: a very small c means a small image window, which can both dramatically increase the probability of faulty detection due to image noise as well as the computing time.

E. Array Shifting

Let the image intensity within a subimage be $f(x, y)$, where x and y are the (integer valued) 2D coordinates. In detecting its displacement, we shift the subimage by a small distance, $(\Delta x, \Delta y)$, relative to the original image to obtain a new subimage, $f_1(x, y)$. The Fourier transform of the original subimage is:

$$F(\xi, \eta) = \int_{-\infty}^{+\infty} \int_{-\infty}^{+\infty} f(x, y) e^{i(\xi x + \eta y)} \, dx dy \tag{3}$$

The transform of the shifted subimage is

$$F_1(\xi, \eta) = \int_{-\infty}^{+\infty} \int_{-\infty}^{+\infty} f(x + \Delta x, y + \Delta y)e^{i(\xi x + \eta y)} \mathrm{d}x\mathrm{d}y = F(\xi, \eta)e^{-i(\xi\Delta x + \eta\Delta y)} \quad (4)$$

where ξ and η are two variables in the (spatial) frequency domain. Equation (4) indicates that in this domain the shifted subimage is proportional, via a simple phase factor, to its counterpart in the original image, $F(\xi, \eta)$. Applying the inverse Fourier transform to $F_1(\xi, \eta)$, the shifted array in real space can be obtained. In software implementation, any discrete form of Fourier transform for $f(x, y)$ with a size of $(2c + 1) \times (2c + 1)$ pixels can be used, such as:

$$F(\xi, \eta) = \sum_{x,y=1}^{2c+1} f(x, y)e^{2\pi i(\xi x + \eta y)/(2c + 1)}, \qquad 1 \le (\xi, \eta) \le (2c + 1) \quad (5)$$

and

$$F_1(\xi, \eta) = e^{-2\pi i(\xi\Delta x + \eta\Delta y)/(2c + 1)} F(\xi, \eta), \qquad 1 \le (\xi, \eta) \le (2c + 1) \quad (6)$$

F. Image Match Searching

To detect the local displacement, two images of the same size are needed: one reference image A and another image B for comparison. The local displacement is identified by comparing the subimages at the same location from the two images. We denote subimages of images A and B by A' and B', respectively. For shifting purposes, the subimage A' includes a 2-pixel peripheral margin [thus having size $(2c + 5) \times (2c + 5)$ pixels] padded with zeros. The idea is to shift A' consecutively in small incremental steps, for example 1/25 pixels, in both horizontal and vertical directions and to search for the best image match with array B'. Here, the best match means the smallest mean square error (MSE) of the intensity of A' relative to B'. The mathematical approach of steepest descent is convenient for converging to the minimal MSE, which yields the local displacement for the best match. A diagram of the searching algorithm is shown in Fig. 3.

G. Synchronous Displacement

Applying the searching algorithm above to two consecutive images, a displacement vector $(Dx, Dy)_{(i,j)}$ can be generated for each subimage at each lattice position (i, j). For $(N + 1)$ consecutive images, N displacements are identified for each subimage at each position, namely, (Dx_k, Dy_k) where $k = 1, 2, ..., N$, and (i, j)

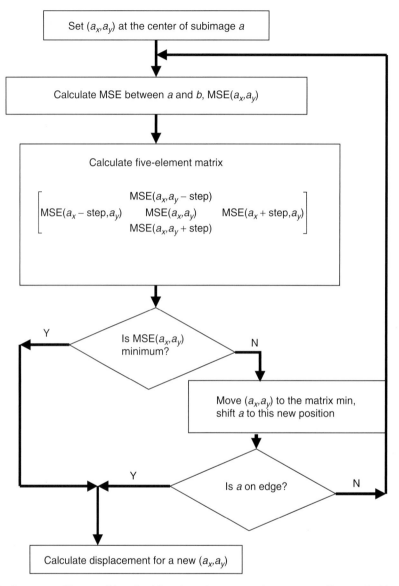

Fig. 3 Image-matching searching algorithm. (a_x, a_y) represents the center coordinates of subimage *a* of size $(2c + 5) \times (2c + 5)$.

is omitted for simplicity. Simple summation suffices to obtain the displacement relative to the first image, (x_k, y_k),

$$x_k = \sum_{m=1}^{k} Dx_m, \qquad y_k = \sum_{m=1}^{k} Dy_m, \qquad k = 1, 2, ..., N \qquad (7)$$

Next, we extract from net displacements the components that occur synchronously with the periodic applied forces, that is, components at the fundamental driving frequency of probing forces. In the horizontal direction, these are given by the amplitude A_x and phase θ_x:

$$[A_x, \theta_x] = \left[\sqrt{s^2 + t^2}, \arctan\left(\frac{t}{s}\right) \right] \qquad (8)$$

where

$$s = \frac{2}{N} \sum_{n=1}^{N} x_n \sin\left(\frac{2\pi(n-1)}{N}\right) \qquad (9)$$

and

$$t = \frac{2}{N} \sum_{n=1}^{N} x_n \cos\left(\frac{2\pi(n-1)}{N}\right) \qquad (10)$$

In the vertical direction, the amplitude and phase $[A_y, \theta_y]$ are calculated similarly from y_n. The net (scalar) displacement magnitude A synchronous with the driving force is given by

$$A = \sqrt{A_x^2 + A_y^2} \qquad (11)$$

Note that the nonsynchronous displacement amplitudes in the frequency domain (i.e., at frequencies other than the fundamental frequency) can be used to estimate the intrinsic resolution of this technique. Using this idea, we have found that a resolution approaching 4–5 nm can be achieved (Hu *et al.*, 2004).

H. Signal-to-Noise Ratio

The SNR is an important parameter in checking the reliability of synchronous detection. The SNR in the horizontal direction can be estimated from x_k ($k = 1, 2, ..., N$) and $[A_x, \theta_x]$. A purely sinusoidal response, including a potential nonzero offset, at the fundamental frequency is given by

$$\hat{x}_k = A_x \sin\left(\frac{2\pi k}{N} + \theta_x\right) + \frac{1}{N}\sum_{n=1}^{N} x_n, \qquad k = 1, 2, ..., N \tag{12}$$

The deviations of observed x_k from this purely sinusoidal response allow an estimate of SNR_x as

$$SNR_x = \frac{A_x/\sqrt{2}}{\sqrt{(1/(N-1))\sum_{n=1}^{N}(x_n - \hat{x}_n)^2}} \tag{13}$$

This is equivalent to estimating the extent to which displacements at the fundamental frequency account for the observed data. The SNR of the array in vertical displacement, SNR_y, can be estimated similarly.

III. Imaging Displacement and Stress Maps in a Live Cell

A. Quantifying Displacement Maps

Here we describe how we have applied the synchronous detection strategy to a living adherent cell to map force propagation in the cytoplasm. An RGD-coated magnetic bead (4.5 μm in diameter) was attached to the apical surface of an adherent cell (Fig. 4A) infected with adenovirus YFP-cytochrome *c* oxidase referred to as YFP-"mitochondria" (Fig. 4B). The bead was twisted by MTC as described previously (Fabry *et al.*, 2001; Wang *et al.*, 1993) and in Chapter 19 by Lele *et al.*, this volume. Before the relative displacements of consecutive images could be quantified, the image was subdivided into subimages. Over the area of each subimage, the standard deviation (SD) of the pixel intensity was calculated (Fig. 4C). When the SD was higher than 10% of the maximum SD in the whole image (e.g., for the scale of Fig. 4C, SD > 0.005), the subimage would be analyzed for its displacement; otherwise it would be regarded as a dark background (Fig. 4D). In the latter case, the displacements would be set to be zero (the dark blue region of the figure). The purpose here is to increase the software efficiency and to ignore subimages with poor quality or with large background noise. The threshold of 10% may be adjusted according to the SNR or the quality of collected images.

The detected displacement maps are shown in Fig. 4E and F. The phase lags of mitochondria displacements relative to the applied torque can be seen in Fig. 4G and H. The corresponding SNR maps are shown in Fig. 4I and J. Considering the area with SNR greater than or equal to 3, the contour diagrams of phase lags can be drawn (Fig. 4K and L). Combining the displacements in both horizontal and vertical directions, a displacement map with directions (depicted by arrows) is shown in Fig. 4M.

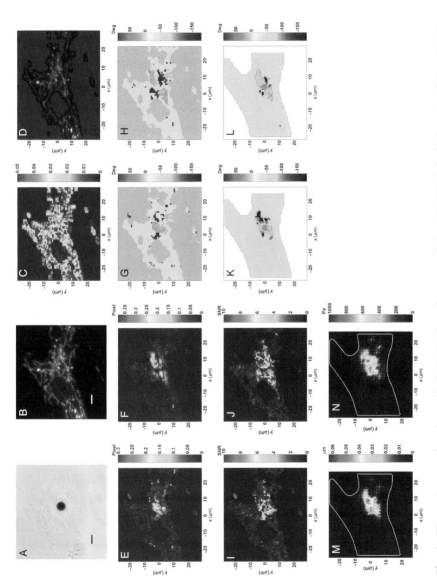

Fig. 4 The synchronous detection method. (A) A 4.5-μm ferromagnetic bead (the black dot in the center) is bound to the apical surface of an adherent airway smooth muscle cell. (B) Fluorescent image of a cell expressing YFP-mitochondria. The image has a size of 360 × 360 pixels, 163 nm/pixel. Scale bar is 5 μm. (C) Map of the standard deviation (SD). (D) The area of the cell selected by the SD value, which will be analyzed by the software. The blue area will be ignored. (E) Displacement map in horizontal (x)-direction obtained, using intensity cross-correlation of fluorescent images as explained in the text, unit for the scale is pixels. (F) Displacement map in vertical (y)-direction. (G) Displacement phase lag map relative to the applied torque in x-direction, unit for the scale is pixels. (H) Phase lag map relative to the applied torque in y-direction, unit for the scale is degrees. (I) Map of the measured SNR for the x-displacement. (J) Map of the measured SNR for the y-displacement. (K) Map of phase lags in x-direction with SNR ≥ 3. (L) Similar conditional map of phase lags in y-direction. (M) Conditional map of directional displacements. Colors represent magnitudes of the displacements. The pink arrow represents the bead center and the direction of the bead lateral movement. (N) Estimated stress map from the displacement map and the measured Young's modulus of the cytoplasm. (See Plate 11 in the color insert section.)

B. Computing Stress Maps from Displacement Maps

Interfacial stresses between the cell and its substrate can be quantified by traction force microscopy, a method that tracks movements of fluorescent beads embedded in an elastomeric gel just beneath the cell base (Butler *et al.*, 2002; Dembo and Wang, 1999). We have adapted this approach to probe the interior of the cell, by tracking movements of the YFP-mitochondria that are synchronized with the movement of the integrin-bound magnetic bead (i.e., the applied load). Using a modified Fourier transform traction cytometry (FTTC) method (Butler *et al.*, 2002), we computed an index of stress analogous to tractions acting on a plane (e.g., the focal plane of the fluorescent image) that would be required to generate the displacement fields such as those illustrated in Fig. 4E and F. This traction field represents an approximation of the stress field inside the cell that acts in the x–y plane at a cell height given by the focal plane. From the lateral bead displacements, we estimated the Young's modulus of the cell (Hu *et al.*, 2003), assuming a Poisson ratio of 0.48 and a uniform elasticity. Figure 4N shows the corresponding stress field calculated from the mitochondrial displacement fields in Fig. 4E and F and the modulus of the cell.

C. Modulation of Stress Distribution Within a Living Cell

Such maps of displacement and stress would provide particularly useful insights into cell biology if they can be perturbed in specific ways. We therefore examined whether induced stress propagation in a normal living cell could be modulated when the actin cytoskeleton or steady state traction stress on the substrate ("prestress") was altered using different treatments. Figure 5B shows that the stress distribution in a cultured smooth muscle cell (Fig. 5A) is dramatically changed when F-actin was disrupted by cytochalasin D. The focusing of stress that was apparent at distal sites before treatment disappeared on the addition of cytochalasin D (Fig. 5B). Similar results were obtained when caldesmon was overexpressed to inhibit actin–myosin contractility in the cell (Fig. 5D); when caldesmon was inhibited by the calcium ionophore A23187, the distal sites with focused stress reappeared (Fig. 5C). In addition, when the cell's prestress was lowered by the relaxing agent isoproterenol, stress focusing at distal sites disappeared (compare Fig. 5E with 5F). Plating the cell nonspecifically on positively charged poly-L-lysine, which does not promote the formation of actin bundles or stress fibers, also leads to loss of remote stress focusing (Fig. 5H). However, treating these cells with histamine elevates the prestress and induces the appearance of stress focusing sites (Fig. 5G). Importantly, these treatments all induce dramatic changes in the stress distribution without immediate effect on the total projected cell area or cell shape. These data not only demonstrate long-range force propagation in the cytoplasm, but also call into question the validity of using continuum models to predict cellular mechanical behaviors at subcellular levels.

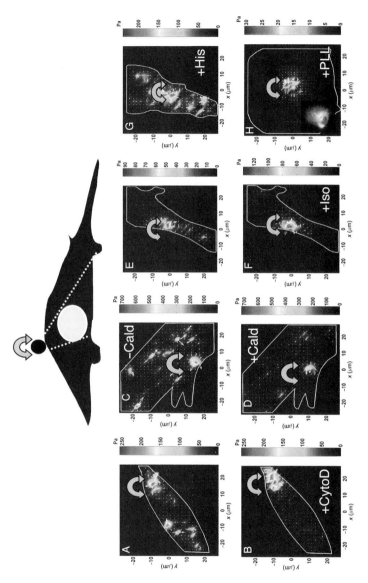

Fig. 5 Modulation of stress propagation by the cytoskeleton. Top: A specific torque of 90 Pa was applied at 0.3 Hz to the apical surface of a cell (the dark blue schematic) via an RGD-coated magnetic bead (the black dot), using a sinusoidal oscillating magnetic field (double green arrow). The stress map of a cultured smooth muscle cell at baseline (A) exhibits stress focusing at several distal sites. When treated with cytochalasin D (+CytoD, 1 μg/ml for 3 min) (B), stresses become localized only near the load. Overexpressing caldesmon (+Cald) to inhibit myosin–actin interactions led to the inhibition of distal stress focusing (D). In contrast, inhibiting caldesmon with the calcium ionophore A23187 (5 μg/ml for 5 min, −Cald) recovered the phenomenon of distal stress focusing (C). A cell that was treated with the contractile relaxing agent isoproterenol (+Iso, 10 μM) for 3 min loses stress focusing at distal sites (F), when compared with its baseline control (E). Elevating the prestress with the contractile agonist histamine (+His, 10 μM for 10 sec) increased the prevalence of stress focusing at distal sites (G; control cell is not shown here), whereas plating the cell on poly-L-lysine (PLL) to inhibit the formation of long actin bundles (and stress fibers) led to the loss of long-range force propagation (H). [Adapted from Hu and Wang (2006) and Hu *et al.* (2003) with permission.] White arrows represent directions and colors represent relative magnitudes of the stresses. (See Plate 12 in the color insert section.)

D. Application of Mechanical Loads in Any Direction Using 3D-MTC

One key issue in elucidating mechanism of mechanotransduction is how to differentiate stress-induced *direct* displacements or deformations, from active remodeling-associated secondary changes in cytoskeletal structures. One potentially effective way is to alter the direction of force application at the same site of loading. The idea is that if the induced displacements are indeed a result of direct stress propagation to the structure, then they should be very sensitive to the direction of loading. On the other hand, if it is a result of stress-induced remodeling of the cytoskeleton, then they should be much less sensitive to the loading direction. There are few existing technologies that can apply a controlled, well-defined mechanical load in different directions. Here, we describe an approach developed recently: 3D-MTC that can apply loading in any specified direction.

Our 3D-MTC device (Fig. 6) contains seven major components: (1) a high-voltage power supply for generating the current in the coils used to magnetize the magnetic particles, (2) three separate bipolar current sources for twisting them, (3) a computer for controlling the twisting, (4) an inverted microscope (Leica, Leica Microsystems, Wetzlar, Germany) for observing the sample, (5) a CCD camera (Hamamatsu, Hamamatsu City, Japan) using tailor-made software for synchronization of image capture with oscillatory magnetic fields, (6) a temperature controller for living cells, and (7) a microscope insert that holds the three pairs of coils for magnetization and twisting and the sample. Components (1)–(3), (6), and (7) are now commercially available from EOL Eberhard (Oberwil, Switzerland) via special order.

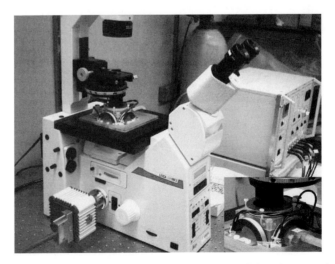

Fig. 6 The 3D-MTC device. Lower right corner is a close-up of the microscope insert with the temperature control tubing on the right corner. [Adapted from Hu *et al.* (2004) with permission.]

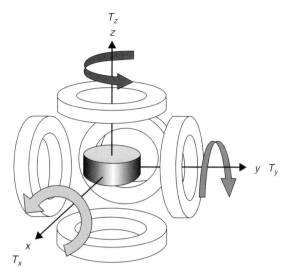

Fig. 7 Schematic of the microscope insert of the 3D-MTC (not to scale). [Adapted from Hu *et al.* (2004) with permission]. Each doughnut represents a coil (the front coil is not drawn for clarity). The gray cylinder in the middle represents the sample dish.

The 3D-MTC uses three pairs of coils oriented along three orthogonal axes, and controls the amplitude, frequency, and phase in each pair independently (Fig. 7). Magnetizing is accomplished by discharging a high-voltage capacitor via one pair of coils selected out of the three pairs of the microscope insert. That way, the axis of magnetization can be chosen to be x, y, or z, depending on the desired direction of twisting. Twisting is done by simultaneously applying computer-controlled bipolar currents to the three pairs of coils. The magnetic field for twisting, which results from superposition of the individual contributions of the three pairs of coils, can therefore be chosen to be along any direction in 3D and any strength within the capability of the electrical current sources [for details of the design and the specifications, see Hu *et al.* (2004)]. Each pair of coils is capable of producing a homogeneous magnetic field in its axial direction, with a nearly zero radial component and negligible field gradients over the diameter of the coil. Thus, in its rotational mode, it can generate pure torques with no linear component of force. The device can magnetize beads with a moment vector by applying a strong short magnetic pulse (1000 gauss for 0.5 msec) via any coil pair. Subsequently, it can apply a weak twisting field vector via another coil pair (0–25 gauss, too weak to remagnetize the beads) to generate a torque **T** on the beads at frequencies up to 1000 Hz. The computer interfaces with the 3D-MTC via a serial port through National Instruments' multiple functional boards. The CCD camera is connected to the computer through IEEE 1394 Firewire interface. A user-friendly interface is controlled via a custom-written software to operate the whole system of the 3D-MTC.

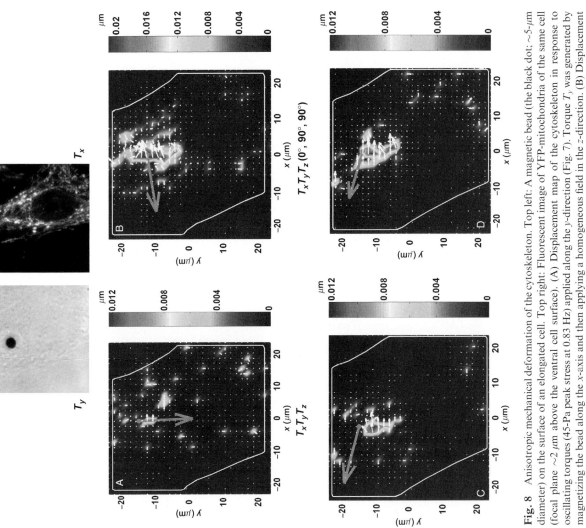

Fig. 8 Anisotropic mechanical deformation of the cytoskeleton. Top left: A magnetic bead (the black dot; ∼5-μm diameter) on the surface of an elongated cell. Top right: Fluorescent image of YFP-mitochondria of the same cell (focal plane ∼2 μm above the ventral cell surface). (A) Displacement map of the cytoskeleton in response to oscillating torques (45-Pa peak stress at 0.83 Hz) applied along the y-direction (Fig. 7). Torque T_y was generated by magnetizing the bead along the x-axis and then applying a homogeneous field in the z-direction. (B) Displacement

E. Mechanical Anisotropic Signaling to the Cytoskeleton and to the Nucleolus

To determine the utility of 3D-MTC in applying mechanical loading in different directions, we mapped the cytoskeletal displacement field of an elongated cell (Fig. 8, top left image) with transfected YFP-mitochodria as markers (Fig. 8, top right image). The load-induced displacement patterns of the cytoskeleton, which constrains the movement of mitochondria, were quite different in response to the applied torques along different directions but of the same magnitude and frequency: Fig. 8A shows the response to torques in the y-direction, T_y, and contrasts with Fig. 8B that shows the response to torques in the x-direction, T_x. Interestingly, when a more complex load was applied, with all three oscillatory twisting fields switched on simultaneously and with different phase lags ($x = 0°$, $y = 90°$, $z = 0°$ in Fig. 8C, or $x = 0°$, $y = 90°$, $z = 90°$ in Fig. 8D), different displacement patterns emerged (compare Fig. 8C with 8D). These results demonstrate that spatial distributions of cytoskeletal deformation depend not only on the magnitude and frequency of force loading but also on the direction of loading.

We next mapped the displacements of the nucleolus, an intranuclear organelle crucial for ribosomal RNA synthesis, in response to the load applied on the cell surface. The nucleolus of an interphase cell is visible under phase-contrast microscopy without staining or GFP tagging (Fig. 9, top left). The displacement map within the nucleolus varied with the loading direction: changing the torque direction by 90° (e.g., from T_y in Fig. 9A to T_x in Fig. 9B) significantly altered the displacement patterns of the nucleolus. As for mitochondria, applying a more complex load by turning on all three twisting fields simultaneously but with different phase lags (thus inducing different patterns of bead rotation) resulted in significantly different nucleolar displacement maps, and therefore different intranucleolar deformation patterns (Fig. 9C and D). Although specific effects on nucleolar function under stress are not known at this time, our results suggest that physiological loads applied at the cell surface might be able to directly alter nuclear functions via the structural pathways of the cytoskeletal and the nuclear matrix (Penman, 1995), thus identifying a potential mechanism independent of chemical pathways mediated by second messengers or cytoplasmic soluble molecules. These results also demonstrate that it is possible to quantify force-induced

---

map of the cytoskeleton in response to torques of the same magnitude applied along the x-direction. Torque T_x was generated by remagnetizing the rotating bead in (A) along the y-axis. Note the drastic differences in displacement patterns between (A) and (B). (C) and (D) are displacement maps of the cytoskeleton in response to equal magnitude torques applied simultaneously along three axes (45-Pa peak stress and 0.83 Hz) with different phase lags: $x = 0°$, $y = 90°$, $z = 0°$ in (C); $x = 0°$, $y = 90°$, $z = 90°$ in (D). The bead was magnetized in the y-direction in both cases. Because T_y was 90° phased-lagged from T_x in (C) and from T_x and T_z in (D), the bead experienced the effect of T_y in both loading conditions. Note the significant differences in displacement maps between (C) and (D). Colors represent the relative magnitudes and small white arrows represent the direction of the displacements. The pink arrow represents the position and direction of the bead center movement. [Adapted from Hu *et al.* (2004) with permission.] (See Plate 13 in the color insert section.)

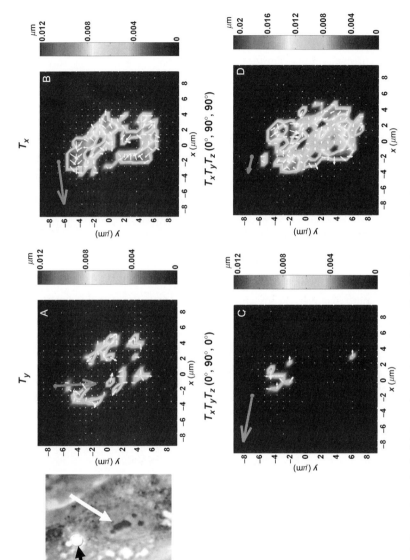

Fig. 9 Anisotropic mechanical deformation to the nucleolus. Top left: Phase-contrast image showing the nucleolus (large white arrow) and the bead (black arrow) of the same cell as in Fig. 8. The distance between the front edge of the bead and the proximal end of the nucleolus is 11.8 μm (not to scale in the displacement maps). Torques were applied similar to those in Fig. 8. Symbols and colors are defined as in Fig. 8. Note the major differences in displacement patterns between (A) and (B) and between (C) and (D). [Adapted from Hu *et al.* (2004) with permission.] (See Plate 14 in the color insert section.)

stress propagation in unperturbed structures or molecules in a living cell as long as the contrast is good.

F. Limitations of 3D–MTC

As mentioned above, the successful application of this technology depends on several issues: (1) The structure of interest should have good contrast relative to its surrounding structures under the microscope. (2) The fluorescent probe or organelle (e.g., nucleolus) should be relatively resistant to irradiation. (3). The specific interactions between the probe (the magnetic bead) and the cell surface receptors need to be quantitated. For example, how many receptors are clustered surrounding the probe? (4) The magnetic bead cannot be precisely placed at a specific desired site on the cell surface (in contrast to AFM or laser tweezers). (5) Like AFM and laser tweezers, the (3D-) MTC can only apply a local load via the apical (dorsal) surface of a cell. (6) The 3D-MTC generates complex stresses in all three dimensions, but z-displacements and out-of-plane rotations/deformations are not measured in the current imaging and analysis protocols. Displacements in the z-direction should be quantified in future studies and be taken into account in computing more realistic stress maps. This potential problem could be alleviated, however, by rotating the bead in the x–y plane (the applied torque is in z-direction) using 3D-MTC, and by applying small surface deformation (<300 nm). (7) Local heterogeneity in stiffness cannot be independently measured and an average stiffness must be used to estimate the stress map. Improvements could include better assessment of local heterogeneity of stiffness.

IV. Future Prospects

A combination of the present synchronous detection approach to map stresses and strains with other existing live cell-imaging technologies and single-cell biochemical assays may reveal new insights into the mechanisms of mechanotransduction. For example, this method could be combined with fluorescent resonance energy transfer (FRET) to reveal possible associations between local mechanical and biochemical activities. In addition to mapping displacements or deformation of cytoplasmic structures/molecules or nuclear structures in response to localized external forces, one can also map friction-related phase lags of different structures. This may shed light on the degree to which local elastic deformation and/or energy dissipation might impact on biochemical activities. With this in mind, we anticipate a bright future for this and related approaches in helping to elucidate the mechanisms of mechanotransduction in living cells.

Acknowledgments

We thank Ben Fabry, Jianxin Chen, and Luc Eberhard for help in developing this technology. This work was supported by NIH GM072744 and HL33009.

References

Bausch, A. V., Ziemann, F., Boulbitch, A. A., Jacobson, K., and Sackmann, K. (1998). Local measurements of viscoelastic parameters of adherent cell surfaces by magnetic bead micro-rheometry. *Biophys. J.* **75,** 2038–2049.

Butler, J. P., Tolic-Norrelykke, I. M., Fabry, B., and Fredberg, J. J. (2002). Estimating traction fields, moments, and strain energy that cells exert on their surroundings. *Am. J. Physiol. Cell Physiol.* **282,** C595–C605.

Crick, F. H. C., and Hughs, A. F. W. (1950). The physical properties of cytoplasm. *Exp. Cell Res.* **1,** 37–80.

Dai, J., and Sheetz, M. P. (1995). Mechanical properties of neuronal growth cone membranes studied by tether formation with laser optical tweezers. *Biophys. J.* **68,** 988–996.

Dembo, M., and Wang, Y. L. (1999). Stresses at the cell-to-substrate interface during locomotion of fibroblasts. *Biophys. J.* **76,** 2307–2316.

Fabry, B., Maksym, G. N., Shore, S. A., Moore, P. E., Panettieri, R. A., Jr., Butler, J. P., and Fredberg, J. J. (2001). Time course and heterogeneity of contractile responses in cultured human airway smooth muscle cells. *J. Appl. Physiol.* **91,** 986–994.

Helmke, B. P., Thakker, D. B., Goldman, R. D., and Davies, P. F. (2001). Spatiotemporal analysis of flow-induced intermediate filament displacement in living endothelial cells. *Biophys. J.* **80,** 184–194.

Hu, S., Chen, J., Butler, J. P., and Wang, N. (2005). Prestress mediates force propagation into the nucleus. *Biochem. Biophys. Res. Commun.* **329,** 423–428.

Hu, S., Chen, J., Fabry, B., Numaguchi, Y., Gouldstone, A., Ingber, D. E., Fredberg, J. J., Butler, J. P., and Wang, N. (2003). Intracellular stress tomography reveals stress focusing and structural anisotropy in the cytoskeleton of living cells. *Am. J. Physiol. Cell Physiol.* **285,** C1082–C1090.

Hu, S., Eberhard, L., Chen, J., Love, J. C., Butler, J. P., Fredberg, J. J., Whitesides, G. M., and Wang, N. (2004). Mechanical anisotropy of adherent cells probed by a three dimensional magnetic twisting device. *Am. J. Physiol. Cell Physiol.* **287,** C1184–C1191.

Hu, S., and Wang, N. (2006). Control of stress propagation in the cytoplasm by prestress and loading frequency. *Mol. Cell Biomech.* **3,** 49–60.

Ingber, D. E. (2006). Cellular mechanotransduction: Putting all the pieces together again. *FASEB J.* **20**(7), 811–827.

Numaguchi, Y., Huang, S., Polte, T. R., Eichler, G. S., Wang, N., and Ingber, D. E. (2003). Caldesmon-dependent switching between capillary endothelial cell growth and apoptosis through modulation of cell shape and contractility. *Angiogenesis* **6,** 55–64.

Penman, S. (1995). Rethinking cell structure. *Proc. Natl. Acad. Sci. USA* **92,** 5251–5257.

Petersen, N. O., McConnaughey, W. B., and Elson, E. L. (1982). Dependence of locally measured cellular deformability on position on the cell, temperature, and cytochalasin B. *Proc. Natl. Acad. Sci. USA* **79,** 5327–5331.

Radmacher, M., Fritz, M., Kacher, C. M., Cleveland, J. P., and Hansma, P. K. (1996). Measuring the viscoelastic properties of human platelets with the atomic force microscopy. *Biophys. J.* **70,** 556–567.

Shroff, S. G., Saner, D. R., and Lal, R. (1995). Dynamic micromechanical properties of cultured rat arterial myocytes measured by atomic force microscopy. *Am. J. Physiol.* **269,** C286–C292.

Valberg, P. A., and Butler, J. P. (1987). Magnetic particle motions within living cells. Physical theory and techniques. *Biophys. J.* **52,** 537–550.

Wang, N., Butler, J. P., and Ingber, D. E. (1993). Mechanotransduction across the cell surface and through the cytoskeleton. *Science* **260,** 1124–1127.

Yanai, M., Butler, J. P., Suzuki, T., Kanda, A., Kurachi, M., Tashiro, H., and Sasaki, H. (1999). Intracellular elasticity and viscosity in the body, leading and trailing regions of locomoting neutrophils. *Am. J. Physiol.* **277,** C432–C440.

CHAPTER 9

Probing Intracellular Force Distributions by High-Resolution Live Cell Imaging and Inverse Dynamics

Lin Ji,[1] Dinah Loerke,[1] Margaret Gardel, and Gaudenz Danuser

The Scripps Research Institute
La Jolla, California 92037

Abstract

Highly coordinated molecular regulation of mechanical processes is central to numerous cell processes. A key challenge in cell biophysics is, therefore, to probe intracellular force distributions and mechanical properties of live cells with high spatial and temporal resolution. This chapter describes a passive (i.e. nonperturbing) approach to map intracellular force distributions with submicron spatial

[1] These authors contributed equally to this work.

resolution, and on a timescale of seconds. On the basis of a continuum mechanical interpretation of the cell cytoskeleton, this approach performs an inverse reconstruction of intracellular forces from cytoskeletal flows measured in high-resolution live cell images acquired by quantitative fluorescent speckle microscopy (qFSM). Our inverse algorithm can robustly reconstruct the relative force distribution even in the absence of a quantitative profile of network elasticity. In addition, we also propose an emerging technique for probing the *in vivo* actin network compliance based on correlation analysis of the same data set. We demonstrate the force reconstruction on migrating epithelial cells, where the reconstructed intracellular force field indicates spatial and temporal coordination of force generation by cytoskeleton assembly, contraction and focal adhesion resistance, and its functional output in the form of cell edge movements. This technique will potentially allow the analysis of intracellular force regulation in numerous other cell functions.

I. Introduction

Many cell functions rely on the precise molecular integration of both force-generating and force-transducing machinery. Examples include cell migration (Lauffenburger and Horwitz, 1996) and cytokinesis (Scholey *et al.*, 2003), where regulated interplay between motor-based contraction, turnover, and transient coupling of cytoskeleton polymer structures to the plasma membrane and the extracellular matrix (ECM) results in dramatic morphological changes of the entire cell. Another example is chromosome movement during cell division (Sharp *et al.*, 2000; Wittmann *et al.*, 2001), where forces across the mitotic spindle generated by multiple motor systems and microtubule assembly and disassembly segregate the replicated chromosome pairs from the mother into the two newly forming daughter cells. Force generation and transduction also play central roles in more local processes, such as the transport of vesicles and organelles throughout the cell (Hirokawa, 1998; Nascimento *et al.*, 2003; Vale and Milligan, 2000), or in small-scale reorganization of the cytoskeleton architecture during internalization and secretion of molecules across the plasma membrane (Conner and Schmid, 2003; Schafer, 2003). Cells respond in a sensitive fashion to extracellular force stimuli either by activation of mechanoreceptors that convert forces into chemical signals (Geiger and Bershadsky, 2001; Geiger *et al.*, 2001) or by global transduction of forces through the cytoskeleton directly to the nucleus, which is thought to modulate gene transcription activity (Ingber, 2003a,b).

To probe force-generation and force-transduction mechanisms and to identify the molecules involved in force regulation, maps of *intracellular* force distributions with high spatial and temporal resolution are required. Several approaches have been devised to probe force development at the cellular scale, many of which are described in this volume. However, most of them are restricted to the measurement of *extracellular* forces, that is the portion of intracellular forces that is coupled to the extracellular domain via transmembrane receptors (see Part *III. Cellular and Embryonic Mechanical Properties and Activities* in this volume). In addition, most of the methods offer limited or no spatial resolution, which precludes the analysis of force regulation across the cell. This applies also to many methods that involve probes embedded inside the cell, for example magnetic beads (see preceding chapters in Part *II. Subcellular Mechanical Properties and Activities* in this volume).

In this chapter, we propose a new approach where forces are extracted from high-resolution live cell fluorescence microscopy movies. The method is built on a set of assumptions about the properties of intracellular materials, which is still under intensive study but has the potential to map intracellular force distributions with submicron spatial resolution and on a timescale of seconds as needed for the dynamic molecular processes in motility. The fundamental idea of this approach is to assess deformations of subcellular structures with high precision, and then to mathematically deduce the intracellular forces that must have been applied to the structure in order to induce the observed deformation. While indirect reconstruction of physical quantities from differential experimental data is widely used in other areas of science and engineering (Kosovichev *et al.*, 2000; Li *et al.*, 2003; Manduca *et al.*, 2001; McIver, 1991; Metherall *et al.*, 1996; Symes and Carazzone, 1991) and has been very narrowly applied to static cell configurations aspirated in micropipettes during fluorescence-imaged microdeformation (Discher *et al.*, 1994), the idea is relatively new to cell biology. In the context of force reconstruction, this approach is commonly referred to as inverse dynamics. Because of the novelty of its application in cell biology, the assumptions involved, and the numerical complexity of the reconstruction procedure, this chapter serves more as an outlook rather than a didactic account of a well-established method. We illustrate the principles of inverse dynamics based on the question of how polymerization, contraction, and adhesion of the actin cytoskeleton together mediate directed protrusion and retraction of the leading edge of a migrating cell. Section I will first introduce the various forces implicated in cell protrusion and then formulate a mechanical model of the actin cytoskeleton. We will also summarize the principles of quantitative fluorescent speckle microscopy (qFSM), whose development was critical in allowing us to apply the ideas of inverse dynamics to subcellular scales. Section II will outline the mathematical ingredients of intracellular force reconstruction by inverse dynamics. Readers who are less interested in these details can directly move on to Section II.E, where we show results of force reconstructions inside protruding cells. In addition, a critical issue in reconstructing forces by inverse dynamics is the characterization of the material properties of the deformed structure. Although data presented in Section II.F suggests that spatial heterogeneity of cytoskeletal stiffness likely plays a minor role in the definition of the intracellular force field, we will conclude the chapter in Section III with a glimpse of a new modality of qFSM, which will potentially allow the probing of material heterogeneity throughout the cell with unprecedented spatial resolution.

═══════════ ## II. Methods

A. Actin Cytoskeleton Mechanics in Cell Protrusion

1. Force Generation During Protrusion

Cell protrusion is the morphodynamic consequence of three integrated force-generating component processes: (1) Actin filament (F-actin) polymerization at the leading edge pushes the cell membrane forward and the growing actin network backward, resulting in cytoskeleton retrograde flow (Cramer, 1997). (2) The filament network transiently anchors to the ECM at focal adhesions, which are specialized multiprotein complexes containing transmembrane receptors (integrins) to ECM ligands and several actin-binding components in the cytoplasmic domain (Geiger, 2001). Adhesive coupling of the cytoskeleton to the ECM provides the mechanical traction for cell migration. Adhesions also create mechanical resistance to cytoskeleton retrograde flow, which allows F-actin assembly at the front to be converted at least partially into forward motion of the cell edge. Accordingly, retrograde flow is slowed down at adhesion sites (Ponti *et al.*, 2004). (3) Contraction of the network by myosin motor proteins can again accelerate F-actin flow. Globally, actomyosin contraction also causes the retraction of the cell rear toward the protruding front (Verkhovsky *et al.*, 1999).

Corresponding to these component processes, three types of intracellular forces act on the actin cytoskeleton (Fig. 1): (1) a boundary force created by actin polymerization pushing against the cell membrane (red arrows); (2) an adhesion resistance force created by cytoskeleton–ECM coupling (orange arrows); and (3) a contraction force (blue arrows). Neglecting further frictional forces between the F-actin and the fluid phase of the cytoplasm, which are thought to be small compared to the three main force classes (Rubinstein *et al.*, 2005), boundary and contraction forces are globally balanced by the resistance forces at adhesion sites. The latter are transmitted to the ECM via the protein layers of adhesion complexes. Assuming that the linkage between the F-actin network and the ECM is 100% efficient, that is, that energy dissipation by protein–protein friction is negligible, the intracellular adhesion resistance force (light green arrows) and the force exerted on the ECM substrate (cyan arrows) are an action–reaction force pair. Thus, intracellular adhesion resistance forces can be approximated by extracellularly probing the forces on the substrate. Several methods have been developed to achieve this. Most prominently, the method of traction force microscopy provides spatiotemporally resolved maps of the adhesive resistance forces exerted on the actin cytoskeleton (Beningo and Wang, 2002; Lee *et al.*, 1994; Munevar *et al.*, 2001; Oliver *et al.*, 1999). Whereas traction forces indicate the location of intracellular force transmission to the extracellular domain, they cannot reveal the individual contributions of polymerization and contraction forces, or their intracellular locations. However, knowledge of precisely this intracellular force distribution is required to establish the mechanisms of molecular regulation of

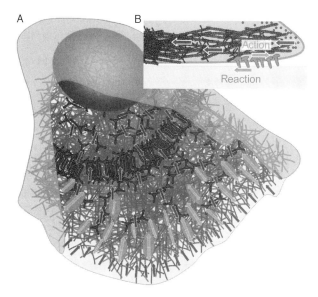

Fig. 1 Schematic representation of intracellular forces in migrating adherent cells. (A) Top view and (B) side view of a protruding cell. Actin polymerization at the leading edge generates a boundary-pushing force (red arrows). At the focal adhesion sites (green), the coupling of the cytoskeleton to the adhesion complexes creates a resistance force (light green arrows), while myosin-driven contractions in the more interior region (blue arrows) can increase the flow speed behind the adhesion sites. The intracellular forces are transmitted to the substrate through the adhesion complexes, creating traction forces (cyan arrows) acting on the substrate. Traction force and adhesion resistance force can be considered an action–reaction force pair that results from the "friction" between the substrate and the actin network. (See Plate 15 in the color insert section.)

cell protrusion and to identify the functions of numerous structural and signaling molecules implicated in cell migration.

In this chapter, we propose an alternative approach to mapping the distribution of intracellular forces acting on the actin cytoskeleton. As discussed above, F-actin polymerization against the cell membrane and intracellular contraction together power the characteristic F-actin retrograde flow observed in the front region of migrating cells (Cramer, 1997; Danuser and Oldenbourg, 2000; Lin and Forscher, 1995; Ponti *et al.*, 2004; Theriot and Mitchison, 1992; Vallotton *et al.*, 2005; Wang, 1985). Transient coupling of the network to adhesions (Suter and Forscher, 2000) as well as changes in the rates of polymerization (Ponti *et al.*, 2004) and contraction (Giannone *et al.*, 2004; Gupton *et al.*, 2002; Vallotton *et al.*, 2004) modulate the flow in space and time. Hence, given the sensitivity of F-actin flow to these spatiotemporal gradients, it should be possible to invert the problem and ask which force distribution would be required to generate the measured flow field. We will show that qFSM can serve the purpose of inverse force reconstruction.

The basic notion of inverse force reconstruction can be readily understood with the analogy of a spring (Fig. 2A). According to Hooke's law, the force F required to extend a spring is proportional to the extension u relative to its initial relaxation length:

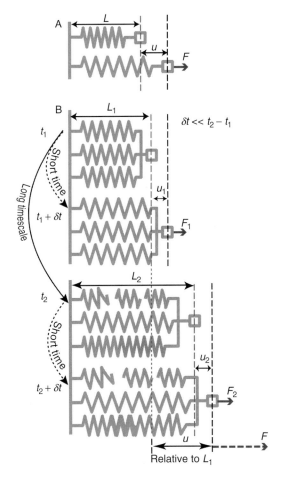

Fig. 2 Elasticity and plasticity are relevant on different timescales. (A) In an elastic spring system, Hooke's law states that $F = ku/L$, with u the elongation, L the resting length, k the elastic spring constant, and F the applied force. When spring constant and resting length are known, the applied force can be calculated from the observed elongation. (B) In an actin network (represented as a system of springs), the elastic properties can change over time due to dynamic remodeling; at long timescales, such as between t_1 and t_2, some structures break while others are newly formed. The individual elastic constants of the springs in this example stay the same, but the resting length L_1 of the entire spring system (dashed blue line) changes to L_2. As a result, the elongation of the system over this timescale does not obey Hooke's law and force reconstruction is impossible. On the other hand, when an extension is fast compared to the time course of remodeling, as from t_1 to $t_1 + \delta t$ or from t_2 to $t_2 + \delta t$, the elastic properties and the relaxation length can be considered to be constant, so that a linear relationship still applies, and $F_1 = ku_1/L_1$ and $F_2 = ku_2/L_2$.

$$k\left(\frac{u}{L}\right) = F \qquad\qquad (1)$$

The proportionality factor k is called the spring constant, which describes the elasticity of the spring. Knowing the spring constant and the relaxation length L, it is straightforward to calculate the exerted force from the measured extension u.

In a two-dimensional (2D) network with linear elasticity, the force can be reconstructed from the deformation along the same basic principles, that is the force is proportional to the degree of deformation. Due to the increase in dimensionality (from 1D to 2D or even 3D), the technicalities of the force reconstruction are more involved than in the simple spring system. However, these problems have been solved in the field of continuum mechanics. Before we proceed to the details of a 2D reconstruction algorithm in Section II.C, we first need to address the question whether such continuum mechanical models apply to the actin cytoskeleton in live cells. In Section II.A.2, we will suggest that the answer to this question is potentially yes, depending on the timescale of the observed deformation. We then give a brief review of qFSM that provides the measurements of actin flow velocities (i.e., deformations) at high spatial resolution and at the appropriate timescales.

2. Mechanical Properties of the Actin Cytoskeleton

In vitro studies have shown that on length scales above the characteristic distance between filament cross-links and branches, F-actin networks behave like a viscoelastic continuum (Shin *et al.*, 2004). The viscous contribution to the mechanical properties of *in vitro* networks is frequency dependent, but can be neglected at frequencies less than 1 Hz (Gardel *et al.*, 2004; MacKintosh *et al.*, 1995), at which high-resolution fluorescence microscopy experiments are typically performed. Although the relationship between force and network deformation becomes nonlinear at high strain rates due to strain stiffening (Gardel *et al.*, 2004; Storm *et al.*, 2005 and Chapter 1 by Janmey *et al.*, this volume), our qFSM data reveal that at the timescale of $\sim$1–10 sec, the intracellular strain induced by transient network deformation rarely exceeds 5%. Thus, for the purpose of force reconstruction on these timescales, *in vitro* F-actin networks can be approximated as a linear elastic material.

Whether F-actin networks in living cells can be approximated as purely elastic is still subject of intensive research, as comprehensively documented in the preceding chapters of this volume (Chapter 1 by Janmey *et al.*, Chapter 7 by Crocker and Hoffman and Chapter 8 by Wang *et al.*, this volume). Currently, the published literature varies in opinion about whether cells behave more viscous- or elastic-like in the frequency range typical for imaging. For 0.1 Hz, that is frame intervals of 10 sec, several groups have estimated the cytoskeletal elastic modulus to be significantly larger than the viscous modulus (Alcaraz *et al.*, 2003; Fabry *et al.*, 2001; Trepat *et al.*, 2004). Others report the opposite, finding a large viscous modulus in the 0.1-Hz frequency range (Kole *et al.*, 2004, 2005; Panorchan *et al.*, 2006) and a dominant elastic modulus only toward higher (1–10 Hz) frequencies.

The variation between *in vivo* studies and their discrepancy from *in vitro* F-actin network results may partly be explained by challenges in defining the elastic and viscous contributions to F-actin mechanical properties in living cells: First, existing live cell rheological measurements do not probe F-actin specifically, but gauge the ensemble behavior of all species of cytoskeletal filaments, membranes, and the cytosol. Presumably, the latter two components make a significant contribution to the viscous behavior only found *in vivo*; this is supported by the fact that the published ratios of elastic to viscous modulus measured in live cells are always much smaller than those documented for *in vitro* networks. Our measurements of F-actin network deformations, however, are molecularly specific, allowing us to neglect contributions by other cytoskeleton components, cytosol, or membranes. We therefore adhere to the more specific data available from *in vitro* networks and consider F-actin networks as essentially elastic at the timescale of high-resolution live cell microscopy. Even when neglecting viscous properties and focusing on values of the elastic modulus, published magnitudes range from <1-Pa (Yanai *et al.*, 2004), 1-Pa levels (Kole *et al.*, 2004; Marion *et al.*, 2005; Wang, 1998; Wang *et al.*, 2001), up to several tens (Tseng *et al.*, 2002; Yamada *et al.*, 2000) or even hundreds of Pa (Bausch *et al.*, 1999). Again, the elastic modulus has also been found to vary between cell types, between individual cells, and between different regions of the same cell. Thus, in order to reconstruct absolute force levels, each experiment requires a careful calibration of the elastic properties.

Second, the most critical distinction between an *in vivo* F-actin cytoskeleton and *in vitro* actin networks is the continuous assembly and disassembly of the former (Diez *et al.*, 2005; McGrath *et al.*, 1998; Ponti *et al.*, 2003, 2005), rendering the material highly plastic. Accordingly, there is no constant global relaxation state of the F-actin network in live cells over the timescale of network remodeling (Wang, 1998); rather, the networks appear to have mainly viscous properties at these timescales. The rates of F-actin turnover vary strongly between cell types and cell regions. Thus, the broad range in the reported ratios of elastic to viscous modulus in cell rheological studies may also be related to variable conditions of cytoskeleton turnover. In most *in vivo* studies, the turnover rate is not known, and its implications for quantitative viscosity measurements are rarely addressed in detail. However, qFSM provides us with precise information on F-actin assembly and disassembly rates in addition to F-actin deformations. This allows us to determine the timescale of network remodeling and to identify the time interval over which network deformations are nonplastic.

The consequences of material plasticity for the reconstruction of forces from F-actin network deformation are illustrated in Fig. 2B, with the analogy of a system of linear springs. For the long timescales at which remodeling takes place, breaking (disassembly) of some springs and formation (assembly) of others lead to a change of the ensemble resting length (relaxation state) and/or the spring constant of the system. Therefore, Hooke's law of a linear relationship between force and spring extension is no longer valid. As an example, if we were to apply Eq. (1) to the spring extension between t_1 and t_2 in Fig. 2B, implicitly assuming identical resting lengths L_1 in both time points, we would significantly overestimate the force level in t_2 (arrow with dashed line), even when the ensemble spring constant were to remain approximately constant. On the other hand, on timescales δt significantly shorter than the remodeling process ($\delta t \ll t_2 - t_1$), plasticity is not manifested. Consequently, Hooke's law still holds for the spring system and, equivalently, linear elastic properties can be assumed for 2D and 3D F-actin networks. Low plasticity conditions apply to the spring extension both between t_1 and $t_1 + \delta t$ and between t_2 and $t_2 + \delta t$ in Fig. 2B. To describe the mechanical properties of such a plastic system at a specific moment, we introduce the concept of *transient elasticity*, which captures the linearity of the relationship between force development and transient F-actin network deformation over short time intervals. Using this concept, the changes in the velocity of F-actin network flow recorded at a short timescale are the microscopic readout of the network deformation induced by *transient* intracellular forces.

Clearly, it is critical to know the timescale of plasticity. In the next section, we show that for the specific cell system studied, the timescale is <40 sec. Hence, F-actin network deformations can be analyzed in the regime of linear elasticity at sampling rates of $\sim$10 sec per frame or faster. To capture variations of transient elasticity at timescales of minutes and longer, it would be critical to measure elasticity parameters with a spatial resolution comparable to the deformation. In Section III, we provide an outlook for a promising approach to intracellular elasticity measurements, which also capitalizes on the information contained in FSM movies. In reference to similar methods using microbeads or organelles as intracellular probes (Chapter 6 by Panorchan *et al.*, Chapter 7 by Crocker and Hoffman, this volume), we refer to this approach as speckle microrheology. The advantage of speckle microrheology is that it is based on the same image data used for flow measurements, thus resolving spatiotemporal heterogeneity in the mechanical properties of *in vivo* F-actin networks at matching temporal and spatial scales.

In summary, this section established that for the purpose of intracellular force reconstruction, *in vivo* F-actin networks can be modeled as linear elastic material over a time window of one to a few seconds. On this timescale, deformations of the F-actin cytoskeleton, as measured with molecular specificity by fluorescence microscopy, result from transient intracellular force changes throughout the cell. By analyzing these changes over multiple time steps, it is possible to monitor the regulation of force generation on a relative scale. To define the force level on an absolute scale, additional measurements of the variable elasticity of the F-actin cytoskeleton are required.

3. Probing Actin Cytoskeleton Dynamics by FSM

The previous sections laid out basic requirements for the measurement of F-actin network deformation, turnover, and transient elasticity in order to reconstruct intracellular force distributions. Network deformations must be derived from spatial gradients in retrograde flow over as short a distance as possible due to spatial variations in the network architecture, yet at length scales sufficiently long such that the network can be described as a continuum of cross-linked and branched filaments without influence of single-filament dynamics. Furthermore, F-actin assembly and disassembly must be monitored to determine a timescale sufficiently short such that the mechanical behavior is determined by transient elasticity instead of network plasticity. Thus, force reconstruction by inverse dynamics would greatly benefit from direct, intracellular measurements of F-actin dynamics at spatial and temporal scales comparable to those of the deformation data. qFSM is a light microscopical modality that can essentially fulfill these requirements. Several reviews of the qFSM technique have been published (Danuser and Waterman-Storer, 2003; Waterman-Storer and Danuser, 2002; Waterman-Storer *et al.*, 1998), the most recent and comprehensive of which may be found in Danuser and Waterman-Storer (2006). Here, we focus on aspects of qFSM relevant to intracellular force reconstruction.

FSM utilizes high-resolution epifluorescence digital light microscopy to analyze the movement, assembly, disassembly, and subunit turnover within macromolecular structures *in vivo* and *in vitro*. It capitalizes on fluorescent analogue cytochemistry in which either purified protein is covalently linked to a fluorophore and microinjected or else the protein is expressed as a GFP fusion construct in living cells. Fluorescently labeled subunits coassemble with endogenous, unlabeled protein, allowing visualization of the localization and organization of structures such as the F-actin cytoskeleton (Fig. 3A). Conventionally, fluorescent signal is maximized by injecting or expressing labeled subunits in high abundance, which limits the ability to report movement or turnover of proteins within assemblies because of the uniform fluorescent labeling. In addition, the fluorescent signal is often blurred by high background fluorescence from unincorporated and out-of-focus fluorescent subunits. These problems are alleviated by FSM.

FSM was discovered by accident when it was noticed that in high resolution, high magnification, and electronically amplified images of very dim cells, due to the injection of apparently too low amounts of X-rhodamine actin, F-actin structures exhibited variations in fluorescence intensity and looked speckled (Fig. 3B). Computational simulations of the random incorporation of subunits into F-actin networks and microtubules (Ponti *et al.*, 2003; Waterman-Storer and Salmon, 1998) revealed that these intensity variations correspond to local fluctuations in the number of fluorescent subunits per resolution-limited image region (Fig. 3C). In conventional modalities of fluorescence microscopy, these fluctuations are obfuscated by the high mean level of fluorescent signal. Injection or expression of very low amounts of labeled subunits both increases local concentration fluctuations and eliminates contributions of diffusible, unincorporated subunits. Thus, a "speckle" represents the image of a circular region of ∼250-nm radius of the focused section of the F-actin network that is significantly higher in fluorophore concentration than its immediately neighboring image regions. For the force analyses of protruding lamellipodial F-actin networks in migrating cells illustrated in this chapter, the network thickness is less than the depth of field of the

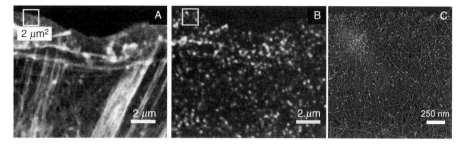

Fig. 3 Speckle formation in FSM. An epithelial cell was microinjected with a low level of X-rhodamine-labeled actin, then fixed and stained with Alexa-488 phalloidin. (A) Phalloidin image showing the organization of F-actin in filament networks and bundles. (B) X-rhodamine image of the same region. In the single FSM image much of the structural information is lost, but time-lapse FSM series contain dynamic information of filament transport and turnover not accessible with higher-level labeling of the cytoskeleton. (C) Close-up of $2 \times 2 \ \mu m^2$ window in panels (A) and (B). The colorized speckle signal is overlaid onto a quick-freeze deep etch image of the same-sized region of the actin cytoskeleton in the leading edge of a fibroblast (kindly provided by Tatiana Svitkina) with a hypothetical fluorophore distribution that could give rise to such speckle pattern. This indicates the scale of FSM in comparison with ultrastructure of the polymer network and illustrates that a small proportion of the total actin fluoresces and that fluorophores from different filaments may contribute to the same speckle. Importantly, the size of a speckle spans several meshes of the network. Accordingly, FSM reports dynamics of actin networks at a scale where single-filament dynamics are homogenized into a continuum. Thus, the tools of continuum mechanics are appropriate to interpret FSM measurements of filament transport and to extract the forces driving it.

microscope (600–800 nm). Therefore, network deformations are captured in one optical section and can be considered effectively 2D.

In theory, optimal contrast is obtained when speckles are formed by a single fluorophore per resolution-limited region (Watanabe and Mitchison, 2002). For F-actin, 80% or more of the speckles should arise from a single fluorophore when the fraction f of labeled monomers in the network is less than 0.01%. In practice, however, the lower bound for f is determined by two factors: (1) The noise level and sensitivity of the imaging system, instability of the microscope, and the dynamics of the observation process may all degrade the feasibility of single fluorophore detection. (2) Very low fractions result in a very low density of speckles and a low spatiotemporal sampling of the underlying network dynamics, which reduces overall spatial resolution. Thus, it is often desirable to image speckles at fractions in the range 0.1–1%. In this range, speckles consist of three to eight fluorophores, which are generally distributed over several filaments (Fig. 3C). However, the associated scale of spatial resolution remains above the mesh size of the network and averaged range of single-filament dynamics. Thus, these speckles probe the network as a locally homogeneous continuum.

The power of FSM lies in its use for monitoring the spatiotemporal dynamics of macromolecular assemblies. In time-lapse FSM of F-actin, speckle trajectories indicate the movement of network patches at a spatial density equivalent to the optical resolution limit. Accordingly, F-actin network flow and flow gradients are sampled at 250- to 500-nm length scale. A $10 \times 10 \ \mu m^2$ cell region can contain up to 40,000 speckles per frame, resulting in several million speckles trackable over an entire movie. To capture this data, we developed algorithms for automated speckle tracking (Ji and Danuser, 2005; Ponti *et al.*, 2005; Vallotton *et al.*, 2003). Dependent on the speckle contrast, tracking is accomplished at the single-speckle level, where trajectories are captured from speckle birth to speckle death (Fig. 4A). As discussed below, these events indicate the location of temporally increased network turnover. Validation of the tracking algorithms has demonstrated that our tracking method have a very high success rate (less than 5% error) in detecting trajectory end points correctly (Ponti *et al.*, 2005). Spatial filtering of speckle trajectories and

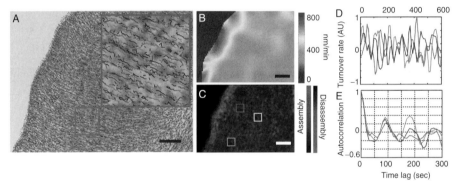

Fig. 4 Data analysis in qFSM. Speckles are detected as statistically significant local maxima in the image signal and tracked by computational single-particle-tracking methods. Generally, an FSM movie contains hundreds of thousands of speckles whose trajectories are reconstructed to obtain high-resolution maps of F-actin transport. (A) Trajectories of single speckles appearing in the first 20 frames of the movie. Time points are color encoded with dark blue for early time points and light green for late time points. Inset: Close-up of the trajectories in a $2 \times 2 \ \mu m^2$ window indicating the high density of data available to extract strain fields in the filament networks. (B) Speed map (color coded between 0 and 800 nm/min) of filament transport generated by interpolation of all speckle trajectories assembled over 60 frames. This data indicates a quasi-steady state distribution of remarkable speed gradients. Fast transport is found next to the cell edge, turning into slower transport at $\sim 2 \ \mu m$ from the edge. (C) Intensity changes in the time points of speckle appearance and disappearance are indicators of the local kinetics of filament assembly and disassembly. Appearance and disappearance events over 60 frames are compiled to generate a quasi-steady state map of filament turnover. A narrow band of average assembly is found next to the cell edge (red) and juxtaposed to a band of average disassembly (green). Together, these two bands overlap with the region of fast filament transport indicated in (B). Behind this region, the filament network assembles and disassembles in random punctate patterns. (D) Three probing windows were randomly selected in (C) and translocated with the filament transport. Assembly and disassembly alternate in these patches at a remarkably stable periodicity. The same oscillation is found in any location of the network (not shown). (E) Autocorrelation of the three turnover time series. The first side maximum occurs at 80 sec, indicating the time over which significant amounts of the network are renewed. Thus, mechanical measurements assuming transient viscoelastic properties of the filament network must be performed one order of magnitude below this timescale. (See Plate 16 in the color insert section.)

display of the flow speed as a scalar map (Fig. 4B) indicate substantial global gradients in network flow at the front of a migrating epithelial cell. Next to the cell edge, speeds of up to 800 nm/min are measured, which decrease to 300–400 nm/min over a very short distance ~2 μm from the cell edge. Clearly, in this region forces counteracting the retrograde transport of actin polymer must be present. As discussed in Section II.C, flow gradients of comparable or even lower magnitude permit the reconstruction of force distributions throughout the entire cell region in the field of view.

The end points of speckle trajectories mark regions where fluorescent monomers exchange actively with subunits in the F-actin network, associated with filament assembly and disassembly. Speckles appear because network assembly yields local accumulation of new fluorescent monomers, or because network disassembly in the immediate surroundings of a newly forming speckle leads to removal of incorporated fluorescent monomers, thereby increasing the contrast between speckle foreground and background. Conversely, speckles disappear because network disassembly dissociates fluorescent monomers from an existing speckle, or because network assembly in the immediate surroundings causes the addition of fluorescent monomers, thereby decreasing the contrast between speckle foreground and background. In addition to the cause of speckle appearance and disappearance, statistical comparison of intensity change at each trajectory's end point with the noise level of the image can indicate the significance of the event in terms of its ability to report true F-actin turnover and the monomer exchange rate (Ponti *et al.*, 2003). This information can be compiled into maps revealing the kinetics of F-actin turnover at a high spatial and temporal resolution (Fig. 4C). For example, the map in Fig. 4C suggests that network polymerization is present in a band subjacent to the cell edge, while fast depolymerization takes place in a band 1–2 μm from the edge. Quantitation of turnover rates in these bands shows that 70–85% of the polymer generated at the leading edge is disassembled after retrograde travel of only 2 μm. This is a direct indicator of the high plasticity of the F-actin network. Given the flow rates of 800 nm/min in this region, we conclude that filaments are not stable for more than 2 min.

Turnover of F-actin is even more dramatic in the region behind the first 2 μm from the edge. Here, static maps display a random pattern of assembly and disassembly puncta, which in time-resolved analyses correspond to local periodic F-actin turnover. Figure 4C depicts the location of three probing windows randomly placed in the network. The windows were displaced through the map to follow the flow field in Fig. 4A over time. Integration of the turnover rates in these moving windows thus defines the assembly and disassembly behavior of a local patch of the F-actin network. The turnover obeys a temporal periodicity of remarkable stability, independent of the location of the window (Fig. 4D). For a noisy, yet periodic signal $s(t)$ with zero mean and variance σ_s^2, sampled in N time points t_i, $i = 1 \ldots 1/4\ N$, the normalized autocorrelation function $C(\tau) = 1/(N\sigma_s^2)\sum_{i=1}^{N} s(t_i - \tau)s(t_i)$ displays side lobes with maxima at time lags $\tau_k = kT$, at multiples of the period T. For the sample signals in Fig. 4D, the maxima of the first side lobe ($k = 1$) locates at $T = 80$ sec (Fig. 4E), indicating that F-actin networks remodel at a timescale of $\sim$80 sec in the epithelial cells studied here. Thus, a sampling interval of 10 sec between frames is at the upper limit to allow the assumption of transient elasticity for the force reconstruction. In practice, we sample at rates between 1 and 10 sec per frame, trading timescales that minimize network plasticity against stronger photobleaching of the cell sample due to repeated illumination.

Besides directed flow, single-speckle trajectories contain a random motion component associated with thermal fluctuations, uncoordinated actomyosin contraction, and noise in the speckle localization. In Section III, we will discuss how one can potentially capitalize on trajectory fluctuations to extract spatially resolved measurements of the transient network elasticity.

In summary, this section introduced qFSM as an appropriate tool to acquire the data of F-actin network dynamics necessary for the reconstruction of intracellular force distributions by an inverse dynamics approach.

B. Force Reconstruction

In this section, we introduce a numerical method to reconstruct intracellular force distributions from F-actin flow fields, assuming transiently linear elastic properties of the network. We will show that spatial heterogeneity in the elastic properties has relatively little influence on the force distribution. Thus, relative force levels can be derived based on flow velocity data only, without knowing the actual elasticity of the actin cytoskeleton. To assess absolute force levels, assumptions must be made about the overall stiffness of F-actin networks. However, this information is not required for analyzing the spatiotemporal force balance, which determines the protrusion and retraction dynamics of the leading edge.

1. Continuum Mechanical Relationship Between Forces and Network Deformation

Under force application, a linear elastic network deforms as illustrated in Fig. 5A. The deformation depends not only on the magnitude of net forces, but primarily on the relative force distribution, that is the location, the direction, and the relative magnitude of the forces exerted on a network patch. Unlike the spring, deformations of a 2D network at a given point can no longer be described by a scalar value, but is measured by a 2×2 matrix referred to as the strain tensor (Landau and Lifshitz, 1986). The strain tensor at any location of the network is calculated from the spatial variation of network displacements around this point. Importantly, nonzero strain is observed only when the displacements in the immediate neighborhood differ in either direction or magnitude or both. Otherwise, homogeneous displacement simply indicates a uniform translational movement of the network with no deformation. This principle is illustrated in Fig. 5A where the flow vectors at the four corners of a small rectangular network patch vary, resulting in deformation of the rectangle at the next time point (solid polygon).

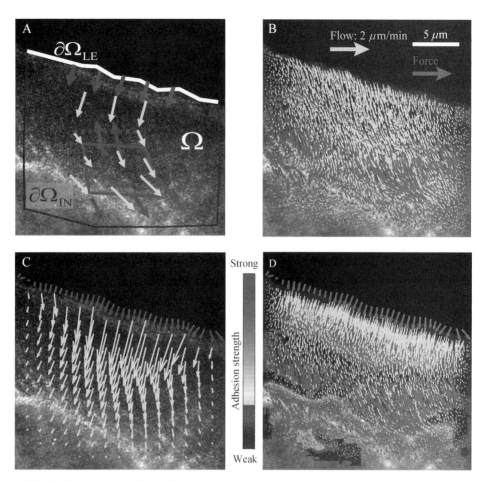

Fig. 5 Forward simulation of force-induced network deformation. (A) The deformed green rectangle illustrates how the three types of intracellular forces (red arrows) are integrated to generate the deformation of the actin cytoskeleton (yellow arrows). (B) Assuming only contraction forces are present (red arrows) with a free leading edge, the simulated flow (yellow arrows) shows an almost uniform flow pattern behind the leading edge. (C) Boundary-pushing force alone can generate huge flows that go deep into the cell, demonstrating the power of actin polymerization at the leading edge. (D) If, in addition to the forces in (B) and (C), adhesions are placed in the appropriate locations (shown as color-coded areas), the integration of all three intracellular forces generates a flow pattern that looks very similar to the experimentally measured flow of the actin network during cell migration. (See Plate 17 in the color insert section.)

Mathematically, the strain tensor is defined as:

$$\boldsymbol{\varepsilon}(\boldsymbol{x}, t) = \tfrac{1}{2}(\triangledown \boldsymbol{u}(\boldsymbol{x}, t) + \triangledown^{\mathrm{T}} \boldsymbol{u}(\boldsymbol{x}, t)) \qquad (2)$$

Here the spatial variation of the flow field $\boldsymbol{u}(\boldsymbol{x}) = [u_1(\boldsymbol{x}); u_2(\boldsymbol{x})]$ at time t is analyzed by applying the partial differential operator $\nabla = [\partial/\partial x_1; \partial/\partial x_2]$ to the flow field at location $\boldsymbol{x} = [x_1, x_2]$.

Network deformations create tension that balances the forces applied to the network. The relationship between network deformation and tension for an elastic material is given by:

$$\boldsymbol{\sigma}_{ij} = \sum_{k,l=1}^{2} c_{ijkl}\boldsymbol{\varepsilon}_{kl}, \qquad i,j = 1,2 \tag{3}$$

In this equation, $\boldsymbol{\sigma}$ denotes the 2×2 stress tensor matrix and $\boldsymbol{c}$ the elasticity tensor of order 4. Equation (3) is very similar to Eq. (1) describing the force–extension relationship for a simple spring. Instead of a scalar force, Eq. (3) contains the stress tensor; instead of a scalar extension, the strain tensor; and instead of a scalar spring constant, the elasticity tensor whose coefficients c_{ijkl} define the material properties of the 2D network. For the simplest case of a linear elastic, locally isotropic 2D network, c_{ijkl} depends on only two parameters (Landau and Lifshitz, 1986). One widely used pair of elastic parameters consists of Young's modulus E and the Poisson ratio γ. Expressing c_{ijkl} in terms of these two parameters, the stress tensor in any location $\boldsymbol{x}$ can be directly calculated from the flow vector $\boldsymbol{u}(\boldsymbol{x})$ by substitution of Eq. (2) into Eq. (3):

$$\boldsymbol{\sigma}(\boldsymbol{x},t) = \frac{E(\boldsymbol{x},t)\gamma(\boldsymbol{x},t)}{(1-2\gamma(\boldsymbol{x},t))(1+\gamma(\boldsymbol{x},t))}(\nabla \cdot \boldsymbol{u}(\boldsymbol{x},t))I + \frac{E(\boldsymbol{x},t)}{2(1+\gamma(\boldsymbol{x},t))}(\nabla\boldsymbol{u}(\boldsymbol{x},t) + \nabla^{\mathrm{T}}\boldsymbol{u}(\boldsymbol{x},t))$$

$$\tag{4}$$

I is the 2×2 identity matrix and $\nabla\cdot$ denotes the divergence of the flow vector. Notice that Eq. (4) incorporates the notions of transient elasticity and network deformation by defining all terms as time-dependent variables. In addition, the material properties are represented by space-dependent variables, which account for spatial heterogeneity of the network.

The balance between network stresses and the intracellular forces $\boldsymbol{F}_{\mathrm{II}}$ and $\boldsymbol{F}_{\mathrm{III}}$ exerted on the network inside the domain Ω (Fig. 5A) is defined as:

$$\nabla \cdot \boldsymbol{\sigma} = \boldsymbol{F}_{\mathrm{II+III}}, \qquad \text{in } \Omega \tag{5}$$

The balance between the boundary force $\boldsymbol{F}_{\mathrm{I}}$ and network stresses along the cell edge $\partial\Omega_{\mathrm{LE}}$ is given by *Neumann boundary conditions*:

$$\boldsymbol{\sigma}\boldsymbol{n} = \boldsymbol{F}_{\mathrm{I}}, \qquad \text{on } \partial\Omega_{\mathrm{LE}} \tag{6}$$

where $\boldsymbol{n}$ is the outward normal to the cell edge.

The goal of force reconstruction is to derive the boundary force $\boldsymbol{F}_{\mathrm{I}}$ at the leading edge $\partial\Omega_{\mathrm{LE}}$, and the forces $\boldsymbol{F}_{\mathrm{II+III}}$ inside the domain Ω from the measured flow field $\boldsymbol{u}(\boldsymbol{x})$ throughout Ω. Before we address an algorithm which achieves this goal, we demonstrate using the framework of Eqs. (4)–(6) that different types of intracellular forces make quite different contributions to the network flow. Hence, it should be possible to turn the table and determine from a flow pattern which forces must be active to drive such a flow field.

2. Forward Simulation of Flow Fields

To perform a simulation as realistic as possible, we assumed contraction, boundary, and adhesion resistance force fields throughout the field of view of an FSM movie for which we will later show the inverse force reconstruction from experimentally measured flows.

In the first example (Fig. 5B), only contraction forces are simulated ($F_{III} \neq 0$). The resulting flow (yellow arrows) displays almost uniform vectors pointing away from the leading edge. In the second example (Fig. 5C), only boundary forces are simulated ($F_I \neq 0$). In the absence of adhesions, the boundary force alone is able to generate retrograde flow deep into the cell. In the last example (Fig. 5D), we added adhesions to both the boundary force and the contraction force. Since resistance forces exerted by adhesions on the network are always opposite to the local flow, the adhesion force field cannot be defined *a priori*. Instead, we modified Eq. (5) by replacing the adhesion resistance force F_{II} with a friction term νu:

$$\nabla \cdot \boldsymbol{\sigma} - \nu \boldsymbol{u} = \boldsymbol{F}_{III}, \qquad \text{in } \Omega \qquad (7)$$

The formulation of adhesions as a friction component implies that two interacting polymer structures, in our case the F-actin network and the actin-binding layers of adhesion complexes, form transient bonds. The faster the two structures slide past each other, the higher is the probability for binding sites to meet and temporarily interact (Howard, 2001). This behavior has been observed qualitatively by comparing the pattern of F-actin flow fields and traction force maps in fast-moving keratocytes (Vallotton *et al.*, 2005). Regions with faster flow displayed proportionally stronger traction forces. The color coding in Fig. 5D indicates variation in the friction coefficient $\nu(x)$. Notice that spatial variations in adhesion forces are caused by spatial variations in the friction coefficient as well as in the magnitude and direction of the flow field. With all three types of intracellular forces acting together, the simulated flow field reproduces the real flow recorded by qFSM in Fig. 6B. The sequence of simulations (Fig. 5B–D) illustrates that the characteristic gradient from fast to slow flow behind the protruding edge of migrating cells occurs by adhesive coupling of the network to the substrate near the cell edge. In other words, adhesion sites modulate the rather uniform flows generated by polymerization and contraction-induced forces.

3. Inverse Reconstruction of Force Fields from Experimental Flow Data

Force reconstruction in the framework of Eqs. (4)–(6) seems straightforward if one can assume a spatially homogenous material. Even when the absolute elastic modulus is unknown, relative force distributions can be extracted throughout the field of view by calculating the stress tensor $\boldsymbol{\sigma}(x,t)$ from the flow field $\boldsymbol{u}(x,t)$ using Eq. (4) and substitution of $\boldsymbol{\sigma}(x,t)$ into Eqs. (5) and (6). The effect of network heterogeneity will be discussed in Section II.F. However, this approach is numerically unstable because the needed second-order differentiation is sensitive to noise

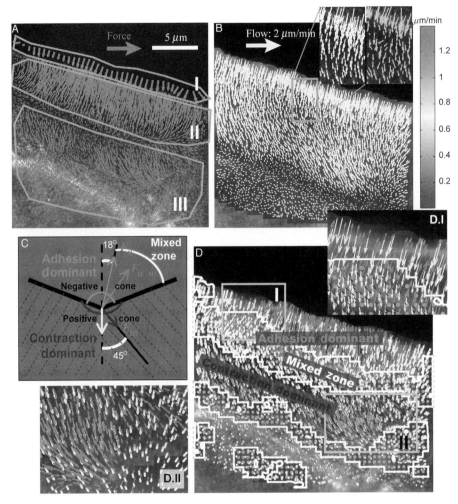

Fig. 6 Example of force reconstruction and the cone rule for the separation of adhesion and contraction. (A) Force field reconstructed from an FSM movie of the actin cytoskeleton. Clearly visible are the three types of intracellular forces: I, boundary-pushing force; II, adhesion resistance force; and III, contraction force. (B) The flow field is recalculated from the reconstructed forces shown in (A) and overlaid onto the speed map of the network flow. Due to the implicit filtering by *Tikhonov regularization*, there is much less random fluctuation in this recalculated flow field (yellow arrows) than in the original flow data (green arrows) shown in the magnified image. (C) The cone rule allows us to separate the adhesion from the contraction in a semiquantitative way. See main text for details. (D) Applying the cone rule to the force distribution shown in (B), we obtain this categorized force map, where the adhesion-dominant region is color coded by the force magnitude, and the mixed zone is marked by gray dots with a white line boundary. In the mixed zone, the separated components of contraction forces, although small (blue arrows in I), are responsible for the consistent bending of the flow away from the edge normal shown in the magnified image I. (See Plate 18 in the color insert section.)

(first the gradient and the divergence of u are calculated, followed by the divergence of σ). In the field of inverse problems, such cases are referred to as ill-posed. In some specialized cases, integration schemes that suitably sum over imaging data can extract traction forces from intrinsically noisy data (Lee *et al.*, 1994), but less restrictive approaches are possible as described here.

In general terms, the goal of solving an inverse problem is to recover physical quantities from data that cannot directly probe them. Examples of inverse problems outside cell biology are as diverse as the detection of oil reservoirs from surface recordings of seismic waves (Symes and Carazzone, 1991) and the nondestructive localization of cracks from optical measurement of the integral strain of loaded material (Yang *et al.*, 2002) or from electrostatic measurements at the boundary (McIver, 1991). Inverse approaches have also been exploited in biomedical diagnostics, for example, for the detection of tumors in mammograms (Li *et al.*, 2003) and blood clots in lungs in electrical impedance tomography (Cheney *et al.*, 1999).

Clearly, the reconstruction of intracellular forces formulates an inverse problem at the macromolecular scale. To circumvent its ill-posed nature, we transform the problem into an optimization problem that searches among all possible force fields for the one minimizing the difference between predicted and experimentally recorded flow. To reduce the effect of measurement noise in the flow data on the minimization, a so-called *regularization term* is added to the objective function acting effectively as a spatial low-pass filter on the measurements. For the mathematically inclined reader, we provide some detail of this solution strategy in the Appendix A.

The algorithm returns the boundary force F_{I} along the cell edge, which reflects the force generated by network polymerization against the plasma membrane, and the domain force $F_{\text{II+III}}$, which includes both adhesion and contraction forces throughout the cell. Figure 6A shows the reconstructed forces F_{I} and $F_{\text{II+III}}$ for a protruding newt lung epithelial cell. The force field agrees well with textbook models of cell migration (Lauffenburger and Horwitz, 1996). At the leading edge, a strong boundary force is generated in response to the forces that push the plasma membrane forward (Fig. 6A.I). Subjacent to the cell edge, a region of forward directed forces is reported, resisting the retrograde flow of the network (Fig. 6A.II). This agrees with traction force maps that suggest strong force development in nascent adhesions at the leading edge (Beningo *et al.*, 2001). A zone of network contraction with opposing retrograde and anterograde force vectors is localized ~10 μm behind the cell edge (Fig. 6A.III).

In this specific case, cell adhesion and contraction are spatially well separated by a region with small forces. It should be noted that, despite the absence of force, network flow in this region is of a magnitude similar to that in the contraction region behind it. Conversely, while anterograde flow at the bottom of the field of view is very slow, the force reconstruction reveals quite strong forces pointing forward. These two observations illustrate that force maps provide a much more sensible means to study mechanical functions in cells. Fast flow does not necessarily indicate strong force or vice versa. The mechanical information in this data is hidden in the spatial variation of flow, which is often difficult to interpret by simple inspection.

The impact of the regularization scheme used to solve the inverse problem is demonstrated in the zoom-ins of Fig. 6B, comparing the original flow field (green arrows) and the flow field recalculated from the reconstructed force distribution (yellow arrows). The latter exhibits much smoother behavior due to the implicit filtering effect of regularization, but preserves the gradient from fast to slow flow, indicative of adhesion at the cell edge.

4. Separation of Contraction and Adhesion Forces

The qualitative distinction of contraction and adhesion in Fig. 6A relied on two assumptions:

1. In adhesive regions, force vectors are directed antiparallel to the flow.
2. In contractile regions, force vectors are directed parallel to the flow.

This behavior can be formulated more quantitatively by the so-called *cone rule* illustrated in Fig. 6C. The rule partitions the direction space around a flow vector into a negative cone (directions opposing the flow vector) and a positive cone (in the direction of the flow vector). Force vectors falling into the negative cone are classified as adhesion-dominated forces, force vectors falling into the positive cone as contraction-dominated forces. The division of the negative and positive cones is not symmetric but defined as the midline between two narrow cones reflecting the expected variation of adhesion and contraction force directions, respectively. For the negative/adhesion cone, we assume an opening half-angle of 18°. Force vectors falling into this cone are strictly classified as adhesion forces. For the positive/contraction cone, we assume an opening half-angle of 45°. Our choice of different openings for the adhesion and contraction cones is based on the observation that adhesion forces must, by definition, exactly oppose the flow direction, while contraction forces do not necessarily need to point in the forward direction of flow. The domain between the narrower (18°) negative/adhesion cone and the division line is identified as the *mixed zone*. Force vectors in this zone are divided into projections onto the two smaller negative/positive cones (Fig. 6C), reflecting the fact that although adhesion forces dominate, the net forces must include a significant contribution from contraction. Forces in the domain between the narrower (45°) positive/contraction zone and the division line may contain minor contributions from adhesion. However, it is equally conceivable that contraction forces alone form an angle with the flow direction greater than 45°. Thus, we do not classify this as a mixed zone. Figure 6D illustrates the force classification according to the cone rule. Adhesion-dominant regions are color coded by the force magnitude. The zoom-in of the mixed zone (Fig. 6D.I) shows how contraction forces in an adhesion-dominated region begin to bend the flow field toward increasingly stronger contraction hubs.

We emphasize the semiquantitative nature of the cone rule. The choice of the two opening angles is to some extent arbitrary and necessarily relies on statistical analyses of force directions. With more noise, the boundaries between adhesion and contraction tend to be more blurred than in this example. Also, it should be noted that the identification of a region as adhesion dominant cannot exclude the possibility of a contribution from contraction forces, even inside the narrow adhesion cone. Nevertheless, the force map categorized by the cone rule can already help the quantitative interpretation of reconstructed forces. For more accurate force identification, it will be necessary to combine the force reconstruction with traction force microscopy and/or the imaging of proteins localized in adhesion sites.

5. Force Maps in Protruding and Retracting Sectors of a Migrating Epithelial Cell

In this section, we demonstrate how intracellular force maps may be used to distinguish mechanisms of cell protrusion and retraction, providing novel insights into force balances at the leading edge of migrating cells (Fig. 7). For this purpose, we select an FSM data set where the observed cell undergoes retraction (left) and protrusion (right) at the same time. Figure 7A displays the reconstructed adhesion forces. Behind the protrusion, strong adhesion resistance is indicated, while such a resistance is absent behind the retracting region. This observation provides direct mechanical evidence that cell protrusion is coupled to the formation of adhesive links between the cytoskeleton and extracellular domain subjacent to the cell edge. Figure 7B shows the contraction forces. Significant contractions are localized immediately behind the retracting portion of the cell edge, suggesting that retraction is promoted by network contraction in the absence of adhesion. The mechanism involved in the contraction of this part of the cytoskeleton is unclear. We have previously found that retrograde flow in the lamellipodium, that is the $\sim$2-μm -wide band next to the cell edge which undergoes rapid polymerization and depolymerization (cf. Fig. 4C), is independent of the activity of myosin II motors (Ponti *et al.*, 2004). However, the present force maps, derived from flow gradients, indicate that another contractile force might be present in order to explain the movement of F-actin in this region. We speculate that our maps might capture the action of an unconventional myosin, or that depolymerization at the base of the lamellipodium could be responsible for this effect. These data show that a detailed mechanical analysis of cytoskeleton dynamics can reveal unexpected processes, which may in turn lead to new and testable hypotheses of the mechanisms of force-dependent cell functions.

Similarly, a comparison of adhesion forces in Fig. 7A with boundary forces in Fig. 7B suggests a strong correlation between the variation of boundary force magnitude and the strength of focal adhesion resistance. Each maximum of boundary force in the protrusion zone is accompanied by a region of strong adhesion force behind the cell edge. It should be noted that this correlation is found at a spatial

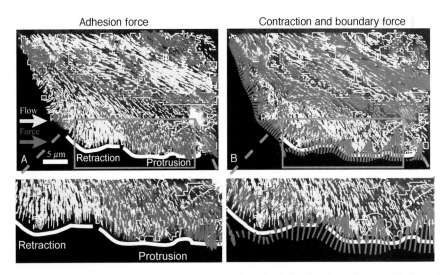

Fig. 7 Intracellular forces as more pronounced probes of cellular functions than speckle flows for addressing unique biological questions. In this example, force maps for adhesion and contraction are shown separately, adhesion forces only in (A), contraction forces and boundary-pushing forces in (B). Five biologically meaningful observations can be made by comparing the two force maps (forces = red arrows, flow vectors = yellow arrows, adhesion sites = color-coded area): (1) adhesion regions that are not easily apparent from the flow map are identified between strong contraction zones; (2) strong adhesion forces are observed only behind the protruding edge and are missing from the retracting portion as shown in (A); (3) a significant amount of active contraction force, as shown in (B), appear just behind the retracting edge, suggesting that contractions provide the main driving force of cell edge retraction; (4) even at the retracting edge, a remaining boundary force can still be detected, raising the question of whether it is still powered by leading edge polymerization; (5) the boundary force along the protruding edge shows a variation in its pattern that correlates with variations in the reconstructed adhesion force magnitude. (See Plate 19 in the color insert section.)

resolution below 1 μm. The close match between the spatial variations in the two force categories suggests the existence of a signaling mechanism that links an F-actin network-coupled ECM adhesive mechanism to F-actin polymerization at the cell edge (Geiger and Bershadsky, 2001). The possibility offered by such reconstructions to pinpoint spatial and temporal relationships between different types of forces will critically contribute to our understanding of how mechanical and chemical signals integrate in cell functions such as migration.

6. Effect of Heterogeneous Network Elasticity

As explained in the introduction, the reconstruction of forces from sample deformation requires knowledge of the sample's material properties. Whereas in the simple spring system the force magnitude defines the only extractable quantity, 2D and 3D reconstructions seek to provide information on the spatial distribution of forces, their local directionality and magnitude. The latter can be defined on an absolute scale only if the material properties are known in every location of the sample. For live cells, this information is currently inaccessible. However, for questions addressing dynamic force regulation, the reconstruction of absolute magnitudes is often unnecessary. Instead, it is sufficient to know where and when tension develops and in which direction relative to other forces. These data could be directly derived from the flow field if the cytoskeleton was homogeneous, as assumed for the calculations illustrated in Figs. 6 and 7. Naturally, the assumption of homogeneous cytoskeleton material properties seems overly simplified. Thus, we ask to what extent reconstructed force distributions still reflect meaningful force relationships, even when homogenous elasticity of the F-actin network has to be assumed in the absence of more complete information.

To address this question, we compared force distributions reconstructed with different assumptions of network elastic properties. We recorded a two-color qFSM movie of X-rhodamine actin (red) and GFP-vinculin (green), one of the major components of adhesions (Fig. 8A). The X-rhodamine signal shows significant heterogeneity, indicating, for instance, the existence of stress fibers. Presumably, stress fibers and other regions of higher F-actin density are stiffer than the rest of the network. To test how such heterogeneity affects force reconstruction, we calculated a map of heterogeneous Young's modulus assuming it is proportional to the F-actin concentration as indicated by X-rhodamine fluorescence (Fig. 8D). The resulting variation range of one order of magnitude is in good agreement with the stiffness heterogeneity reported by high-resolution AFM mapping (Chapter 15 by Radmacher, this volume). We then reconstructed the force fields from F-actin flow, assuming homogenous elasticity (Fig. 8B and C) or heterogeneous elasticity (Fig. 8E and F). The two force fields display very similar patterns. In both cases, the identified adhesion-dominant regions (Fig. 8B and E) correlate well with the location and intensity of the GFP-vinculin signal (Fig. 8A, green). Closer comparison of force vectors in the two zoom-in panels of Fig. 8 suggests that the distributions of both the direction and relative magnitudes are similar. Thus, it appears that the outcome of force reconstruction is relatively insensitive to elasticity variations. A possible explanation for this robust behavior is the high resolution of the qFSM-measured flow data, which warrants the determination of stable flow gradients with submicron resolution. At this length scale, variations in the average network elasticity of an interlinked network of F-actin are small compared to the magnitude of the flow gradient. Consequently, the numerical solution of the inverse problem is dominated by the strain tensor for which high-quality information is available.

C. Probing Heterogeneous Network Elasticity with Speckle Microscopy

Up until this point, this chapter has focused on F-actin flow as the feature of interest in qFSM movies. However, it is obvious that single-speckle trajectories contain both directed and random motion components, which vary with the network location (Fig. 9A). Thus, the random motion component itself could be

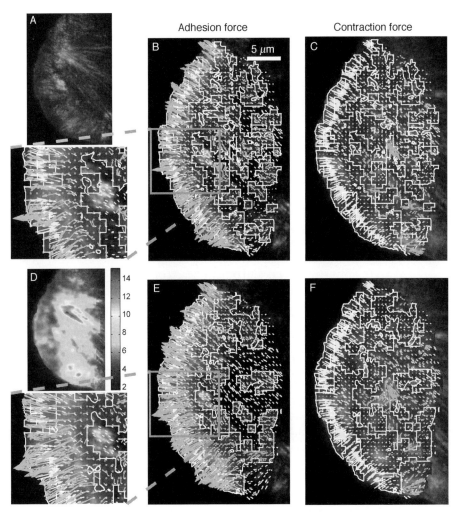

Fig. 8 Sensitivity of force reconstruction to the heterogeneity of network elasticity. This example shows that, even in the absence of an accurate *in vivo* elasticity map of the actin cytoskeleton, the relative force distribution in terms of force locations and directions can still be robustly reconstructed. Row I (A, B, and C): The contraction and adhesion forces reconstructed with the assumption that the actin network elasticity is homogeneous with a constant Young's modulus of 2 Pa. Forces are reconstructed from the actin flow of a two-channel FSM movie of actin (red) and vinculin (green). One frame of the overlaid images is shown in (A). Row II (D, E, and F): The contraction and adhesion forces reconstructed with the assumption that the network elasticity is heterogeneous, with Young's modulus varying over almost one decade as shown by the color map in (D). Note the similarity between the two sets of results. (See Plate 20 in the color insert section.)

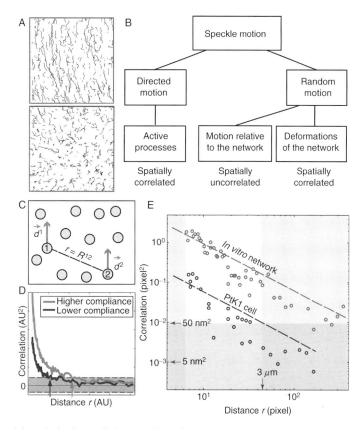

Fig. 9 Deriving relative intracellular compliance from correlated speckle movement. (A) Examples of speckle trajectories in different regions of a cell; trajectories are a superposition of varying contributions of directed and random motion. (B) The directed motion component is generated by active processes in the cell such as polymerization or contraction forces. The random motion component has two origins. One is "positional noise," which is the apparent or real motion of speckles relative to their surrounding network. The other source is random small-scale deformation of the polymer network itself. Unlike positional noise, polymer fluctuations are spatially correlated since they are transmitted through the material. (C) Principle of two-point microrheology: in an elastic medium, a force displacing probe 1 is transmitted through the medium to also displace probe 2; the correlation of the motion of neighboring probes $D(r,\tau)$ is the displacement vectors' dot product, a function of the distance r and the time interval of displacement τ. (D) Simulated correlation function $D(r)$ in an ideal 3D medium (where it represents the $1/r$ strain field decay), with added "measurement noise." The magnitude of the correlation function increases with the compliance of the material; thus, the cutoff distance at which the signal amplitude drops below noise threshold (shaded gray area) is smaller for low compliance (blue arrow) than for high compliance (red arrow). (E) Double logarithmic plot of the measured correlation $D(r)$ (for $\tau = 0.06$ sec) in an *in vitro* actin network (~ 0.1 Pa) and in a PtK1 cell. For low-compliance samples such as cells, we expect useful measurements only in a small range of distances (white window). For distances that are too large (yellow-shaded area on the right), the correlation signal drops below the noise threshold (gray, in this example 50 nm^2). For distances that are too small (yellow band on the left), the measured correlation can become biased, since not all displacement directions are detected with equal probability close to the optical resolution limit—the measurement thus favors relative movements away from each other.

indicative of spatial differences in F-actin dynamics. In this section, we provide an outlook of how random speckle motion could potentially be exploited to probe the spatiotemporal heterogeneity of network elasticity, seeking the analogy to intracellular two-particle microrheology (Chapter 7 by Crocker and Hoffman, this volume).

Random motion of speckles has a number of sources (Fig. 9B). First, speckle images are very noisy, leading to substantial fluctuations of the apparent speckle positions. Second, even though speckles are considered fiduciary markers, their locations within the F-actin network may fluctuate. Since speckles are clusters of several fluorophores distributed over one or multiple filaments, loss or addition of a fluorophore can lead to a shift in the centroid of the intensity distribution, resulting in a shift of the speckle position relative to the local architecture of filaments. Alternatively, random motion may reflect physical movements of fluorophores within the speckle-forming cluster of fluorophores, induced, for example, by the fluctuation of a locally unlinked filament in the network. Superimposed onto these random motions of a speckle relative to the network, the network itself is also subject to small-scale deformations, causing speckles to move with it. In a cell, both thermal and nonthermal forces can cause such deformations. How the associated forces are transmitted through the network is determined by the network viscoelastic properties.

How might the two types of random speckle motion be distinguished? Speckle fluctuations relative to the network are spatially uncorrelated. Network fluctuations, on the other hand, are spatially correlated since force-induced deformations can propagate through the network. Thus, a local force displacing one speckle will also displace its neighbors to a certain degree. The pairwise correlated displacement decays as a function of the speckle-to-speckle distance. The magnitude of the decaying correlations is characteristic of the material. It should be noted that this approach assumes that the random motion component has previously been separated from the directed motion component (Fig. 9B), which, of course, can also be spatially correlated. The "randomness" of the presumably random component of speckle trajectories can be validated by analysis of the mean-squared displacement (MSD) of individual speckles as a function of time. The random motion component should be diffusive or subdiffusive, while a superdiffusive MSD curve indicates directed motion.

1. Speckles as Probes for Two–Point Microrheology

Because cells actively react to externally applied forces, it is essential to develop techniques to measure intracellular material properties that are nonpertubative to the parameters of interest. Relating the characteristics of the passive, random movement of probes in a viscoelastic material to its mechanical properties is the realm of passive microrheology (MacKintosh and Schmidt, 1999; Mason and Weitz, 1995), which classically uses embedded colloidal probes such as beads (Chapter 6 by Panorchan *et al.* and Chapter 7 by Crocker and Hoffman, this volume). While this approach usually necessitates knowledge of the probe size and its viscous drag on the surrounding material, two-particle microrheology overcomes this limitation (Crocker *et al.*, 2000; Levine and Lubensky, 2000) by considering the pairwise correlated motion of neighboring particles (Chapter 7 by Crocker and Hoffman and Chapter 11 by Lammerding *et al.*, this volume). Two-particle microrheology using embedded colloidal tracers has been applied to recover the viscoelastic parameters of *in vitro* actin networks (Gardel *et al.*, 2003; Valentine *et al.*, 2004).

As fiduciary markers, fluorescent speckles lack a meaningful size and a well-defined Stokes drag. However, since these parameters are irrelevant to two-particle measurements, we can extract mechanical information of F-actin networks by examining the pairwise correlated motion of randomly driven speckle fluctuations. Additionally, speckles have the clear advantage that they are molecularly specific cytoskeletal markers; this allows us to directly ascribe the measured material properties to, for example, the F-actin or the intermediate filament network. Actin speckles are also fairly uniformly distributed throughout the cell, while microinjected or endocytosed particles may not locate to all places of interest. This advantage, in combination with the high speckle density, also lends the technique a high degree of spatial resolution. Particle-based measurements typically operate with some tens of probes inside a cell, yielding some hundreds of probe pairs per frame and requiring on the order of 10^4 frames for averaging. Speckle images typically contain several thousand speckles, yielding 10^6–10^7 probe pairs in a single frame, so that even movies of a few hundred frames produce enough data to allow spatial binning.

In practice, we measure the average pairwise correlated movement of neighboring speckles as a function of their distance from each other. If an individual displacement vector of a probe is

$$d(t, \tau) = (x(t + \tau) - x(t), y(t + \tau) - y(t))$$

at time t and for a lag time τ (Fig. 9C), then the correlation is the dot product

$$D_{\alpha\beta}^{ij}(r, \tau) = \langle d_\alpha^i(t, \tau) d_\beta^j(t, \tau) \delta[r - R^{ij}(t)] \rangle_{i \neq j, t}$$

of the displacement vectors, where i, j are the indexes for the two particles and α, β are the indexes for the coordinate axes. Since the distance-dependent correlation represents spatial decay of the strain field, it is expected to decay approximately as $1/r$ where r is the distance (see simulated trace in Fig. 9D). Thus, we fit the correlation function with the general expression $D(r, \tau) = A(\tau)/r$ and interpret the correlation amplitude A as a linear relative measure of local material compliance. If the compliance is increased in a certain region of the cell (i.e. the material is softer), this is reflected by larger speckle displacements, and thus by a larger magnitude of $A(\tau)$ in that region. It should be mentioned that in materials at thermal equilibrium such as *in vitro* actin networks, the magnitude of the thermal forces deforming the network is known, so that the compliance can be directly related to an absolute value of the elastic modulus. This is more difficult in live cells since these are governed by forces of both thermal and nonthermal origin (Lau *et al.*, 2003), for example the random activation of motors. In spite of this limitation, speckle microrheology can still map out spatial heterogeneities in the effective compliance, which is all the more important since it represents a nonperturbing measurement. Figure 9D shows two simulated traces representing the $1/r$—decay of the correlation function for two different compliances, with superimposed noise, which comprises both error in the position/displacement measurement and random contributions to the correlation from position fluctuations of speckles relative to the network. This figure demonstrates that due to the noise floor (represented by the gray-shaded area), the correlation measurement deteriorates very quickly with increasing speckle-to-speckle distance. For a given noise level, the "cutoff" distance, that is the distance at which the decaying amplitude disappears in the noise, depends on the material compliance. The noise level, in turn, is partly determined by the typical error of measuring individual speckle positions/displacements in noisy images and partly by the amount of averaging in the experiment. Figure 9E shows two examples of experimentally measured correlation functions of speckle displacements (the estimated noise level again shaded in gray), in a soft cross-linked *in vitro* actin network with an elastic modulus of $\sim$0.1 Pa and in a PtK1 epithelial cell. The functions (both for $\tau = 60$ msec) are plotted on a double logarithmic scale so that the expected $1/r$ decay corresponds to a straight line (dashed line). The correlation amplitude of the intracellular measurements (blue) is more than an order of magnitude smaller than the *in vitro* network (red); notably, both data sets decay relatively precisely with $1/r$. Since the estimated noise threshold is in the range of 5–50 nm^2 in these experiments, the cutoff distance for intracellular measurements may be on the order of a few micrometers (yellow-shaded area in Fig. 9E).

While the results show that speckle microrheology can be successfully applied as an *in vitro* assay and that relative compliances can also be measured in cells, more experiments will be needed to quantify the exact range of the corresponding elastic moduli that are accessible in cells with this method. The highest published estimates for the stiffness of an adherent cell's lamellipodium as measured by currently available techniques are in the range of some hundreds of Pascals, and the thermal displacements predicted by this stiffness would actually be too small to be detected above the estimated noise floor. However, it is possible that a true intracellular microrheological technique would probe a much softer elasticity and/or that the measured correlations are produced by nonthermal displacements of much larger magnitude. Assuming a linear relationship, the measured cellular correlation amplitude A can yield maps of the relative F-actin compliance with a spatial resolution in the single micron range (Fig. 10). Despite the high noise level, these maps indicate a variation of the relative compliance by a factor of about 5 over distances of <2 μm. These data can provide an invaluable input for the force reconstruction by inverse dynamics, by constraining the elasticity in Eq. (4); additionally, speckle microrheology can be used for the analysis of intracellular material heterogeneity implicated in other dynamic cell functions.

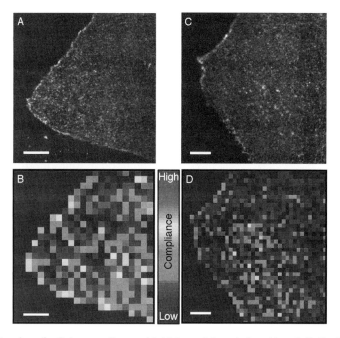

Fig. 10 Mapping of cellular compliance with high spatial resolution. (A and C) Confocal speckle image of contact-inhibited PtK1 cells, pixel size 67 nm. (B and D) Spatial maps of the relative cellular compliance (i.e., amplitude A of the correlation function). Panel A: spatial binning 1.3 μm. Panel B: spatial binning 1 μm. In all images, scale bars are 5 μm. (See Plate 21 in the color insert section.)

III. Summary

Innumerable cell functions depend on the precise coordination of intracellular forces, frequently in conjunction with local structural reorganization. In order to study these processes and to elucidate the regulatory pathways involved, intracellular force measurements need to be made with high resolution, both in space and time. In this chapter, we proposed a force reconstruction method based on inverse dynamics, which allows us to derive continuous intracellular force fields with second scale time resolution and submicron spatial resolution from the deformation of the F-actin cytoskeleton. The method rests on two pillars: First, nonplastic, that is elastic, deformations of F-actin structures can be measured by qFSM if measurements are performed at timescales between 1 and 10 sec per frame. Second, the intracellular forces that must have been applied to the network to induce the observed elastic deformations can be deduced from the microscopy data by a continuum mechanical model of the F-actin cytoskeleton. Relative force distributions can be extracted even in the absence of a precise spatial profile of the network elasticity, at the timescale of network deformation observed in FSM. In its future development, the force reconstruction method can be easily extended to 3D, and viscosity can be incorporated into the model for those cell systems where it is proven to be significant. However, major challenges will have to be tackled regarding the implementation of high-resolution imaging that allows the measurement of deformations over submicron distances. The basic idea of reconstructing forces from structural deformation by using inverse dynamics and regularization techniques is applicable to other macromolecular assemblies once the mathematical relation between force application and the microscopically observable deformation is known. Additionally, we also propose an emerging speckle-based method that will allow us to probe spatial heterogeneity of *in vivo* actin network elasticity with high spatiotemporal resolution. Both methods are still in their infancy, and it will take significant further development to generalize them for everyday use in cell biology laboratories. However, the various experimental results presented in this chapter indicate that the idea of force analysis from motion in high-resolution live cell microscopy defines a promising approach to studying cell mechanics, complementary to many of the other methods described in this volume. The particular strength of the proposed method is that force fields can be reconstructed at the timescale of seconds and throughout the entire microscopic field of view. This approach will potentially allow us to analyze the spatial regulation of forces inside living cells that is absolutely critical to understanding mechanical cell functions.

IV. Appendix

A. Solution of the Inverse Problem

In this section, we offer some technical detail of the force reconstruction algorithm. The cellular domain of interest (Ω in Fig. 5A) is bounded by the natural cell edge (denoted by $\partial\Omega_{LE}$ in Fig. 5A) together with the artificially drawn interior boundary (denoted by $\partial\Omega_{IN}$ in Fig. 5A). Furthermore, we also make use of so-called *Dirichlet boundary conditions* in addition to the Neumann boundary conditions defined in Eq. (6). They are either defined on the leading edge boundary

$$u = h, \qquad \text{on } \partial\Omega_{LE} \tag{A.1}$$

or on the interior boundary

$$u = h, \qquad \text{on } \partial\Omega_{IN} \tag{A.2}$$

or on the entire boundary of Ω

$$u = h, \qquad \text{on } \partial\Omega \tag{A.3}$$

dependent on the targeted solution. Here, h denotes the measured flow vectors on the boundary.

We reconstruct the boundary and domain forces by formulating two separate optimization problems. The boundary force is a function of only one independent variable, whereas the domain force is a function of two independent variables. To define the objective functions of the two optimization problems, we first establish two forward maps, going from the forces to the flow fields they generate. Let L_Ω and $L_{\partial\Omega}$ denote the function spaces for the domain forces and the boundary forces, respectively, and H the function space for the flow vectors. The forward map A from the domain force space to the flow vector space

$$A : F \in L_\Omega \mapsto u \in H \tag{A.4}$$

and the forward map B from the boundary force space to the flow vector space

$$B : F \in L_{\partial\Omega} \mapsto u \in H \tag{A.5}$$

are defined by the solution u to the partial differential equation (PDE) [Eq. (5)] that satisfies the following two different homogenization conditions:

HA. The solution u satisfies the Dirichlet boundary conditions (A.3.), where the boundary flow h is set to zero.

HB. The solution u satisfies the PDE [Eq. (5)] with the domain force F_{II+III} set to zero, and satisfies the Neumann boundary condition (6) on $\partial\Omega_{LE}$ and the Dirichlet boundary condition (A.2) on $\partial\Omega_{IN,}$ where the boundary flow h is set to zero.

Under these two conditions both maps are linear and of infinite dimensions. More importantly, they are compact operators whose eigenvalues approach zero, which makes their direct inversion sensitive to noise. Therefore, instead of direct inversion of A and B, we formulate the reconstruction of the domain force F_{II+III} and the boundary force F_I, given the measured flow data u as two least square optimization problems:

$$F_{II+III} = \underset{F \in L_\Omega}{\arg\min} \, ||AF - u_{II+III}||^2 + \alpha||F||^2 \tag{A.6}$$

and

$$F_I = \underset{F \in L_{D\Omega}}{\arg\min} \, ||BF - u_I||^2 + \alpha||F||^2 \tag{A.7}$$

In these equations the norm $||\cdot||$ represents the L_2 norm. In addition, u_{II+III} and u_I denote the homogenized flow derived from the measured flow field u, so that the homogenization conditions introduced to define the two maps A and B are satisfied. The calculation of u_{II+III} and u_I is performed in the two inverse algorithms summarized at the end of this section.

The factor α in (A.6) and (A.7) sets the level of regularization introduced to reduce the effect on the force reconstruction of noise in the flow data. In the field of inverse problems, this kind of extension of the objective function is called *Tikhonov regularization*. With this term, the objective function is penalized for forces of irregularly big magnitudes, which may occur with the calculation of flow gradients from noisy speckle trajectories. The value of the regularization parameter $\alpha(>0)$

determines the balance between finding the force field which fits the flow data best (targeted by the first term) and the suppression of noise in the flow data (targeted by the second term). The choice of α is critical to obtaining a meaningful force reconstruction. Various parameter choice strategies have been developed to identify the "optimal" range of α values, some of which require *a priori* knowledge of the noise level. The most practical approach is the L-curve method (Engl and Grever, 1994), which is an *a posteriori* strategy not requiring *a priori* knowledge of the noise level. In most applications, the range of optimal α values is robust and can span two decades. For our inverse problem, we adopted the L-curve method to identify the optimal value range of the α parameter. The final choice of α within the optimal range, however, depends on how much detail in the force field is needed to address the biological question asked. A smaller α value offers more detail, but is also prone to more noise, while a bigger α value yields a smoother and thus globally more discernible force pattern.

Numerical solutions to (A.6) and (A.7) can be found in the finite subspaces that approximate L_Ω and $L_{\partial\Omega}$. Let $L_\Omega^{(m)}$ and $L_{\partial\Omega}^{(n)}$ denote two such finite subspaces of dimensions m and n, respectively, and A_m and B_n the corresponding finite matrix representations of A and B, respectively. The discrete solutions to (A.6) and (A.7) are then given by inversion of two linear equations:

$$(aI + A_m^{\mathrm{T}} A_m)\boldsymbol{F}_{\mathrm{II+III}} = A_m^{\mathrm{T}} \boldsymbol{u}_{\mathrm{II+III}} \tag{A.8}$$

and

$$(aI + B_n^{\mathrm{T}} B_n)\boldsymbol{F}_{\mathrm{I}} = B_n^{\mathrm{T}} \boldsymbol{u}_{\mathrm{I}} \tag{A.9}$$

which is straightforward once A_m and B_n are calculated. Here, A_m^{T} and B_n^{T} denote the transposes of A_m and B_n, respectively.

Following the definition of A and B, their finite representations A_m and B_n are calculated columnwise, where each column corresponds to one basis function in $L_\Omega^{(m)}$ and $L_{\partial\Omega}^{(n)}$, respectively. Each basis function in $L_\Omega^{(m)}$ is used as the domain force $\boldsymbol{F}_{\mathrm{II+III}}$, and each column in A is obtained by solving the elastic equation [Eq. (5)] subject to the homogenized boundary condition HA in every location of Ω where a flow vector is measured. Columns of B are calculated analogously, but each basis function in $L_{\partial\Omega}^{(n)}$ is used as the boundary force $\boldsymbol{F}_{\mathrm{I}}$ on $\partial\Omega_{\mathrm{LE}}$ and the elastic equation [Eq. (5)] solved subject to the homogenization condition HB.

The choices of $L_\Omega^{(m)}$ and $L_{\partial\Omega}^{(n)}$ are arbitrary, as long as they can represent forces with sufficiently fine resolution and are compatible with the resolution of the flow measurements. We chose to use the shape functions of the finite elements of the meshed 2D cellular domain Ω as the domain force space $L_\Omega^{(m)}$ and B-splines as the 1D boundary force space $L_{\partial\Omega}^{(n)}$.

In summary, we propose two inverse algorithms for the reconstruction of the domain force $\boldsymbol{F}_{\mathrm{II+III}}$ and the boundary force $\boldsymbol{F}_{\mathrm{I}}$, respectively.

Inverse Algorithm A:

- Establish the finite subspace $L_\Omega^{(m)}$ of the function space of domain forces.
- Use each basis function in $L_\Omega^{(m)}$ as the domain force $F_{\text{II+III}}$ and solve Eq. (5) subject to HA. The solution evaluated at the positions of flow measurements defines one column of the finite forward map A_m.
- Calculate $u_{\text{II+III}}$ by subtracting from the recorded flow data u the solution to Eq. (5) that satisfies the Dirichlet boundary condition with h set to the measured flow velocities on $\partial\Omega$ and the domain force $F_{\text{II+III}}$ set to zero.
- Solve Eq. (A.8) using the regularization parameter α as determined by the L-curve method.

Inverse Algorithm B:

- Establish the finite subspace $L_{\partial\Omega}^{(n)}$ of the function space of boundary forces.
- Use each basis function in $L_{\partial\Omega}^{(n)}$ as the boundary force F_{I} on $\partial\Omega_{\text{LE}}$ and solve Eq. (5) subject to HB. The solution evaluated at the positions of flow measurements defines one column of the finite forward map B_n.
- Calculate u_{I} as the difference between the two solutions of Eq. (5) described next. For both solutions, the domain force reconstructed by application of the Inverse Algorithm A is used. They also satisfy the same Dirichlet boundary condition Eq. (A.2) on $\partial\Omega_{\text{IN}}$, with h set to the measured flow velocities on $\partial\Omega_{\text{IN}}$. The difference is that the first solution satisfies the Dirichlet boundary condition Eq. (A.1) also on $\partial\Omega_{\text{LE}}$ with the recorded flow velocities, whereas the second solution satisfies the Neumann boundary condition Eq. (6) on $\partial\Omega_{\text{LE}}$ with a free edge, that is $F_{\text{I}} = 0$ in Eq. (6).
- Solve Eq. (A.9) using the regularization parameter α as determined by the L-curve method.

Acknowledgments

The authors would like to thank James Lim for his contribution of Fig. 1. The research compiled in this chapter was funded in parts by NIH grants R01 GM67230 and U54 GM64346.

References

Alcaraz, J., Buscemi, L., Grabulosa, M., Trepat, X., Fabry, B., Farre, R., and Navajas, D. (2003). Microrheology of human lung epithelial cells measured by atomic force microscopy. *Biophys. J.* **84,** 2071–2079.

Bausch, A. R., Moller, W., and Sackmann, E. (1999). Measurement of local viscoelasticity and forces in living cells by magnetic tweezers. *Biophys. J.* **76,** 573–579.

Beningo, K. A., Dembo, M., Kaverina, I., Small, J. V., and Wang, Y. L. (2001). Nascent focal adhesions are responsible for the generation of strong propulsive forces in migrating fibroblasts. *J. Cell Biol.* **153,** 881–887.

Beningo, K. A., and Wang, Y.-L. (2002). Flexible substrata for the detection of cellular traction forces. *Trends Cell Biol.* **12,** 79–84.

Cheney, M., Isaacson, D., and Newell, J. C. (1999). Electrical impedance tomography. *SIAM Rev.* **41**, 85–101.

Conner, S. D., and Schmid, S. L. (2003). Regulated portals of entry into the cell. *Nature* **422**, 37–44.

Cramer, L. P. (1997). Molecular mechanism of actin-dependent retrograde flow in lamellipodia of motile cells. *Front. Biosci.* **2**, 260–270.

Crocker, J. C., Valentine, M. T., Weeks, E. R., Gisler, T., Kaplan, P. D., Yodh, A. G., and Weitz, D. A. (2000). Two-point microrheology of inhomogeneous soft materials. *Phys. Rev. Lett.* **85**, 888–891.

Danuser, G., and Oldenbourg, R. (2000). Probing f-actin flow by tracking shape fluctuations of radial bundles in lamellipodia of motile cells. *Biophys. J.* **79**, 191–201.

Danuser, G., and Waterman-Storer, C. M. (2003). Fluorescent speckle microscopy: Where it came from and where it is going. *J. Microsc.* **211**, 191–207.

Danuser, G., and Waterman-Storer, C. M. (2006). Quantitative fluorescent speckle microscopy of cytoskeleton dynamics. *Annu. Rev. Biophys. Biomol. Struct.* **35**, 361–387.

Diez, S., Gerisch, G., Anderson, K., Muller-Taubenberger, A., and Bretschneider, T. (2005). Subsecond reorganization of the actin network in cell motility and chemotaxis. *Proc. Natl. Acad. Sci. USA* **102**(21), 7601–7606.

Discher, D. E., Mohandas, N., and Evans, E. A. (1994). Molecular maps of red cell deformation: Hidden elasticity and *in situ* connectivity. *Science* **266**, 1032–1035.

Engl, H. W., and Grever, W. (1994). Using the L-curve for determining optimal regularization parameters. *Numerische Mathematik* **69**, 25–31.

Fabry, B., Maksym, G. N., Butler, J. P., Glogauer, M., Navajas, D., and Fredberg, J. J. (2001). Scaling the microrheology of living cells. *Phys. Rev. Lett.* **87**(14), 148102-1–4.

Gardel, M. L., Shin, J. H., MacKintosh, F. C., Mahadevan, L., Matsudaira, P., and Weitz, D. A. (2004). Elastic behavior of cross-linked and bundled actin networks. *Science* **304**, 1301–1305.

Gardel, M. L., Valentine, M. T., Crocker, J. C., Bausch, A. R., and Weitz, D. A. (2003). Microrheology of entangled F-actin solutions. *Phys. Rev. Lett.* **91**(15), 158302-1–4.

Geiger, B. (2001). Cell biology. Encounters in space. *Science* **294**, 1661–1663.

Geiger, B., and Bershadsky, A. (2001). Assembly and mechanosensory function of focal contacts. *Curr. Opin. Cell Biol.* **13**, 584–592.

Geiger, B., Bershadsky, A., Pankov, R., and Yamada, K. M. (2001). Transmembrane extracellular matrix-cytoskeleton crosstalk. *Nat. Rev. Mol. Cell Biol.* **2**, 793–805.

Giannone, G., Dubin-Thaler, B. J., Doeberbeiner, H.-G., Kieffer, N., Bresnick, A. R., and Sheetz, M. P. (2004). Periodic lamellipodial contractions correlate with rearward actin waves. *Cell* **116**, 431–443.

Gupton, S. L., Salmon, W. C., and Waterman-Storer, C. M. (2002). Converging populations of F-actin promote breakage of associated microtubules to spatially regulate microtubule turnover in migrating cells. *Curr. Biol.* **12**, 1891–1899.

Hirokawa, N. (1998). Kinesin and dynein superfamily protein and the mechanism of organelle transport. *Science* **279**, 519–526.

Howard, J. (2001). "Mechanics of Motor Proteins and the Cytoskeleton." Sinauer Associates Inc., Sunderland, MA.

Ingber, D. E. (2003a). Tensegrity I. Cell structure and hierarchical systems biology. *J. Cell Sci.* **116**, 1157–1173.

Ingber, D. E. (2003b). Tensegrity II. How structural networks influence cellular information processing networks. *J. Cell Sci.* **116**, 1397–1408.

Ji, L., and Danuser, G. (2005). Tracking quasi-stationary flow of weak fluorescent signals by adaptive multi-frame correlation. *J. Microsc. Oxford* **220**, 150–167.

Kole, T. P., Tseng, Y., Huang, L., Katz, J. L., and Wirtz, D. (2004). Rho kinase regulates the intracellular micromechanical response of adherent cells to rho activation. *Mol. Biol. Cell* **15**, 3475–3484.

Kole, T. P., Tseng, Y., Jiang, I., Katz, J. L., and Wirtz, D. (2005). Intracellular mechanics of migrating fibroblasts. *Mol. Biol. Cell* **16**, 328–338.

Kosovichev, A. G., Duvall, T. L., and Scherrer, P. H. (2000). Time-distance inversion methods and results (invited review). *Solar Phys.* **192**, 159–176.

Landau, L. D., and Lifshitz, E. M. (1986). "Theory of Elasticity" Pergamon Press, Oxford, UK.

Lau, A. W. C., Hoffman, B. D., Davies, A., Crocker, J. C., and Lubensky, T. C. (2003). Microrheology, stress fluctuations, and active behavior of living cells. *Phys. Rev. Lett.* **91**(19), 198101-1–4.

Lauffenburger, D. A., and Horwitz, A. F. (1996). Cell migration: A physically integrated molecular process. *Cell* **84**, 359–369.

Lee, J., Leonard, M., Oliver, T., Ishihara, A., and Jacobson, K. (1994). Traction forces generated by locomoting keratocytes. *J. Cell Biol.* **127**, 1957–1964.

Levine, A. J., and Lubensky, T. C. (2000). One- and two-particle microrheology. *Phys. Rev. Lett.* **85**, 1774–1777.

Li, A., Miller, E. L., Kilmer, M. E., Brukilacchio, T. J., Chaves, T., Stott, J., Zhang, Q., Wu, T., Chorlton, M., Moore, R. H., Kopans, D. B., and Boas, D. A. (2003). Tomographic optical breast imaging guided by three-dimensional mammography. *Appl. Opt.* **42**, 5181–5190.

Lin, C. H., and Forscher, P. (1995). Growth cone advance is inversely proportional to retrograde f-actin flow. *Neuron* **14**, 763–771.

MacKintosh, F. C., Kaes, J., and Janmey, P. A. (1995). Elasticity of semiflexible biopolymer networks. *Phys. Rev. Lett.* **75**, 4425–4428.

MacKintosh, F. C., and Schmidt, C. F. (1999). Microrheology. *Cur. Opin. Colloid Interface Sci.* **4**, 300–307.

Manduca, A., Oliphant, T. E., Dresner, M. A., Mahowald, J. L., Kruse, S. A., Amromin, E., Felmlee, J. P., Greenleaf, J. F., and Ehman, R. L. (2001). Magnetic resonance elastography: Non-invasive mapping of tissue elasticity. *Med. Image Anal.* **5**, 237–254.

Marion, S., Guillen, N., Bacri, J. C., and Wilhelm, C. (2005). Acto-myosin cytoskeleton dependent viscosity and shear-thinning behavior of the amoeba cytoplasm. *Eur. Biophys. J.* **34**, 262–272.

Mason, T. G., and Weitz, D. A. (1995). Optical measurements of frequency-dependent linear viscoelastic moduli of complex fluids. *Phys. Rev. Lett.* **74**, 1250–1253.

McGrath, J. L., Tardy, Y., Dewey, C. F., Jr., Meister, J. J., and Hartwig, J. H. (1998). Simultaneous measurements of actin filament turnover, filament fraction, and monomer diffusion in endothelial cells. *Biophys. J.* **75**, 2070–2078.

McIver, M. (1991). An inverse problem in electromagnetic crack detection. *IMA J. Appl. Math.* **47**, 127–145.

Metherall, P., Barber, D. C., Smallwood, R. H., and Brown, B. H. (1996). Three-dimensional electrical impedance tomography. *Nature* **380**, 509–512.

Munevar, S., Wang, Y., and Dembo, M. (2001). Traction force microscopy of migrating normal and H-ras transformed 3T3 fibroblasts. *Biophys. J.* **80**, 1744–1757.

Nascimento, A. A., Roland, J. T., and Gelfand, V. I. (2003). Pigment cells: A model for the study of organelle transport. *Annu. Rev. Cell Dev. Biol.* **19**, 469–491.

Oliver, T., Dembo, M., and Jacobson, K. (1999). Separation of propulsive and adhesive traction stresses in locomoting keratocytes. *J. Cell Biol.* **145**, 589–604.

Panorchan, P., Lee, J. S. H., Kole, T. P., Tseng, Y., and Wirtz, D. (2006). Microrheology and ROCK signaling of human endothelial cells embedded in a 3D matrix. *Biophys. J.* **91**(9), 3499–3507 (doi: 10.1529/biophysj.106.084988).

Ponti, A., Machacek, M., Gupton, S. L., Waterman-Storer, C. M., and Danuser, G. (2004). Two distinct actin networks drive the protrusion of migrating cells. *Science* **305**, 1782–1786.

Ponti, A., Matov, A., Adams, M., Gupton, S., Waterman-Storer, C. M., and Danuser, G. (2005). Periodic patterns of actin turnover in lamellipodia and lamellae of migrating epithelial cells analyzed by quantitative fluorescent speckle microscopy. *Biophys. J.* **89**, 3456–3469.

Ponti, A., Vallotton, P., Salmon, W. C., Waterman-Storer, C. M., and Danuser, G. (2003). Computational analysis of F-actin turnover in cortical actin meshworks using fluorescent speckle microscopy. *Biophys. J.* **84**, 3336–3352.

Rubinstein, B., Jacobson, K., and Mogilner, A. (2005). Multiscale two-dimensional modeling of a motile simple-shaped cell. *SIAM J. Multiscale Model. & Simul.* **3**, 413–439.

Schafer, D. A. (2003). Actin puts on the squeeze. *Nat. Cell Biol.* **5**, 693–694.

Scholey, J. M., Brust-Mascher, I., and Mogilner, A. (2003). Cell division. *Nature* **422**, 746–752.

Sharp, D. J., Rogers, G. C., and Scholey, J. M. (2000). Microtubule motors in mitosis. *Nature* **407**, 41–46.

Shin, K. H., Gardel, M. L., Mahadevan, L., Matsudaira, P., and Weitz, D. A. (2004). Relating microstructure to rheology of a bundled and cross-linked F-actin network *in vitro*. *Proc. Natl. Acad. Sci. USA* **101**, 9636–9641.

Storm, C., Pastore, J. J., MacKintosh, F. C., Lubensky, T. C., and Janmey, P. A. (2005). Nonlinear elasticity in biological gels. *Nature* **435**, 191–194.

Suter, D. M., and Forscher, P. (2000). Substrate-cytoskeletal coupling as a mechanism for the regulation of growth cone motility and guidance. *J. Neurobiol.* **44**, 97–113.

Symes, W. W., and Carazzone, J. J. (1991). Velocity inversion by differential semblance optimization. *Geophysics* **56**, 654–663.

Theriot, J., and Mitchison, T. (1992). Comparison of actin and cell surface dynamics in motile fibroblasts. *J. Cell Biol.* **119**, 367–377.

Trepat, X., Grabulosa, M., Puig, F., Maksym, G. N., Navajas, D., and Farre, R. (2004). Viscoelasticity of human alveolar epithelial cells subjected to stretch. *Am. J. Physiol. Lung Cell. Mol. Physiol.* **287**, L1025–L1034.

Tseng, Y., Kole, T. P., and Wirtz, D. (2002). Micromechanical mapping of live cells by multiple-particle-tracking microrheology. *Biophys. J.* **83**, 3162–3176.

Vale, R. D., and Milligan, R. A. (2000). The way things move: Looking under the hood of molecular motor proteins. *Science* **288**, 88–95.

Valentine, M. T., Perlman, Z. E., Gardel, M. L., Shin, J. H., Matsudaira, P., Mitchison, T. J., and Weitz, D. A. (2004). Colloid surface chemistry critically affects multiple particle tracking measurements of biomaterials. *Biophys. J.* **86**, 4004–4014.

Vallotton, P., Danuser, G., Bohnet, S., Meister, J. J., and Verkhovsky, A. (2005). Retrograde flow in keratocytes: News from the front. *Mol. Biol. Cell* **16**, 1223–1231.

Vallotton, P., Gupton, S. L., Waterman-Storer, C. M., and Danuser, G. (2004). Simultaneous mapping of filamentous actin flow and turnover in migrating cells by quantitative fluorescent speckle microscopy. *Proc. Natl. Acad. Sci. USA* **101**, 9660–9665.

Vallotton, P., Ponti, A., Waterman-Storer, C. M., Salmon, E. D., and Danuser, G. (2003). Recovery, visualization, and analysis of actin and tubulin polymer flow in live cells: A fluorescence speckle microscopy study. *Biophys. J.* **85**, 1289–1306.

Verkhovsky, A. B., Svitkina, T. M., and Borisy, G. G. (1999). Network contraction model for cell translocation and retrograde flow. "Cell Behaviour: Control and Mechanism of Motility," pp. 207–222. Portland Press, London.

Wang, N. (1998). Mechanical interactions among cytoskeletal filaments. *Hypertension* **32**, 162–165.

Wang, Q., Chiang, E. T., Lim, M., Rogers, R., Janmey, P. A., Shepro, D., and Doerschuk, C. M. (2001). Changes in the biomechanical properties of neutrophils and endothelial cells during adhesion. *Blood* **97**, 660–668.

Wang, Y. (1985). Exchange of actin subunits at the leading edge of living fibroblasts: Possible role of treadmilling. *J. Cell Biol.* **101**, 597–602.

Watanabe, Y., and Mitchison, T. J. (2002). Single-molecule speckle analysis of actin filament turnover in lamellipodia. *Science* **295**, 1083–1086.

Waterman-Storer, C. M., and Danuser, G. (2002). New direction of fluorescent speckle microscopy. *Curr. Biol.* **12**, R633–R640.

Waterman-Storer, C. M., Desai, A., Bulinski, J. C., and Salmon, E. D. (1998). Fluorescent speckle microscopy, a method to visualize the dynamics of protein assemblies in living cells. *Curr. Biol.* **8**, 1227–1230.

Waterman-Storer, C. M., and Salmon, E. D. (1998). How microtubules get fluorescent speckles. *Biophys. J.* **75,** 2059–2069.

Wittmann, T., Hyman, A., and Desai, A. (2001). The spindle: A dynamic assembly of microtubules and motors. *Nat. Cell Biol.* **3,** E28–E34.

Yamada, S., Wirtz, D., and Kuo, S. C. (2000). Mechanics of living cells measured by laser tracking microrheology. *Biophys. J.* **78,** 1736–1747.

Yanai, M., Butler, J. P., Suzuki, T., Sasaki, H., and Higuchi, H. (2004). Regional rheological differences in locomoting neutrophils. *Am. J. Physiol. Cell Physiol.* **287,** C603–C611.

Yang, Z. L., Liu, G. R., and Lam, K. Y. (2002). An inverse procedure for crack detection using integral strain measured by optical fibers. *Smart Mater. Struct.* **11,** 72–78.

CHAPTER 10

Analysis of Microtubule Curvature

Andrew D. Bicek,[*] Erkan Tüzel,[†,‡] Daniel M. Kroll,[§] and David J. Odde[*]

[*]Department of Biomedical Engineering
University of Minnesota
Minneapolis, Minnesota 55455

[†]School of Physics and Astronomy
University of Minnesota
Minneapolis, Minnesota 55455

[‡]Supercomputing Institute
University of Minnesota
Minneapolis, Minnesota 55455

[§]Department of Physics
North Dakota State University
Fargo, North Dakota 58105

METHODS IN CELL BIOLOGY, VOL. 83
Copyright 2007, Elsevier Inc. All rights reserved.

0091-679X/07 $35.00
DOI: 10.1016/S0091-679X(07)83010-X

═══════ ## Abstract

The microtubule cytoskeleton in living cells generate and resist mechanical forces to mediate fundamental cell processes, including cell division and migration. Recent advances in digital fluorescence microscopy have enabled the direct observation of bending of individual microtubules in living cells, which has enabled quantitative estimation of the mechanical state of the microtubule array. Although a variety of mechanisms have been proposed, the precise origins of microtubule deformation in living cells remain largely obscure. To investigate these mechanisms and their relative importance in cellular processes, a method is needed to accurately quantify microtubule bending within living cells. Here we describe a method for quantification of bending, using digital fluorescence microscope images to estimate the distribution of curvature in the microtubule. Digital images of individual microtubules can be used to obtain a set of discrete x–y coordinates along the microtubule contour, which is then used to estimate the curvature distribution. Due to system noise and digitization error, the estimate will be inaccurate to some degree. To quantify the inaccuracy, a computational model is used to simulate both the bending of thermally driven microtubules and their observation by digital fluorescence microscopy. This allows for direct comparison between experimental and simulated images, a method which we call model convolution microscopy. We assess the accuracy of various methods and present a suitable method for estimating the curvature distribution for thermally driven semiflexible polymers. Finally, we discuss extensions of the method to quantify microtubule curvature in living cells.

═══════ ## I. Introduction

Living cells respond to mechanical signals from their environment. A potential cellular-based sensory apparatus for mechanical signal transduction is the cytoskeleton, a filamentous network composed of microtubules, actin filaments, and intermediate filaments. Together, these structures provide shape and mechanical integrity for the cell. In addition, they mediate motor-based transport of membrane-bound organelles and vesicles, generate force for cell locomotion, and are essential for cell division. Since force is generated within the cytoskeleton via molecular motors and can be locally accommodated and dissipated (Brangwynne *et al.*, 2006; Heidemann *et al.*, 1999; Odde *et al.*, 1999; Waterman-Storer and Salmon, 1997), the cytoskeleton may act as a mechanosensitive element by responding to force (Putnam *et al.*, 1998). Therefore, by measuring the characteristic shapes of cytoskeletal structures in the cell and comparing, if possible, to *in vitro* shapes of the isolated structures, it may be possible to gain insight into the intracellular mechanical stresses.

One type of cytoskeletal filament in particular, the microtubule, is believed to be important for vesicle trafficking and transport, organelle positioning, chromosome segregation, and cell shape and integrity. Because of its resistance to bending and deformation under compressive loads, an analysis of microtubule shape holds promise for force determinations. Microtubules are linear polymers composed of the $\alpha\beta$-heterodimeric protein tubulin, which assembles head-to-tail to form a protofilament. In a typical microtubule, there are 13 protofilaments that form a tube of 25-nm outer diameter and 16-nm inner diameter (Desai and Mitchison, 1997). The tubular structure increases the second moment of the cross-sectional area (I; a measure of the distribution of mass relative to the centroid of the cross section) compared to a solid rod of the same mass, thereby increasing its resistance to bending from external forces. The flexural rigidity, EI, (elastic modulus multiplied by the second moment of the cross-sectional area) of microtubules has been estimated *in vitro*, and the reported values range from 1×10^{-24} N m^2 to 200×10^{-24} N m^2 (1–200 pN μm^2) depending on the experimental conditions and the measurement technique (Cassimeris *et al.*, 2001; Felgner *et al.*, 1997, 1996; Fygenson *et al.*, 1997; Gittes *et al.*, 1993; Janson and Dogterom, 2004; Kis *et al.*, 2002; Kurachi *et al.*, 1995; Kurz and Williams, 1995; Mickey and Howard, 1995; Takasone *et al.*, 2002; Venier *et al.*, 1994). Assuming that the microtubule is an isotropic solid continuum, I (second moment of the cross-sectional area) is estimated to be on the order of 10^{-32} m^4 based on the known dimension of the tubular structures, and thus the elastic modulus is approximately $E \cong 1 \times 10^9$ Pa, which is similar to Plexiglas$^\circledR$ and indicates that microtubules are indeed stiff filaments.

Dynamic microtubules observed *in vivo* often stochastically switch between alternate states of roughly constant growth and shortening, a phenomenon known as dynamic instability (Desai and Mitchison, 1997; Mitchison and Kirschner, 1984). Dynamic instability presumably allows microtubules to rapidly explore a variety of arrangements within the cytoplasm, with preferred spatial arrangements arising via spatially selective protection of microtubules against depolymerization (Kirschner and Mitchison, 1986). Understanding the mechanisms that provide spatially selective stabilization of microtubules is a key issue in understanding cell polarization. In particular, previous studies have focused largely on the chemical origin of microtubule-stabilization as mediated by microtubule-associated proteins, but recent studies also point to a significant mechanical basis of stabilization, where compressive forces acting on microtubules *in vitro* slow microtubule assembly and promote catastrophe, the abrupt transition from growing to shortening (Dogterom and Yurke, 1997; Janson and Dogterom, 2004; Janson *et al.*, 2003). Conversely, stretching forces applied to living cells have been found to induce microtubule extension (Zheng *et al.*, 1993; Kaverina *et al.*, 2002; Putnam *et al.*, 1998).

The extent to which compressive forces play a role in controlling microtubule assembly *in vivo* is unclear, but it is clear that microtubules are under mechanical stress, as evidenced by the direct observation of the curving of individual fluorescently tagged microtubules in living cells. Much like classical rods and tubes in macroscopic structures, microtubules are capable of bending and breaking (Gupton *et al.*, 2002; Odde *et al.*, 1999; Waterman-Storer and Salmon, 1997). Bending enables the relatively stiff microtubules to store elastic strain energy. This is evident in cilia and flagella, where the energy from the bent microtubule is used to propel the cell relative to the surrounding fluid. In addition, microtubule bending could more generally affect microtubule dynamic instability, and thereby change the microtubule organization within the cell. By increasing the elastic strain energy through bending, the growth and shortening dynamics of the microtubule may be altered, thereby creating a mechanical mechanism that is capable of regulating and reorganizing the microtubule network over time (Odde *et al.*, 1999). In addition, microtubule breaking exposes the labile core of microtubules, which tends to promote disassembly from the site of breaking (Odde *et al.*, 1999; Waterman-Storer and Salmon, 1997).

II. Rationale

Microtubule bending is commonly observed in fluorescent images of microtubules *in vivo*; however, the mechanisms which cause microtubule bending are still largely unknown. Specific sources of bending have been identified, including thermal forces (Cassimeris *et al.*, 2001; Gittes *et al.*, 1993; Kurz and Williams, 1995; Mickey and Howard, 1995; Venier *et al.*, 1994), microtubule polymerization (Dogterom and Yurke, 1997; Janson and Dogterom, 2004; Janson *et al.*, 2003), and actomyosin contractility (Gupton *et al.*, 2002; Waterman-Storer and Salmon, 1997). The extent to which these mechanisms contribute to the deformation of microtubules in living cells is still largely unknown. In addition, analysis of microtubule bending may serve as a useful tool for measuring net mechanical stress in different regions of a cell.

A number of groups have investigated the mapping of microtubules, actin filaments, and DNA shapes into discrete x–y coordinates in order to estimate the flexural rigidity (Gittes *et al.*, 1993; Janson and Dogterom, 2004; Kurz and Williams, 1995; Ott *et al.*, 1993; Venier *et al.*, 1994) and microtubule curvature *in vivo* (Odde *et al.*, 1999). While the analysis techniques vary, the underlying data collection techniques are similar in that they first require collection of digitized x–y coordinates from a raw image as the initial step, followed by estimation of the mechanical deformation.

The resistance to microtubule bending is characterized by the flexural rigidity, which is both a material and geometrical property of the microtubule. There are many ways to estimate EI, but it is essential to know the manner in which the forces are applied in order to make an accurate estimate. One approach is to simply exploit thermal forces and measure quantities such as end fluctuations (Cassimeris *et al.*, 2001), end-to-end fluctuations (Van Noort *et al.*, 2003), shape fluctuations

(Gittes *et al.*, 1993; Kurz and Williams, 1995; Mickey and Howard, 1995; Venier *et al.*, 1994), or tangent correlations (Ott *et al.*, 1993). An alternative approach is to use applied forces, including hydrodynamic flow (Kurz and Williams, 1995; Venier *et al.*, 1994), optical trapping (Felgner *et al.*, 1997, 1996; Kurachi *et al.*, 1995; Takasone *et al.*, 2002), and atomic force microscopy (Kis *et al.*, 2002).

Once *EI* is known, the deformation of the microtubule can, in principle, be used to estimate the distribution of load on the microtubule. However, to date there has not been, to our knowledge, an estimate of *EI in vivo*. The only information available therefore is the deformation of the microtubule, characterized by the curvature distribution (Odde *et al.*, 1999). It is important to note that *EI* and the curvature distribution are directly related (discussed below) when the only forces are thermal. However, microtubule bending in living cells is unlikely to be driven solely by thermal forces. The curvature distribution therefore serves, as the principal characterization of microtubule bending, and so the methods used to estimate it will be the focus of this chapter.

There are various ways to estimate local curvature and construct a distribution. One method involves fitting a circle to a specific part of the microtubule to estimate the curvature, since for a given circle the curvature is the reciprocal of the radius (Cassimeris *et al.*, 2001; Gupton *et al.*, 2002; Waterman-Storer and Salmon, 1997). A similar, yet simpler approach is to use three adjacent points along the microtubule's contour to calculate the angle change with respect to the arc length, namely the three-point method (Odde *et al.*, 1999). This method provides a simple estimate of local curvature and does not require knowledge of the microtubule's position or curvature at its ends, that is the boundary conditions. Alternatively, microtubule shapes can be fit to a set of basis functions, such as cosines, sines, and polynomials, and the curvature calculated analytically. In this chapter, we focus on estimating the curvature distribution using both the three-point and the shape-fitting methods.

In order to determine the accuracy of a given method, it is necessary to know the underlying microtubule shape. This presents an immediate problem because the image used for collecting the x–y coordinates of the microtubule is noisy, digitized, and blurred by diffraction from a circular aperture in the microscope. To address this problem, we used a computational model to simulate both the bending of a thermally driven semiflexible polymer, as well as the subsequent digital imaging via fluorescence microscopy, an approach that we call model-convolution microscopy (Sprague *et al.*, 2003). This allows us to construct essentially exact simulated data sets (where the position of the fluorescent object, the deformed microtubule, is known to the computer's decimal accuracy), which can then be used to validate the accuracy of each method in terms of its ability to properly reproduce the (known) curvature distribution. Although the method is general, we will analyze the particular case of thermally driven microtubules as observed *in vitro*. From the insight we gain through the simulations, we are able to determine the pitfalls associated with estimating the curvature distribution from experimental data. Finally, we discuss extending the method to analyze microtubule deformation in living cells.

III. Raw Data Collection

Modern optical microscopes with high numerical aperture (NA) objectives coupled with scientific grade digital cameras are routinely used to gather digital images of microstructure within the cell. Both differential interference-contrast (DIC) microscopy and fluorescence microscopy are commonly used to visualize microtubules, with DIC usually dominating *in vitro* observations and fluorescence microscopy dominating *in vivo* applications. While the details of each type of microscopy ultimately affect the quality of the collected images, both methods provide digital images of microtubules as sources of raw data. An example of an image taken with our microscope of

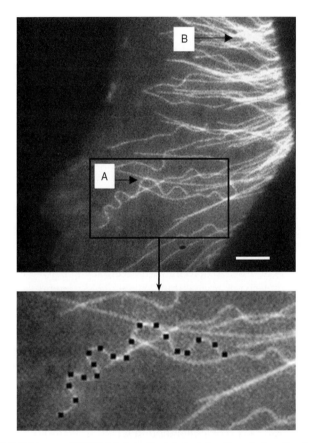

Fig. 1 Typical digital image of EGFP-labeled microtubules taken from the lamella of an epithelial cell. The image was acquired using a 60×, 1.4-NA Plan APO oil immersion objective with a 2.5× projection lens. Note that the deformation observed in living cells is highly variable. The arrows indicate regions where crossover (A) and high density (B) of microtubules makes determination of the microtubule shape ambiguous. The inset shows typical $x–y$ coordinates collected from a highly curved microtubule. To remain in focus, the z-coordinates must remain within about 0.5 μm of each other, and so the microtubules that are in focus all along their length can be approximated as deforming in the $x–y$ plane only. Scale bar is 4 μm.

fluorescently labeled microtubules in a living epithelial cell stably transfected with GFP-tubulin (Rusan *et al.*, 2001) is shown in Fig. 1.

Once digital images of microtubules have been collected, individual microtubules need to be identified. While this is usually trivial with *in vitro* data, it can be difficult with *in vivo* images because microtubules are rarely visible along their entire length and frequently cross over each other. In addition, the high density of microtubules in some regions of the cell body makes identifying single microtubules very difficult (Fig. 1). To deal with these problems, we collect data from the edges of cells, typically in the lamella of epithelial cells, where individual microtubules can clearly be distinguished. Data is collected from the plus-end of a microtubule back to a point where the position of the microtubule becomes difficult to determine visually. Finally, data is excluded from microtubules that ambiguously cross over each other.

A. Point-Click Method

On defining a microtubule of interest, the x–y coordinates of the microtubule are extracted from the image. Since the depth of field of a 1.4-NA lens is $\sim$0.5 μm, we assume that if the entire length of a microtubule ($\sim$10–30 μm) is in focus, its position is essentially limited to a plane (i.e., the focal plane). A simple method to visually extract the positional information is to use image-processing software such as ImageJ (public domain, http://rsb.info.nih.gov/ij/) or MetaMorph (Molecular Devices Corporation, Downingtown, Pennsylvania) to view the digital images and record the x–y coordinates that correspond to positions on the microtubule at regular intervals. This is commonly done by hand with a mouse-controlled cursor, by moving the pointing arrow over the position of the microtubule and then clicking the button to record the pixel coordinates. Most image-processing software has a feature to automatically collect this data. For example, in MetaMorph, the "measure pixel" function is set up to record the x–y coordinates from any position in the image by "clicking" on the feature of interest. These data points are then exported to MatLab 7.1 (The Mathworks, Natick, Massachusetts) for later analysis. While this method is simple, it is prone to errors from visually aligning the mouse pointer with the center of the projected image of the microtubule.

B. Semiautomated Methods

Semiautomated image-processing algorithms can make the recording of the coordinates more precise. Janson and Dogterom (2004) presented a semiautomated method for collection of digitized coordinates in a DIC image of an *in vitro* microtubule. Briefly, to maximize contrast, they selected microtubules orthogonal to the direction of greatest optical shear in the DIC microscope. They then obtained line scans perpendicular to the microtubule across every pixel column in a region and convolved the line scans with one period of a sine function which mimics the shadow-cast appearance of DIC images. This one-dimensional sine wave convolution effectively acts as a pattern recognition filter and amplifies the true signal of the microtubule, which increases the accuracy of the raw data collection process. However, this method is specific to DIC microscopy as it capitalizes on the contrast pattern inherent to DIC microscopy, and would need to be modified for fluorescent images, that is with a Gaussian function or a Bessel's function squared, to mimic the point spread function of light. In principle, such semiautomated methods for collecting positional data from within fluorescently labeled cells could be developed to increase positional accuracy.

C. Data Collection Errors

Accurate collection of raw data is important because the methods described above for extracting positional data effectively discretize the microtubule from a "continuum" into a set of discrete coordinate values. The coordinates are then used to estimate the curvature, and the degree of uncertainty in these coordinates directly affects the associated uncertainty in the estimate. In Sections III.C.1 and III.C.2, we discuss sources of error attributed to collecting coordinates from digital images.

1. Digitization Error

A charge-coupled device (CCD) camera is an array of pixels (picture elements) that detect light and transfer it to a digital array (image). Since each pixel sums the light intensity projected onto it, the resulting image array will lose spatial information with increasing pixel size. This leads to the first type of error associated with collecting data from a digital image, which we will refer to as "digitization error." For a single image, the digitization of data determines the smallest detectable deformation of the microtubule; according to the Nyquist limit, it is twice the pixel size.

Because the microtubule is only 25 nm in diameter, optical microscopes utilize high-NA lenses, typically 1.30 or higher, to maximize the resolution (typically ~200–250 nm). In addition, high magnification allows for projecting the image onto more pixels, which results in better sampling and less digitization error. However, there is a practical limit to the magnification, due to concerns of signal-to-noise ratio. When the magnification is increased, the intensity of the signal from the image is spread over more pixels and eventually the noise in each pixel starts to mask the signal. Therefore, to achieve high-quality images of fluorescently labeled microtubules, a typical experimental setup might include a 60×, 1.4-NA objective with a 2.5× projection lens for a resulting camera magnification of 150×. The digital image is obtained by a cooled CCD camera which has a physical pixel dimension of ~8 μm. In this setup, the resulting pixel dimension in the image, which we will term the image pixel size, is ~50 nm. When extracting microtubule coordinates in this setup, the digitization errors from the image pixel size (50 nm) are large compared to the diameter of the microtubule (25 nm). As an example, a perfectly straight microtubule aligned diagonally with respect to the pixel array will look like a step function when digitized. In this extreme case, the calculated radius of curvature when using every point could be as small as 32 nm.

2. Measurement Error

Another source of error is introduced by the method used to select the x–y coordinates of the microtubule. If an individual coordinate selected from an image deviates from the true position on the microtubule, the resulting estimate of curvature will likely be overestimated. This can be understood by thinking of a straight microtubule. If the selected coordinates for a perfectly straight microtubule do not fall exactly on a line, the curvature of that microtubule will artificially increase, introducing error in data collection. Throughout this chapter, we will call these deviations from the true position of the microtubule "measurement error." Note that the impact of this error on the curvature estimate will be additive to the errors introduced by digitization (discussed further in Section IV.B.4).

IV. Validation Strategy

To assess the accuracy of any particular curvature estimation method, it is necessary to have test cases where the actual coordinates of the microtubule contour are known with high precision. In order to establish these test cases, we utilize a computer algorithm to generate a large ensemble of equilibrium semiflexible polymer configurations, which simulate microtubules. Using these simulated polymers allows us to eliminate the problem of experimental error and perform a quantitative comparison of different analysis methods. In addition, we can simulate the spread of light due to diffraction and the uncertainties in estimating the position of the microtubules due to digitization and measurement errors. We then determine how much each of these errors affects the final curvature distribution estimate. Also, since microtubules observed in living cells often appear more curved than microtubules *in vitro*, we vary the shape undulation of simulated microtubules by changing the persistence length, the characteristic distance over which the tangent angles become uncorrelated, over four orders of magnitude. This allows us to test the ability of various methods to estimate the curvature over a wide range of shapes.

For a thermally driven polymer, the average angle spanned by the tangent angles at any two points along the curve $s_1 = 0$ and $s_2 = s$ is given by

$$\langle \vec{t}(s) \cdot \vec{t}(0) \rangle = \langle \cos \left[\theta(s) - \theta(0) \right] \rangle = \exp\left(-\frac{s}{2L_p} \right) \tag{1}$$

in two dimensions, where $\vec{t}$ is the tangent vector along a curve parameterized by s (Fig. 2). Since the choice of origin is arbitrary, we chose, for simplicity, the left end of the microtubule to be the origin ($s_1 = 0$). The characteristic distance at which this average angular span, or "tangent correlation," decays to $1/\sqrt{e}$ geometrically

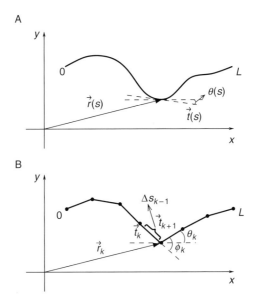

Fig. 2 Continuum versus discrete representations of a polymer. (A) A polymer of length L is described by the curve $\vec{r}(s)$ in the x–y plane, where s denotes the position along the contour. The variables $\vec{t}(s)$ and $\theta(s)$ are the tangent vector and the angle it makes with the horizontal axis at a point s, respectively. (B) The continuum curve in (A) is discretized. The tangent vector and the angle with the x-axis are now defined at coordinates $\vec{r}_k = (x_k, y_k)$. The variable ϕ_k denotes the angle between consecutive tangent vectors $\vec{t}_k$ and $\vec{t}_{k+1}$, and Δs_k is the spacing between neighboring coordinates $\vec{r}_k$ and $\vec{r}_{k+1}$.

defines the persistence length, L_p. In the case where curvature is driven by thermal forces alone, equipartition of energy can be used to show that L_p is related to the temperature, T, and flexural rigidity by (Howard, 2001)

$$L_p = \frac{EI}{k_B T} \qquad (2)$$

where k_B is Boltzmann's constant.

In three dimensions, the tangent correlations decay twice as quickly as that shown in Eq. (1), due to an additional angular degree of freedom (Howard, 2001). Note that Eq. (1) is not universal, but rather depends on the polymer model used as well as the forces acting on the system. For example, it was shown that tangent correlations exhibit oscillatory decay for polymer models with a nonzero diameter (Marenduzzo *et al.*, 2005). One might speculate that for systems with additional nonthermal forces, the tangent correlations will have nonexponential decay. For example: (1) extensional forces suppress fluctuations and compressional forces tend to introduce buckles (Fygenson *et al.*, 1997) and (2) lateral confinement forces tend to suppress fluctuations (Brangwynne *et al.*, 2006).

A. Modeling of Semiflexible Polymers

Our approach is to model the microtubule as a discrete chain embedded in a solvent, which may be described using a stochastic model for fluid flow (Ihle and Kroll, 2001; Malevanets and Kapral, 1999, 2000), often referred to as stochastic rotation dynamics (SRD). SRD is an attractive simulation tool for the coarse-grained modeling of a fluctuating solvent, in particular colloidal and polymer suspensions, since it provides the correct hydrodynamic interactions between embedded particles and fully incorporates thermal fluctuations. For details regarding the algorithm and its transport properties, the reader is referred to a series of articles (Ihle and Kroll, 2003a,b; Ihle *et al.*, 2004, 2005; Tüzel *et al.*, 2003, 2006).

In the continuum approach, the conformation of a semiflexible polymer such as a single microtubule can be described by a curve $\vec{r}(s)$, where s is the contour position along the curve (Fig. 2A). Semiflexible polymers are often assumed to be inextensible so that the only relevant potential energy is the bending energy, given by

$$U_{b} = \frac{1}{2} EI \int_{0}^{L} \kappa^2 \mathrm{d}s = \frac{1}{2} EI \int_{0}^{L} \left(\frac{\mathrm{d}\vec{t}(s)}{\mathrm{d}s} \right)^2 \mathrm{d}s = \frac{1}{2} EI \int_{0}^{L} \left| \frac{\mathrm{d}\theta(s)}{\mathrm{d}s} \right|^2 \mathrm{d}s \qquad (3)$$

where $\kappa \equiv |\,\mathrm{d}\theta(s)/\mathrm{d}s\,|$ is the curvature, L is the total length of the polymer, $\vec{t}(s)$ is the unit tangent vector at a point s, and $\theta(s)$ is the angle of the tangent vector relative to the x-axis (Fig. 2A). The correct discretization for the bending energy in Eq. (3) is (Klapper and Qian, 1998)

$$U_{b} = \frac{1}{2} EI \sum_{k=1}^{N-1} \frac{(\vec{t}_{k+1} - \vec{t}_k)^2}{(\vec{r}_{k+1} - \vec{r}_k)^2} \cdot \Delta s_k \qquad (4)$$

where N is the total number of nodes (x–y coordinate pairs) on the discrete chain in the simulation. Here the magnitude of the distance between two consecutive coordinate pairs in a chain is given by $\Delta s_k \equiv |\vec{r}_{k+1} - \vec{r}_k|$. Figure 2B shows the discretized position and tangent vectors. Coupling to the solvent is described by including the polymer nodes in the SRD collision step (see Malevanets and Yeomans, 2000 for details).

The time evolution of the polymer between collisions is determined by solving the resulting Newton's equations of motion using the velocity Verlet algorithm (Frenkel and Smit, 2002). We will refer to this procedure as a hybrid SRD-MD simulation. A typical snapshot of a single semiflexible polymer in solvent is shown in Fig. 3. This approach has been used previously to study the behavior of polymers in solution. In particular, Malevanets and Yeomans (2000) studied the dynamics of short polymer chains and Kikuchi *et al.* (2002) investigated polymer collapse in the presence of hydrodynamic interactions using this approach. SRD has been used to investigate the effects of hydrodynamic interactions (Falck *et al.*, 2003; Ripoll *et al.*, 2004; Winkler *et al.*, 2004) on the behavior of rod-like colloids

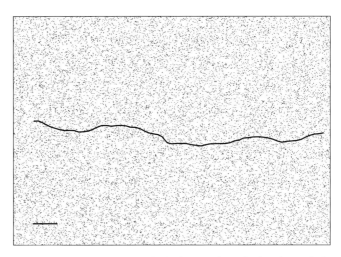

Fig. 3 A snapshot from a hybrid SRD-MD simulation showing a single polymer chain surrounded by solvent particles. Total number of solvent particles is 4.6×10^4, and the simulation box size is 96×96 pixels. For clarity, a short chain of 64 nodes is shown, the scale bar is 5 pixels and $L/L_p = 6.3$.

and flexible polymer chains in solution. Others have also used SRD as a tool to study colloids (Falck *et al.,* 2004; Hecht *et al.,* 2005; Padding and Louis, 2004) and vesicles in shear flow (Noguchi and Gompper, 2004). In this study, SRD provides a heat bath for the polymers.

B. Generation of Simulated Data

Using hybrid SRD-MD, we simulated the bending of thermally driven polymers to test each shape parameterization method. Since microtubules in a cell appear to bend much more than their counterparts from *in vitro* experiments, we are unsure if the persistence length (L_p) for microtubules in a cell is the same as the L_p estimated from *in vitro* experiments. If the persistence lengths are different, then either the *EI* is different or additional forces are acting on the microtubules within the cell that increase the apparent flexibility. To examine this effect, we simulated polymers with a wide range of persistence lengths, L_p. In particular, we studied polymers of length L for which $0.03 \leq L/L_p \leq 62$. This range of persistence lengths allowed us to test the shape parameterization methods across a wide range of flexibility regimes in an attempt to find a robust method for curvature distribution estimation that works for all persistence lengths, or at least to determine where each method breaks down (Fig. 4). Note that when L/L_p is much smaller than one, the filament is essentially straight, and when L/L_p is larger than one, the filament is highly flexible. For most biologically relevant polymers, L/L_p is of order 1, and is described properly using semiflexible chain models (Storm *et al.,* 2005).

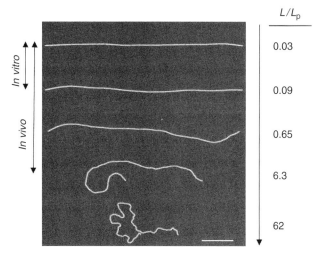

Fig. 4 Representative examples of simulated thermally driven polymers having $L/L_p = 0.03, 0.09,$ 0.65, 6.3, 62 from top to bottom, respectively. Each chain has been convolved with the point spread function and Gaussian white noise to create a simulated fluorescent image of the polymer. We observe microtubules within the cell with shapes similar to the first four images. Scale bar is 5 μm.

1. Simulation Conversion Factor

We simulated $M = 500$ thermally driven polymers for different values of L/L_p, with representative examples shown in Fig. 4. To convert between physical length and the dimensionless simulation units, we set the bond length (distance between consecutive coordinates) in the polymer chain to be equal to 1 pixel. The reason for this is twofold. First, the pixel is a pseudo-dimensionless unit that can be directly compared to any experimental setup. Second, the pixel is the smallest unit of resolution in the experimental data. Therefore, it is a natural choice for the conversion factor. The polymers in our simulations consist of 622 nodes which correspond to a length of 621 pixels. If we take 50 nm as a pixel size, an approximate value for a typical high-resolution microscopy system, then our polymer length is ~31 μm. This length is typical for *in vitro* measurements of microtubule shapes (Gittes *et al.*, 1993; Janson and Dogterom, 2004), and the simulated polymers can always be truncated to a shorter length that might more closely match the length of microtubules observed *in vivo*.

2. Model Convolution Microscopy

The position of each coordinate in the hybrid SRD-MD simulations is accurate to decimal precision of 1×10^{-7} so that the simulated data can be thought of as pure and devoid of any measurement error. This pure data provides the best-case scenario for each shape parameterization method used to estimate the curvature distribution. However, in any experimental system, such as the one described in Section III.C.1, the digital image is always contaminated with noise from the detector and optical aberration from the microscope. In addition, the spread of light due to a circular aperture blurs the image and makes the microtubules appear much wider than they actually are. When taken together, these sources of noise and blur make accurate collection of x–y coordinates difficult. To simulate noise and blur, we convolved our pure data with the characteristics of the microscope system to approximate experimental data, a method called model convolution microscopy (Sprague *et al.*, 2003). Our procedure was as follows:

a. We took 35 independent simulations of polymers with $L/L_p \cong 0.02$ and projected the pure data onto a 5-nm two-dimensional grid.

b. The grid was filled in to make the simulated chains appear 25-nm wide, corresponding to the width of a microtubule.

c. The projected polymer data was convolved with the point spread function, effectively making every location within the polymer a point source of light.

d. The signal and background levels were normalized to experimentally measured levels typical for EGFP-labeled microtubules imaged from LLC-PK1-α epithelial cells.

e. The fine grid was coarse grained to a larger grid size of 50 nm, approximately corresponding to the size of pixels in our microscope system.

f. The noise due to collecting the image with a digital camera was estimated by calculating the variance of the intensity for a typical background region in an experimental image, and by adding Gaussian white noise with a similar variance.

This procedure resulted in an image of a simulated polymer that is directly comparable to an image collected experimentally from a fluorescently labeled microtubule *in vivo* with two important exceptions: (1) we know the precise position of the underlying polymer and (2) we know its persistence length (Fig. 4).

3. Estimation of Digitization Error

Whenever data are binned (i.e., pixelated), there is loss of information. Therefore, digitization errors are inherent in all digital images. Since we set the units for our pure coordinate pairs in the simulation to be equivalent to pixels in a digital image, the correct digitization is achieved by simply rounding the pure coordinate values to integer values corresponding to the nearest pixel location. This is analogous to the binning action that a CCD camera performs on every pixel in an experimental image.

4. Estimation of Measurement Error

We used very stiff simulated polymers ($L/L_p = 0.02$) and rotated them so that their end points were both on the x-axis. We applied model convolution microscopy to the polymers and obtained simulated images. In practice these simulated images appeared very straight, and it was therefore reasonable to assume that the measurement errors from point clicking were limited to deviations in the y-direction only.

We first collected x–y coordinates from simulated images using the point-click method. We then overlaid the estimated pixel locations onto the digitized microtubule image to determine the magnitude of the deviations in pixel units in the transverse (y) direction only, which are an estimate of our measurement error. The distribution of a one-dimensional measurement error is shown in Fig. 5 in units of pixels, with one pixel corresponding to 50 nm for the experimental system described in III.C.1. The histogram indicates an error (standard deviation) of 52 nm. (Note that for any other pixel size, another corresponding histogram needs to be constructed.)

The measurement of a highly curved microtubule is expected to have deviations in the x-direction as well as the y-direction. So, one can use the one-dimensional error as a lookup table independent of direction by assuming the deviations in x and y are statistically independent and equal. Therefore, the root-mean-squared (RMS) magnitude of a two-dimensional measurement error is ~74 nm. When compared to an earlier estimation of measurement error by Gildersleeve *et al.* (1992)

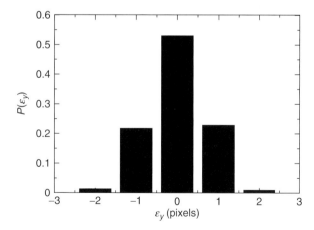

Fig. 5 Histogram of the magnitude of measurement errors in y-direction, namely ε_y, calculated using data collected from 35 simulated polymers. The RMS error is 52 nm and 97% of the time the clicking error is 0 or 1 pixels. The total number of data points used to construct the histogram is 1463. Note that one pixel corresponds to 50 nm in the experimental system, and therefore, for any other image pixel size, the histogram needs to be reconstructed.

(RMS deviation of 162 nm), our system is roughly twice as accurate (note: Gildersleeve *et al.* were tracking microtubule tips using DIC, which is more difficult than tracking the position of the microtubule using fluorescence). Since the values in the lookup table have units of pixels, they can be directly added to the simulated polymer coordinates. Therefore, to recreate the measurement error associated with the point-click method in our simulated polymers, we randomly picked two values from the measurement error lookup table and added each value to the x and y coordinates, respectively.

5. Coarse-Graining

Since each position estimate contains errors as described above, the coordinate data from a microtubule must be collected at some larger interval to reduce the effects of measurement error on the estimation of curvature. For example, if the point-click method is used to collect the coordinates from every adjacent pixel, the magnitude of the measurement error will dominate and will result in a noisy estimate of the x–y coordinates. Further, if this data is then used to estimate a curvature distribution, the effect of the measurement noise will substantially alter the associated curvature distribution toward that of a more flexible polymer. Therefore, when the RMS measurement error approaches the size of a pixel, the x–y coordinates should be sampled at some lower spatial frequency to reduce the effects of measurement error, a practice we use for coarse-graining. At the other extreme, if the data is collected at too low a spatial frequency, sharp fluctuations in the microtubule's shape between sampling points will be missed, and the resulting curvature estimate will underestimate the bending. Therefore, some optimum must exist between these two extremes that provides the most accurate estimate of the curvature distribution.

In order to investigate the effect of coarse-graining, we skipped coordinates in the simulated images, effectively increasing the distance between nodes. We use the resulting average spacing, $\Delta \tilde{s}$, as a measure of the coarse-graining. Because the uncertainty in our collected coordinates is a maximum of 2 pixels in any direction, and since we do not want overlapping data points, the minimum spacing between coordinates for data collected by the point-click method should be at least 5 pixels. In the experimental system described by Odde *et al.* (1999), their coordinate spacing was ∼0.5 μm, which corresponds to a spacing of $\Delta \tilde{s} = 10$ pixels. Note that $\Delta \tilde{s}$ should always be smaller than the persistence length, since using spacing larger than the persistence length will miss the correlations in the polymer shape along the length and result in significant entropic contributions to the polymer conformation (De Gennes, 1979; Doi and Edwards, 2004).

Applying the processes described above to the simulation data results in three sets of simulated data: pure, digitized, and digitized with measurement error. Figure 6 shows all three sets of data from representative simulated polymers at two different values for L/L_p. Note for a stiff polymer, $L/L_p = 0.03$ (Fig. 6A), the

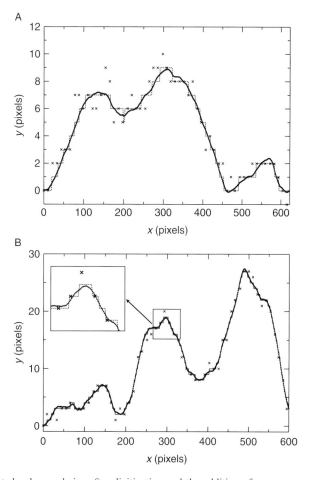

Fig. 6 Simulated polymer chains after digitization and the addition of measurement errors. Bullets show pure data, dashed lines show the effect of digitization, and crosses show the data after digitization and measurement error addition. (A) $L/L_p = 0.03$ and (B) $L/L_p = 0.65$. The inset in (B) shows a close-up of a highly curved region.

relative effects of the digitization and measurement errors are larger than for a more flexible polymer, $L/L_p = 0.65$ (Fig. 6B).

C. Validation of Semiflexible Polymer Simulation

The hybrid SRD-MD simulations correctly reproduce the statistics of thermally driven polymers. We measured tangent correlations (Ott *et al.*, 1993) and end-to-end fluctuations (Van Noort *et al.*, 2003), and found that the results all agree with analytical predictions of polymer theory (not shown). However, we also wanted to validate the simulation's ability to correctly incorporate noise and reproduce experimental data. Our strategy to ensure that the model and noise simulation were working properly was to apply a bending mode analysis to our simulated data to determine whether we could recreate the results of Gittes *et al.* (1993) and Howard (2001).

A bending mode analysis can be used to measure the flexural rigidity of *in vitro* microtubules by expressing the polymer's shape as a superposition of a large number of Fourier cosine modes (Gittes *et al.*, 1993). Using a Fourier series expansion, the shape $\theta(s)$ of a microtubule can be expressed by

$$\theta(s) = \sqrt{\frac{2}{L}} \sum_{n=0}^{\infty} a_n \cos\left(\frac{n\pi s}{L}\right) \tag{5}$$

where a_n denotes the amplitude of the nth cosine mode. For a discrete chain, s will be replaced by s_k^{mid} and the sum is limited to $N - 1$. The spacing between coordinates is given by

$$\Delta s_k = \sqrt{(x_{k+1} - x_k)^2 + (y_{k+1} - y_k)^2} \tag{6}$$

which yields for the position along the contour

$$s_k^{\mathrm{mid}} = \frac{1}{2}\Delta s_k + \sum_{i=1}^{k-1} \Delta s_i \tag{7}$$

Using the orthogonality of cosine functions, one can show that

$$a_n = \sqrt{\frac{2}{L}} \int_0^L \theta(s) \cos\left(\frac{n\pi s}{L}\right) ds \cong \sqrt{\frac{2}{L}} \sum_{k=1}^{N} \theta_k \Delta s_k \cos\left(\frac{n\pi s_k^{\mathrm{mid}}}{L}\right) \tag{8}$$

Here the tangent angle is given by

$$\theta_k = \tan^{-1}\left(\frac{y_{k+1} - y_k}{x_{k+1} - x_k}\right) \tag{9}$$

The discrete coordinates from the polymers are used to solve for the mode amplitudes using Eq. (8), and the variance of mode amplitudes var(a_n) computed from all polymers is used to estimate the flexural rigidity. In theory, each cosine mode yields an independent measure of flexural rigidity. For a more detailed description of the method see Gittes *et al.* (1993).

We computed the mode amplitudes for 500 simulated polymers and calculated the variance. Using equipartition of energy, the variance of the mode amplitudes can be expressed as

$$\text{var}(a_n) = \langle a_n^2 \rangle = \frac{k_B T}{EI} \left(\frac{L}{n\pi} \right)^2 \qquad (10)$$

Note that the mean of the mode amplitudes is equal to zero since the simulated polymers have no intrinsic curvature. As can be seen from Eq. (10), the variance for each mode should scale as $1/n^2$ where n is the mode number. Plotting the variance of mode amplitudes against the mode number (Fig. 7A) shows good agreement with theory.

Next, we simulated the effects of digitization and measurement error as discussed in Sections IV.B.3 and IV.B.4, respectively. In addition, the data was

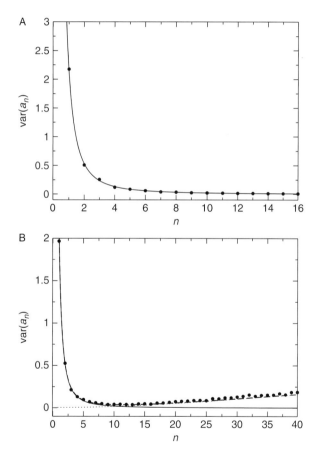

Fig. 7 Validation of hybrid SRD-MD simulation. Variance of the mode amplitude a_n as a function of the mode number n is shown in bullets (A) for the pure simulation data and (B) for the digitized data with measurement error. Solid line is a plot of Eq. (10). Dotted and dashed lines represent theoretical predictions given by Eqs. (11) and (12), respectively. The results match well the relationship for experimental data presented by Gittes *et al.* (1993).

coarse-grained at an interval of $\Delta \tilde{s} = 10$ pixels. To compare our simulated noisy data to theory, we used the relation

$$\langle a_n^2 \rangle^{\text{noise}} = \frac{4}{L} \langle \varepsilon_y^2 \rangle \left[1 + (N-1) \sin^2 \left(\frac{n\pi}{2N} \right) \right] \tag{11}$$

for the variance of noise reported by Gittes *et al.* (1993). Here ε_y is the magnitude of the measurement error in pixels, from our measurement error lookup table (Fig. 5). The noise variance can be directly added to the theoretical mode variance, which yields

$$\text{var} \, (a_n)^{\text{measured}} = \frac{1}{L_p} \left(\frac{L}{n\pi} \right)^2 + \frac{4}{L} \langle \varepsilon_y^2 \rangle \left[1 + (N-1) \sin^2 \left(\frac{n\pi}{2N} \right) \right] \tag{12}$$

for the measured variance. We plotted the measured variance, calculated using the digitized data with measurement error, against the mode number and the results are shown in Fig. 7B. The agreement with Eq. (12) is excellent indicating that we are able to correctly reproduce the experimental results of Gittes *et al.* (1993) using the hybrid SRD-MD simulations. In Section V, we discuss methods to estimate the curvature distribution.

V. Curvature Estimation Methods

A. Three-Point Method

A simple method was used by Odde *et al.* (1999) to determine the curvature at which microtubules broke, as well as the curvature distribution of all microtubules in the lamella of fibroblasts. They collected x–y coordinate data from fluorescent images every 500 nm along the length of microtubules in Swiss 3T3 fibroblasts. The curvature (κ) was calculated at each coordinate by taking three adjacent points and computing the change in the angle (ϕ_k) (Fig. 2) over the average arc length of the two adjacent segments to yield

$$\kappa \approx \left| \frac{2\phi_k}{\Delta s_{k-1} + \Delta s_k} \right| \tag{13}$$

which is an approximation of the curvature for small angle changes and small bond lengths.

B. Shape-Fitting Method

We thought it natural to extend the bending mode analysis (Gittes *et al.*, 1993) to estimate the curvature distribution. Our strategy is again to map the shape, $\theta(s)$ of the microtubule to a sum of cosine waves, and calculate curvature at the nodes along the microtubule's length and then produce the curvature distribution. Differentiating Eq. (5) with respect to s, we have

$$\kappa = \left| \frac{\mathrm{d}\theta(s)}{\mathrm{d}s} \right| = \left| \sqrt{\frac{2}{L}} \sum_{n=0}^{\infty} a_n \left(\frac{n\pi}{L} \right) \sin \left(\frac{n\pi s}{L} \right) \right| \tag{14}$$

for the curvature. Note that Eqs. (5) and (14) are given in terms of the contour length, s, and in the discrete approximation, s is replaced by s_k^{mid} as shown in Eq. (7).

Alternatively, one can use other analytical functions to map the shape of a microtubule. For example, sine and cosine transforms or polynomials can be fit to the shape and used to estimate $\theta(s)$ or the curvature $\mathrm{d}\theta(s)/\mathrm{d}s$. While the selection of the analytical function seems arbitrary, a complete set of basis functions (Arfken and Weber, 2001) is needed to correctly reconstruct the microtubule's shape.

C. Constructing the Curvature Distribution

The curvature distribution can be estimated by creating a histogram of the discrete curvature values obtained from either the three-point method or the shape-fitting method. It is important to note that the number of bins used in the histogram should not exceed the number of discrete values that curvature can take. For stiff microtubules ($L/L_p < 0.65$) and digitized data collected for every pixel this may be as low as three bins.

To determine the accuracy of the curvature distribution, it should be compared to theory. For the case of a thermally driven polymer, the local energy should be distributed exponentially according to Boltzmann's law. Since energy is proportional to the curvature squared, the curvature distribution should be normally distributed, and for polymers with zero mean curvature and unit bond length, will be given by

$$P(\kappa) = \sqrt{\frac{2L_p}{\pi}} e^{(-L_p \kappa^2)/2} \tag{15}$$

The extra factor of $\sqrt{2}$ in the prefactor is because of the curvature distribution being a half Gaussian, which is due in turn to the absolute value in Eq. (5). This equation shows that the variance of the normal distribution of κ equals the reciprocal of persistence length, L_p. The cumulative distribution function, $C(\kappa)$, can be derived from Eq. (15) using the following relation

$$C(\kappa) = \int_0^\kappa P(\kappa') \, \mathrm{d}\kappa' \tag{16}$$

Using Eqs. (15) and (16) it can be shown that

$$C(\kappa) = \mathrm{erf} \left(\sqrt{\frac{L_p}{2}} \kappa \right) \tag{17}$$

where erf is the error function.

The empirical cumulative distribution function is easily generated from the experimental data and it can be compared to Eq. (17). First, begin by sorting all the discrete curvature values from lowest to highest. The position of each data point in the sorted list is indicated by its index from 0 to $K-1$ where $K = M \cdot N$ is the total number of data points. These curvature values define the x-position for each data point in the distribution. Next, the index for each sorted data point is rescaled between 0 and 1. This effectively performs the integration in Eq. (16), and the y-data is generated. Finally, by plotting the normalized index as a function of the curvature values, the empirical cumulative distribution is generated (Press *et al.*, 1992).

VI. Results

A. Three-Point Method

1. Pure Data

For the case of pure data, the three-point method for calculating curvature accurately reproduced the curvature distribution for the entire range of L/L_p investigated. Typical results are shown in Fig. 8, where it can be seen that the data are in good agreement with the predictions of Eqs. (15) and (17).

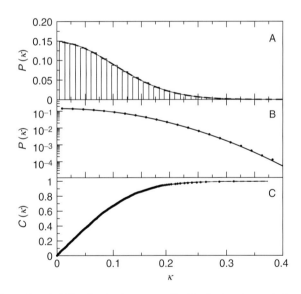

Fig. 8 The estimated probability density (curvature distribution) and cumulative distribution functions for $L/L_p = 6.3$ calculated using pure data. (A) A histogram of the curvature values, κ. (B) Semilog plot of the curvature distribution. The profile is parabolic as expected from a Gaussian distribution. (C) Empirical cumulative distribution function calculated using the sorted curvature data. Solid lines in (A) and (B) show theoretical predictions of Eq. (15), and solid line in (C) shows the theoretical cumulative distribution given by Eq. (17). Dashed lines show the fitted distributions.

We also investigated coarse-graining the data, which is similar to what might be done in a typical experiment with "point-clicked" data. In this case, we found that the variance of the curvature distribution depends on the coarse-graining level. The estimated curvature distribution from simulated polymers with $L/L_p = 6.3$ using $\Delta \tilde{s} = 2$ pixels is shown in Fig. 9. The solid lines in Fig. 9A and B show predictions of Eq. (15), while the solid line in Fig. 9C shows Eq. (17). Comparison with theory shows that the curvature distribution overestimates the small curvatures and underestimates the large curvatures. This is due to smoothing the sharp curvature fluctuations when coarse-graining. Note that Eq. (15) is derived under the assumption that the *local* energy (i.e., energy per unit length) is Boltzmann distributed, and since the total energy of a polymer chain is constant, the local energy must be rescaled on coarse-graining. The rescaling of the local energy can be represented by an effective persistence length, $\tilde{L}_p$. This can be understood in the following way. Let us start with a polymer chain of N nodes and total length L. Coarse-graining the polymer chain by taking every other coordinate results in a chain of $N/2$ nodes with twice the distance between coordinates. If the spacing between the coordinates in the coarse-grained chain is scaled back to the original node spacing, then the total length will be reduced by a factor of 2. This means the ratio L/L_p is reduced by the same factor, which is equivalent to having an effective persistence length twice the value of L_p. This scaling argument is intended for illustration; in

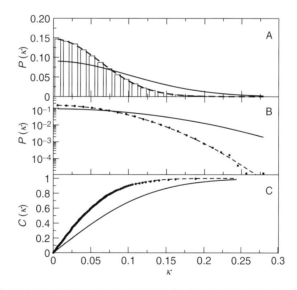

Fig. 9 The estimated probability density (curvature distribution) and cumulative distribution functions for $L/L_p = 6.3$ for a coarse-graining level of $\Delta \tilde{s} = 2$ pixels, calculated using pure data. (A) A histogram of the curvature values, κ. (B) Semilog plot of the curvature distribution. (C) Empirical cumulative distribution function calculated using the sorted curvature data. Solid lines in (A) and (B) show theoretical predictions of Eq. (15) using the correct persistence length L_p, and solid line in (C) shows theoretical cumulative distribution given by Eq. (17). Dashed lines show the fitted distributions.

practice, there might be additional dependencies on parameters besides node spacing and a discussion of these is beyond the scope of this chapter.

We found that the effective persistence length scales linearly with the average spacing of coarse-graining, $\Delta\tilde{s}$. The bullets in Fig. 10 show the effective persistence

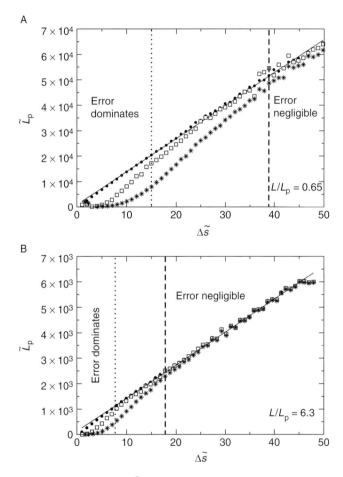

Fig. 10 The effective persistence length $\tilde{L}_p$ as a function of coarse-grained spacing $\Delta\tilde{s}$. The scaling relation is shown for (A) $L/L_p = 0.65$ and (B) $L/L_p = 6.3$, with $L = 621$ pixels. The bullets, empty squares, and stars show results for pure, digitized, and digitized with measurement error data, respectively. The solid line is a fit to the scaling relation for pure data which is used to estimate $\tilde{L}_p$ for a given $\Delta\tilde{s}$. The region to the left of the dotted line corresponds to an error dominated regime (for digitized data with measurement error) where one is mainly measuring the noise contribution to the curvature. The region to the right of the vertical dashed line corresponds to a regime where the noise contribution to the curvature is negligible, and the minimum coarse-graining interval, in principle, is given by this dashed line. However, in practice a value for $\Delta\tilde{s}$ may need to be chosen between the two vertical lines, which will result in some impact of noise on the curvature distribution.

length, $\tilde{L}_p$, as a function of $\Delta\tilde{s}$, for $L/L_p = 0.65$ and 6.3. The slope and intercept of the line fitted to the pure data in the figure could be used to estimate the effective persistence length for a given coarse-graining level, $\Delta\tilde{s}$. If this estimated $\tilde{L}_p$ is substituted for L_p in Eq. (15), the resulting curvature distribution is in excellent agreement with the simulation data.

2. Digitized Data

For the case of digitized data, the three-point method had trouble estimating the correct curvature distribution when every data point was used. This is because for a given stiff polymer, that is $L/L_p = 0.03$, the coordinate change from one pixel to the next is small (typically 0 or 1), and the resulting pixel is either directly next to the last pixel or perhaps one pixel above or below. Thus, for digitized data, the resulting calculated curvature values using the three-point method are only one of three values. Furthermore, when the curvature distribution is constructed from this data, the resulting histogram has only three bins, which leads to an imprecise estimate of the curvature distribution.

Alternatively, a more precise estimate of the position could be used, such as a semiautomated method described in Section III.B, to increase the precision of data from integer values of pixel coordinates, to interpolated decimal values corresponding to a position with subpixel precision. With this high precision data, the distribution of angles would not be limited to a few values even if coordinate data was collected from every pixel.

Therefore, to apply the three-point method correctly to digitized data, the coordinates should be coarse-grained, effectively increasing the number of values curvature can take. This means that, just as in the case for pure data described above, the effect of coarse-graining will change the variance of the curvature distribution. The squares in Fig. 10 show the effective persistence length, $\tilde{L}_p$ as a function of $\Delta\tilde{s}$, for $L/L_p = 0.65$ and 6.3. Whereas the scaling of $\tilde{L}_p$ for pure data was completely linear (bullets in Fig. 10), the plot for digitized data has a nonlinear region for small $\Delta\tilde{s}$ before crossing over to a linear scaling regime at larger spacing. This nonlinear region corresponds to an error dominated regime, and the width varies for each value of L/L_p thereby giving us the minimum coarse-graining interval (coordinate spacing). Note that this procedure provides a quantitative basis to the minimum coarse-graining estimate discussed in Section IV.B.5.

In practice, the observed images of microtubules may not be long enough to sample at large coarse-graining intervals. Therefore, microtubules can be concatenated, as long as the angles at the microtubule ends are matched with each other. This can be accomplished by taking the last two coordinates of one microtubule and aligning them with the first two coordinates of another, with the cost of losing two coordinates (where they overlap) per microtubule.

3. Digitized Data with Measurement Error

Finally, since the data collected from a digital image is also usually biased with measurement error, the real test for the three-point method was to see if it can simultaneously handle both sources of error. Once again the method has trouble when analyzing every pixel of data, but this now has more to do with the magnitude of the measurement error applied to every pixel. This suggests that data should not be collected for every pixel, but rather at some larger interval, similar to the digitized case. The stars in Fig. 10 show the effective persistence length, $\tilde{L}_p$ as a function of $\Delta\tilde{s}$, for $L/L_p = 0.65$ and 6.3. Again the minimum spacing can be determined, and an effective persistence length can be estimated for a given spacing.

The three-point method works well for highly curved polymers as typically observed from *in vivo* microtubule data; however, the method has trouble estimating the curvature of very stiff polymers for $L/L_p = 0.65$ due to noise. To correctly use the method in this stiff regime, the coordinates must be collected more precisely, and/or the data must be coarse-grained to reduce the effect of digitization and measurement error.

The shape of a thermally driven equilibrium polymer with $L/L_p = 6.3$ (Fig. 4) might resemble a highly curved *in vivo* microtubule (shown in Fig. 1 inset). To test how the three-point method performs on polymers with this persistence length, we collected data using the point-click method at a spacing of $\Delta\tilde{s} = 20$, and the resulting curvature distribution is shown in Fig. 11. The solid line shows the agreement of simulation data with Eq. (15) when using the effective persistence length, $\tilde{L}_p$. Note that the variance of the curvature distribution in Fig. 11 is different from the variance in Fig. 8. A precise knowledge of L_p is needed to rescale the curvature distribution in Fig. 11 and to match the variance in Fig. 8. While the fitted line in Fig. 10 can be used to estimate L_p by extrapolating the line to $\Delta\tilde{s} = 1$, we urge caution in applying this technique to experimental data, as small errors in the slope may lead to large discrepancies in the estimation of L_p. If one chooses to rescale the data to obtain the curvature distribution with the correct variance, a more accurate method for estimating L_p should be used, that is a bending mode analysis.

B. Shape-Fitting Method

We found that the Fourier shape-fitting method can only be applied to estimate the curvature distribution if the correct number of modes is utilized. However, the correct number of modes to use is different for each value of L/L_p, amount of noise, and coarse-graining level investigated. In addition, using the maximum number of modes $(N - 1)$ does not necessarily give the correct curvature distribution, as one might expect. This is due to the fact that the sine function in Eq. (13) gives zero curvature values when it is evaluated at the nodes for certain mode numbers.

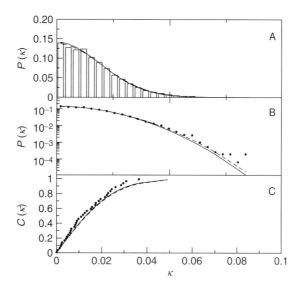

Fig. 11 The estimated probability density (curvature distribution) and cumulative distribution functions for $L/L_p = 6.3$ for a coarse-graining level of $\Delta \bar{s} = 20$ pixels, for digitized data with measurement error. (A) A histogram of the curvature values, κ. (B) Semilog plot of the curvature distribution. (C) Empirical cumulative distribution function calculated using the sorted curvature data. Solid lines show theoretical predictions of Eqs. (15) and (17), whereas dashed lines show the fitted distribution. Using the effective persistence length, $\tilde{L}_p$, in Eqs. (15) and (16) rescales the theoretical predictions to match the simulation data.

This happens with increasing frequency for the higher modes, and effectively adds zeros to the curvature distribution, resulting in an overestimate of low curvatures and underestimate of high curvatures. Without the knowledge of the true curvature distribution, which is the case in the analysis of the experimental data, the correct number of modes is difficult to determine *a priori*. We therefore conclude that this method is of limited value for determining the curvature distribution for unknown conditions in living cells.

We also investigated using polynomials or sine transforms for fitting shapes. We found a similar problem with using polynomials because an optimal order for the polynomial needs to be determined, and this order depends on the shape of each polymer. In addition, the polynomial method has problems measuring the curvature at the ends of the polymers due to poor fitting. The sine transform also has difficulty matching the shape at the ends due to the fact that sine functions do not match the angles at the ends. This edge effect, also known as ringing, is magnified when taking the derivative of the Fourier transform to obtain the curvature. We therefore recommend not using shape-fitting approaches to determine the curvature distribution, especially if: (1) additional nonthermal forces act on the microtubules or (2) the flexural rigidity is not known *a priori*.

VII. Discussion

The results given in this chapter have all been obtained using thermally driven polymer simulation data, and it remains unclear whether *in vivo* data would yield similar results. The shape of the curvature distribution for a thermally driven polymer is Gaussian. Therefore, estimating the curvature distribution serves as an important tool to determine whether the polymer is only subject to thermal forces. If the estimated curvature distribution from a cell is not Gaussian in shape, then the associated bending must be influenced by other factors such as molecular motors, polymerization forces, or actomyosin contractility. However, if the shape is Gaussian, it is still possible that nonthermal sources of bending contribute to the curvature distribution.

For systems where the curvature distribution is not Gaussian, alternative scenarios can be investigated in the framework provided by hybrid SRD-MD simulations. In particular, the effect of molecular motors, elastic forces in the surrounding cytoplasm, polymerization forces, as well as hydrodynamic effects due to the presence of other microtubules can be investigated using this approach. Furthermore, by comparing the experimental curvature distribution with the simulated distribution, we can potentially identify and eliminate models for microtubule bending.

Using shape-fitting methods for estimating curvature distributions may lead to large discrepancies. In particular, results obtained using Fourier cosine transforms depend sensitively on the number of modes used, even for data that is not coarse-grained. In addition, both sine transforms and polynomial methods suffer from end effects. All shape-fitting methods effectively interpolate between the nodes, which results in spurious curvature values in the curvature distribution. We therefore do not recommend the use of fitting functions for estimating the curvature distribution. By comparison, the three-point method is capable of producing a model-independent curvature distribution and is therefore a suitable method for estimating the curvature distributions in living cells.

For noisy data, coarse-graining is necessary in order to prevent digitization and measurement errors from dominating the curvature distribution. The disadvantage of coarse-graining is that the correct variance is lost due to smoothing. However, the curvature distribution still allows for comparison between different experiments, assuming the data collection is done at the same coarse-graining level.

When collecting data from living cells, the position and curvature at the microtubule's ends are not easy to determine. This is due in part to a limited field of view in the microscope, as well as high density of microtubules in the cell body. Even if the microtubules could be traced back to the centrosome, it is still unclear as to whether the ends are clamped or hinged. This is another reason we believe it is necessary to use a local curvature estimation method, such as the three-point method, when analyzing data in living cells.

Our investigation has allowed us to identify a number of important considerations for data collection. First, the minimum spacing between data points along microtubules is important and it should be determined experimentally. Second, the errors associated with digitization can be large, especially when the image pixel size is larger than the diameter of the microtubule. This error could be reduced by using higher magnification to project the microtubule image over more pixels, effectively reducing the pixel size in the image. However, this will reduce the signal to noise, which will then require one to increase the exposure time or to average over more data points. Perhaps a better approach would be to estimate the microtubule position to subpixel precision, which could be done through the use of a semiautomated method.

VIII. Conclusions

Given an experimental set of discrete coordinates collected from images of microtubules, the first question to be asked is whether the data can be explained by applying a theoretical model. The simplest approach is to assume that thermal forces dominate and use a thermally driven polymer model. To check this hypothesis, we suggest using the three-point method as described in Section V.A to estimate the curvature distribution, making sure the data collection is done at the appropriate coarse-graining interval. Then, if the estimated curvature distribution is Gaussian in shape, one can interpret the results using a thermal model.

More likely, the bending of microtubules in living cells involves additional nonthermal forces, otherwise one would have to assume that they are much less rigid *in vivo* to match the observed bending. The three-point method provides a simple approach for estimating the curvature of microtubules, independent of the forces that influence bending. The difference in curvature distributions between experiments at similar coarse-graining intervals may then be used as a tool to provide insight into how different cellular processes affect curvature. We believe this approach will complement the existing techniques for investigating microtubule deformation and could potentially turn microtubules into intracellular force transducers.

In order to use microtubules as force transducers, it is important to have an accurate measure of the flexural rigidity *in vivo*. Since the flexural rigidity of microtubules *in vivo* is to the best of our knowledge unknown, one could try for a more realistic "*in vivo*" estimate of *EI* by using cell extracts. Assuming that forces attributed to molecular motors can be silenced, then one could use any of the methods referenced in Section II to estimate flexural rigidity from cell extracts. Once *EI* is known, the deformation of the microtubules, as measured by the curvature distribution can, in principle, be used to estimate the distribution of loads on these microtubules.

Acknowledgments

We thank Dr. Patricia Wadsworth, University of Massachusetts Amherst for kindly supplying us the LLC-PK1-α cell line.

D.M.K. and E.T. acknowledge support from the National Science Foundation under Grant No. DMR-0513393 and ND EPSCoR through NSF grant EPS-0132289. D.J.O. acknowledges support from NSF grants BES 9984955, BES 0119481, and NIGMS R01GM71522.

References

Arfken, G. B., and Weber, H. J. (2001). "Mathematical Methods for Physicists." Academic Press, San Diego.

Brangwynne, C. P., Mackintosh, F. C., Kumar, S., Geisse, N. A., Talbot, J., Mahadevan, L., Parker, K. K., Ingber, D. E., and Weitz, D. A. (2006). Microtubules can bear enhanced compressive loads in living cells because of lateral reinforcement. *J. Cell Biol.* **173**(5), 733–741.

Cassimeris, L., Gard, D., Tran, P. T., and Erickson, H. P. (2001). XMAP215 is a long thin molecule that does not increase microtubule stiffness. *J. Cell Sci.* **114**, 3025–3033.

Doi, M., and Edwards, S. F. (2004). "The Theory of Polymer Dynamics." Oxford University Press, Oxford.

De Gennes, P. G. (1979). "Scaling Concepts in Polymer Physics." Cornell University Press, Ithaca, New York.

Desai, A., and Mitchison, T. J. (1997). Microtubule polymerization dynamics. *Annu. Rev. Cell Dev. Biol.* **13**, 83–117.

Dogterom, M., and Yurke, B. (1997). Measurement of the force-velocity relation for growing microtubules. *Science* **278**, 856–860.

Falck, E., Lahtinen, J. M., Vattulainen, I., and Ala-Nissila, T. (2004). Influence of hydrodynamics on many-particle diffusion in 2D colloidal suspensions. *Eur. Phys. J. E* **13**, 267–275.

Falck, E., Punkkinen, O., Vattulainen, I., and Ala-Nissila, T. (2003). Dynamics and scaling of 2D polymers in a dilute solution. *Phys. Rev. E* **68**, 050102(R).

Felgner, H., Frank, R., Biernat, J., Mandelkow, E. M., Mandelkow, E., Ludin, B., Matus, A., and Schliwa, M. (1997). Domains of neuronal microtubule-associated proteins and flexural rigidity of microtubules. *J. Cell Biol.* **138**, 1067–1075.

Felgner, H., Frank, R., and Schliwa, M. (1996). Flexural rigidity of microtubules measured with the use of optical tweezers. *J. Cell Sci.* **109**, 509–516.

Frenkel, D., and Smit, B. (2002). "Understanding Molecular Simulation. From Algorithms to Applications." Academic Press, Boston.

Fygenson, D. K., Elbaum, M., Shraiman, B., and Libchaber, A. (1997). Microtubules and vesicles under controlled tension. *Phys. Rev. E* **55**, 850–859.

Gildersleeve, R. F., Cross, A. R., Cullen, K. E., Fagen, A. P., and Williams, R. C., Jr. (1992). Microtubules grow and shorten at intrinsically variable rates. *J. Biol. Chem.* **267**(12), 7995–8006.

Gittes, F., Mickey, B., Nettleton, J., and Howard, J. (1993). Flexural rigidity of microtubules and actin filaments measured from thermal fluctuations in shape. *J. Cell Biol.* **120**(4), 923–934.

Gupton, S. L., Salmon, W. C., and Waterman-Storer, C. M. (2002). Converging populations of F-actin promote breakage of associated microtubules to spatially regulate microtubule turnover in migrating cells. *Curr. Biol.* **12**, 1891–1899.

Hecht, M., Harting, J., Ihle, T., and Herrmann, H. J. (2005). Simulation of claylike colloids. *Phys. Rev. E* **78**, 011408.

Heidemann, S. R., Kaech, S., Buxbaum, R. E., and Matus, A. (1999). Direct observations of the mechanical behavior of the cytoskeleton in living fibroblasts. *J. Cell Biol.* **145**, 109–122.

Howard, J. (2001). "Mechanics of Motor Proteins and the Cytoskeleton." Sinauer Associates, Sunderland, MA.

Ihle, T., and Kroll, D. M. (2001). Stochastic rotation dynamics: A Galilean-invariant mesoscopic model for fluid flow. *Phys. Rev. E* **63**, 020201(R).

Ihle, T., and Kroll, D. M. (2003a). Stochastic rotation dynamics. I. Formalism, galilean invariance, and green-kubo relations. *Phys. Rev. E* **67**, 066705.

Ihle, T., and Kroll, D. M. (2003b). Stochastic rotation dynamics. II. Transport coefficients, numerics, and long-time tails. *Phys. Rev. E* **67**, 066706.

Ihle, T., Tüzel, E., and Kroll, D. M. (2004). Resummed Green-Kubo relations for a fluctuating fluid-particle model. *Phys. Rev. E* **70**, 035701(R).

Ihle, T., Tüzel, E., and Kroll, D. M. (2005). Equilibrium calculation of transport coefficients for a fluid-particle model. *Phys. Rev. E* **72**, 046707.

Janson, M. E., De Dood, M. E., and Dogterom, M. (2003). Dynamic instability of microtubules is regulated by force. *J. Cell Biol.* **161**(6), 1029–1034.

Janson, M. E., and Dogterom, M. (2004). A bending mode analysis for growing microtubules: Evidence for a velocity-dependent rigidity. *Biophys. J.* **87**, 2723–2736.

Kaverina, I., Krylyshkina, O., Beningo, K., Anderson, K., Wang, Y. L., and Small, J. V. (2002). Tensile stress stimulates microtubule outgrowth in living cells. *J. Cell Sci.* **115**(11), 2283–2291.

Kikuchi, N., Gent, A., and Yeomans, J. M. (2002). Polymer collapse in the presence of hydrodynamic interactions. *Eur. Phys. J. E* **9**, 63–66.

Kirschner, M., and Mitchison, T. (1986). Beyond self-assembly: From microtubules to morphogenesis. *Cell* **45**, 329–342.

Kis, A., Kasas, S., Babic, B., Kulik, A. J., Benoit, W., Briggs, G. A. D., Schonenberger, C., Catsicas, S., and Forro, L. (2002). Nanomechanics of microtubules. *Phys. Rev. Lett.* **89**, 248101.

Klapper, I., and Qian, H. (1998). Remarks on discrete and continuous large-scale models of DNA dynamics. *Biophys. J.* **74**, 2504–2514.

Kurachi, M., Hoshi, M., and Tashiro, H. (1995). Buckling of a single microtubule by optical trapping forces—direct measurement of microtubule rigidity. *Cell Motil. Cytoskeleton* **30**, 221–228.

Kurz, J. C., and Williams, R. C. (1995). Microtubule-associated proteins and the flexibility of microtubules. *Biochemistry* **34**, 13374–13380.

Malevanets, A., and Kapral, R. (1999). Mesoscopic model for solvent dynamics. *J. Chem. Phys.* **110**, 8605–8613.

Malevanets, A., and Kapral, R. (2000). Solute molecular dynamics in a mesoscale solvent. *J. Chem. Phys.* **112**, 7260–7269.

Malevanets, A., and Yeomans, J. M. (2000). Dynamics of short polymer chains in solution. *Europhys. Lett.* **52**, 231–236.

Marenduzzo, D., Micheletti, C., Seyed-allaei, H., Trovoto, A., and Maritan, A. (2005). Continuum model for polymers with finite thickness. *J. Phys. A: Math. Gen.* **38**, L277–L283.

Mickey, B., and Howard, J. (1995). Rigidity of microtubules is increased by stabilizing agents. *J. Cell Biol.* **130**, 909–917.

Mitchison, T., and Kirschner, M. (1984). Dynamic instability of microtubule growth. *Nature* **312**, 237–242.

Noguchi, H., and Gompper, G. (2004). Fluid vesicles with viscous membranes in shear flow. *Phys. Rev. Lett.* **93**, 258102.

Odde, D. J., Ma, L., Briggs, A. H., DeMarco, A., and Kirschner, M. W. (1999). Microtubule bending and breaking in living fibroblast cells. *J. Cell Sci.* **112**, 3283–3288.

Ott, A., Magnasco, M., Simon, A., and Libchaber, A. (1993). Measurement of the persistence length of polymerized actin using fluorescence microscopy. *Phys. Rev. E* **48**(3), R1642–R1645.

Padding, J. T., and Louis, A. A. (2004). Hydrodynamic and Brownian fluctuations in sedimenting suspensions. *Phys. Rev. Lett.* **93**, 220601.

Putnam, A. J., Cunningham, J. J., Dennis, R. G., Linderman, J. J., and Mooney, D. J. (1998). Microtubule assembly is regulated by externally applied strain in cultured smooth muscle cells. *J. Cell Sci.* **111**, 3379–3387.

Press, W. H., Flannery, B. P., Teukolsky, S. A., and Vetterling, W. T. (1992). "Numerical Recipes in C: The Art of Scientific Computing," 2nd edn. Cambridge University Press, New York.

Ripoll, M., Mussawisade, K., Winkler, R. G., and Gompper, G. (2004). Low-Reynolds-number hydrodynamics of complex fluids by multi-particle-collision dynamics. *Europhys. Lett.* **68**, 106–112.

Rusan, N. M., Fagerstrom, C. J., Yvon, A. M., and Wadsworth, P. (2001). Cell cycle dependent changes in microtubule dynamics in living cells expressing green fluorescent protein-alpha tubulin. *Mol. Biol. Cell* **12**(4), 971–980.

Sprague, B. L., Pearson, C. G., Maddox, P. S., Bloom, K. S., Salmon, E. D., and Odde, D. J. (2003). Mechanisms of microtubule-based kinetochore positioning in the yeast metaphase spindle. *Biophys. J.* **84**(6), 3529–3546.

Storm, C., Pastore, J. J., MacKintosh, F. C., Lubensky, T. C., and Janmey, P. A. (2005). Non-linear elasticity in biofilament gels. *Nature* **435,** 191–196.

Takasone, T., Juodkazis, S., Kawagishi, Y., Yamaguchi, A., Matsuo, S., Sakakibara, H., Nakayama, H., and Misawa, H. (2002). Flexural rigidity of a single microtubule. *Jpn. J. Appl. Phys.* **41,** 3015–3019.

Tüzel, E., Strauss, M., Ihle, T., and Kroll, D. M. (2003). Transport coefficients for stochastic rotation dynamics in three dimensions. *Phys. Rev. E* **68,** 036701.

Tüzel, E., Ihle, T., and Kroll, D. M. (2006). Dynamic correlations in stochastic rotation dynamics. Pre-print, http://lanl.arxiv.org/cond-mat/0606628.

Van Noort, J., van der Heijden, T., de Jager, M., Wyman, C., Kanaar, R., and Dekker, C. (2003). The coiled-coil of the human Rad50 DNA repair protein contains specific segments of increased flexibility. *Proc. Natl. Acad. Sci. USA* **100**(13), 7581–7586.

Venier, P., Maggs, A. C., Carlier, M.-F., and Pantaloni, D (1994). Analysis of microtubule rigidity using hydrodynamic flow and thermal fluctuations. *J. Biol. Chem.* **269**(18), 13353–13360.

Waterman-Storer, C. M., and Salmon, E. D. (1997). Actomyosin-based retrograde flow of microtubules in the lamella of migrating epithelial cells influences microtubule dynamic instability and turnover and is associated with microtubule breakage and treadmilling. *J. Cell Biol.* **139,** 417–434.

Winkler, R. G., Mussawisade, K., Ripoll, M., and Gompper, G. (2004). Rod-like colloids and polymers in shear flow: A multi-particle-collision dynamics study. *J. Phys.: Condens. Matter* **16,** 3941–3954.

Zheng, J., Buxbaum, R. E., and Heidemann, S. R. (1993). Investigation of microtubule assembly and organization accompanying tension-induced neurite initiation. *J. Cell Sci.* **104**(4), 1239–1250.

CHAPTER 11

Nuclear Mechanics and Methods

Jan Lammerding,[*] **Kris Noel Dahl,**[†] **Dennis E. Discher,**[‡] **and Roger D. Kamm**[§]

[*]Department of Medicine, Cardiovascular Division
Brigham and Women's Hospital/Harvard Medical School
Boston, Massachusetts 02115

[†]Biomedical Engineering and Chemical Engineering
Carnegie Mellon University
Pittsburgh, Pennsylvania 15213

[‡]Biophysical Engineering Laboratory
University of Pennsylvania
Philadelphia, Pennsylvania 19104

[§]Departments of Biological Engineering and Mechanical Engineering
Massachusetts Institute of Technology
Cambridge, Massachusetts 02139

METHODS IN CELL BIOLOGY, VOL. 83

0091-679X/07 $35.00
DOI: 10.1016/S0091-679X(07)83011-1

Abstract

The role of the nucleus in protecting and sequestering the genome is intrinsically mechanical, and disease-causing structural mutants in lamins and other components underscore this function. Various methods to measure nuclear mechanics, isolated or *in situ*, are outlined here in some detail.

I. Introduction

The nucleus is generally the largest organelle of a eukaryotic cell and is literally the defining feature. The nucleus is also among the stiffest organelle in a cell. The primary constituent of the nucleus is chromatin—a complex of nucleic acids (DNA) and proteins, primarily histones. In the present postgenomic era, epigenetic factors that regulate DNA expression are increasingly understood to be important to cell state, and among the epigenetic factors of likely importance are the structural properties of the nucleus and its response to force. The mechanics of isolated DNA during bending, torsion, and extension has been well studied at the molecular level (Cui and Bustamante, 2000), in recognition of the fact that the physical properties of DNA are critical to processes such as replication and transcription, histone wrapping, chromatin compaction, and more. However, in higher-order eukaryotes, chromosomal DNA is packed within a nucleus from over a meter in length to only micrometers in dimension (Fig. 1), raising questions about chromatin mechanics *in situ*. Moreover, the nucleus not only houses DNA and other genetic factors but also establishes a protected nucleoplasmic microenvironment that modulates DNA access and gene expression.

The nuclear envelope can also contribute significantly to the mechanical behavior of the nucleus (Dahl *et al.*, 2004, 2005; Rowat *et al.*, 2005). The nuclear envelope is composed of a double lipid bilayer, the inner and outer nuclear membranes, and the nuclear lamina. The nuclear lamina plays important roles in structure, stability, and gene regulation (Goldman *et al.*, 2002; Gruenbaum *et al.*, 2005) and consists of a dense network of protein—composed principally of lamins—that underlies the inner nuclear membrane. Lamins are nuclear intermediate filament proteins that form coiled-coil dimers, which assemble further into stable strings and higher-order networks (Aebi *et al.*, 1986; Panorchan *et al.*, 2004). However, the *in vivo* structure of the nuclear lamina, especially the organization of the lamin network, is not completely understood. The inner and outer nuclear membranes join at the nuclear pores and enclose a perinuclear space (Fig. 1). Lending further to the complexity of the nuclear envelope, the outer nuclear membrane and the perinuclear space are continuous with the endoplasmic reticulum (ER), providing a reservoir of membrane that can, in principle, be drawn on during nuclear deformation. Transcription and other regulatory factors, as well as products of gene expression (RNA), shuttle between the nucleus and the cytoplasm through nuclear pores. Ions, metabolites, and small proteins diffuse freely through the nuclear pore complex, while proteins larger than about 40 kDa that lack a

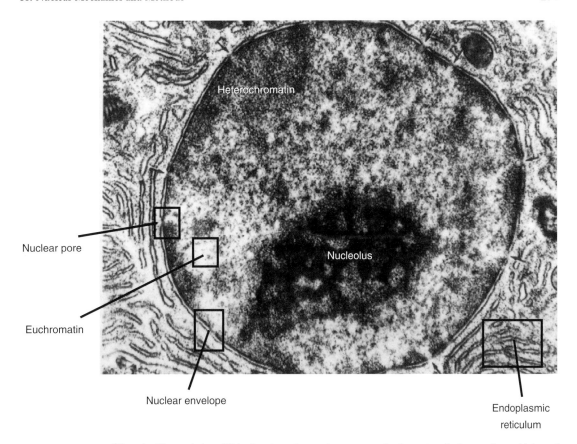

Fig. 1 Transmission EM showing the major structural elements of the nucleus. (Adapted from NUS Histonet, WWW Electronic Guide to Histology for Medicine and Dentistry, online at http://www.med.nus.edu.sg/ant/histonet/txt/tacsem/tac01.sem.html. TEM image courtesy of Dr. P. Gopalakrishnakone.)

nuclear localization sequence are excluded (Tran and Wente, 2006). The cytoplasm is literally crowded with such excluded proteins and other organelles. Equally important is the fact that the nucleus is not suspended or floating within this cytosol; rather it is physically coupled to the cytoskeleton through protein complexes (i.e., Sun proteins, nesprins, and so on) that cross the perinuclear space and connect to the inner nuclear lamina (Crisp *et al.*, 2006; Grady *et al.*, 2005; Padmakumar *et al.*, 2005; Starr and Han, 2002, 2003; Zhang *et al.*, 2005).

Nuclear structure and mechanics not only affect the overall cellular mechanical properties—as a large and relatively stiff inclusion—but they also play a critical role in many cellular functions. Nuclear organization directly affects accessibility of chromatin to transcriptional regulators. In addition, since extracellular mechanical stimuli are transmitted to the nucleus, the nucleus has been implicated as a mechanosensor and thus important for mechanotransductive signaling (Ingber *et al.*, 1987), although the contributions of specific nuclear components to possible mechanosensing have remained unclear. Nuclear size and shape vary between different cell types, ranging from textbook spheroids to multilobed nuclei in neutrophils and ruffled shapes in aged and some diseased cells (Haithcock *et al.*, 2005; Muchir *et al.*, 2004; Scaffidi and Misteli, 2006). How such diverse nuclear morphology and the associated complex arrangements of components, which may reflect the state of nuclear functions, respond to stresses and strains are questions with significant implications for a number of topics in cell mechanics, mechano-transduction, and nuclear structure–function.

The discovery that mutations in genes encoding lamins and associated nuclear envelope proteins are responsible for a wide range of human diseases has propelled research on the nuclear envelope to the forefront. Diseases collectively referred to as laminopathies include Hutchinson–Gilford progeria syndrome, Emery-Dreifuss muscular dystrophy, Dunnigan-type familial partial lipodystrophy, limb-girdle muscular dystrophy, and dilated cardiomyopathy with conduction system defects (Bione *et al.*, 1994; Bonne *et al.*, 1999; Cao and Hegele, 2000; Chen *et al.*, 2003; De Sandre-Giovannoli *et al.*, 2002, 2003; Eriksson *et al.*, 2003; Fatkin *et al.*, 1999; Muchir *et al.*, 2000; Novelli *et al.*, 2002; Shackleton *et al.*, 2000). Mechanical studies of diseased nuclei have yielded important clues as to the role of nuclear anatomy in pathological deformity. For example, lamin A/C-deficient cells have mechanically weak nuclei (Lammerding *et al.*, 2004), while nuclei from patients with Hutchinson–Gilford progeria syndrome, where there is an increase in lamin A at the nuclear envelope, show a stiffer, less resilient lamina (Dahl *et al.*, 2006). Two hypotheses, neither mutually exclusive, have emerged to explain how mutations in lamin proteins could cause the diverse disease phenotypes reflected in the laminopathies (Broers *et al.*, 2004a; Burke and Stewart, 2002; Hutchison and Worman, 2004). The "gene-regulation hypothesis" proposes that perturbed gene regulation may underlie the development of different disease phenotypes, while the "structural hypothesis" proposes that mutations in the lamin proteins render the nucleus more fragile, causing cell death and eventually disease in mechanically stressed tissues (Broers *et al.*, 2004a; Burke and Stewart, 2002; Hutchison, 2002). Nuclear mechanics and structure could affect both of these functions, through either direct physical/structural functions or the regulation of chromatin organization, access to transcriptional regulators, or coupling of the nucleus to the cytoskeleton, and so on. In addition to these physiological and pathological functions, detailed knowledge of nuclear mechanics is also important for somatic cell nuclear transfer (SCNT), in which the nucleus from adult somatic cells is transferred to the denucleated, unfertilized egg (Wakayama *et al.*, 1998). While SCNT has the potential to revolutionize many levels of biological research from cloning to nuclear reprogramming, the techniques are notoriously inefficient (Wilmut and Paterson, 2003) and may benefit from a better fundamental understanding of the mechanics of nuclei. In this chapter, we describe some of the current methods being used to probe and understand nuclear mechanics, focusing on methods that provide measures not only of nuclear shape but also of stress, strain, and structural failure.

II. Experimental Methods for Probing Nuclear Mechanical Properties

Nuclei can be studied both within cells and as isolated organelles. Advantages of studying nuclei *in situ* include maintaining the balance of salts and molecules that diffuse or shuttle between the cytoplasm and the nucleoplasm through the large nuclear pores. Furthermore, nuclei *in situ* maintain membrane connections between the double bilayered nuclear envelope and the ER with potentially important implications for membrane flow and lateral diffusion of membrane proteins. Experiments with isolated nuclei provide complementary information. Isolation allows one to clearly distinguish of properties of the nucleus from those of the rest of the cell and also to detect specific physical interactions while minimizing visual artifacts. Isolation also allows stress to be applied directly to the nucleus rather than having complicated redistributions of stress to the cell cytoskeleton. In some cases, nuclear lamina defects associated with lamin mutations have been measured under extreme stresses that might exceed the average physiological stress and might not even be sustainable by the cellular cytoskeleton (Dahl *et al.*, 2006). Understanding the failure modes of nuclear structures under such conditions is nonetheless very informative. For example, imposing direct forces to isolated nuclei replicates the stresses and strains imposed on nuclei during SCNT. In the following sections, we present experimental methods for isolating nuclei from living cells and for probing the mechanics of nuclei both in isolation and in intact cells. Although generalizations can be tenuous in cellular mechanics, several of the methods below have provided measures of nuclear mechanical properties that indicate the nucleus of most differentiated cells is about 5- to 10-fold stiffer than the cell cytoplasm. As a very rough but perhaps usefully tactile sense of the mechanical properties presented, the nucleus is about as stiff as a tender steak while the cytoplasm has a stiffness that might be approximated by a jellyfish.

A. Isolation of Individual Nuclei

Spheroidal nuclei of many cell types can be readily isolated by both single-cell mechanical extraction and bulk methods (Dahl *et al.*, 2005; Deguchi *et al.*, 2005; Guilak *et al.*, 2000). Since extremely fragile nuclei have not yet been isolated successfully, and since the nuclear envelope breaks down during mitosis, the method applies only to mechanically stable interphase nuclei.

An individual nucleus can be isolated from a cell by taking advantage of the nucleus' relative rigidity and tenuous connections to the rest of the cell. In some cases, a micropipette is used to simply tear through the cell membrane to access the nucleus (Rowat *et al.*, 2005). In other cases, the micropipette is used to mechanically rupture the cell (Guilak *et al.*, 2000). Alternatively, nuclei can be isolated in bulk and then probed mechanically. A number of protocol methods are available for nuclear isolation (Caille *et al.*, 2002; Dahl *et al.*, 2005; Deguchi *et al.*, 2005). In general, the cell membrane is disrupted mechanically with hypotonic swelling or chemically with digitonin or other surfactants that perturb the plasma membrane but not nuclear membranes. The cell is then opened with mechanical homogenization, and cellular contents are separated from the nuclei by ultracentrifugation through a sucrose gradient. One detailed protocol for bulk isolation, described in detail (Dean and Kasamatsu, 1994) and shown to yield nuclei suited to micromechanical characterization (Dahl *et al.*, 2005), is briefly described as follows:

1. Cells are grown to confluence. Typically, starting with 10^7 cells (a T75-flask or 100-mm dish) yields 10^5 nice looking nuclei. Nuclei can be isolated similarly from transfected cells, 24–48 h after transfection.

2. Cells are rinsed twice with phosphate buffered saline (PBS) and once with 10-mM HEPES, 1-mM dithiothreitol (DTT), pH 7.5.

3. Cells are scraped into 1 ml of the HEPES/DTT solution described above and allowed to equilibrate with the hypotonic solution for 10 min on ice.

4. Cells are broken open with 25 strokes in a Dounce homogenizer.

5. The lysate is added to one-fifth volume of 5× STKMC (1.25-M sucrose, 250-mM Tris, pH 7.6, 125-mM KCl, 15-mM $MgCl_2$, 15 mM-$CaCl_2$, 10 $\mu g/ml$ each of the protease inhibitors pepstatin and leupeptin) and incubated for 10 min on ice. The salt in the resulting 1× STKMC condenses the nuclei and aids in their separation later.

6. An appropriate amount of 2.3-M sucrose in TKMC (50-mM Tris, pH 7.6, 25-mM KCl, 3-mM $MgCl_2$, 3-mM $CaCl_2$, 2 $\mu g/ml$ each leupeptin and pepstatin) is added to the cell lysate suspension in order to achieve a final sucrose concentration of 1.6 M.

7. The 1.6-M STKMC suspension is layered onto a 150-μl cushion of 2.3-M STKMC and spun at 166,000 × g in an ultracentrifuge (Optima TLX, Beckman Coulter, Palo Alto, CA).

8. Cellular debris collects at the interface between the 1.6- and 2.3-M STKMC, while the denser nuclei collect at the bottom of the tube.

At this point nuclei can be resuspended into any media, although adding bovine serum albumin helps break up clumps and facilitates manipulation later. Nuclei can be counted relatively accurately on a hemacytometer. In addition to their use in mechanical and other biophysical measurements (such as FRAP, fluorescence recovery after photobleaching), isolated nuclei can be used for parallel experiments. For example, immunofluorescence imaging of nuclei bound to poly-L-lysine-coated coverslips can clarify nuclear localization of proteins that are sometimes masked by strong cytoplasmic signals. Similarly, Western blots of lysates of a known number of isolated nuclei can be used to compare normal and diseased nuclei and to quantify disease-induced perturbations in protein composition.

These characterizations seem likely to be of increasing importance as the molecular bases for nuclear structure, function, and disease are increasingly clarified.

Subsequent to isolation, it is important to assess the quality of the nuclei by comparison with nuclei in intact cells, and to make sure that the nuclear envelope remains intact while avoiding excess membranes such as the ER or cytoskeletal structures. DNA stains such as Hoechst dyes or, better, GFP-lamins allow fluorescence visualization for assessment of nuclear morphology. The concentrations of salts can dramatically affect nuclear mechanics since divalent and, to a lesser extent, monovalent cations affect chromatin condensation (Aaronson and Woo, 1981; Dahl *et al.*, 2005). With changes in salt, nuclear volume changes significantly and may induce mechanical stress on the nuclear lamina. It is unclear exactly what effect these salt concentrations have on protein–protein interactions within the nucleus. Therefore, while nuclei isolated in buffers with low salt tend to swell and resemble closely the contours and size of nuclei inside live cells, it is impossible to know how the mechanical properties might be affected. To accurately approximate nuclear mechanics inside the cell, it is usually desirable to mimic the intracellular ion concentrations. The difficulty is that the ion concentrations inside the cytoplasm and within organelles have not been reproducibly determined, and values for ions such as calcium range from submillimolar to millimolar and can vary greatly as a function of disease (Dobi and Agoston, 1998). Also, it is difficult to predict the shift in nuclear salt homeostasis after isolation from the cell. Some studies have deliberately examined mechanics of isolated nuclei at extreme salt conditions to determine the maximum possible range of mechanical responses. This strategy proves effective since the chromatin condensation and dilation appears to shift the load within the nucleus from the chromatin to the lamina (Dahl *et al.*, 2005).

B. Micropipette Aspiration Experiments

While removal of a nucleus from the native cytoplasm increases the uncertainty about effects of the chemical environment, stress, and strain can be determined more directly with isolated nuclei. Micropipette aspiration has been the most common method for measuring the mechanics of isolated nuclei (Fig. 2), where the scale of deformation is on the order of the whole nucleus. At this length scale, the nucleus can be considered a relatively course-grained homogeneous material. Also, a key advantage of micropipette aspiration is the ability to visualize the responses of stressed subnuclear structures during the deformation, using various optical methods especially fluorescence and polarization optics. Here, we briefly summarize the methods involved, though more complete descriptions of micropipette methods can be found elsewhere (Evans, 1989).

1. Micropipettes of 2- to 8-μm diameter are pulled from 1-mm borosilicate capillaries (WPI, Sarasota, Florida) with commercial pipette pullers (Sutter Instrument Company, Novato, California) that have been developed for a host of micropipette needs such as microinjection and patch clamping.

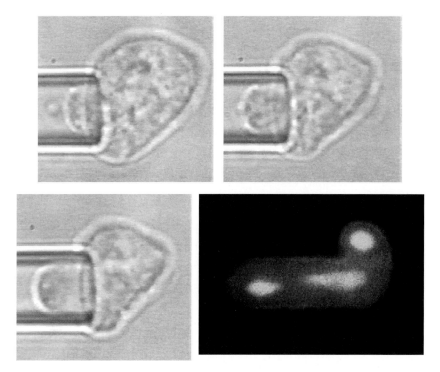

Fig. 2 An isolated epithelial nucleus is aspirated into a micropipette of 6-μm inner diameter at a constant applied pressure of 7 kPa at 2, 14, and 132 sec. While nuclear and subnuclear deformation can be seen in bright field, GFP-labeled proteins allow detailed assessments of connectivity and resilience (Dahl *et al.*, 2005; Rowat *et al.*, 2005). Fluorescence image shows a nucleus coexpressing GFP-lamin A and the nucleolar protein GFP-B23 (courtesy of D. Pajerowski).

2. Quantitative studies of nuclear properties by aspiration require a pipette of relatively constant tip diameter over at least 10 μm of length. This can generally be achieved at high pulling rates with some practice and optimization.

3. The micropipette is filled with water, most easily by a method that entails fully submerging the micropipettes (tied to a glass rod and tips up) in boiling water. The elevated temperature expands the air within, so that, on cooling, the decrease in air pressure draws the water into the micropipette.

4. The water-filled micropipette is connected to standard micropipette holders (Narishige, Japan) via a waterline to a water-filled reservoir. Nuclear isolation buffer is then aspirated into the micropipette to ensure solution compatibility.

5. Since nuclei are typically stiffer than other parts of the cell, negative pressures of ~0.1–10 kPa are required for aspiration (Dahl *et al.*, 2005; Deguchi *et al.*, 2005; Guilak *et al.*, 2000). They are readily generated by syringe suction. A pressure

transducer or manometer in parallel or within the apparatus is required for monitoring the applied pressure.

To determine from nuclear aspiration the effective stress versus strain behavior (Chapter 1 by Janmey *et al.* and Chapter 2 by Kandow *et al.*, this volume), the applied pressure is set and the increase in the projection length determined as a function of time. Nuclei can be viscoelastic, recovering significantly after a deformation, or they can appear more plastic or fluid-like, showing largely irreversible deformations. The relationship between plastic rearrangements and phenotypic plasticity has yet to be assessed. For studies of either viscoelasticity or plasticity, the increase in the projection length into the micropipette as a function of time is measured while applying a step in suction pressure. Using this technique in conjunction with high-resolution microscopy, several groups have been able to examine lamina rearrangement and identify roles for subnuclear structures in nuclear mechanics (Dahl *et al.*, 2005; Rowat *et al.*, 2005). Many questions remain such as the effects of differentiation, aging, as well as disease and treatments on nuclear plasticity in relation to nuclear shape.

Isolated nuclei exhibit viscoelasticity and creep in response to micropipette aspiration at constant pressures. While pressure versus length curves can be revealing, viscoelastic models appear to succinctly capture the time dependence and differences between nuclei. Among the models applied to the nucleus, some are purely elastic (Deguchi *et al.*, 2005), while others involve three-parameter spring-dashpot (Guilak *et al.*, 2000) or power-law rheological responses that are equivalent to an infinite series of Kelvin solids (Dahl *et al.*, 2005). The latter applies if micropipette aspiration (or the equivalent for AFM) yields a plot of pressure/length that increases with time as t^{α}. The relationship of these responses to changes in nuclear components, including the chromatin, remains a major challenge.

C. Substrate Strain Experiments

Culturing of cells on elastic membranes allows the cells and their nuclei to be strained as the substrate is stretched. Tracking of particles on the cells allows such strains to be measured. The method was first outlined by Barbee *et al.* (1994) for the case of isotropic stretching of the substrate, and it was later refined for uniaxial stretching by Caille *et al.* (1998). Similar techniques have been employed to measure nuclear mechanics in mouse embryo fibroblasts lacking specific nuclear envelope proteins (Lammerding *et al.*, 2004, 2005). In these experiments, the substrates underlying the attached cells are subjected to either uniaxial or biaxial strains. Substrate strains are transmitted to the cytoskeleton through focal adhesion complexes, and induced nuclear and cytoskeletal strains can be quantified to infer mechanical properties of the nucleus relative to the cytoskeleton. The advantages of this experimental technique are several fold. It does not require the isolation of nuclei with the needed identification of physiological buffer conditions. Also, in preserving the normal nuclear and cytoskeletal architecture and the interconnections, the nucleus is strained in a relatively physiological fashion. In addition, the method can be applied to many cells simultaneously. One drawback is that cells must spread well and not detach during strain application; nonadherent or weakly adherent cells are not suitable for these experiments. Another drawback is that measurements depend on pathways of force transmission through the cytoskeleton (e.g., contractile state of actin–myosin) and only reflect nuclear stiffness relative to cytoskeletal stiffness.

1. Preparation

Thin, transparent silicone membranes (Silastic, Specialty Mfg, Pineville, North Carolina) are generally used as the substrate for these experiments, due to their translucent optical characteristics and their elastic material properties. The typical membrane thickness used is 0.15–0.17 mm, which is similar to a No. 1 coverslip. In order to promote cell attachment, it is necessary to coat the membranes with appropriately chosen extracellular proteins, for example fibronectin, laminin, or collagen. Since silicone membranes tend to be hydrophobic, adsorption of protein is generally successful. It is recommended to perform preliminary experiments with different coating proteins and concentrations to find conditions maximizing cell adhesion and spreading. In one set of experiments, for example, mouse skeletal myoblasts as well as human and mouse fibroblasts were found to adhere well to fibronectin-coated silicone membranes: 2 μg/ml of fibronectin in phosphate-buffered saline solution, incubated at 37 °C for 5 h or alternatively at 4 °C overnight (Lammerding *et al.*, 2004, 2005).

2. Plating and Culturing Cells on the Membranes

In order to induce uniform strains in each cell, cells should be plated at low densities ($\approx$30–60% confluence at the time of experiments), as this minimizes cell-to-cell interactions and allows cells to spread freely. However, too low of a cell density is often detrimental to cell viability, and so the exact plating density must be determined empirically; for mouse embryo fibroblasts, $\approx$1000–2000 cells/cm^2 are recommended. Cells are plated in full-growth media (containing serum) to achieve firm cell attachment, but are subsequently serum starved for 48 h to minimize the number of mitotic cells, as measurements in these cells would not accurately reflect mechanical properties of the interphase nucleus. In the case of myoblasts, cells are induced to differentiate into myotubes by serum- and growth factor withdrawal for 5 days to obtain well-spread, multinucleated myotubes that are well suited for the strain experiments.

3. Experimental Methods

Depending on the particular strain device used, cells are subjected to either uniaxial (Caille *et al.*, 1998) or biaxial (isotropic) strain (Barbee *et al.*, 1994; Lammerding *et al.*, 2004, 2005). The particular preference depends on the cell type (isotropic cell shape vs elongated cells) and the physiological problem under investigation, but both methods are suitable to infer nuclear mechanical properties. Commercially available strain devices such as the Stageflexer (Flexcell International, Hillsborough, North Carolina) can be mounted on an inverted epifluorescence or confocal microscope so that cells can be imaged before, during, and after the induction of strain at high magnification (at least 40×). The magnitude of the strain can generally be regulated by the particular strain device. While larger strain magnitudes lead to larger nuclear deformation and, thus, improve the precision of the measurements (assuming constant optical resolution for all conditions), excessive substrate strain can lead to cell damage and partial cell detachment. Since cellular sensitivity to strain can be cell-type specific, it is important to perform pilot experiments with various strain magnitudes and to carefully evaluate cell morphology during and after the strain application. Drastic changes in the cell outline or morphology (viewed by phase-contrast microscopy) during the experiments indicate excessive strain and potential detachment.

In order to measure induced nuclear deformations, it is necessary to fluorescently label the nucleus. Commonly used, cell-permeable reagents are Hoechst 33342 or SYTO 13 (both Invitrogen, Carlsbad, California) that intercalate into double-stranded DNA. The corresponding substrate strain can be quantified by tracking small impurities in the silicone membrane or small markers placed in or on top of the membranes (embedded FluoSpheres, Invitrogen, Carlsbad, California). Induced cytoskeletal strain can be quantified using microbeads internalized and embedded in the cytoskeleton (FluoSpheres, Invitrogen, Carlsbad, California; $\approx$0.2-μm diameter). The 0.2-μm bead size reflects a compromise: the beads must be small enough to be internalized by the cell, yet not so small that they freely diffuse through the spaces of the cytoskeleton and are therefore unsuitable as markers for cytoskeletal strain (Caille *et al.*, 1998).

4. Analysis of Results

Experimental analysis requires computation of the applied substrate strain and the induced nuclear and cytoskeletal strains. Strains are generally computed based on images acquired before and during the induction of strain, while residual strains are computed based on comparison between images acquired before the induction and after the release of strain on the substrate. To compute substrate and cytoskeletal strains, it is necessary to track the (centroid) positions of distinct markers, for example impurities in the silicone membrane or microbeads internalized and embedded in the cytoskeleton. The actual strain values are then computed either based on triangulation, with a simple MatLab program, or using the scaling factor of the linear affine image transformation in NIH Image (which deforms an image by the translation, rotation, and scaling) that best fits control points from the base image (prestrain) to the input image (full-strain). Similar methods can be applied to compute nuclear strain, exploiting brightly fluorescent speckles in the chromatin rather than internalized beads. Alternatively, one can quantify nuclear strain using an image analysis algorithm that minimizes the normalized cross-correlation function between the base and the input images, by applying image rotation, translation, and scaling transformation to the input image (Lammerding *et al.*, 2005). This technique works especially well for nuclei that only have relatively few distinct fluorescent speckles. Finally, nuclear strain can also be computed from direct measurements of nuclear dimensions based on thresholded fluorescence images, that is, after setting pixel intensities that are lower than a threshold to zero. However, this technique is more prone to artifacts caused by changes in fluorescence intensity over time.

Regardless of the specific technique used to analyze substrate, cytoskeletal, and nuclear strain, it is important to visually inspect the images of the cells during and following strain application to exclude cells that become damaged or partially detached as a result of the strain. A useful metric to determine cellular damage/detachment can be the residual nuclear strain, as cell damage/detachment is often found to result in decreased nuclear size following the removal of applied forces (Lammerding *et al.*, 2005).

To compare nuclear mechanics between different cells or experiments, it is best to normalize nuclear strain to the applied substrate strain or cytoskeletal strain, as the actual substrate strain can vary slightly from experiment to experiment. Caille *et al.* (1998) found that the induced cytoskeletal strain closely resembles the applied substrate strain, especially in regions farther away from the nucleus, suggesting that it might be sufficient to limit the analysis to the ratio of nuclear strain to substrate strain. However, additional experiments (e.g., immunofluorescence labeling of cytoskeletal components, magnetic bead microrheology) should be performed to verify that differences in nuclear deformation are not caused by disruptions in cytoskeletal morphology and mechanics (Lammerding *et al.*, 2004).

D. Compression Experiments

Uniaxial compression experiments to investigate cellular and nuclear deformations under load were first conducted in chondrocytes in articular cartilage explants (Guilak, 1995) and embedded in alginate constructs (Knight *et al.*, 2002). Subsequently, compression experiments on single cells and isolated nuclei have been used to study nuclear mechanics in more detail (Broers *et al.*, 2004b; Caille *et al.*, 2002; Thoumine and Ott, 1997). In these later experiments, single cells or isolated nuclei located on a rigid, transparent substrate (e.g., coverslips) on an inverted microscope are subjected to controlled, uniaxial displacement normal to the substrate surface, while simultaneously recording the uniaxial compression force and the induced nuclear and cellular deformations. The compressive force is uniformly applied to the entire cell surface in contact with the microindenter or microplate, unlike conventional atomic force microscopy (AFM) that applies a highly localized force (Section II.E and Chapter 15 by Radmacher, this volume). In contrast to the cellular strain experiments described above (Section II.C), compression experiments generally do not require firm cell adhesion, but nuclear deformations in intact cells are still affected by the cytoskeletal architecture surrounding the nucleus.

1. Experimental Methods for Cellular/Nuclear Compression

a. General Consideration

The current compression experiments can be grouped into different categories depending on the sample (e.g., cells embedded in extracellular matrix or gels, single adherent cells, single nonadherent cells, or isolated nuclei) and the mode of compression (unconfined or confined expression that allows or not displacements perpendicular to the direction of compression). Experiments on living cells in their three-dimensional (3D) environment (e.g., cartilage explants, alginate constructs) preserve the natural mechanical environment of the cell and provide a more realistic model of cellular and nuclear deformations under physiological conditions, but the surrounding tissue makes observation during force application difficult. Here, we summarize experiments on single cells and isolated nuclei that avoid this limitation. These experiments generally consist of unconfined compression through uniaxial force application, allowing the cell or nucleus to deform freely in the perpendicular directions during compression. These experiments can be carried out on adherent (i.e., spread) cells or on non- or only weakly adherent cells. Most spread cells have a relatively flat topology with a central elevation at the site of the nucleus, resembling a fried egg. These cells often have a very structured cytoskeleton consisting of actin stress fibers and networks of microtubules and intermediate filaments that will affect nuclear deformations. Even in uncompressed, spread cells, the nucleus often has a flattened shape as a result of cell spreading and physical linkage to the spread cytoskeleton. In contrast, uncompressed, nonadherent cells have an approximately spherical shape with a centrally located, approximately spherical nucleus. Isolated nuclei resemble round cells in shape, but they are significantly stiffer than intact cells (Caille *et al.*, 2002). In the following sections, we briefly discuss the two most common experimental techniques used in cellular and nuclear compression experiments.

b. Microplate Manipulation

Cellular compression by microplate manipulation was first described by Thoumine *et al.* (Thoumine and Ott, 1997; Thoumine *et al.*, 1999). In this experimental setup (Fig. 3), cells are attached to a rigid microplate (cross-sectional tip dimensions $\approx 3 \times 30 \ \mu m^2$) and are subsequently compressed using a thinner, more flexible microplate (cross-sectional tip dimensions $\approx 1 \times 10 \ \mu m^2$, stiffness $\approx 10^{-9}$ to $10^{-8} \ nN/\mu m$) mounted on a micromanipulator (Thoumine *et al.*, 1999). The magnitude of the applied force can be determined based on the deflection of the flexible plate during the experiments and is calibrated using microneedles of known stiffness or micropipette

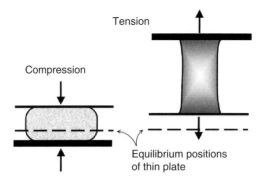

Fig. 3 Microplate manipulation of a single cell or nucleus adhering between a thin, flexible plate and a thick, stiff plate.

aspiration (Thoumine *et al.*, 1999). Generally, the error in force measurements in the microplate experiments is around 10%. In order to visualize plate deflection on an inverted microscope, the microplate assembly is rotated at 90° once a single cell adheres to the tip of the rigid plate, thus providing a side view of the cell sandwiched between the two microplates. The microplates are pulled from rectangular borosilicate bars (100 mm in length, 1 mm in width, and thicknesses of 0.075 mm and 1 mm; VitroCom, Mountain Lakes, New Jersey) using a micropipette puller. The microplates are coated with fibronectin to foster cell adhesion or treated with aminosilane (Sigma, St. Louis, Missouri) plus glutaraldehyde for even stronger, covalent adhesion. Isolated nuclei will adhere to uncoated microplates (Thoumine *et al.*, 1999). An alternative approach to custom-made microplates is to use commercially available tip-less AFM probes (e.g., Veeco). In this technique, single cells or nuclei are placed on a flat glass coverslip and compressed using the tip-less probe mounted on an AFM. The applied force and compression can be directly obtained from the AFM setup. These experiments are described in more detail in Section II.E.

c. Compression Using a Microindenter

The microindenter technique was first developed by Peeters *et al.* (2003) and was used to measure nuclear mechanics and cellular sensitivity to compression in lamin A/C-deficient mouse embryo fibroblasts (Broers *et al.*, 2004b). The compression device consists of a thin glass microindenter (tip diameter 500 μm) that is controlled by a piezo-actuator and coupled to a force transducer, enabling direct quantification of the applied compression force. A tilting mechanism allows parallel alignment of the microindenter tip with the experimental chamber containing the cells. The compression occurs parallel to the optical axis, so that lateral displacements can be directly visualized on an inverted microscope, while vertical displacements require 3D reconstruction. Due to the vertical setup of the microindenter and the piezo-actuator and force transducer, cells cannot be monitored with transmitted light microscopy. However, fluorescently labeled cells or nuclei can be imaged on an inverted microscope during the compression experiments. The cells are plated sparsely so that only single cells are compressed when lowering the microindenter.

d. Compression Parameters

Both aforementioned compression devices work by controlling the applied displacement and quantifying the force required for a given compression. This compression force depends on the extent of compression and, due to the viscoelastic properties of most biological tissues, on the indentation speed (Chapter 1 by Janmey *et al.* and Chapter 2 by Kandow *et al.*, this volume). The compression is generally defined relative to the initial cell/nucleus height (Section II.D.3), and most experiments on isolated cells use compression of up to 50–70%, corresponding to forces on the order of 0.1–10 μN (Broers *et al.*, 2004b; Caille *et al.*, 2002). In contrast, experiments on cells embedded in alginate gels or cartilage explants generally apply $\approx$15–20% compression, as these experiments typically aim to emulate the physiological load placed on these cells (Guilak, 1995, 2000; Knight *et al.*, 2002). Regardless of the chosen experimental techniques, compression velocities are generally on the order of 1–10% height change per second, such that the final compression is achieved in less than a minute to avoid artifacts caused by cellular remodeling or adaptation. Higher compressions ($\approx$60–70%) can lead to cellular damage (e.g., rupture of the plasma membrane or protrusion of chromatin from the nucleus) and should be avoided when measuring nuclear elasticity. On the other hand, the characteristic drop in the compression force associated with plasma membrane rupture allows easy detection of this event, and comparison of such burst forces between different cells (e.g., lamin A/C-deficient fibroblasts vs wild-type controls) can provide important insights into cellular sensitivity to mechanical stress (Broers *et al.*, 2004b).

2. Preparation of Cells and Nuclei

Single cells can be probed when adherent to a flat surface (i.e., spread fibroblasts or endothelial cells; Broers *et al.*, 2004b; Caille *et al.*, 2002) or in a rounded state (i.e., detached cells or nonadherent cells; Caille *et al.*, 2002). Spread cells are obtained by incubating cells on appropriately treated (e.g., fibronectin-coated) microplates or coverslips for a minimum of 1 h at 37°C. The extent of spreading is determined by the cell type, the adhesiveness of the substrate, and the incubation time; therefore, each condition should be empirically determined based on the cell type under investigation. In order to visualize the nucleus, cells are stained with a DNA-intercalating fluorescent dye such as Hoechst 33342. While the lowest detectable dye concentrations should be used and a two-fold higher concentration should be used to check for dye-induced artifacts (as with any labeling), this strategy also allows for the visualization of chromatin fragments protruding from the damaged nucleus under large compression (Broers *et al.*, 2004b). The cells in turn can be labeled with CellTracker or other fluorescent markers that are retained in the cytoplasm. Adding propidium iodide (2 μg/ml; Sigma, St. Louis, Missouri) to the medium in the experimental chamber can additionally be used to detect rupture of the plasma membrane, as the nonpermeable dye rapidly stains the nucleus after membrane rupture. The compression technique can also be applied to isolated nuclei (Section II.A). The isolated nuclei are resuspended in an appropriate buffer and are manipulated similar to the intact cells (Broers *et al.*, 2004b; Caille *et al.*, 2002; Thoumine and Ott, 1997; Thoumine *et al.*, 1999). It is important to note that the buffer used for the final suspension of isolated nuclei can significantly affect the nuclear mechanical behavior, as ion concentrations can lead to nuclear swelling or contraction as described earlier (Dahl *et al.*, 2004, 2005; Rowat *et al.*, 2005).

3. Analysis of Compression Experiments

The aforementioned experiments impose a controlled uniaxial displacement (strain) on the cell or nucleus and measure the corresponding reaction force, either with an in-line force transducer (Broers *et al.*, 2004b) or based on the deflection of a flexible microplate (Caille *et al.*, 2002). Cellular or nuclear deformation along the direction of the force is generally defined as the relative reduction in height, that is, $\Delta H = H/H_0 - 1$, where H is the deformed height, H_0 the initial height, and the result is expressed in percentage (Caille *et al.*, 2002). The cell/nucleus height can be directly determined from the phase-contrast side-view images (Caille *et al.*, 2002) or 3D reconstruction of confocal image slices (Broers *et al.*, 2004b). The resulting plots of compression force versus axial deformation can then be used for direct comparison between different cell types (e.g., lamin A/C-deficient fibroblasts vs wild-type controls). Alternatively, Caille *et al.* (2002) have used a finite element model representing the intact cell or isolated nucleus to obtain numerical values for the cytoskeletal and nuclear Young's moduli. The finite element model assumes homogeneous, incompressible, isotropic, and hyperelastic materials for the nucleus and the cytoskeleton. Further details are beyond the scope of this chapter, but Young's moduli are obtained by varying the material properties of the model (e.g., $E_{cytoskeleton}$, $E_{nucleus}$) to best fit the model predictions to the experimental force–deformation curves.

In addition to deformation along the direction of applied force, lateral displacements of the cells/nuclei perpendicular to the force direction can also be quantified by image analysis of fluorescently labeled cells/nuclei. The extent of isotropic or anisotropic nuclear deformations perpendicular to the axis of compression allows inferences to be drawn on intracellular architecture, specifically cytoskeletal–nuclear coupling, as physical connection between an anisotropic cytoskeleton often result in anisotropic nuclear deformations (Broers *et al.*, 2004b). Nuclei in wild-type fibroblasts and myoblasts show anisotropic nuclear deformations that depend on the orientation with respect to initial cell morphology (parallel or perpendicular to a cell's long axis) but are independent of the initial nuclear shape (Broers *et al.*, 2004b; Peeters *et al.*, 2003). This anisotropy is lost in lamin A/C-deficient fibroblasts, indicating uncoupling of the nucleus from the cytoskeleton (Broers *et al.*, 2004b). Similarly, cytochalasin D treatment of chondrocytes in cartilage explants significantly altered induced changes in nuclear shape and height, indicative of disrupted nuclear–cytoskeletal coupling (Guilak, 1995).

Compression experiments can also be used to measure membrane sensitivity to mechanical load. Propidium iodide is impermeable to the intact plasma membrane but rapidly intercalates into DNA once inside the cells, fluorescently labeling the nucleus. Thus, when added to the media in the experimental chamber, propidium iodide will only stain the nuclei in cells whose integrity of plasma membrane has been compromised under compression. On the other hand, cell-permeable nuclear stains, such as Hoechst 33342, can label chromatin in intact cells and can be used to monitor protrusion of chromatin through ruptured nuclear envelopes (Broers *et al.*, 2004b).

E. Indentation by AFM

AFM has also been used to measure nuclear rheology. While it is similar to the compression methods described above, AFM measurements typically probe the nucleus on a smaller length scale. The distal end of an AFM cantilever has a pyramidal probe tip with a radii of curvature as small as 25 nm for so-called sharpened tips (Veeco, Woodbury, New York). Because of the localized nature of force application, AFM measurements of nuclear rheology are generally performed with isolated nuclei rather than intact cells. Forces measured are typically in the range of nanonewton while indentations are in the range of 50 nm to several *μ*m.

Cell nuclei are first isolated following the methods of Section II.A. They are then allowed to settle onto a rigid substrate which can either be treated to minimize adhesion [e.g., with poly-HEMA (Sigma, St. Louis, Missouri) (Folkman and Moscona, 1978)] or with poly-L-lysine to promote it (Dahl *et al.*, 2005). AFM studies, thus far, have only been conducted on adherent nuclei using sufficient poly-L-lysine—determined experimentally—to immobilize while avoiding the strong adhesion the tends to increase the tension in the structures of the nuclear envelope (Hategan *et al.*, 2003).

Methods differ depending on the nature of the AFM tip used in the experiments. Typically, the sharpened tip is used, but a cantilever with an attached spherical microbead of radius 2.5 *μ*m has also been used as the probe in some experiments (Dahl *et al.*, 2005). A tip-less probe, which consists of only the flat surface of the AFM cantilever, might be used as described above in Section II.C for nuclear compression. To obtain accurate measurements of indentation depth and applied force, cantilevers are used with stiffness in the range of 10–100 pN/nm. Force is then determined by the stiffness of the cantilever and the deflection of the cantilever arm from the resting (zero force) value, and indentation of the nucleus is taken as the vertical displacement of the cantilever from the point of contact, subtracting off the cantilever deflection.

Various testing protocols can be used, each yielding different information. In one method, the AFM probe is brought down onto the nucleus at a constant velocity to cause a given degree of indentation, and then the tip is retracted. Results using this approach will depend on the rate and maximum depth of indentation.

This approach can be challenging, however, as it requires accurate determination of the point of contact; errors in identifying the point of contact can be minimized using the method of Domke *et al.* (2000). An alternative approach, less sensitive to these errors, is to first bring the probe down to an initial degree of indentation, then to oscillate the probe about this position while monitoring tip deflection and force (Mahaffy *et al.*, 2000).

1. Method of Analysis

Because the probe is often considered small compared to the radius of curvature of the nucleus, the Hertzian contact analysis is most frequently used to analyze the results (Dahl *et al.*, 2005). In this method, the probe is assumed to be either a cone with half angle α or a sphere of radius a, and the nucleus is assumed to be elastic (Young's modulus E), homogeneous, and isotropic, with a flat surface of infinite extent. Under these assumptions, the predicted force (F)–indentation (δ) relationship takes on one of the following two forms:

$$F = \frac{2E \tan \alpha}{\pi(1 - v^2)} \delta^2 \tag{1}$$

for a conical tip or

$$F = \frac{4}{3} \sqrt{a\delta^3} \frac{E}{1 - v^2} \tag{2}$$

for a spherical tip. v is the Poisson ratio of the nucleus, which is assumed to be 0.5 as for an incompressible material. One of these two expressions is then fit to the experimental results to determine the value of Young's modulus that gives the best fit. A correction factor for finite diameter of the nucleus has been included in a study (Dahl, 2005), which also shows that the apparent E can depend on the rate at which the AFM tip indents the nucleus. As with results from micropipette analyses, typical magnitudes for E are in the range of 1–10 kPa.

The expressions above are based on numerous further assumptions. For example, they fail to distinguish between the nuclear envelope and the nucleoplasm and, therefore, give, as best, an average measure of nuclear stiffness.

F. Particle-Tracking Microrheology

As described elsewhere in this volume (Chapter 6 by Panorchan *et al.* and Chapter 7 by Crocker and Hoffman), particle-tracking microrheology (PTM) has flourished in recent years because (1) particle tracking has the highest spatial resolution (nanometers) among the various mechanical measurement methods, (2) PTM has the potential to cause the least possible disturbance to the cell or nucleus and is also referred to as "passive microrheology," and (3) PTM can, in principle, provide a complete characterization of the elastic and viscous properties of the material over several decades in frequency. Numerous studies have been published demonstrating the use of single, double, and multiple particle-tracking methods to determine the rheological properties of the cytoskeleton (Lau *et al.*, 2003), the membrane cytoskeleton under strain (Lee and Discher, 2001), and reconstituted actin matrices (Gardel *et al.*, 2004; Palmer *et al.*, 1999; Shin *et al.*, 2004; Xu *et al.*, 1998), and one group has also applied this technique to the nucleus (Tseng *et al.*, 2004). Determinations of nucleoplasmic viscosity and elasticity by PTM are on the soft side of measurements with other methods described above. The mean viscosity is about 500 Pa, which is 500-fold more viscous than water. At the same time, the average elasticity is reported to be about 200 Pa, which is about 10- to 100-fold lower than what were reported by micropipette and AFM methods.

1. Potential Limitations

Particle tracking and fluctuation methods are subject to several unique complicating factors. First, the act of microinjection and the presence of foreign bodies within the nucleus might adversely affect cell viability or nuclear integrity, or lead to a biological response that could alter stiffness. In addition, results can differ depending on the size of the microparticles, the size of the nucleus, or the size of the nuclear domain probed (Lee and Discher, 2001), all of which affect the length scale of measurements. PTM results also depend on the surface treatment, which affects the extent to which the particles are tethered to the surrounding structure.

G. Microneedle-Imposed Extension

Nuclear and cytoskeletal mechanics can also be probed with a micropipette manipulation technique developed by Ingber and coworkers (Maniotis *et al.*, 1997). Cells are plated on glass bottom cell culture dishes coated with low concentrations of fibronectin (0.1 μg/ml), since highly spread cells are very stiff and less amenable to micromanipulation. Subsequently, a micropipette needle (narrowed at the tip to about 100 nm) is carefully inserted into the cell at one of two specific distances (5 and 10 μm) from the nucleus and moved a fixed distance away from the nucleus. The induced nuclear and cytoskeletal deformations are then imaged at 30 frames/sec with a charge-coupled device (CCD) camera (Fig. 4). The induced nuclear strain can be directly computed from changes in nuclear dimensions in the phase-contrast images and additionally by triangulation using intranuclear phase-dense particles (e.g., nucleoli). Similarly, the induced cytoplasmic strain is computed by tracking cytoplasmic phase-dense particles. Alternatively, fluorescently labeled cellular components (e.g., mitochondria) can be used as reference markers. The ratio of nuclear to cytoskeletal stiffness can then be inferred from comparison of the nuclear to cytoskeletal strain ratio at the two distances (Maniotis *et al.*,

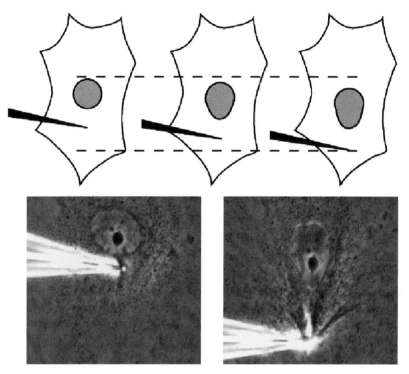

Fig. 4 Microneedle extension through cytoskeletal force application. A micropipette is inserted into the cytoplasm adjacent to the nucleus and subsequently moved a fixed distance from the nucleus. Induced nuclear and cytoplasmic strain are measured based on displacements of phase-dense particles.

1997). For the differentiated cells studied (endothelial cells), the nucleus appeared about 9 ± 4-fold stiffer than the cytoplasm. Disruption of the cytoskeleton with various drugs eliminated any transfer of stress to the nucleus as expected.

III. Discussion and Prospects

Perhaps, because of the wide range of methods used to probe nuclear mechanics, the reported elastic and viscoelastic properties differ widely. While part of this might reflect differences between cells or cell types, or to the effects of nuclear isolation as discussed below; two additional factors likely exert a strong influence on the measurements: the model used to interpret the results and fundamental differences in the structures being probed by different experimental methods.

In most of the published reports of nuclear mechanics, a model is used to infer structural properties from the unprocessed data, which might consist of indentation depth versus applied force (as in the cases of AFM indentation or compression) or nuclear length versus time for a given level of pressure in micropipette aspiration. While it is useful to extract parameters that can be compared between different methods, there is an inherent danger in using any single existing model to infer properties, as it is almost certainly true that none can fully capture all of the details of nuclear structure with a reasonable degree of precision. The simplest model represents the nucleus as a homogeneous elastic sphere or half-space. However, since the nucleus is neither purely elastic nor homogenous, such representation can at best be viewed as gross approximations. Viscoelastic models begin to address the viscous (e.g., rate-dependent) aspects of deformation, but these, too, tend to be simplified and approximate. Among the various viscoelastic models employed, the latest evidence suggests that nuclei exhibit power law rheology (i.e., the existence of an essentially infinite spectrum of relaxation timescales; Dahl *et al.*, 2005), but this, too, is likely to be only an approximation. Some models are beginning to acknowledge the membrane as a separate and important structural element (Dahl *et al.*, 2004; Rowat *et al.*, 2005), although substantial differences remain in the details of these models.

In view of these different models, it comes as little surprise that experimental methods which emphasize the importance of one structure over another (e.g., the nuclear lamina over the nucleoplasm) will yield different results. While several approaches can be used to distinguish various experimental techniques, a parameter that is particularly useful is the characteristic length scale of the measurement. For example, whole nucleus compression or micropipette aspiration experiments probe the nucleus on a very different length scale from AFM or microbead rheology. Even within a single method such as microbead rheology, beads of different sizes can, in principle, yield different results depending on the extent to which the particles are free to diffuse through the nucleoplasm, or the extent to which they probe local versus spatially averaged structures over a larger domain. In addition to these differences in characteristic length scale between various experimental approaches, it is also important to consider the chosen mode of force application. Experiments that exert tensile stress (e.g., substrate strain or micropipette aspiration) on the nucleus might probe different stress-bearing components compared to experiments applying compressive forces (e.g., AFM), and even the same structural components can respond differently to tensile and compressive loads.

There are several reasons to expect that nuclei within intact cells would exhibit different properties from those isolated by one of the techniques described above. When in its native environment, the nucleus is tethered to the cytoskeleton via a collection of interconnected proteins that have been termed the linker of nucleoskeleton and cytoskeleton (LINC) complex (Crisp *et al.*, 2006). Members of the family of nesprin proteins reside in the outer nuclear membrane and are known to have binding sites for actin (Crisp *et al.*, 2006). Sun1 and Sun2 are integral membrane proteins that span the inner nuclear membrane with binding sites for nesprin on the outside, within the perinuclear space between the inner and outer membranes, and for A-type lamins on the inside. Thus, the nesprin–Sun–lamin complex provides a means by which the nucleoskeleton and cytoskeleton can be linked structurally, and the effect of nuclear isolation on the structural integrity of this complex is unknown. *In situ*, intracellular forces might be transmitted to the nucleus through this specific complex, while force application on isolated nuclei might be transmitted through nonspecific contact sites. In addition, the outer leaflet of the nuclear membrane is continuous with the ER, and the extent to which the ER remains tethered to the nucleus depends on the specific isolation procedure. This can in turn affect the ability of the nuclear envelope to expand in procedures that involve substrate stretching. Finally, the structural organization of the nucleus is highly dependent on the intranuclear concentration of ions, proteins, and ATP, and the conditions for isolated nuclei to best recapitulate the intracellular environment are not obvious. The plethora of experimental conditions is likely to account for some of the reported variation in nuclear mechanics between intact cells and isolated nuclei.

An explicit or implicit objective of all studies of nuclear mechanics is to ultimately link mechanical behavior to nuclear biology. While compression of chondrocytes *in vitro* corresponds to physiological loading conditions in cartilage, the physiological relevance of cellular compression for other cell types (e.g., fibroblasts) is less clear. Previous experiments show that cellular responses to mechanical stimulation can differ significantly between stretching and compression application (Wille *et al.*, 2004). In addition, the response could be cell-type specific, as different cell types are characterized by unique cellular structure and organization (e.g., muscle cells) that can significantly influence the overall mechanical behavior. Nonetheless, many nuclear structural features are preserved among many cell types studied *in vitro*, and findings for a specific cell type (e.g., fibroblast) can indicate general characteristics representative for many different cells. For example, we found that fibroblasts from lamin A/C-deficient mouse embryo fibroblasts display similar deficiencies in nuclear mechanics when subjected to substrate strain as myoblast and myotubes from the same mice (Lammerding, unpublished observation).

IV. Outlook

Currently, a number of experimental methods exist to probe the mechanics of nuclei in isolation and in living cells. However, several challenges remain to improve our understanding of nuclear mechanics *in vivo*, especially how nuclear structure and mechanics interact with cellular biological functions. Future experiments must be directed at elucidating how individual nuclear components contribute to nuclear mechanics, and how specific human diseases can interfere with biological functions. To achieve this, a combination of sophisticated experimental methods to measure precisely mechanical properties along with molecular biology tools to manipulate specific components of the nuclear architecture will be necessary. Insights from these experiments will not only help to reveal the molecular mechanism underlying the diverse nuclear envelope diseases (laminopathies) but will also improve our understanding of normal cellular architecture and functions.

References

Aaronson, R. P., and Woo, E. (1981). Organization in the cell nucleus: Divalent cations modulate the distribution of condensed and diffuse chromatin. *J. Cell Biol.* **90**(1), 181–186.

Aebi, U., Cohn, J., Buhle, L., and Gerace, L. (1986). The nuclear lamina is a meshwork of intermediate-type filaments. *Nature* **323**, 560–564.

Barbee, K. A., Macarak, E. J., and Thibault, L. E. (1994). Strain measurements in cultured vascular smooth muscle cells subjected to mechanical deformation. *Ann. Biomed. Eng.* **22**(1), 14–22.

Bione, S., Maestrini, E., Rivella, S., Mancini, M., Regis, S., Romeo, G., and Toniolo, D. (1994). Identification of a novel X-linked gene responsible for Emery-Dreifuss muscular dystrophy. *Nat. Genet.* **8**(4), 323–327.

Bonne, G., Di Barletta, M. R., Varnous, S., Becane, H. M., Hammouda, E. H., Merlini, L., Muntoni, F., Greenberg, C. R., Gary, F., Urtizberea, J. A., Duboc, D., Fardeau, M., *et al.* (1999). Mutations in the gene encoding lamin A/C cause autosomal dominant Emery-Dreifuss muscular dystrophy. *Nat. Genet.* **21**(3), 285–288.

Broers, J. L., Hutchison, C. J., and Ramaekers, F. C. (2004a). Laminopathies. *J. Pathol.* **204**(4), 478–488.

Broers, J. L., Peeters, E. A., Kuijpers, H. J., Endert, J., Bouten, C. V., Oomens, C. W., Baaijens, F. P., and Ramaekers, F. C. (2004b). Decreased mechanical stiffness in LMNA–/– cells is caused by defective nucleo-cytoskeletal integrity: Implications for the development of laminopathies. *Hum. Mol. Genet.* **13**(21), 2567–2580.

Burke, B., and Stewart, C. L. (2002). Life at the edge: The nuclear envelope and human disease. *Nat. Rev. Mol. Cell Biol.* **3**(8), 575–585.

Caille, N., Tardy, Y., and Meister, J. J. (1998). Assessment of strain field in endothelial cells subjected to uniaxial deformation of their substrate. *Ann. Biomed. Eng.* **26**(3), 409–416.

Caille, N., Thoumine, O., Tardy, Y., and Meister, J. J. (2002). Contribution of the nucleus to the mechanical properties of endothelial cells. *J. Biomech.* **35**(2), 177–187.

Cao, H., and Hegele, R. A. (2000). Nuclear lamin A/C R482Q mutation in canadian kindreds with Dunnigan-type familial partial lipodystrophy. *Hum. Mol. Genet.* **9**(1), 109–112.

Chen, L., Lee, L., Kudlow, B. A., Dos Santos, H. G., Sletvold, O., Shafeghati, Y., Botha, E. G., Garg, A., Hanson, N. B., Martin, G. M., Mian, I. S., Kennedy, B. K., *et al.* (2003). LMNA mutations in atypical Werner's syndrome. *Lancet* **362**(9382), 440–445.

Crisp, M., Liu, Q., Roux, K., Rattner, J. B., Shanahan, C., Burke, B., Stahl, P. D., and Hodzic, D. (2006). Coupling of the nucleus and cytoplasm: Role of the LINC complex. *J. Cell Biol.* **172**(1), 41–53.

Cui, Y., and Bustamante, C. (2000). Pulling a single chromatin fiber reveals the forces that maintain its higher-order structure. *Proc. Natl. Acad. Sci. USA* **97**, 127–132.

Dahl, K. N., Engler, A. J., Pajerowski, J. D., and Discher, D. E. (2005). Power-law rheology of isolated nuclei with deformation mapping of nuclear substructures. *Biophys. J.* **89**(4), 2855–2864.

Dahl, K. N., Kahn, S. M., Wilson, K. L., and Discher, D. E. (2004). The nuclear envelope lamina network has elasticity and a compressibility limit suggestive of a molecular shock absorber. *J. Cell Sci.* **117**(Pt. 20), 4779–4786.

Dahl, K. N., Scaffidi, P., Islam, M. F., Yodh, A. G., Wilson, K. L., and Misteli, T. (2006). Distinct structural and mechanical properties of the nuclear lamina in Hutchinson-Gilford progeria syndrome. *Proc. Natl. Acad. Sci. USA* **103**(27), 10271–10276.

De Sandre-Giovannoli, A., Bernard, R., Cau, P., Navarro, C., Amiel, J., Boccaccio, I., Lyonnet, S., Stewart, C. L., Munnich, A., Le Merrer, M., and Levy, N. (2003). Lamin A truncation in Hutchinson-Gilford progeria. *Science* **300**(5628), 2055.

De Sandre-Giovannoli, A., Chaouch, M., Kozlov, S., Vallat, J. M., Tazir, M., Kassouri, N., Szepetowski, P., Hammadouche, T., Vandenberghe, A., Stewart, C. L., Grid, D., and Levy, N. (2002). Homozygous defects in LMNA, encoding lamin A/C nuclear-envelope proteins, cause autosomal recessive axonal neuropathy in human (Charcot-Marie-Tooth disorder type 2) and mouse. *Am. J. Hum. Genet.* **70**(3), 726–736.

Dean, D. A., and Kasamatsu, H. (1994). Signal- and energy-dependent nuclear transport of SV40 Vp3 by isolated nuclei. Establishment of a filtration assay for nuclear protein import. *J. Biol. Chem.* **269**(7), 4910–4916.

Deguchi, S., Maeda, K., Ohashi, T., and Sato, M. (2005). Flow-induced hardening of endothelial nucleus as an intracellular stress-bearing organelle. *J. Biomech.* **38**(9), 1751–1759.

Dobi, A., and Agoston, D. V. (1998). Submillimolar levels of calcium regulates DNA structure at the dinucleotide repeat (TG/AC)n. *Proc. Natl. Acad. Sci. USA* **95**(11), 5981–5986.

Domke, J., Dannohl, S., Parak, W. J., Muller, O., Aicher, W. K., and Radmacher, M. (2000). Substrate dependent differences in morphology and elasticity of living osteoblasts investigated by atomic force microscopy. *Colloids Surf. B Biointerfaces* **19**(4), 367–379.

Eriksson, M., Brown, W. T., Gordon, L. B., Glynn, M. W., Singer, J., Scott, L., Erdos, M. R., Robbins, C. M., Moses, T. Y., Berglund, P., Dutra, A., Pak, E., *et al.* (2003). Recurrent *de novo* point mutations in lamin A cause Hutchinson-Gilford progeria syndrome. *Nature* **423**(6937), 293–298.

Evans, E. A. (1989). Structure and deformation properties of red blood cells: Concepts and quantitative methods. *Methods Enzymol.* **173**, 3–35.

Fatkin, D., MacRae, C., Sasaki, T., Wolff, M. R., Porcu, M., Frenneaux, M., Atherton, J., Vidaillet, H. J., Jr., Spudich, S., De Girolami, U., Seidman, J. G., Seidman, C., *et al.* (1999). Missense mutations in the rod domain of the lamin A/C gene as causes of dilated cardiomyopathy and conduction-system disease. *N. Engl. J. Med.* **341**(23), 1715–1724.

Folkman, J., and Moscona, A. (1978). Role of cell shape in growth control. *Nature* **273**, 345–349.

Gardel, M. L., Shin, J. H., MacKintosh, F. C., Mahadevan, L., Matsudaira, P., and Weitz, D. A. (2004). Elastic behavior of cross-linked and bundled actin networks. *Science* **304**(5675), 1301–1305.

Goldman, R. D., Gruenbaum, Y., Moir, R. D., Shumaker, D. K., and Spann, T. P. (2002). Nuclear lamins: Building blocks of nuclear architecture. *Genes Dev.* **16**(5), 533–547.

Grady, R. M., Starr, D. A., Ackerman, G. L., Sanes, J. R., and Han, M. (2005). Syne proteins anchor muscle nuclei at the neuromuscular junction. *Proc. Natl. Acad. Sci. USA* **102**(12), 4359–4364.

Gruenbaum, Y., Margalit, A., Goldman, R. D., Shumaker, D. K., and Wilson, K. L. (2005). The nuclear lamina comes of age. *Nat. Rev. Mol. Cell Biol.* **6**(1), 21–31.

Guilak, F. (1995). Compression-induced changes in the shape and volume of the chondrocyte nucleus. *J. Biomech.* **28**(12), 1529–1541.

Guilak, F. (2000). The deformation behavior and viscoelastic properties of chondrocytes in articular cartilage. *Biorheology* **37**(1–2), 27–44.

Guilak, F., Tedrow, J. R., and Burgkart, R. (2000). Viscoelastic properties of the cell nucleus. *Biochem. Biophys. Res. Commun.* **269**(3), 781–786.

Haithcock, E., Dayani, Y., Neufeld, E., Zahand, A. J., Feinstein, N., Mattout, A., Gruenbaum, Y., and Liu, J. (2005). Age-related changes of nuclear architecture in *Caenorhabditis elegans*. *Proc. Natl. Acad. Sci. USA* **102**(46), 16690–16695.

Hategan, A., Law, R., Kahn, S., and Discher, D. E. (2003). Adhesively-tensed cell membranes: Lysis kinetics and atomic force microscopy probing. *Biophys. J.* **85**(4), 2746–2759.

Hutchison, C. J. (2002). Lamins: Building blocks or regulators of gene expression? *Nat. Rev. Mol. Cell Biol.* **3**(11), 848–858.

Hutchison, C. J., and Worman, H. J. (2004). A-type lamins: Guardians of the soma? *Nat. Cell Biol.* **6**(11), 1062–1067.

Ingber, D. E., Madri, J. A., and Folkman, J. (1987). Endothelial growth factors and extracellular matrix regulate DNA synthesis through modulation of cell and nuclear expansion. *In Vitro Cell. Dev. Biol.* **23**, 387–394.

Knight, M. M., van de Breevaart Bravenboer, J., Lee, D. A., van Osch, G. J., Weinans, H., and Bader, D. L. (2002). Cell and nucleus deformation in compressed chondrocyte-alginate constructs: Temporal changes and calculation of cell modulus. *Biochim. Biophys. Acta* **1570**(1), 1–8.

Lammerding, J., Hsiao, J., Schulze, P. C., Kozlov, S., Stewart, C. L., and Lee, R. T. (2005). Abnormal nuclear shape and impaired mechanotransduction in emerin-deficient cells. *J. Cell Biol.* **170**(5), 781–791.

Lammerding, J., Schulze, P. C., Takahashi, T., Kozlov, S., Sullivan, T., Kamm, R. D., Stewart, C. L., and Lee, R. T. (2004). Lamin A/C deficiency causes defective nuclear mechanics and mechanotransduction. *J. Clin. Invest.* **113**(3), 370–378.

Lau, A. W., Hoffman, B. D., Davies, A., Crocker, J. C., and Lubensky, T. C. (2003). Microrheology, stress fluctuations, and active behavior of living cells. *Phys. Rev. Lett.* **91**(19), 198101.

Lee, J.-C., and Discher, D. E. (2001). Deformation-enhanced fluctuations in the red cell skeleton with theoretical relations to elasticity, connectivity, and spectrin unfolding. *Biophys. J.* **81**(6), 3178–3192.

Mahaffy, R. E., Shih, C. K., MacKintosh, F. C., and Kas, J. (2000). Scanning probe-based frequency-dependent microrheology of polymer gels and biological cells. *Phys. Rev. Lett.* **85**, 880–883.

Maniotis, A. J., Chen, C. S., and Ingber, D. E. (1997). Demonstration of mechanical connections between integrins, cytoskeletal filaments, and nucleoplasm that stabilize nuclear structure. *Proc. Natl. Acad. Sci. USA* **94**(3), 849–854.

Muchir, A., Bonne, G., van der Kooi, A. J., van Meegen, M., Baas, F., Bolhuis, P. A., de Visser, M., and Schwartz, K. (2000). Identification of mutations in the gene encoding lamins A/C in autosomal dominant limb girdle muscular dystrophy with atrioventricular conduction disturbances (LGMD1B). *Hum. Mol. Genet.* **9**(9), 1453–1459.

Muchir, A., Medioni, J., Laluc, M., Massart, C., Arimura, T., van der Kooi, A. J., Desguerre, I., Mayer, M., Ferrer, X., Briault, S., Hirano, M., Worman, H. J., *et al.* (2004). Nuclear envelope alterations in fibroblasts from patients with muscular dystrophy, cardiomyopathy, and partial lipodystrophy carrying lamin A/C gene mutations. *Muscle Nerve* **30**(4), 444–450.

Novelli, G., Muchir, A., Sangiuolo, F., Helbling-Leclerc, A., D'Apice, M. R., Massart, C., Capon, F., Sbraccia, P., Federici, M., Lauro, R., Tudisco, C., Pallotta, R., *et al.* (2002). Mandibuloacral dysplasia is caused by a mutation in LMNA-encoding lamin A/C. *Am. J. Hum. Genet.* **71**(2), 426–431.

Padmakumar, V. C., Libotte, T., Lu, W., Zaim, H., Abraham, S., Noegel, A. A., Gotzmann, J., Foisner, R., and Karakesisoglou, I. (2005). The inner nuclear membrane protein Sun1 mediates the anchorage of Nesprin-2 to the nuclear envelope. *J. Cell Sci.* **118**(Pt. 15), 3419–3430.

Palmer, A., Xu, J., Kuo, S. C., and Wirtz, D. (1999). Diffusing wave spectroscopy microrheology of actin filament networks. *Biophys. J.* **76**(2), 1063–1071.

Panorchan, P., Wirtz, D., and Tseng, Y. (2004). Structure-function relationship of biological gels revealed by multiple-particle tracking and differential interference contrast microscopy: The case of human lamin networks. *Phys. Rev. E Stat. Nonlin. Soft Matter. Phys.* **70**(4 Pt. 1), 041906.

Peeters, E. A., Bouten, C. V., Oomens, C. W., and Baaijens, F. P. (2003). Monitoring the biomechanical response of individual cells under compression: A new compression device. *Med. Biol. Eng. Comput.* **41**(4), 498–503.

Rowat, A., Foster, L., Nielsen, M., Weiss, M., and Ipsen, J. (2005). Characterization of the elastic properties of the nuclear envelope. *J. Roy. Soc. Interface* **2**, 63–69.

Scaffidi, P., and Misteli, T. (2006). Lamin A-dependent nuclear defects in human aging. *Science* **312** (5776), 1059–1063.

Shackleton, S., Lloyd, D. J., Jackson, S. N., Evans, R., Niermeijer, M. F., Singh, B. M., Schmidt, H., Brabant, G., Kumar, S., Durrington, P. N., Gregory, S., O'Rahilly, S., *et al.* (2000). LMNA, encoding lamin A/C, is mutated in partial lipodystrophy. *Nat. Genet.* **24**(2), 153–156.

Shin, J. H., Gardel, M. L., Mahadevan, L., Matsudaira, P., and Weitz, D. A. (2004). Relating microstructure to rheology of a bundled and cross-linked F-actin network in vitro. *Proc. Natl. Acad. Sci. USA* **101**(26), 9636–9641.

Starr, D. A., and Han, M. (2002). Role of ANC-1 in tethering nuclei to the actin cytoskeleton. *Science* **298**(5592), 406–409.

Starr, D. A., and Han, M. (2003). ANChors away: An actin based mechanism of nuclear positioning. *J. Cell Sci.* **116**(Pt. 2), 211–216.

Thoumine, O., and Ott, A. (1997). Time scale dependent viscoelastic and contractile regimes in fibroblasts probed by microplate manipulation. *J. Cell Sci.* **110**(17), 2109–2116.

Thoumine, O., Ott, A., Cardoso, O., and Meister, J. J. (1999). Microplates: A new tool for manipulation and mechanical perturbation of individual cells. *J. Biochem. Biophys. Methods* **39**(1–2), 47–62.

Tran, E. J., and Wente, S. R. (2006). Dynamic nuclear pore complexes: Life on the edge. *Cell* **125**(6), 1041–1053.

Tseng, Y., Lee, J. S., Kole, T. P., Jiang, I., and Wirtz, D. (2004). Micro-organization and visco-elasticity of the interphase nucleus revealed by particle nanotracking. *J. Cell. Sci.* **117**(10), 2159–2167.

Wakayama, T., Perry, A. C., Zuccotti, M., Johnson, K. R., and Yanagimachi, R. (1998). Full-term development of mice from enucleated oocytes injected with cumulus cell nuclei. *Nature* **394**(6691), 369–374.

Wille, J. J., Ambrosi, C. M., and Yin, F. C. (2004). Comparison of the effects of cyclic stretching and compression on endothelial cell morphological responses. *J. Biomech. Eng.* **126**(5), 545–551.

Wilmut, I., and Paterson, L. (2003). Somatic cell nuclear transfer. *Oncol. Res.* **13**(6–10), 303–307.

Xu, J., Schwarz, W. H., Kas, J. A., Stossel, T. P., Janmey, P. A., and Pollard, T. D. (1998). Mechanical properties of actin filament networks depend on preparation, polymerization conditions, and storage of actin monomers. *Biophys. J.* **74**(5), 2731–2740.

Zhang, Q., Ragnauth, C. D., Skepper, J. N., Worth, N. F., Warren, D. T., Roberts, R. G., Weissberg, P. L., Ellis, J. A., and Shanahan, C. M. (2005). Nesprin-2 is a multi-isomeric protein that binds lamin and emerin at the nuclear envelope and forms a subcellular network in skeletal muscle. *J. Cell Sci.* **118**(Pt. 4), 673–687.

PART III

Cellular and Embryonic Mechanical Properties and Activities

CHAPTER 12

The Use of Gelatin Substrates for Traction Force Microscopy in Rapidly Moving Cells

Juliet Lee

Department of Molecular and Cell Biology
University of Connecticut
Storrs, Connecticut 06269

Abstract

The study of traction forces generated by rapidly moving cells requires the use of substrates that are highly elastic because these cells typically generate weaker traction forces than slower moving cells. Gelatin substrates are soft enough to allow deformation by rapidly moving cells such as fish epidermal keratocytes and *Dictyostelium discoideum* amoebas. In addition, gelatin substrates are thin ($\sim$30–40 μm) and transparent, allowing them to be used in combination with high-resolution calcium imaging. Importantly, the responsiveness of gelatin substrates allows changes in traction force generation to be detected within seconds, corresponding to the timescale of calcium transients. Here we describe the manufacture and application of gelatin substrates to study the role of mechanochemical signaling in the regulation of keratocyte movement. We show how patterns of traction force generation can be analyzed from a time series of traction vector maps, and how to interpret them in relation to cell movement. In addition, we discuss how the gelatin traction force assay is being used to study the mechanics of *Dictyostelium* cell motility, and future applications such as the study of neuronal path finding.

I. Introduction

Cells can respond to a wide range of mechanical stimuli such as fluid flow (Malek and Izumo, 1996), mechanical stretching (Pender and McCulloch, 1991), and even their own area of spreading (Chen *et al.*, 1997). In addition, cells can respond to physical properties of their microenvironment (Chicurel *et al.*, 1998; Janmey, 1998). For example, moving cells can sense the rigidity of their substrate (Lo *et al.*, 2000), surface topology (Dunn and Brown, 1986), and changes in cytoskeletal tension (Lee *et al.*, 1999).

Cell movement requires the generation of mechanical forces (Lauffenburger and Horwitz, 1996; Schwarzbauer, 1997) including a protrusive force at the front cell edge, together with contractile forces to shift the cell body forward and to facilitate retraction at the rear. In addition, the strength of adhesion between cell and its substrate must be regulated so that newly formed adhesions at the cell front strengthen, while older ones at the rear disassemble (Sheetz *et al.*, 1998). Although much is known about the molecular basis of various subprocesses of cell migration, such as protrusion, retraction, contractile force generation, and adhesion, we still do not understand how these processes are coordinated spatially or temporally. The generation of mechanical force is believed to play a central role in integrating different molecular processes to produce movement at the cellular level (Lauffenburger and Horwitz, 1996; Wang and Ingber, 1994).

The first study of traction forces generated by moving fibroblasts employed a thin polymerized film of silicone that wrinkled in response to the contractile forces exerted by moving cells via cell–substrate adhesions (Harris *et al.*, 1980). Since then a variety of traction force assays have been developed, as reviewed by Beningo and Wang (2002). These include elastic nonwrinkling films of silicone (Dembo *et al.*, 1996; Lee *et al.*, 1994; Oliver *et al.*, 1998), polyacrylamide gels (Wang and Pelham, 1998), micromachined cantilevers (Galbraith and Sheetz, 1997), UV-cured wrinkling elastomers (Burton *et al.*, 1999), micropatterned silicone (Balaban *et al.*, 2001), or polyacrylamide gels (Wang *et al.*, 2002), gelatin substrates (Doyle and Lee, 2002), and silicone posts (Tan *et al.*, 2003). Details of some of these traction force assays are the subject of other chapters in this volume. However, the basic idea behind most traction force assays is that the sensor deforms (e.g., bends in the case of cantilevers or silicone posts) in proportion to the magnitude of the traction forces exerted by motile cells. Given the material properties of an elastic substrate, such as the Young's modulus or the bending coefficient of a cantilever, it is then possible to determine the magnitude and orientation of traction stresses at discrete points beneath the cell.

An important consideration before using any traction force assay is to match the sensitivity of the sensor with the magnitude of traction forces. For example, rapidly moving cell types such as fish epidermal keratocytes are generally unable to deform polyacrylamide substrates, but can deform the more compliant silicone or gelatin. Conversely, slower moving fibroblastic cell types, which form prominent stress fibers and focal adhesions, tend to rip weakly cross-linked silicone substrates (J. Lee, unpublished observation). Since fibroblastic cells are perhaps the most common subject of cell motility studies, the majority of traction force assays have been developed for these slower moving cell types. However, it is equally important to understand mechanical events underlying the rapid movement of cells such as fish epidermal keratocytes and amoeboid cells, since they may utilize traction forces differently from fibroblastic cells.

II. Rationale

The simple shape and rapid movement of fish epidermal keratocytes make them a particularly useful model system for studying how cytoskeletal functions are organized to produce cellular movement (Lee *et al.*, 1993; Svitkina *et al.*, 1997). In addition, the relationship between traction force production and movement has been studied extensively in keratocytes for more than a decade (Burton *et al.*, 1999; Galbraith and Sheetz, 1999; Lee *et al.*, 1994; Oliver *et al.*, 1998,1999). Furthermore, keratocyte movement is dependent on transient increases in intracellular calcium $[Ca^{2+}]_i$ triggered by the activation of stretch-activated calcium channels (SACs), when retraction at the rear is impeded. The ensuing calcium transient induces retraction that helps to maintain a rapid mode of movement (Lee *et al.*, 1999). Thus, SAC activity may represent a critical mechanism for maintaining the high degree of coordination between protruding and retracting cell edges that is required for rapid movement. To investigate how calcium transients coordinate traction force generation in different regions of moving cells, it was necessary to develop a traction force assay that is compatible with high-resolution calcium imaging. Although nonwrinkling silicone substrates have previously been combined with calcium imaging (Lee *et al.*, 1999), they were not suitable for use with traction force microscopy because changes in substrate deformation cannot be detected within the time frame of a calcium transient. This problem was compounded by the fact that the maximum substrate deformation was less than 1 μm.

Gelatin substrates possess all of the characteristics necessary to combine traction force microscopy with calcium imaging (Doyle and Lee, 2002). They are transparent and nonfluorescent (Fig. 2A). In addition, the thickness as prepared with this proto-col (~40 μm) is well within the working distance of most 100× microscope objectives. An equally important feature is that gelatin substrates are highly responsive, allowing submicron bead displacements to be detected in less than 1 sec. The main limitation of gelatin substrates is that experiments must be conducted well below the melting temperature, which is typically 30 °C.

III. Methods

A. Preparation of Gelatin Substrates

The following steps produce a thin (~30 μm), highly compliant substrate, whose top surface is embedded with small (~0.2 μm) marker beads. These beads will be used for the detection of strain caused by traction forces exerted on the surface. For rapidly moving, weakly adherent cells, a gelatin concentration of 3% is recommended which has a Young's modulus of ~3 kPa. However, this concentra-tion can be increased or decreased to make stiffer or softer gels with a Young's modulus ranging from 2.2 to 16 kPa. If the gelatin substrate is being used for the first time, we recommend that several substrates be made over a range of concen-trations to determine an optimal concentration that allows both normal cell movement and detection of marker bead displacement.

1. Assemble Rappaport chambers by attaching a glass cylinder (diameter = 22 mm, height = 8 mm) to a 22 × 22 mm^2 glass coverslip (#0 thickness), using Sylgard 184 (Dow Corning, Midland, Michigan) silicone elastomer (Fig. 1A).

2. To make a 3% gelatin gel, powdered gelatin (Nabisco, Parsippany, New Jersey) is dissolved in prewarmed (~40 °C) Ca^{2+}- and Mg^{2+}- free Fish Ringer's solution (112-mM NaCl, 2-mM D-glucose, 2-mM KCl, 2.4-mM NaHCO$_3$) for about 3 min. The gelatin solution should become clear with no observable particulates, otherwise brief vortexing should remove them.

3. Transfer a 400 μl of liquid gelatin into the Rappaport chamber and allow it to solidify in a humid environment at 4 °C for 2–24 h. Humidity is necessary to pre-vent gels from drying out, which will effectively increase the gelatin concentration and thus the rigidity of the substrate.

4. Dilute a solution of fluorescent microbeads (0.2-μm diameter, Molecular Probes, California) 1:100 in distilled water, vortex for about 1 min, then cool to 4 °C. Red or orange beads are used in conjunction with the calcium indicator (Calcium Green-1 dextran, Molecular Probes, California).

5. Add 400 μl of the cooled bead solution to the solidified gelatin, then aspirate off immediately and leave this to dry for 1 h at 4 °C, in an environment room with good air circulation (Fig. 1B). At this time, it is useful (but not necessary) to make a small indentation with a pipette tip, at the extreme edge of the gelatin, which will serve as an indicator for the melting state of the gelatin in the next step.

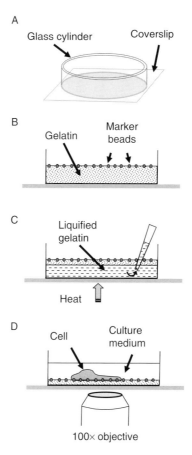

Fig. 1 Diagrams illustrating the key steps in the manufacture and use of the gelatin substrate. (A) Components of a glass Rappaport chamber. (B) Cross-section of a Rappaport chamber containing 400-μl solidified gelatin with a layer of marker beads dried on to the top surface (step 5 of the protocol). (C) Diagram showing step 6 of the protocol in which the lower layer of gelatin is heated and aspirated off. (D) The completed substrate for the traction force assay.

6. Briefly warm the chamber with gelatin on a hot plate set at ~62 °C for 5–15 sec to liquefy only the lower layer of gelatin, then carefully aspirate ~330 μl of the solution from the bottom of the chamber using a small pipette tip, being careful not to disturb the top surface (Fig. 1C). Using the indentation as an indicator, remove the chamber from the hot plate as soon as the indentation disappears. This is a critical step, because if the gelatin liquefies completely, the marker beads will sink into it. Alternatively, if the lower layer of gelatin does not liquefy sufficiently, it will be difficult to remove a sufficient volume such that the thickness of the substrates falls within the working distance of the microscope objective, making it impossible to image cells on top.

7. Remove the chamber from the hot plate and rapidly cool for 30 sec by placing on a level metal sheet, precooled to −20 °C.

8. Let the gelatin substrate sit at 4 °C for 1 h to allow complete solidification. The chamber may be stored for up to 72 h under 0.5 ml of serum-supplemented culture medium.

9. Calibrate each substrate prior to use, by measuring Young's modulus as described in another chapter of this volume, since this value will be needed for calculation of traction stresses. This measurement should be performed after the gelatin has equilibrated to room temperature for 1 h because gelatin substrates are temperature sensitive. In addition, it is often useful to repeat these measurements after an experiment, to check for any changes in Young's modulus, especially if variations in room temperature are suspected. The elasticity gelatin substrates can be varied between ~2.2 and 16 kPa with a high degree of reproducibility by altering gelatin concentration. At gelatin concentrations of 2.5% and above and at temperatures below 26 °C, substrates behave as an ideal elastic material (Doyle and Lee, 2002). This is supported by the observation that during cell movement beads return immediately to their original undisplaced positions, following release by the cell. Furthermore, when a microneedle is used to displace a single marker bead, for a distance double that made normally by the cell (~12 μm) and for 10 times the duration (5 min), beads return instantaneously to within ~4% (11.52 μm) of their original positions, following release.

B. Plating Cells onto a Gelatin Substrate

Some trial and error may be necessary to plate cells at a density such that there is only one cell per field of view. Otherwise if cells are too close, traction forces generated by neighboring cells will interfere. Although the following protocol has been optimized for fish epidermal keratocytes, it should be applicable to other rapidly moving cell types.

1. Bring the gelatin substrate to room temperature. Before plating cells, place ~200 μl of serum-rich culture medium, supplemented with 30% FBS, on the gelatin substrate and incubate for 30 min.

2. Epidermal sheets are obtained from explants of ~4 fish scales placed on a 22-mm^2 square coverslip within a 35-mm plastic Petri dish, as described previously (Lee et al., 1994). Wash explants twice with Ca^{2+}- and Mg^{2+}-free PBS (CMF-PBS) and once with a few drops of trypsin-EDTA (0.05% trypsin containing 0.53-mM EDTA, Life Technologies, Grand Island, New York). Carefully aspirate all the fluid from the Petri dish.

3. Immediately add 130 μl of a 1:1 mixture of CMF-PBS and trypsin-ETDA to the coverslip. Monitor under a microscope until cells begin to round up.

4. Immediately add 1 ml of serum-rich culture medium. Pipette the solution up and down several times (~5) to remove as many adherent cells as possible.

5. Transfer the cell-containing medium to a 1.5-ml Eppendorf tube, then *quickly* empty it onto the gelatin substrate. Note if the medium is emptied too slowly it will tend to "bead-up" on the substrate, making it difficult to spread the medium without agitating the chamber and possibly rupturing the gelatin.

6. Allow the cells to attach to the gelatin substrate for 45–60 min at room temperature before experimentation (Fig. 1D).

C. Data Collection and Analysis

The methods used for image acquisition and data analysis have been described in detail elsewhere (Doyle and Lee, 2005; Doyle *et al.*, 2004), so only the key points will be covered here. Keratocyte cultures grown in plastic Petri dishes (35-mm diameter) were loaded with both the calcium indicator Calcium Green™-1 dextran (3000 MW) and a calcium insensitive, Texas red fluorescent dextran (3000 MW, Molecular Probes, Eugene, Oregon) using the Influx™ pinocytic cell-loading reagent (Molecular Probes) as previously described (Doyle and Lee, 2002) and then replated onto the gelatin substrate, embedded with blue fluorescent marker beads. Triple fluorescence imaging was performed using a Zeiss 100×/1.4 emersion oil objective on a Zeiss Axiovert 200M inverted microscope (Carl Zeiss MicroImaging, Inc., Thornwood, New York) equipped with a DG-4 filter-changer (Sutter Instruments Co., Novato, California). Three fluorescence images of Calcium Green-1, the Texas red dextran, and the marker beads were acquired using FITC, Texas Red, and Cy-5/DAPI excitation filters, respectively, every ~1.8 sec. Emitted fluorescence was collected using a DAPI/FITC/Texas Red/Cy-5 quad band-pass filter set (Chroma Technology Corp., Vermont). An Orca II CCD camera (Hamamatsu Corp., Bridgewater, New Jersey) controlled by Openlab software (Improvision, Massachusetts) on an Apple G4 platform was used for image acquisition.

To detect changes in $[Ca^{2+}]_i$, measurements of the average fluorescence intensity were made over the cell body region in sequential, background-subtracted images. Morphometric analyses were performed using Metamorph analysis software (Molecular Devices Corp., California). Cell speed was calculated using the x–y coordinates of the cell centroid. The distance between every 11th centroid was calculated (~every 20 sec), from which a running average of 5 was obtained. Cell shape was measured for each frame, in terms of a shape factor according to the following equation: shape factor $= (4\pi A)/P^2$, where A is the cell area and P is the cell perimeter. This provides a measure between 0 and 1, for how close cell shape resembles a circle, where 0 is a straight line and 1 is a circle.

A qualitative impression of the relative magnitude and direction of traction stresses can be obtained by generating a "streak" image, in which pixels of maximum brightness from each image of the marker beads in a series, are superimposed in a single image (Fig. 2B). This type of analysis provides a useful means for detecting: (1) any drift of marker beads, (2) the maximum bead displacement, (3) whether bead displacements are occurring outside of the field of view, and

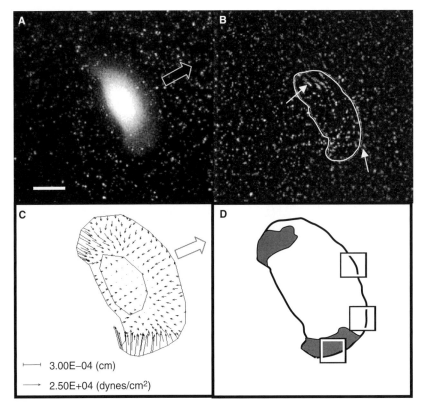

Fig. 2 Fluorescence images of the traction force assay and resulting vector and magnitude maps of traction stresses generated by a moving keratocyte. (A) Merged dual fluorescence image of a keratocyte loaded with a fluorescent calcium indicator, Calcium Green-1 dextran (excitation $\lambda = 500$ nm) moving on a gelatin substrate embedded with red fluorescent marker beads (excitation $\lambda = 560$ nm). (B) A fluorescence "streak" image showing the total bead displacements associated with a single calcium transient of 18-sec duration. Bead displacements (arrows) are clearly seen on either side of the cell (white outline) with the largest total displacements occurring on the left side. (C) A vector map generated by the LIBTRC software of traction stresses beneath a different cell. Strong tractions are oriented inward with respect to the left and right cell margins. Small inward directed tractions are also visible beneath the leading lamellipodium and at the rear cell edge. (D) A traction magnitude or "mag map" in which regions of high traction stress, equal to or greater than the 90th percentile, are colored gray. The boxes show several sampling regions on the right side of the cell. Block arrows represent direction of cell movement. Scale bar is 10 μm.

(4) the ability of all displaced beads to return to their original positions. Streak images also provide a good indication of the magnitude and direction of the traction stresses generated beneath the cell but only for cell types that maintain a simple relatively constant shape during movement such as fish keratocytes.

It is important to realize that some bead displacements may represent the response of the substrate, rather than the result of locally applied traction stresses.

For example, moving keratocytes generate relatively strong "pinching" tractions at their lateral cell edges in a direction perpendicular to the direction of cell movement that causes the substrate to be pushed outward with respect to the front and back of the cell (parallel to cell motion). To the naive observer, it may appear as if marker beads are being pushed outward by the leading cell edge. However, a given bead displacement may not accurately represent the size or direction of tractions generated by adjacent regions of the cell because the experimentally observed pattern of bead displacements are the result of a complex integration of traction stresses generated beneath the entire ventral cell surface.

To determine the magnitude and direction of the traction stresses generated at specific locations of the ventral cell surface, we use custom software LIBTRC (Dembo and Wang, 1999; Marganski *et al.*, 2003), that may be obtained from Dr. Micah Dembo. The first step is to obtain the substrate displacement field for each time point, which requires an image of beads in their displaced positions and a reference or null image of beads in their undisplaced positions, after the cell has moved away. Displacement of the substrate is calculated by comparing the positions of marker beads between the disturbed and reference image, using a correlation-based optical flow algorithm (Marganski *et al.*, 2003). Using the traced cell outline as a guide, a custom algorithm is then used to generate a mesh of ~200 quadrilateral elements tessellating the interior of the cell. The most likely traction vector at each node of this mesh is then estimated by fitting the displacement data using the formulas of Boussinesque relating substrate displacement to delta function forces acting at the substrate surface (Dembo and Wang, 1999).

D. Trouble-Shooting Guide

Since gelatin substrates are made "by hand," practice is required before they are suitable for an experiment. Some common problems that are likely to be encountered by those unfamiliar with the technique are described below.

1. Marker beads are not in the same plane of focus.

This usually occurs when attempting to liquefy the lower layer of gelatin. If the Rappaport chamber is heated for a few seconds too long, the gelatin will liquefy, allowing the beads to sediment from the top surface. The most effective way to avoid this problem is to observe when the small indentation, previously made at the edge of the gelatin, disappears. If the indentation disappears too quickly, then lower the temperature of the hot plate. We usually make substrates in triplicate and choose the best ones for an experiment. These substrates should have an evenly distributed monolayer of beads at the top surface of the gelatin and none of the defects described below.

2. Marker beads are too sparse or are clumped together.

The concentration of marker beads may have to be adjusted to obtain a dense, uniform distribution (~5 beads/μm^2) since a greater number of markers allow substrate deformation to be determined with a greater resolution. Beads in solution have a tendency to clump when dried onto the gelatin. Although they can still serve as markers for deformation, the loss of spatial resolution is undesirable. To prevent bead aggregation, we sonicate the stock solution for ~10 min before use, using a cleaning sonicator (Laboratory Supplies Company, Inc., Hicksville, New York) at full power.

3. The surface of the gelatin is uneven, furrowed or cracked.

Imperfections in the surface of the gelatin usually arise during the brief period of heating and aspiration of the lower layer of liquefied gelatin. If aspiration is too rapid or the gelatin is insufficiently solated, the top surface may pucker into ripples or furrows. This can also occur if the surface of the hot plate or the precooled metal sheet (step 7 in the procedure) is not completely level. In addition, gelatin substrates will crack if they dry out, so it is important to keep them in a humid environment and/or under the culture medium. Likewise, it is important not to let the marker bead solution sit for more than 1 h while waiting for them to dry onto the gelatin surface.

4. Beads detach from the gelatin substrate.

Although not a common problem, this can occur when the gelatin begins to disintegrate, which can happen if there is more than a 5 °C rise in room temperature or if the substrate is more than ~3 days old.

5. Marker bead displacements are either undetectable or too large.

Since the magnitude of average traction stress varies among individual keratocytes, we look at several moving cells to determine whether any of them can produce detectable bead displacements. Displacements may be difficult to see in a movie sequence at a slow speed; increasing the playback speed may make small displacements more noticeable. Otherwise, the substrate may be too rigid and requires downward adjustment of gelatin concentration.

Conversely, a gelatin substrate is considered too compliant when maximal bead displacements are greater than 10 μm. This is particularly problematic if the displacements drive the beads out of focus or beyond the field of view. The gelatin concentration is adjusted in 0.5% increments until the displacements are readily detectable and the maximal bead displacements are smaller than 10 pixels.

6. Displaced beads do not return immediately to their original undisplaced positions.

Typically, this indicates that the substrate is not behaving as an elastic material and therefore should not be used in the traction force assay. Although this is usually not a problem, it is likely to occur with more compliant gelatin substrates, or when the temperature rises above 26 °C.

7. Marker beads are not in the same focal plane as the cell.

While beads on the surface may be recorded selectively by proper focusing, it is helpful to have the beads on the same focal plane as the cell particularly when using an objective of high magnification and/or high numerical aperture (NA).

Sinking of beads into the substrate may be avoided by reducing the time and/or temperature for heating the gelatin, or increasing the concentration of gelatin.

8. Drift in x-, y-, or z-planes. Small drifts in the x- and y-directions may be corrected by software (e.g., LIBTRC per below), although it is advisable to limit any drift to no more than 10 pixels. Similarly, a small amount of focus drift can be tolerated as long as the pattern of marker beads remains recognizable by the computer program for mapping substrate deformation (Marganski $et\ al.$, 2003).

9. The cell stops moving.

As mentioned earlier, in most cases one can simply wait for cells to move out of the field before collecting the null-force image of marker beads. Cells that fail to move away must be removed by force. Weakly adhesive cells such as keratocytes may be removed by applying a gentle jet of culture medium through a pipette or by exposure to intense fluorescence illumination to kill the cell.

IV. Applications of the Gelatin Traction Force Assay to Study Mechano-signal Transduction in Moving Keratocytes

SACs have been shown to play an important role in regulating the movement of keratocytes (Lee $et\ al.$, 1999). To learn more about the role of mechano-chemical signaling, it is necessary to perform dual calcium and traction force microscopy. The optical and physical properties of 3% gelatin substrates are well suited for this purpose (Doyle and Lee, 2002). Changes in traction forces can be monitored during individual calcium transients over a range of $\sim$5–30 sec (Fig. 2B). Since the method used to perform dual calcium and traction force imaging have been described in detail elsewhere (Doyle $et\ al.$, 2004), we will focus here on the interpretation of data obtained from this type of experiment.

To generate vector maps that reveal the distribution and magnitude of traction stresses (Fig. 2C), we use the custom traction mapping software, LIBTRC (Dembo and Wang, 1999; Marganski $et\ al.$, 2003). Additionally, to quantify the temporal relationship between SAC-induced calcium transients and traction stress, we take the value of the 90th percentile traction stress at each time point, and correlate this with the fluorescence intensity of the calcium indicator (Fig. 3B). The 90th percentile value was used because it provides a more sensitive measure of changes in traction stress than the average traction stress, which is sensitive to the spreading area of the cell, or the maximal traction stress, which may reflect outlying events.

Our results indicate that a calcium transient leads to an immediate increase in traction stress, which is maintained, despite the return of $[Ca^{2+}]_i$ to baseline, until retraction occurs. To examine the spatial distribution of traction stress, contour plots of traction stress magnitude referred to here as mag maps are generated. Ranges of traction stress magnitude are rendered in different colors (Doyle $et\ al.$, 2004), which allow easy visualization of changes in size and distribution of traction

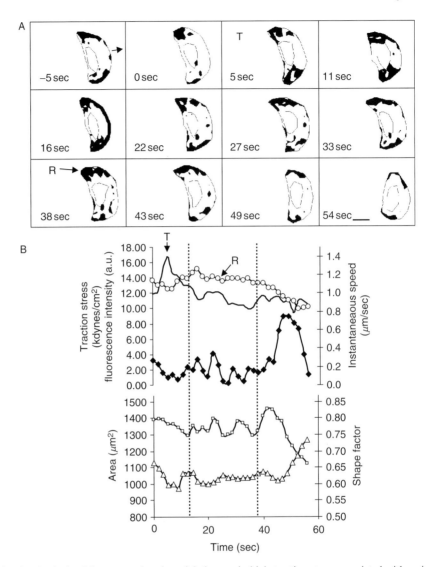

Fig. 3 Analysis of the temporal and spatial changes in high traction stress associated with a single calcium transient in a moving keratocyte. (A) Sequence of mag maps at ∼5-sec intervals, corresponding to the time period in (B). The regions of high traction enlarge along the outer cell margins following a calcium transient (marked as T), then decrease on retraction (marked as R). (B) Plots of calcium indicator fluorescence (solid line), 90th percentile traction stress (line with open circles), instantaneous cell speed (solid rhomboids), cell area (open squares), and shape factor (open triangles). Decreasing values of shape factor indicate that the cell is elongating, while increasing values represent a rounding of cell shape. A single calcium transient, with a peak (T), occurs during the signal transduction phase and ends at ∼13 sec (dotted line). In the following response phase, retraction (R) occurs at ∼38 sec (dotted line) and ends by 54 sec.

stresses. To quantify changes in the distribution of traction stress in relation to a calcium transient, we binarize mag maps by coloring only regions of high traction stress, which we define as being equal or greater than the average 90th percentile stress for the entire series (Fig. 2D). The presence high traction stress is sampled at the front, rear, and sides of the cell within equal-sized square sampling windows (Fig. 2D). Analysis of the frequency of high traction stress appearing within the sampling window indicates that high traction stresses are present at the rearmost lateral edges of the cell for the entire observation period, and for 27% and 17% of the time at the sides and front edge of the lamellipodium, respectively (Fig. 3A). The relationship between traction force generation and lamellar dynamics may be further inferred by comparing these values with the average time that the cell undergoes protrusion or retraction in each region.

To gain insight into the interrelationship between $[Ca^{2+}]_i$, traction stress, and cell movement, we compare the plot of 90th percentile traction stress with changes in cell speed, shape, and area during a single calcium transient (Doyle and Lee, 2005; Fig. 3B). The following events occur with a high degree of consistency among keratocytes. Prior to a calcium increase, which we term the "sensing" phase, a decrease in cell speed usually accompanies an increase in cell elongation, indicating that cell movement is inhibited by the failure to retract (data not shown). During the subsequent "transduction" phase, a calcium transient is followed immediately by a rise in traction stress. However, cell speed remains reduced and the cell is still elongated indicating that movement is still impeded. In addition, the degree of cell elongation at this stage suggests that cytoskeletal tension is being maintained. During the "response phase," retraction is marked by a drop in traction stress and an increase in speed, as cytoskeletal tension is released and the cell resumes movement.

V. Other Applications and Future Directions

Cells that exhibit rapid amoeboid movement, such as *Dictyostelium discoideum*, leukocytes, and highly metastatic transformed cells, typically generate weaker traction forces compared with slow moving cells such as fibroblasts, which have been measured in the majority of past traction force studies. The gelatin substrate is sufficiently sensitive to detect the weak traction stress ($\sim 7 \times 10^2$ Pa) of wild-type (WT) *D. discoideum* and even myosin II null mutants that generate tractions an order of magnitude lower (Lombardi *et al.*, 2007). Furthermore, the existence of many cytoskeletal mutants makes *Dictyostelium* a particularly useful model system to investigate the mechanism of traction mechanics. Studying the spatiotemporal patterns of traction stress in various mutants of *Dictyostelium* should provide insight into the role of different cytoskeletal components in generating and coordinating traction forces. In addition, gelatin substrate has enabled the study of GFP-myosin II redistribution in relation to traction force production in moving *Dictyostelium* (Lombardi *et al.*, unpublished data). Similar studies may be applied to other fast moving cell types such as leukocytes, metastatic cells, and neurons during pathfinding. However, if cells need to be maintained at 37 °C, preparation of gelatin substrates can be modified to increase their thermal stability (Bigi *et al.*, 2001).

The transparency of gelatin substrates permits its use in combination with a variety of fluorescence and photomanipulation techniques. Cells expressing fluorescent proteins or biosensors may be used to allow simultaneous observation of the dynamics of molecular machinery and generation of traction forces. Measurements of traction forces may be further combined with fluorescence resonance energy transfer (FRET), fluorescence recovery after photobleaching (FRAP), photoactivation, and chromophore-assisted laser inactivation (CALI) to allow real-time observation and manipulation of protein activities in relation to traction force generation. The highly elastic nature of gelatin substrates will be particularly advantageous for detecting small, rapid changes in traction stress in relation to localized molecular events.

VI. Summary

The use of gelatin substrates to study traction forces generated by rapidly moving cells represents an important addition to the repertoire of assays currently available to investigate the mechanics of cell motility. The major advantage of gelatin substrate is that it is sensitive enough to detect changes in the spatiotemporal pattern of traction stresses over short timescales, for example, during individual SAC-induced calcium transients.

In addition to the preparation of gelatin substrate and its application to keratocytes, we show how to extract information from a time series of traction maps and how to interpret these data in relation to the mechanochemical regulation of keratocyte movement. We further discuss future applications of this assay, including its use in conjunction with fluorescence imaging and photomanipulation techniques and applications to other rapidly moving cell types. Together these approaches will play an important role in furthering our understanding of cellular mechanosensing.

Acknowledgments

The work described herein was supported by a National Science Foundation grant MCB-0114231 to J.L.

References

Balaban, N. Q., Schwarz, U. S., Riveline, D., Goichberg, P., Tzur, G., Sabanay, I., Mahalu, D., Safran, S., Bershadsky, A., Addadi, L., and Geiger, B. (2001). Force and focal adhesion assembly: A close relationship studied using elastic micropatterned substrates. *Nat. Cell Biol.* **3**, 466–472.

Beningo, K. A., and Wang, Y.-L. (2002). Flexible substrates for the detection of cellular traction forces. *Trends Cell Biol.* **12,** 79–84.

Bigi, A., Cojazzi, G., Panzavolta, S., Rubini, K., and Roveri, N. (2001). Mechanical and thermal properties of gelatin films at different degrees of glutaraldehyde crosslinking. *Biomaterials* **22,** 763–768.

Burton, K., Park, J. H., and Taylor, D. L. (1999). Keratocytes generate traction forces in two phases. *Mol. Biol. Cell* **10**(11), 3745–3769.

Chen, C. S., Mrksich, M., Huang, S., Whitesides, G. M., and Ingber, D. E. (1997) Geometric control of cell life and death. *Science* **276**(5317), 1425–1428.

Chicurel, M. E., Chen, C. S., and Ingber, D. E. (1998). Cellular control lies in the balance of forces. *Curr. Opin. Cell Biol.* **10,** 232–239.

Dembo, M., Oliver, T., Ishihara, A., and Jacobson, K. (1996). Imaging the traction stresses exerted by locomoting cells with the elastic substrate method. *Biophys. J.* **70,** 2008–2022.

Dembo, M., and Wang, Y.-L. (1999). Stresses at the cell-to-substrate interface during locomotion of fibroblasts. *Biophys. J.* **76,** 2307–2316.

Doyle, A. D., and Lee, J. (2002). Simultaneous, real-time imaging of intracellular calcium and traction force production. *Biotechniques* **22**(2), 358.

Doyle, A. D., and Lee, J. (2005). Cyclic changes in keratocyte speed and traction stress arise from the Ca^{2+}-dependent regulation of cell adhesiveness. *J. Cell Sci.* **118**(Pt. 2), 369–379.

Doyle, A., Marganski, W., and Lee, J. (2004). Calcium transients induce spatially coordinated increases in traction force during the movement of fish keratocytes. *J. Cell Sci.* **117**(Pt. 11), 2203–2214.

Dunn, G. A., and Brown, A. F. (1986). Alignment of fibroblasts on grooved surfaces described by a simple geometric transformation. *J. Cell Sci.* **83,** 313–340.

Galbraith, C. G., and Sheetz, M. P. (1997). A micromachined device provides a new bend on fibroblast traction forces. *Proc. Natl. Acad. Sci. USA* **94**(17), 9114–9118.

Galbraith, C. G., and Sheetz, M. P. (1999). Keratocytes pull with similar forces on their dorsal and ventral surfaces. *J. Cell Biol.* **147**(6), 1313–1323.

Harris, A. K., Wild, P., and Stopak, D. (1980). Silicone rubber substrates: A new wrinkle in the study of cell locomotion. *Science* **208**(4440), 177–179.

Janmey, P. A. (1998). The cytoskeleton and cell signaling: Component localization and mechanical coupling. *Physiol. Rev.* **78,** 763–781.

Lauffenburger, D. A., and Horwitz, A. F. (1996). Cell migration: A physically integrated molecular process. *Cell* **84**(3), 359–369.

Lee, J., Ishihara, A., Oxford, G., Johnson, B., and Jacobson, K. (1999). Regulation of cell movement is mediated by stretch-activated calcium channels. *Nature* **400,** 382–386.

Lee, J., Ishihara, A., Theriot, J. A., and Jacobson, K. (1993). Principles of locomotion for simple-shaped cells. *Nature* **362,** 167–171.

Lee, J., Leonard, M., Oliver, T., Ishihara, A., and Jacobson, K. (1994). Traction forces generated by locomoting keratocytes. *J. Cell Biol.* **127,** 1957–1964.

Lo, C., Wang, H. B., Dembo, M., and Wang, Y. L. (2000). Cell movement is guided by the rigidity of the substrate. *Biophys. J.* **79,** 144–152.

Lombardi, M. L., Knecht, D. A., Dembo, M., and Lee, J. (2007). Traction force microscopy in dictyostelium reveals distinct roles for myosin II motor and actin-crosslinking activity in polarized cell movement. *J. Cell Sci.* (in press).

Malek, A. M., and Izumo, S. (1996). Mechanism of edothelial shape change and cytoskeletal remodeling in response to fluid shear stress. *J. Cell Sci.* **109,** 713–726.

Marganski, W. A., Dembo, M., and Wang, Y.-L. (2003). Measurements of cell-generated deformations on flexible substrates using correlation-based optical flow. *Methods Enzymol.* **161,** 197–211.

Oliver, T., Dembo, M., and Jacobson, K. (1999). Separation of propulsive and adhesive traction stresses in locomoting keratocytes. *J. Cell Biol.* **145**(3), 589–604.

Oliver, T., Jacobson, K., and Dembo, M. (1998). Design and use of substrates to measure traction forces exerted by cultured cells. *Methods Enzymol.* **298,** 497–521.

Pender, N., and McCulloch, C. A. (1991). Quantitation of actin polymerization in two human fibroblast sub-types responding to mechanical stretching. *J. Cell Sci.* **100,** 187–193.

Schwarzbauer, J. E. (1997). Cell migration: May the force be with you. *Curr. Biol.* **7,** R292–R294.

Sheetz, M. P., Felsenfeld, D. P., and Galbraith, C. (1998). Cell migration: Regulation of force on extracellular-matrix-integrin complexes. *Trends Cell Biol.* **8,** 51–54.

Svitkina, T. M., Verkhovsky, A. B., McQuade, K. M., and Borisy, G. (1997). Analysis of the actin-myosin II system in fish epidermal keratocytes: Mechanism of cell body translocation. *J. Cell Biol.* **139**(2), 397–415.

Tan, J. L., Tien, J., Pirone, D. M., Gray, D. S., Bhadriraju, K., and Chen, C. S. (2003). Cells lying on a bed of microneedles: An approach to isolate mechanical force. *Proc. Natl. Acad. Sci. USA* **100,** 1484–1489.

Wang, N., and Ingber, D. E. (1994). Control of cytoskeletal mechanics by extracellular matrix, cell shape, and mechanical tension. *Biophys. J.* **66,** 2181–2189.

Wang, N., Ostuni, E. O., Whitesides, G. M., and Ingber, D. E. (2002). Micropatterning tractional forces in living cells. *Cell Motil. Cytoskeleton* **52,** 97–106.

Wang, Y.-L., and Pelham, R. J. (1998). Preparation of a flexible porous polyacrylamide substrate for mechanical studies of cultured cells. *Methods Enzymol.* **298,** 489–496.

CHAPTER 13

Microfabricated Silicone Elastomeric Post Arrays for Measuring Traction Forces of Adherent Cells

Nathan J. Sniadecki[*] and Christopher S. Chen[*,†]

[*]Department of Bioengineering
University of Pennsylvania
Philadelphia, Pennsylvania 19104

[†]Department of Physiology
University of Pennsylvania
Philadelphia, Pennsylvania 19104

METHODS IN CELL BIOLOGY, VOL. 83
Copyright 2007, Elsevier Inc. All rights reserved.

0091-679X/07 $35.00
DOI: 10.1016/S0091-679X(07)83013-5

Abstract

Nonmuscle cells exert biomechanical forces known as traction forces on the extracellular matrix (ECM). Spatial coordination of these traction forces against the ECM is in part responsible for directing cell migration, for remodeling the surrounding tissue scaffold, and for the folds and rearrangements seen during morphogenesis. The traction forces are applied through a number of discrete adhesions between a cell and the ECM. We have developed a device consisting of an array of flexible, microfabricated posts capable of measuring these forces under an adherent cell. Functionalizing the top of each post with ECM protein allows cells to attach and spread across the tops of the posts. Deflection of the tips of the posts is proportional to cell-generated traction forces during cell migration or contraction. In this chapter, we describe the microfabrication, preparation, and experimental use of such microfabricated post array detector system (mPADs).

I. Introduction

Aside from bursts of locomotion during embryonic development or wound healing, most nonphagocytic cells in tissues lead a relatively nonmotile existence. Yet, underlying this apparently stationary state, cells can generate biomechanical forces that are important for cytokinesis (Scholey *et al.*, 2003), cortical contraction (Adelstein, 1982), cytoskeletal rearrangement (Ingber, 1993, 2003), adhesion remodeling (Geiger *et al.*, 2001), and sensing the elasticity of the microenvironment (Discher *et al.*, 2005). Cells generate force through fibrous bundles of myosin acting on actin microfilaments in the cytoskeleton (Huxley, 2004). In muscle cells, the actin–myosin bundles are known as myofibrils, while in nonmuscle cells, they are known as stress fibers. Myosin moves in a stepwise, walking cycle along actin microfilaments. During each power stroke, it exerts pulling forces of 3–4 pN on actin filaments of opposite polarities to shorten the total length of the fibrous bundle (Brenner, 2006; Finer *et al.*, 1994). The ends of many of these microfilaments are linked to focal adhesions (FAs), which are transmembrane "spot-welds" that attach cells to the extracellular matrix (ECM) through integrin receptors (Chapter 5 by Spatz and Geiger, this volume) or to adherens junctions (AJs), which are intercellular patches of cadherins that hold cells together (Fig. 1A). Contraction of the actin–myosin bundle thus leads to the exertion of concentrated stress at these loci of attachment. At FAs, stress acts on the matrix lattice and creates "traction forces" to propel a cell forward, to induce strain in the matrix fibers (e.g., collagen), and perhaps to remodel the matrix. Measuring traction forces generated by cells can thus provide insight into how intracellular forces regulate and are regulated by other cell functions.

Several techniques have been developed to measure these forces and biochemical activities associated with them. Traction forces of locomotive fibroblasts and other cell types were first observed as wrinkles in a thin, flexible film of silicone rubber (Harris *et al.*, 1980). The numbers and lengths of wrinkles or buckles gave a rough gauge of traction forces. This simple yet elegant tool provided insights into

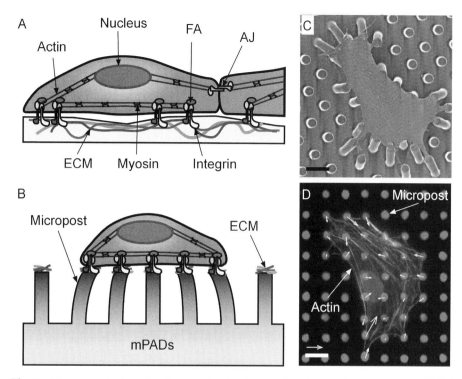

Fig. 1 Measurement of traction forces with microfabricated post array detector system (mPADs). (A) Myosin-driven sliding of actin microfilaments (stress fibers) causes forces to be loaded at focal adhesions (FAs), which attach the cell to the ECM through integrin receptors and at adherens junctions (AJs), which mediate cell–cell adhesions. (B) Forces at FAs, known as traction forces, are measured with the mPADs through the deflection of individual microposts. (C) Electron micrograph of a cell attached to the top of the microposts. Scale bar, 10 μm. (D) Vector map of traction forces is obtained by measuring the deflection of each micropost (DiI) due to the contraction of actin microfilament bundles (Phalloidin-Alex Fluor 488) within a cell (nucleus, Hoescht 33258). Scale bar, 10 μm; arrow, 20 nN. (See Plate 22 in the color insert section.)

the molecular pathways that regulated traction forces, stress fiber formation, and FA assembly (Chrzanowska-Wodnicka and Burridge, 1996; Helfman *et al.*, 1999). To better assess film distortions and allow measurements of traction forces, small latex beads were embedded into nonwrinkling silicone films for the quantification of film distortion during cell migration (Lee *et al.*, 1994). Subsequently, fluorescent microbeads were embedded into polyacrylamide gels, which could be cross-linked to different degrees in order to "tune" the stiffness of the elastic substrate for the range of traction forces generated by a particular cell type (Dembo and Wang, 1999). Arrays of fluorescent beads imprinted onto elastomeric substrates afforded even great precision than randomly seeded beads for the measurement of traction forces at individual FAs (Balaban *et al.*, 2001). Additionally, microfabricated cantilevers that deflect parallel to the plane of cell migration have provided measurements of transient tractions forces underneath small areas of a migrating cell (Galbraith and Sheetz, 1997): as a cell migrated over each microcantilever, the cantilever's deflection allowed a simple calculation of the local traction force.

To expand on this general approach, we have developed a microfabricated system that uses an array of vertical cantilevers to measure the traction forces at multiple locations on a cell (Tan *et al.*, 2003; Fig. 1B and C). During migration or contraction of a cell, its traction forces bend several posts such that each tip deflection is linearly proportional to the local force (Fig. 1D). However, while this technique can be straightforward to trained engineers, it requires many techniques that may not be familiar to a typical cell biologist. The goal of this chapter is to describe these techniques in sufficient detail such that they can be reproduced reliably and with minimal outside resources. In this chapter, we will describe (1) the fabrication steps used to construct the microfabricated post array detector system (mPADs), (2) the methods used to calibrate micropost stiffness, (3) the modification of mPADs for cell attachment, (4) the techniques in microscopy and image analysis for measuring micropost deflections, and (5) some applications with the mPADs for understanding cell mechanics.

II. Microfabrication of the Micropost Arrays

The mPADs are composed of a transparent silicone rubber called polydimethylsiloxane (PDMS), which can be modified with different surface treatments to allow or prevent deposition of ECM proteins onto the posts. Each micropost is cylindrical in shape, and the diameters (D), spacing (S), and heights (L) within an array are kept strictly uniform (within a fraction of a micrometer) in order to accurately measure traction forces ($D = 3 \, \mu m$, $S = 6$–$9 \, \mu m$, and $L = 8$–$12 \, \mu m$, respectively). Due to the microscale dimensions of the microposts, it is necessary to fabricate the devices with techniques developed in the semiconductor industry for the fabrication of integrated circuits (ICs) and microelectrical mechanical systems (MEMS). The first step involves photolithography of a hard master made from the photoresist SU-8. The second step consists of molding the master in PDMS to create soft replicas of the arrays of microposts.

The master can be fabricated as either a positive or a negative replica of the micropost arrays—positive being microposts of SU-8 and negative being holes within a film of SU-8. For a positive master, two casting steps in PDMS are required to generate the PDMS mPADs (Fig. 2). The first casting generates a negative mold, into which PDMS is cast a second time to create the array of microposts. With this process, multiple molds can be made in batch from the same SU-8 master. For a negative master, PDMS is directly cast onto the master to form the final product. This process eliminates one step from the positive master approach, but with a drawback that PDMS tends to get clogged into the SU-8 holes, which can lead to degradation of a negative master. In our laboratory, we have chosen to fabricate arrays of microposts with a positive master and we describe this technique here. In discussing the fabrication, we first provide a general introduction to photolithography, then detail the steps specific to fabricating the mPAD master in SU-8, and finally describe the replication process in PDMS.

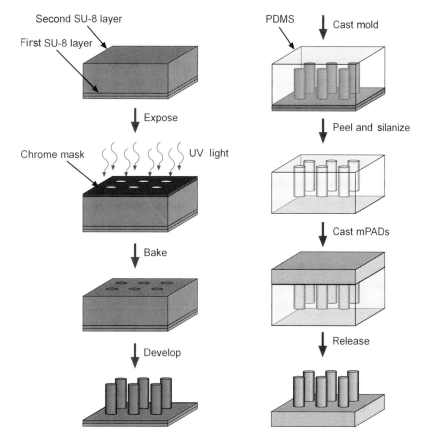

Fig. 2 Steps involved in the fabrication of mPADs. The process involves two photolithography steps of the SU-8 film and two soft lithography steps with PDMS.

A. Standard Photolithography

While we present below a concise introduction to the technique of photolithography as it pertains to fabricating mPADs, there are several excellent sources that describe this process in greater detail (Campbell, 2001; Jaeger, 2002; Madou, 1997). Standard photolithography involves the exposure of photosensitive material through a photomask. In MEMS or IC fabrication, the photosensitive material is photoresist such as SU-8, which is a polymer that can be patterned depending on where transparent patterns on the photomask allow light to pass through. In the case of positive photoresist, ultraviolet (UV) light causes scission in the polymer, rendering it more soluble in the high-pH developer solution than the unexposed regions. For negative photoresist such as SU-8, UV light enables cross-linking of the polymer in the exposed region, making it insoluble to the organic developer.

Each commercially available resist has its own recommended steps to follow in its use, but a general procedure is as follows. First, photoresist polymer is poured

onto a wafer and—in order to generate a film of uniform thickness on the wafer—
the wafer is spun at 1000–4000 rpm using a photoresist spin coater, which is a
centrifuge-like device with a vacuum chuck and adjustable rotational speed to
allow control of the uniformity and thickness of the film. The polymer film on the
wafer is "soft baked" between 90 and 100°C to evaporate out solvent from
the polymer film. The wafer is then overlaid with the photomask in an exposure
system, commonly known as a mask aligner, in order to correctly position the
patterns to the wafer. After exposure according to the manufacturer's specification, the wafer is placed into a developer solution to wash away the soluble regions,
leaving behind the insoluble patterns in the photoresist. Expertise with spin coat-
ers, mask aligners, and photolithography is often available within electrical engineering and material science departments, or in a core microfabrication facility at
many universities.

Photomasks contain microscopic features that are designed like blueprints with
a computer-aided design (CAD) tool. A photomask consists of a soda lime glass or
quartz plate with a patterned chromium layer, which absorbs visible to deep UV
light and casts shadows of designed pattern on the photoresist. For feature sizes as
small as 1–2 μm, such masks can be readily obtained from commercial outsourcing
services or at local university facilities. The patterns on the chrome mask are
"written" using optical pattern generators or e-beam mask writers. Once a design
is laid out and fabricated, a single mask can be used repeatedly to generate the
same pattern on tens of thousands of silicon wafers.

A less expensive approach but with a trade-off of larger feature sizes and lower
resolution is a film photomask. Here, a transparency film is printed with minimum
feature sizes of 10–20 μm using a high-resolution laser printer or photoplotter that
is often available at custom print shops (graphic printing or reprography). The film
is fixed to a glass plate with clear tape along the edges and shadows the photoresist
like the metal layer of a chrome mask. A film mask is not sufficient for generating
the mPADs master because 1- to 3-μm feature sizes are needed for the microposts.
However, it is an inexpensive approach to generate a master for PDMS stamps
used in microcontact printing ECM proteins onto the tops of the microposts as
described below (also in Chapter 19 by Lele *et al.*, this volume).

B. SU-8 Photolithography for Micropost Arrays

Currently, mPADs masters are made with SU-8 (MicroChem Corp., Newton,
Massachusetts), which is a negative, epoxy-type, near-UV photoresist. Unlike
common photoresists used as a protective layer for subsequent etching or deposition of materials, SU-8 has found widespread use in MEMS applications as a
structural layer. Due to its high viscosity, SU-8 can be spun as thick as 2 mm and
can achieve features with height-to-width aspect ratios as great as 25 with standard
photoexposure systems. SU-8 is a highly functionalized molecule with eight epoxy
groups (1,2-epoxide), where photogeneration of acid initiates cross-linking sites
and renders it insoluble and mechanically rigid (4 GPa; Lorenz *et al.*, 1997). It is an

excellent material for making the mPADs master due to its ability to produce the high aspect ratio required for the microposts, its compatibility with standard exposure systems, and its mechanical strength for subsequent soft lithography.

The lithography of the mPADs master involves two lithography steps to create layers of SU-8 resist (Fig. 2). The first layer is unpatterned and UV exposed without a mask (flood exposure) to create a base for the microposts. The second layer is patterned with a dark-field chrome mask with arrays of $D = 3$-μm holes in the metal film, spaced $S = 9$-μm center-to-center to create the micropost arrays. The double layer of SU-8 ensures good adhesion between the posts and the underlying silicon wafer.

Before the first layer is spun on, the test grade, n-type, $\langle 100 \rangle$ silicon wafer (Silicon Quest International, Santa Clara, California) is dehydrated at 175°C for 30 min on a hot plate (Model 721A, Barnstead, Dubuque, Iowa) and ozone cleaned (Model 342, Jelight, Irvine, California) for 10 min to prime the wafer surface for optimal attachment of SU-8. SU-8 2002 is spun onto the wafer at 500 rpm for 5 sec followed by a ramp up to 2000 rpm for 30 sec to create a base layer of 2-μm thickness in a spin coater (WA400B-6NPP-LITE, Laurell, North Wales, Pennsylvania). The SU-8 film is soft baked at 65°C for 1 min and then 95°C for 2 min on a hot plate to dry the solvent out of the film. Next, the SU-8 film is exposed to UV light (365 nm) in an MJB-3 mask aligner (Karl Suss, Munich, Germany) without a photomask at 70 mJ of total light energy (=power × time). After the exposure, the wafer is postexposure baked with the same temperature conditions as the soft-baked step, in order to thermally drive the cross-linking reaction. The SU-8 film is then allowed to cool for 1 h at room temperature before spinning the next layer of photoresist.

For the second layer of SU-8, which will form the microposts, SU-8 2010 is spun at 500 rpm for 5 sec followed by a ramp up to 4000 rpm for 30 sec, yielding a 10-μm thick film. The film is soft baked with the same conditions as before. To pattern arrays of microposts, a dark-field photomask (Advanced Reproductions, North Andover, Massachusetts) is loaded into the MJB-3 mask aligner, placed into hard contact with the photoresist on the wafer, and exposed to 80 mJ of total light energy. The wafer is then postexposure baked with the same conditions as before to cross-link the pattern. Afterward, it is allowed to slowly cool to room temperature to reduce cracking in the SU-8 film.

To develop the pattern, the wafer is placed in a glass dish containing propylene glycol methyl ether acetate (PGMEA) to dissolve away the unexposed SU-8. (Note that PGMEA is an organic solvent and should be handled in a chemical fume hood.) After 2 min in the developer, the wafer is transferred to a second glass dish containing PGMEA for 5 sec to dilute away the dissolved SU-8. Next, the wafer is transferred into a third dish containing isopropyl alcohol (IPA) for 20 sec to dissolve the PGMEA. After IPA, the wafer is quickly transferred into the fourth and fifth dishes containing hexane for 5 sec each to remove the IPA. The wafer is removed from the last dish and rapidly dried with nitrogen. Hexane has a lower surface tension than IPA and helps reduce micropost collapse due to capillary forces. The master is now fully fabricated and inspection under a metallurgical microscope should reveal arrays of microposts. Hard baking at 150°C for 20 h is encouraged to increase the mechanical strength of the master. Although these steps have been described in detail, results will vary depending on the equipment used. We strongly suggest that these process parameters be optimized according to specific facilities.

C. Soft Lithography

PDMS (Sylgard 184, Dow Corning, Midland, Michigan) is an optically clear, biologically inert, silicone rubber that closely matches the contour of a micro- or nanofabricated mold when cured. This material has been used to make inexpensive microfluidic devices (Duffy et al., 1998), microlens (Chen et al., 2004; Xia et al., 1996), or stamps for microcontact printing (Tien and Chen, 2001; Xia and Whitesides, 1998; Chapter 19 by Lele et al., this volume). Here, it is used to replicate the micropost structures in SU-8 through a double casting procedure. A negative mold is cast from the master to create an inverse of the mPADs, which are arrays of holes. The second casting into the negative mold results in the array of PDMS microposts on which cells can be seeded. The advantage of the double casting process is that one master can be used to create over several hundred PDMS micropost arrays for fabrication cost savings, which reduces variability between "identical" devices and provides a large supply of substrates.

PDMS is mixed at a 10:1 base polymer to curing agent ratio and allowed to degas for 1 h. The master is placed in an aluminum weight boat and PDMS is poured onto the master to form a layer ~1-cm thick. The polymer is rapidly cured in a 110°C convection oven for 10 min. The polymer is allowed to cool for 10 min before cutting away the aluminum boat and gently peeling the master out of the negative mold. This casting procedure is repeated in order to generate a large batch of negative masters for the second casting.

The surfaces of the negative molds need to be passivated with a fluorinated silane in order to prevent the liquid PDMS from permanently bonding to the PDMS negative mold during the second casting. First, the molds are placed in a plasma etcher (SPI Supplies, West Chester, Pennsylvania) for 2 min to activate the PDMS surface groups. Afterward, the molds are loaded into a desiccator and a few drops of (tridecafluoro-1,1,2,2-tetrahyrooctyl)-1-trichlorosilane (United Chemical Technologies, Bristol, Pennsylvania) are placed on a glass slide inside the desiccator. The chamber is evacuated overnight to allow the silane gas to diffuse over the negative molds and covalently bond to the PDMS surface groups. Extreme caution should be taken with trichlorosilanes as they readily react with water vapor to generate hydrochloric gas, which is toxic if inhaled.

After passivation, PDMS can be cast into the negative molds and released without permanent bonding. PDMS is mixed at a 10:1 ratio and degassed for 30 min. A thin layer of PDMS is applied to the negative mold and then a glass slide or cover glass pretreated in a plasma etcher is placed on top to sandwich the film.

The mold is placed into a 110 °C oven and cured for 20 h to ensure maximum cross-linking of the PDMS polymer. The mPADs, bound to the top cover glass or slide, is then peeled away from the negative mold and the excess PDMS runoff is trimmed with a razor blade. Large batches of substrates can be prepared and stockpiled for multiple experiments.

D. Troubleshooting and Helpful Suggestions

In lithography of SU-8, it is essential to make sure that there is good contact between the mask and SU-8 film in order to generate sharply defined posts. Due to its viscosity, SU-8 has more edge beading than most other photoresists. The excess SU-8 beads at the edge of the wafer, which introduces topographic features on the wafer and prevents its intimate conformal contact with the mask. The gaps between the mask and wafer result in light diffraction and pattern loss. Spinning SU-8 at higher speeds assists in material removal at the wafer edge. Additionally, edge beading can be removed by spraying the edge of a spinning wafer with PGMEA in a syringe. Another common problem with SU-8 processing is enlarged feature dimensions at the top of the film, often called T-topping. We suggest using a UV filter (U-360, Hoya Optics, San Jose, California) to reduce the amount of deep UV light (<350 nm) transmitted to the film to reduce the T-topping effect. Finally, we strongly encourage the optimization of process conditions that we have detailed. Screening a range of lithography condition—exposure times, lamp intensities, hot plate temperatures, and baking times—will help identify the appropriate parameters to enable repeatable SU-8 fabrication success.

III. Characterization of Micropost Spring Constant

When using the mPADs as a force sensor, each tip deflection reports the local traction force and the relationship is proportional to the spring constant of a micropost. An estimate of the spring constant can be obtained through classical relationships describing beam bending, which closely matches small deflections of the microposts. We assume that microposts have uniform material properties and dimensions across the entire array so that equivalent deflections reflect equivalent forces. Calibration is, however, required to obtain empirically the spring constant of the posts against the deflection of a calibrated, pulled glass microneedle.

A. Beam-Bending Theory

A micropost can be regarded as a cantilever beam, fixed at one end and loaded with forces at the other end, undergoing pure bending and negligible shearing (Fig. 3A). From the theory of slender beam bending, the relationship between force, F, and tip displacement, x, for a cylindrical beam is given by

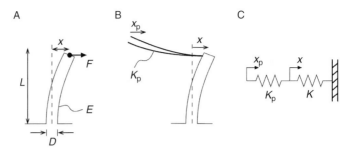

Fig. 3 Calculating the micropost stiffness. (A) Each micropost can be regarded as simple cantilever beam where the traction force, F, is linearly proportional to the tip deflection, x, as determined by the modulus of elasticity, E, the diameter, D, and height of the micropost, L. (B) Glass micropipettes can be used to empirically obtain the micropost stiffness by bring a calibrated tip with known stiffness, K_p, into contact with the posts and deflecting it by a known distance, x_p, while measuring the tip displacement of the post. (C) Equivalent two springs in series model used to calculate the micropost stiffness.

$$F = \frac{3\pi E D^4}{64 L^3} x \qquad (1)$$

where E is the modulus of elasticity of PDMS (2.0–2.5 MPa), D is the diameter, and L is the height of the micropost. The diameter of the posts is measured optically with a calibrated microscope eyepiece reticle. The height is measured with a profilometer on the silicon master. The spring constant of the micropost is therefore $K = 3\pi E D^4/64 L^3$. Calculation of this relationship provides the conversion from tip displacement (micrometers) to traction forces (nanonewtons) when cells are cultured on the array. A key assumption in the derivation of the spring constant equation is that vertical deflection of the micropost is small compared to horizontal deflection. For large tip deflections, a nonlinear relationship between force and deflection must be considered.

B. Measurement of Micropost Stiffness

The spring constant of the posts can be calibrated against the known spring constant of a glass micropipette prepared with a puller (World Precision Instruments, Sarasota, Florida). A crystal of p-nitrophenol was carefully placed on the end of the micropipette tip and the vertical deflection of the tip under the weight of the crystal ($58 \pm 32\ \mu g$) was recorded with a metallurgical microscope and measuring reticle. The microscope was horizontally mounted to observe the deflection of the tip. To determine the exact mass of the crystal, it was dissolved into bicarbonate buffer (50-mM Na_2CO_3, 50-mM $NaHCO_3$) and the transmittance of the solution at 400 nm was measured in a spectrophotometer (Lowry and Passonneau, 1972). The transmittance value was compared against a standard curve of transmission for known p-nitrophenol concentrations.

The calibrated micropipette can then be mounted on a micromanipulator and positioned into contact with a micropost, while under observation on an inverted phase microscope (Fig. 3B). The spring constant of the microposts can be found from the displacement of the micropost, x, against a prescribed translation of the micropipette, x_p, by the relationship for two springs in series,

$$K = \frac{K_p(x_p - x)}{x} \tag{2}$$

where K_p is the spring constant of the glass micropipette (Fig. 3C). We generally find close agreement between this direct measurement of K and calculated K as described above based on properties of PDMS and dimension of the posts.

IV. Analysis of Traction Forces Through Micropost Deflections

The mPADs are microcontact printed with ECM on the top surface to enable cells to attach and spread. This process involves the transfer of ECM proteins from a PDMS stamp to the tops of the microposts. The shaft and base are coated with a nonadhesive surfactant to confine cell attachment to the top. With proper coating, nearly all cell types are able to attach to the microposts and spread across the plane of tips. Spreading will be limited, of course, if the spacing S between posts becomes too large. For analysis of forces and identification of cellular structures responsible, the cells are fixed and immunofluorecently stained as one would do with cells on microscope cover glass. The samples are imaged in a fluorescent microscopy and images of the microposts are analyzed for deflections in MatLab (The Mathworks, Inc., Natick, Massachusetts).

A. Substrate Preparation

The methods of making and preparing micropatterned stamps for controlling cell adhesion area have been described elsewhere (Tan *et al.*, 2004; Tien and Chen, 2001; Chapter 19 by Lele *et al.*, this volume). Here, we discuss the general transfer of ECM onto the tops of all posts with a flat PDMS stamp (Fig. 4). First, PDMS and catalyst is mixed at a 30:1 ratio, allowed to degas, and then poured onto a silanized silicon wafer. The polymer is cured at 60°C for 1 h and then peeled from the wafer. The flat stamps are formed by cutting the PDMS into smaller pieces that match the mPAD array (typically 1×1 cm^2). In a biosafety cabinet, human fibronectin (BD Biosciences, San Jose, California) is prepared at 50 μg/ml in DI water. Aliquots of 50–100 μl are spread across the flat surface of the stamps and the protein is allowed to hydrophobically adsorb for 1 h to fully saturate the surface. Excess fibronectin is rinsed off in DI water and the stamp is dried with nitrogen. The PDMS surfaces of the mPADs are rendered hydrophilic with UV ozone treatment for 7 min in a UV Ozone cleaner (Jelight, Irvine, California). The flat

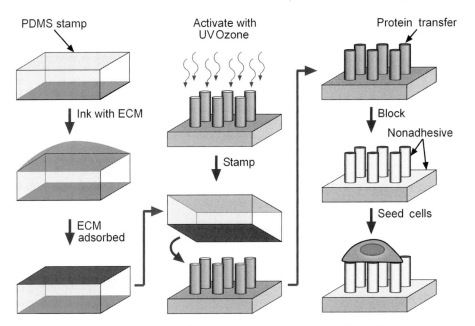

Fig. 4 Preparation of mPADs for seeding cells. ECM proteins are first applied to the top surface of a PDMS stamp by adsorption. Next, the mPADs are "activated" with UV ozone to change the PDMS surface from hydrophobic to hydrophilic, in order to allow transfer of proteins from the stamp to the tops of the microposts. The mPADs are then blocked with Pluronics to prevent nonspecific protein adsorption to unprinted areas. Finally, cells are seeded onto the microposts and allowed to spread for at least 10 h.

stamps are then placed over the microposts and pressed gently to form intimate contact with the microposts. The physical contact allows fibronectin to be transferred to the microposts by hydrophilic interactions.

After microcontact printing of the ECM protein, the mPADs undergo a series of sterilizing and washing steps. First, the mPADs are submerged in 100% ethanol to fully wet the microposts followed by 70% ethanol to sterilize the PDMS. The substrates are then washed in successive dishes containing DI water. The PDMS microposts are then stained with a lipophilic tracer (DiI, 1,1′-dioleyl-3,3,3′, 3′-tetramethylindocarbocyanine methanesulfonate, Invitrogen, Carlsbad, California) at 5 μg/ml for 1 h. The mPADs are washed with DI water and PBS to remove excess DiI and then submerged in 0.2% Pluronics F-127 (BASF, Ludwigshafen, Germany) for 30 min to prevent protein adsorption and cell adhesion to areas that were not stamped with ECM.

The substrates now have ECM on the tips and Pluronics blocking on the sides and base of the microposts. The mPADs are washed in successive dishes containing PBS and then placed into a standard tissue culture dish filled with the appropriate culture medium for the cell type. An mPADs substrate cast onto a standard 22×22 mm^2 cover glass easily fits inside a 35-mm tissue culture dish. Cells are harvested and seeded typically at a 1:10 ratio into the new dish using normal tissue culture procedures (Freshney, 2005). The seeding ratio affects the distance between cells and should be optimized to ensure a large likelihood of single cells for force analysis. After seeding, the mPADs are place in an incubator for at least 10 h to allow the cells to attach and spread on the microposts. The cells are then ready for experiments.

B. Staining and Microscopy of Micropost Arrays

Before microscopic imaging of the mPADs, the cells can be fixed and stained to identify specific structures, organelles, or proteins. We use 4% paraformeldahyde in PBS as our fixing solution to cross-link proteins, 0.05–0.2% Triton X-100 in PBS to permeabilize the membrane, and then antibodies or other fluorescent agents for immunofluorescent and direct staining. The samples can then be mounted onto a glass slide with a microscope cover glass overlaid on top for fluorescent and phase-contrast microscopy.

The first published technique for determining tip deflections compared a single fluorescent image of the grid deformed by traction forces, acquired at the focal plane of the tips of the posts, with an equally spaced grid that approximates the ideal, undeflected positions of the microposts (Tan et al., 2003). The difference in positions, between the observed circular tops of the posts and the corresponding circles in the ideal grid, determined the magnitude and direction of tip deflection due to the traction forces.

We have since advanced the measurements by imaging the entire length of posts from top to base (Lemmon et al., 2005). This method provides a more precise measure of micropost deflections because the base image of the posts approximates the undeflected position of the tips. It accurately shows any deviations in micropost position regularity due to defects in photomask design or microfabrication, which may be incorporated into the analysis for improved accuracy as described below. We use high-magnification objectives (60×) with oil to acquire our top and bottom images of the microposts.

C. Image Analysis Techniques

To calculate the direction and magnitude of the tip deflections, we use an image analysis routine developed with MatLab's image-processing toolbox (Lemmon et al., 2005). The original code imports the acquired microscope images, performs a localized thresholding algorithm to calculate the centroids of the fluorescent microposts. This routine is repeated for both top and bottom images and generates respective matrices of centroid positions. The difference in positions (in pixels) between the top and bottom images is then converted to traction forces (nanonewtons) by converting pixel to micrometer and multiplying the displacement by the spring constant of the microposts. The calculated deflections can be used to generate vector plots of the resulting cell-generated forces. If MatLab is not accessible, a similar imaging process can be implemented with other software programs such as IPLab (Scanalytics, Rockville, Maryland) or IGOR Pro (Wavemetrics, Lake Oswego, Oregon). We freely share our MatLab code on request.

V. Experimental Applications of Microposts and Discussion

On the mPADs, cells exert an average traction force per micropost on the order of tens of nanonewtons. Often, a range of three decades of traction forces (1–100 nN) is observed to be exerted by a cell attached to the posts, with largest forces at the perimeter and smallest ones in the interior. We consider the lower limit on force resolution to be equal to the variance in the apparent forces reported on the microposts that are not attached to cells and have no net force exerted on them. Using current techniques, this limit is ~3.2 nN.

One modification of the technique involves functionalizing only a subset of the microposts with ECM protein, in order to constrain cells to adhere to a specific geometry (square, rectangle, triangle, and so on) or total area (2×2 posts, 3×3 posts, and so on). In this case, instead of using a flat stamp, ECM can be loaded onto a stamp with microfeatures that have the desired printing dimensions and transferred to the tops of the microposts for patterning cells. This technique has been used to demonstrate that cell area positively correlates with average traction force per post (Tan et al., 2003). In scaling up, larger stamp features may be used to produce patterned clusters of cells on the mPADs to study the cooperative effect of traction forces within a monolayer of cells (Nelson et al., 2005).

The spring constant of the microposts can be adjusted by changing the height or diameter of the posts to measure the relationship between traction force generation and underlying substrate stiffness. In addition, while the procedure described above generates posts with a center-to-center distance $S = 9\ \mu m$, it can be used down to a resolution of ~3 μm. Spatial resolution has been further increased by shortening S to 1.6, by the use of deep reactive ion etching (DRIE) to generate the high aspect ratio features (Du Roure et al., 2005; Saez et al., 2005). However, because DRIE is not as widely available and can be expensive, the cost-to-benefit ratio should be considered when deciding on which fabrication process to use. Through screening a range of aspect ratios of the posts, such variations in post geometry indicate that there is a strong correlation between substrate stiffness and traction force generation (Saez et al., 2005). We have looked at traction forces with respect to cell types and have found that there is also phenotypic specificity in force generation for endothelial cells, epithelial cells, fibroblasts, and smooth muscle cells (Lemmon et al., 2005).

Analysis of the spatiotemporal dynamics of traction forces is also permissible with the mPADs. A fluorescent microscope equipped with a chamber for regulating live cell conditions (temperature, CO_2, and humidity) can be employed to record the traction force over time during isometric contraction or directed migration of a cell. To analyze these images, a single bottom image is acquired and used as the reference image for subsequent frames of the tip deflections in the top image. Attention should be made to accurately register the frames to the reference image because shifting in the sample position relative to the objective is common due to thermal fluctuations of the microscope or transmittance of vibration during operation. Additionally, one can fabricate mPADs onto a microscope cover glass and assemble them as the bottom of PDMS chambers for high-resolution measurements with oil immersion objectives.

It should be pointed out that the microposts provide a different topography to cells from the planar surface of glass or plastic tissue culture dishes. Such topography may introduce additional signals that could alter cell behavior, and requires further study. This possibility also raises the interesting question of whether flat surfaces or regularized discrete features are better models for the complex fibrous ECM that cells typically encounter *in vivo*. One important observation is that cells on mPADs appear qualitatively similar to those on glass coverslips in cell shape and organization of stress fibers and FAs, and generate similar traction forces as do cells cultured on flat polyacrylamide substrates (Lemmon *et al.*, 2005). Thus, while observations should always be checked by comparing different systems, these data suggest that the mPADs can provide meaningful insight into the mechanics of cultured cells.

Acknowledgments

The authors would like to thank John L. Tan and Michael T. Yang for their helpful discussions. This work was supported in part by the National Institutes of Health (grants EB00262 and HL073305), the Department of Defense Multidisciplinary University Research Initiative, DARPA, and the Nano/Bio Interface Center through the National Science Foundation NSEC DMR-0425780. NJS is supported by the National Institutes of Health Ruth Kirschstein National Research Service Award Postdoctoral Fellowship.

References

Adelstein, R. S. (1982). Calmodulin and the regulation of the actin-myosin interaction in smooth muscle and nonmuscle cells. *Cell* **30**(2), 349–350.

Balaban, N. Q., Schwarz, U. S., Riveline, D., Goichberg, P., Tzur, G., Sabanay, I., Mahalu, D., Safran, S., Bershadsky, A., Addadi, L., and Geiger, B. (2001). Force and focal adhesion assembly: A close relationship studied using elastic micropatterned substrates. *Nat. Cell Biol.* **3**(5), 466–472.

Brenner, B. (2006). The stroke size of myosins: A reevaluation. *J. Muscle Res. Cell Motil.* **27**(2), 173–187.

Campbell, S. A. (2001). "The Science and Engineering of Microelectronic Fabrication," 2nd edn., Vol. xiv, pp. 603. The Oxford series in electrical and computer engineering. Oxford University Press, New York.

Chen, J., Wang, W., Fang, L., and Varahramyan, K. (2004). Variable-focusing microlens with micro-fluidic chip. *J. Micromech. Microeng.* **14**, 675–680.

Chrzanowska-Wodnicka, M., and Burridge, K. (1996). Rho-stimulated contractility drives the formation of stress fibers and focal adhesions. *J. Cell Biol.* **133**(6), 1403–1415.

Dembo, M., and Wang, Y. L. (1999). Stresses at the cell-to-substrate interface during locomotion of fibroblasts. *Biophys. J.* **76**(4), 2307–2316.

Discher, D. E., Janmey, P., and Wang, Y. L. (2005). Tissue cells feel and respond to the stiffness of their substrate. *Science* **310**(5751), 1139–1143.

Duffy, D. C., McDonald, J. C., Schueller, O. J. A., and Whitesides, G. M. (1998). Rapid prototyping of microfluidic system in poly(dimethylsiloxane). *Anal. Chem.* **70**(23), 4974–4984.

Du Roure, O., Saez, A., Buguin, A., Austin, R. H., Chavrier, P., Siberzan, P., and Ladoux, B. (2005). Force mapping in epithelial cell migration. *Proc. Natl. Acad. Sci. USA* **102**(7), 2390–2395.

Finer, J. T., Simmons, R. M., and Spudich, J. A. (1994). Single myosin molecule mechanics: Piconewton forces and nanometre steps. *Nature* **368,** 113–119.

Freshney, R. I. (2005). "Culture of Animal Cells: A Manual of Basic Technique," 5th edn., p. 642. Wiley-Liss, Hoboken, NJ.

Galbraith, C. G., and Sheetz, M. P. (1997). A micromachined device provides a new bend on fibroblast traction forces. *Proc. Natl. Acad. Sci. USA* **94**(17), 9114–9118.

Geiger, B., Bershadsky, A., Pankov, R., and Yamada, K. M. (2001). Transmembrane extracellular matrix—cytoskeleton crosstalk. *Nat. Rev. Mol. Cell Biol.* **2**(11), 793–805.

Harris, A. K., Wild, P., and Stopak, D. (1980). Silicone rubber substrata: A new wrinkle in the study of cell locomotion. *Science* **208**(4440), 177–179.

Helfman, D. M., Levy, E. T., Berthier, C., Shtutman, M., Riveline, D., Grosheva, I., Lachish-Zalait, A., Elbaum, M., and Bershadsky, A. D. (1999). Caldesmon inhibits nonmuscle cell contractility and interferes with the formation of focal adhesions. *Mol. Biol. Cell* **10**(10), 3097–3112.

Huxley, H. E. (2004). Fifty years of muscle and the sliding filament hypothesis. *Eur. J. Biochem.* **271**(8), 1403–1415.

Ingber, D. E. (1993). Cellular tensegrity: Defining new rules of biological design that govern the cytoskeleton. *J. Cell Sci.* **104**(Pt. 3), 613–627.

Ingber, D. E. (2003). Tensegrity I. Cell structure and hierarchical systems biology. *J. Cell Sci.* **116**(Pt. 7), 1157–1173.

Jaeger, R. C. (2002)."Introduction to Microelectronic Fabrication," 2nd edn., Vol. xiv, p. 316. Prentice Hall, Upper Saddle River, NJ.

Lee, J., Leonard, M., Oliver, T., Ishihara, A., and Jacobson, K. (1994). Traction forces generated by locomoting keratocytes. *J. Cell Biol.* **127**(6 Pt. 2), 1957–1964.

Lemmon, C. A., Sniadecki, N. J., Alom Ruiz, S., Tan, J. T., Romer, L. H., and Chen, C. S. (2005). Shear force at the cell-matrix interface: Enhanced analysis for microfabricated post array detectors. *Mech. Chem. Biosyst.* **2**(1), 1–16.

Lorenz, H., Desont, M., Fahrni, N., LaBianca, N., Renaud, P., and Vettiger, P. (1997). SU-8: A low cost negative resist for MEMS. *J. Micromech. Microeng.* **7**, 121–124.

Lowry, O. H., and Passonneau, J. V. (1972)."A Flexible System of Enzymatic Analysis," Vol. xii, p. 291. Academic Press, New York.

Madou, M. J. (1997). "Fundamentals of Microfabrication," p. 589. CRC Press, Boca Raton, FL.

Nelson, C. M., Jean, R. P., Tan, J. L., Liu, W. F., Sniadecki, N. J., Spector, A. A., and Chen, C. S. (2005). Emergent patterns of growth controlled by multicellular form and mechanics. *Proc. Natl. Acad. Sci. USA* **102**(33), 11594–11599.

Saez, A., Buguin, A., Silberzan, P., and Ladoux, B. (2005). Is the mechanical activity of epithelial cells controlled by deformations or forces? *Biophys. J.* **89**(6), L52–L54.

Scholey, J. M., Brust-Mascher, I., and Mogilner, A. (2003). Cell division. *Nature* **422**(6933), 746–752.

Tan, J. L., Liu, W., Nelson, C. M., Raghavan, S., and Chen, C. S. (2004). Simple approach to micropattern cells on common culture substrates by tuning substrate wettability. *Tissue Eng.* **10**(5–6), 865–872.

Tan, J. L., Tien, J., Pirone, D. M., Gray, D. S., Bhadriraju, K., and Chen, C. S. (2003). Cells lying on a bed of microneedles: An approach to isolate mechanical force. *Proc. Natl. Acad. Sci. USA* **100**(4), 1484–1489.

Tien, J., and Chen, C. S. (2001). Microarrays of cells. *In* "Methods in Tissue Engineering" (A. Atala and R. Lanza, eds.), pp. 113–120. Academic Press, San Diego.

Xia, Y., Kim, E., Zhao, X. M., Rogers, J. A., Prentiss, M., and Whitesides, G. M. (1996). Complex optical surfaces formed by replica molding against elastomeric masters. *Science* **273**(5273), 347–349.

Xia, Y. N., and Whitesides, G. M. (1998). Soft lithography. *Angew. Chem. Int. Ed.* **37**(5), 551–575.

CHAPTER 14

Cell Adhesion Strengthening: Measurement and Analysis

Kristin E. Michael and Andrés J. García

Woodruff School of Mechanical Engineering
Petit Institute for Bioengineering and Bioscience
Georgia Institute of Technology
Atlanta, Georgia 30332

Abstract

Cell adhesion to the extracellular matrix is a dynamic process involving numerous focal adhesion components, which act in coordination to strengthen and optimize the mechanical anchorage of cells over time. A method for systematically analyzing the cell adhesion strengthening process and the components involved in this process is described here. The method combines an adhesion strength assay based on applying fluid shearing to a population of cells and quantitative biochemical analyses.

I. Introduction

Cell adhesion to the extracellular matrix (ECM) provides tissue structure and integrity as well as triggers signals that regulate complex biological processes such as cell cycle progression and tissue-specific cell differentiation (Danen and Sonnenberg, 2003; Garcia and Reyes, 2005; Walker *et al.*, 2005; Wehrle-Haller and Imhof, 2003). Hence, cell adhesion is critical to numerous physiological and pathological processes, including embryonic development, cancer metastasis, and wound healing, as well as biotechnological applications such as host responses to implanted devices and integration of tissue-engineered constructs (Garcia, 2005; Hubbell, 1999; Jin and Varner, 2004; Wilson *et al.*, 2005). For example, genetic deletion of adhesion components, including ECM proteins, integrin receptors, and focal adhesion (FA) proteins, results in lethality at early embryonic stages (Furuta *et al.*, 1995; George *et al.*, 1993; Stephens *et al.*, 1995; Xu *et al.*, 1998). These deletion studies clearly demonstrate that cell–ECM adhesion is essential to development. The cell adhesion process involves binding of integrin surface receptors to ECM proteins, clustering of these bound integrins, association with the actin cytoskeleton, and subsequent strengthening of the integrin–actin cytoskeleton interaction via complexes of proteins known as FAs (Hynes, 2002; Zamir and Geiger, 2001; Zimerman *et al.*, 2004). Due to the close association between biochemical and biophysical processes within adhesion complexes, mechanical analyses can provide important new insights into structure–function relationships involved in regulating the adhesion process.

Cell spreading and migration are often used as indirect indicators of adhesion strength. However, these multistep, dynamic processes exhibit complex dependencies on adhesion strength (Palecek *et al.*, 1997), and therefore do not provide sufficiently direct or sensitive metrics of adhesion. This lack of quantitative understanding of the regulation of adhesion strength limits the interpretation of functional studies of structural and signaling adhesive components. Furthermore, it is increasingly evident that mechanotransduction between cells and their environment regulates gene expression and cell fate (Engler *et al.*, 2004; Mammoto *et al.*, 2004; McBeath *et al.*, 2004; Polte *et al.*, 2004; Wozniak *et al.*, 2003). Direct measurements of cell–ECM adhesion strength are therefore necessary for understanding these mechanosensory functions.

II. The Cell Adhesion Process

Cell adhesion to ECM components, such as collagen, laminin, and fibronectin, is primarily mediated by integrin receptors. Integrins are heterodimeric, transmembrane receptors consisting of noncovalently associated α and β subunits (Hynes, 2002). Following activation, via changes in conformation, integrins bind their

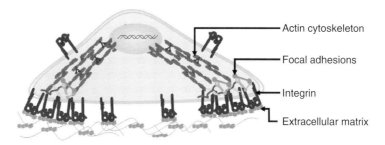

Fig. 1 Cell adhesion involves binding integrin receptors to extracellular matrix proteins, clustering of integrins, and association of these integrins with the cytoskeleton forming complexes called focal adhesions (FAs). FAs provide mechanical linkage to the cytoskeleton and trigger signaling pathways that regulate cell survival, growth, and differentiation. (See Plate 23 in the color insert section.)

target ligands and cluster together. The cytoplasmic domains of bound and clustered integrins form supramolecular complexes with FA proteins, such as talin, FAK, vinculin and paxillin, and become associated with the actin cytoskeleton (Beningo *et al.*, 2001; Galbraith *et al.*, 2002; Zamir and Geiger, 2001; Zimerman *et al.*, 2004; Fig. 1). Through these initial adhesive complexes, the cell generates myosin-mediated contractile forces resulting in enhanced, more mature FA complexes and a strengthened cell–ECM interaction (Galbraith *et al.*, 2002). Most FA proteins contain multiple binding sites for other adhesion components. Thus, FA complexes can assemble in multiple structural configurations and the regulation of such complex interactions among FA components is central to the function of these complexes in cell adhesion (Zamir and Geiger, 2001).

Cytoskeletal tension and FA assembly also drive changes in cell shape and spreading. In addition, coordinated adhesive interactions and force generation are central to cell migration (Lauffenburger and Horwitz, 1996; Sheetz *et al.*, 1998). During migration, cellular processes extend and attach to ECM at the leading edge, allowing actomyosin contractile forces to be transmitted to the ECM as "traction forces" (Chapter 12 by Lee, this volume). As the cell pulls itself forward with these traction forces, it extends more processes to sustain the forward motion. Once the cell has moved sufficiently forward relative to the initial attachment point, the attachment at the rear is released and adhesion components are recycled for reuse at the front of the cell (Palecek *et al.*, 1996), or else proteolytically cleaved. Therefore, migration involves coordinated regulation of adhesive interactions along the cell–ECM interface, and the resulting distribution of traction forces is highly nonuniform and position dependent (Dembo and Wang, 1999).

III. Measurement Systems for Adhesion Characterization

Cell migration and spreading are complex processes that involve multiple steps and are highly regulated (Lauffenburger and Horwitz, 1996). Although changes in cell spreading and migration are commonly studied to gain insights into adhesive interactions and FA function, obtaining a fundamental, mechanistic understanding of how FA components regulate these processes is difficult. Moreover, the highly complex dependency of migration and spreading on adhesion strength, down to the position of an adhesive complex within a migrating cell, makes the interpretation of these results with regard to adhesion challenging (DiMilla *et al.*, 1991; Palecek *et al.*, 1997, 1999). Due to the close association between biochemical and biophysical processes within adhesion complexes, mechanical analyses can provide insights into structure–function relationships in the regulation of the adhesion process. Mechanical analyses have identified two stages of cell adhesion, consisting of initial integrin–ligand binding followed by rapid strengthening (Choquet *et al.*, 1997; Garcia *et al.*, 1998b; Lotz *et al.*, 1989; McClay *et al.*, 1981). This strengthening response arises from (1) cell spreading, which increases in cell–substrate contact area, (2) clustering, which entails receptor recruitment to anchoring sites, and (3) FA assembly, which involves integrative interactions with cytoskeletal elements that lead to enhanced force distribution among bound receptors through local membrane stiffening (Gallant *et al.*, 2005; Lotz *et al.*, 1989).

Several mechanical methods have been developed to quantify adhesion strength, including hydrodynamic shear force, centrifugation, and micromanipulation assays (Doroszewski *et al.*, 1977; Garcia and Gallant, 2003; Garcia *et al.*, 1997; Lawrence *et al.*, 1987; McClay *et al.*, 1981; Mohandas *et al.*, 1974). Wash assays provide an easy method to apply hydrodynamic shear force to detach cells and compare the number of attached cells after force application. But, the amount of force applied in wash assays is nonuniform and uncontrolled, limiting the reproducibility and interpretation of these results. Centrifugation and micromanipulation measurements do apply controlled forces; however, these assays apply a single force at one time, making them tedious to extend over a range of forces, and are often limited in the overall range of forces than can be applied. Hydrodynamic techniques, based on systems such as parallel plate, rotating disk(s), and radial flow between parallel disks, can be configured to apply a controlled, wide range of forces and provide consistent, reliable measurements of adhesion strength. On the basis of characterized flow patterns in a specialized flow chamber, fluid flow over adherent cells generates detachment shear forces. The adhesion strength is typically reported as the shear stress (force/area) at the flow chamber wall (τ_w) that produces a prescribed level of cell detachment (e.g., 50% detachment). Although these assays provide sensitive and reproducible measurements of adhesion strength, they are often limited by the ability to apply sufficient forces for detachment of cells with strengthened adhesion. Furthermore, complex detachment mechanics are associated with these hydrodynamic assays due to variability in cellular morphology (spread vs rounded cells) as well as nonuniform distributions of adhesive structures, thus limiting the interpretation of results.

Sophisticated mechanical experimental systems have recently been developed to further elaborate structure–function relationships in cell adhesion. Using laser tweezers to trap ECM-coated beads and apply small forces (nN) in a short time period (sec), Sheetz and colleagues have shown that mechanical force itself can be converted into a signal to strengthen initial integrin–ECM adhesions in advancing lamellapodia (Galbraith *et al.*, 2002; Giannone *et al.*, 2003). These studies elegantly demonstrate that applied force can enhance maturation of initial FA complexes and provide insights into mechanical mechanisms at the leading edge of migrating cells. Similarly, cantilever-based techniques analogous to atomic force microscopy have been developed to study cellular force application or generation (Chen and Moy, 2000; Micoulet *et al.*, 2005). These methods resolve forces at localized points on the cell during short time periods, thereby demonstrating maturation of initial FAs induced by externally applied or internally generated forces. Extending our understanding of force generation at edges of adherent cells, traction force microscopy techniques incorporate embedded beads in or extended posts on compliant substrates to measure forces applied by the cell to the substrate, based on the mechanical theory of material deformation or beam deflection (Balaban *et al.*, 2001; Dembo and Wang, 1999; Reinhart-King *et al.*, 2005; Tan *et al.*, 2003). Notably, Balaban *et al.* (2001) discovered a direct correlation between local generated force and the orientation, intensity, and area of FAs, suggesting that internally generated force and extent of FA maturation combine to a constant net stress of ~ 5.5 nN/mm^2 at adhesive complexes. Although traction force microscopy techniques provide an excellent means for studying force generated by cells, these techniques are presently limited to 2D measurements and are, therefore, only able to resolve forces in the tangential (parallel) plane to the surface. This limitation obscures the interpretation of results at the center of contact. In addition, traction force measurements, while highly relevant to migration, do not provide direct measurements of adhesion strength. In all, these recent studies have enhanced our understanding of various adhesion-related functions; however, all of these methods examine single-cell interactions with substrates and focus on a small population of FA complexes at a specific time in FA development, and none of these methods provide direct information on the strength of the adhesive interaction.

We have developed a robust approach for studying adhesion strengthening, which incorporates a hydrodynamic adhesion strength assay and quantitative biochemical assays to analyze the mechanisms of the strengthening process (Gallant *et al.*, 2005; Garcia *et al.*, 1997, 1998a). Using population-based assays, we can obtain reproducible measurements of adhesion strength and FA assembly. These measurements may be performed over multiple time points to extend our characterization of strengthening from initial adhesion to steady state adhesion. By studying detachment of cells, our analysis focuses on the maximum force that a cell can resist before failure of attachment, a common parameter used in mechanics. This experimental failure strength measurement can then be extended to theoretical mechanical analyses to enhance our understanding of the strength of adhesion complexes under various conditions and time points.

IV. Hydrodynamic Assay for Quantifying Adhesion Strength

Cell adhesion strength is quantified using a submerged spinning disk device, which applies a well-defined range of hydrodynamic forces to adherent cells on the disk surface (Garcia *et al.*, 1998a). Using this device, a large range of forces sufficient to detach cells can be applied in the laminar flow regime, before turbulence obscures the generation of regular, controlled hydrodynamic forces as commonly occurs in other devices such as parallel plate flow chambers. These large forces are often necessary to detach cells with strengthened adhesions (e.g., spread fibroblasts).

A. Experimental Design

Cells are seeded on a 25-mm diameter coverslip/disk and maintained under the appropriate conditions at 37°C until reaching the desired experimental point (see micropatterning paragraph below for details). Seeded coverslips are mounted on the spinning device, which imparts detachment forces due to fluid shear stress. The flow patterns in the spinning disk device approximate the flow around an infinite disk spinning in an infinite fluid (Garcia *et al.*, 1997). Due to the "no-slip" condition between the fluid and the solid surface, a velocity gradient is generated that imparts a shear stress at the surface of the disk. This shear stress τ (force/area) increases linearly with radial position r along the surface of the disk as given by Eq. (1). Fluid density ρ, viscosity μ, and rotational velocity ω all remain constant for each disk spun.

$$\tau = 0.8r\sqrt{\rho\mu\omega^3} \tag{1}$$

Therefore, the device generates a range of detachment forces such that no force is imparted at the center of the disk ($r = 0$), whereas large forces are applied at the edge of the disk.

Uniformly distributed, adherent cells are subject to spinning for 5 min in serum-free media (PBS +2-mM dextrose) at room temperature and subsequently fixed and stained. Cells are counted in 60 fields across four axes of radial position. Radial position is then converted to shear stress based on the experimental values of the parameters in Eq. (1), and the resulting fraction of adherent cells postspin (f) versus shear stress is fit to a sigmoidal curve [Eq. (2)].

$$f(\tau) = \frac{1}{1 + e^{b(\tau - \tau_{50})}} \tag{2}$$

τ_{50} is the shear stress for 50% cellular detachment (Fig. 3) and represents the mean adhesion strength for a population of cells. In order to achieve appropriate values of τ_{50} for a given population of cells, appropriate input values for a constant rotational velocity ω must be determined to impart shear stresses such that cells remain attached at the center of the disk and less than 10% of the cells remain attached at the edge of the disk. This rotational velocity can range from 500 to

4000 rpm with 3500 rpm being ideal for detaching steady state adherent NIH3T3 cells. Shifts in the resulting adhesion profile represent changes in the adhesion strength, and the force applied to the cells by the experimental shear stress can be determined as illustrated in the modeling section of this text (Section VI).

Because the geometry of an object in flow is critical to the amount of force applied to that object (Munson *et al.*, 2006), we use micropatterning techniques to maintain uniform cell geometry across all conditions of interest. Under normal plating conditions, cells spread across the surface over time, and various cellular conditions can modulate the extent of cell spreading. This modulation of cell spreading thereby modulates the amount of force applied, thus making interpretation of strengthening results difficult. By confining the amount of area available to cells for spreading, we can maintain uniform cellular geometry across various conditions and time points. Moreover, the micropatterned substrates control the size and position of the adhesive area, an important parameter in interpreting contributions to adhesion strength (Gallant *et al.*, 2005).

We are able to generate adhesive islands for attachment of rounded, single cells using photolithography techniques in conjunction with microcontact printing of alkanethiol self-assembled monolayers on gold (Chapter 19 by Lee *et al.*, this volume; Gallant *et al.*, 2002; Fig. 2). Alkanethiols consist of a long-chain carbon backbone with a thiol group at one terminus and a functional group at the other terminus. These molecules self-assemble from solution onto gold surfaces (through strong coordination of S to Au) to form stable, well-packed, and ordered monolayers presenting the end-group of interest (Bain *et al.*, 1989; Ulman *et al.*, 1989). In our assay, well-defined arrays of CH_3-terminated alkanethiol ($HS-(CH_2)_{11}-CH_3$) circles or "islands" are stamped onto a gold-coated glass coverslips using a PDMS stamp (Chapter 19 by Lee *et al.*, this volume). The remaining exposed gold is then filled in with a triethylene glycol-terminated alkanethiol ($HS-(CH_2)_{11}-(EG)_3OH$), which resists protein adsorption and cell attachment. The patterned coverslip is coated with human plasma fibronectin (pFN), which adsorbs passively onto the

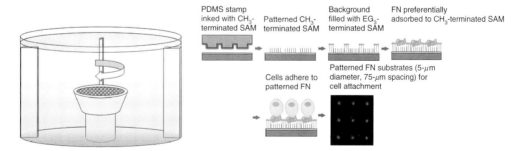

Fig. 2 Our adhesion strength assay incorporates a spinning disk device, which applies a well-defined range of hydrodynamic shear stress to cells. The surface for cell adhesion is micropatterned to restrict contact area and generate uniform cellular geometries. This technique provides reproducible measurements of the mean adhesion strength for a population of cells. (See Plate 24 in the color insert section.)

CH$_3$-terminated alkanethiol. The substrate is then blocked with 1% heat-denatured borine serum albumin (BSA) to prevent nonspecific cellular binding, and incubated in PBS to elute proteins that are weakly bound to the triethylene glycol surface. This process results in patterned areas of adsorbed pFN (Fig. 2) in an array of circular islands varying in diameter, currently from 2 to 20 μm, that are spaced 75-μm apart to promote single-cell attachment to each island. On 2-, 5-, and 10-μm islands, the island size is much smaller than the diameter of cell allowing cells to remain rounded. On 20-μm islands, the cells form a hemisphere over the island due to the circular pattern of FN. Cells are seeded on these patterned coverslips in complete media and uniformly adhere across the disk surface such that one cell attaches to one island and remains rounded. The uniform, rounded shape of the cell allows for consistent application of force across all cells on the surface and is used to determine the force applied at the cell–ECM bond as shown in Section VI. With an empirically derived cell density of ~200 cells/mm^2, cells uniformly distribute on patterned substrates placed in a tissue culture dish. Presently, we have patterned several cell types using this technique, including NIH3T3, MC3T3-E1, and C2C12 cell lines and primary fibroblasts.

Measurements of cell adhesion strength at various time points after plating yield a kinetic adhesion profile that shows an exponential rise to maximum [Eq. (3)], and fits the solution of a simple first-order differential equation that characterizes the biological strengthening process.

$$\tau_{50}(t) = \tau_{\infty}(1 - e^{-k_s t}) \qquad (3)$$

This curve fitting generates two parameters: τ_{∞} is equal to the maximum adhesion strength achieved at steady state and has units of shear stress, and k_s characterizes the strengthening rate and is given by the time required to reach 67% of the maximum strength (Fig. 3). Using these two parameters, we can determine if

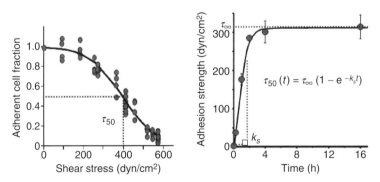

Fig. 3   The adhesion assay yields cell counts from 60 fields subject to a range of applied shear stresses. The adherent cell fraction profile is fit with a sigmoidal curve (left), and the shear stress at which 50% of the cells remain attached is considered the mean adhesion strength for the population of cells. Measurements of adhesion strength over time can be regressed using an exponential rise to maximum curve, to yield the steady state adhesion strength and the strengthening rate (right).

the experimental treatment modulates the steady state adhesion strength or the rate to reach this steady state strength.

B. Interpretation of Adhesion Strength Results

Changes in either steady state strength or strengthening rate provide information on the functional mechanism of a particular component or process in adhesion strengthening (Fig. 4). For instance, if the localization of a protein were to modulate preferentially the steady state strength, then this protein must in some way act to enhance the overall mechanical coupling of the ECM to the cell. This enhancement could in turn be due to an increased number of bonds, enhanced position/distribution of bonds, or enhanced coupling of bonds to the cytoskeleton (Section VI). Coupling of FA proteins and integrins has also been shown to modulate the affinity of the integrin–ECM bond. This modulation of affinity is likely to cause adhesion strengthening through an increase in the number of bonds, rather than an increase in the force required to disrupt individual bonds, because the enhanced affinity increases the probability of bond formation rather than the strength of each bond. However, proteins that modulate preferentially the strengthening rate likely do so by enhancing the rate of mechanical coupling. This enhancement could be due to an enhanced recruitment of strengthening molecules to the site of interaction or altered localization of bond positions for a more optimal distribution of load. It should be noted that enhancements of bond strength and formation rate are not mutually exclusive, as some adhesion components may be able to modulate both parameters. As an example of this type of analysis, increasing the number of bound integrins via an increase in FN surface density preferentially modulates steady state adhesion strength rather than the strengthening rate (Gallant *et al.*, 2005). This preferential modulation suggests that an increased

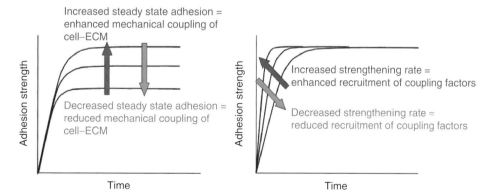

Fig. 4 Different responses of adhesion strengthening to molecular activity. Preferential modulation of either steady state strength or strengthening rate can give insight into the functional mechanism of a particular protein in adhesion strengthening.

density of FN–integrin bonds increases the overall adhesion strength by increasing the number of bonds rather than increasing the rate of bond formation or optimizing the position of the bonds. How an adhesive component modulates the strength of adhesion may in turn be determined by investigating changes in the size, position, and composition of adhesive complexes following the activity of that component.

V. Quantitative Biochemical Methods for Adhesion Analysis

Within adhesion complexes, we have identified two key factors that contribute significantly to mechanical strength: integrin binding and FA complexes (Gallant *et al.*, 2005). These factors contribute to strengthening through both increasing the number of molecules bound to the FA complex and changing the position and distribution of localized molecules. We have therefore developed biochemical techniques to quantify both of these contributing factors.

A. Quantification of Bound Integrin

We quantify the number of bound integrins using a cross-linking, extraction, and reversal technique (Garcia and Boettiger, 1999). This technique employs a membrane-impermeable, homobifunctional cross-linker to couple primary amine groups in the integrin and ECM ligand. Equivalent results have been obtained for sulfo-BSOCOES (13-Å spacer arm) and DTSSP (12-Å spacer arm) cross-linkers (both from Pierce, Rockford, Illinois). Taking advantage of the fact that most ECM proteins are detergent insoluble, the bulk of the cellular components, including unbound receptors, is then extracted using an ionic detergent (SDS). After washing, dithiothreitol (DTT) is used to cleave the disulfide bonds in the remaining cross-linkers thus releasing the bound integrins, which can be collected and quantified by Western blotting (Fig. 5).

B. Wet-Cleaving Assay for Localized FA Protein Quantification

We have modified the wet-cleaving technique developed by Brands and Feltkamp for quantification of the amount of specific FA proteins, for example vinculin and talin, localized to FA complexes (Brands and Feltkamp, 1988; Keselowsky and Garcia, 2005; Fig. 6). This wet-cleaving assay mechanically disrupts the cellular membrane to release and remove cytosolic elements that are not associated with adhesion complexes. It involves placing a piece of nitrocellulose membrane on top of cells, allowing nonspecific interactions between the nitrocellulose membrane and the dorsal cell membrane. The nitrocellulose membrane is then ripped from the substrate, thereby rupturing the cellular membrane and releasing cytosolic proteins and cellular melieu into a protease-inhibiting solution. The released cellular fraction is washed away and the materials that remain on the substrate

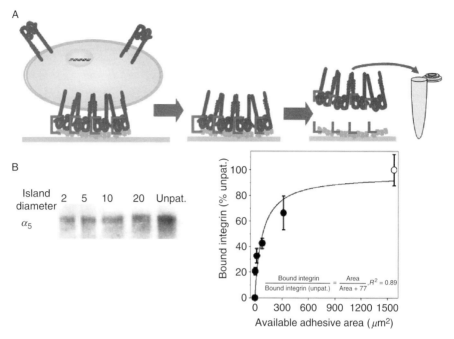

Fig. 5 Quantification of bound integrins using a cross-linking, extraction and reversal procedure. (A) This method involves addition of a reversible, cell-impermeable cross-linker to capture integrins bound to matrix proteins. The cellular milieu is then removed with an ionic detergent. The bound fraction is collected after reversal of the cross-linker and analyzed by Western blotting. (B) Quantification of Western blotting results is achieved using image intensity analysis. Representative data for cells patterned on different island sizes (●) and unpatterned surfaces (○) reveal that integrin binding increases nonlinearly with adhesive area. The relationship is accurately described by a hyperbolic curve (Gallant *et al.*, 2005). (See Plate 25 in the color insert section.)

are collected with a detergent-containing solution for quantification by Western blotting. By controlling the cleaving conditions (overlay time and volume of liquid at the cell–nitrocellulose interface), cells can be ruptured at different planes relative to the underlying surface. With an overlay time of 1 min, most cells rupture close to the cell–ECM interface.

C. Immunofluorescence Staining and Quantification

The distribution of bound integrins and localization of FA proteins to the adhesion area is visualized and quantified by immunofluorescence staining. Analysis of such images is particularly useful for quantifying the area occupied by adhesion components in cells on micropatterned substrates. Bound integrins are visualized by modifying the cross-linking and extraction technique such that after washing away the unbound cellular components, the remaining surface is blocked

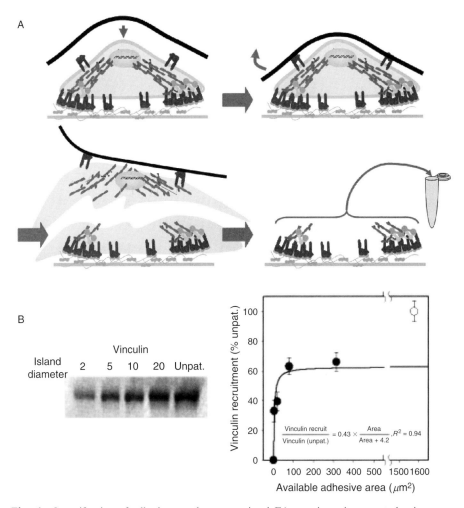

Fig. 6 Quantification of adhesive membrane-associated FA proteins using a wet-cleaving assay. (A) Using nonspecific interactions between nitrocellulose membrane and the dorsal cell membrane, the cell is mechanically disrupted, unbound cytosolic molecules are washed away, and substrate bound molecules are removed by detergent conditions and collected for quantification by Western blotting. (B) Quantification of Western blotting results is completed using image intensity analysis. The representative data for cells patterned on different island sizes (●) revealed that vinculin localization also increases nonlinearly with the adhesive area. The relationship is accurately described by a hyperbolic curve (Gallant *et al.*, 2005). Vinculin localization on unpatterned surfaces (○) is given as a reference. (See Plate 26 in the color insert section.)

and labeled with primary and secondary antibodies, without reversing the cross-linker. FA proteins are visualized using standard immunofluorescence staining techniques, including permeabilization of the membrane, fixation, blocking, and labeling with primary and secondary antibodies. Once the images have been collected using a fluorescence microscope, the amount of adhesion area occupied by FA molecules can be quantified using image analysis software. Our laboratory quantifies this value by defining an area of interest and calculating the area occupied by fluorescent molecules divided by the total area of interest using ImagePro Plus software (MediaCybernetics, Maryland). Immunofluoresence staining and biochemical quantification techniques provide complementary information, such as the total amount of adhesive components versus component localization and area of occupation, for mechanistic analysis of adhesion strengthening.

VI. Simple Mathematical Modeling of Adhesion Strengthening Mechanics

Mathematical modeling of adhesion strengthening mechanics can help interpret experimental results and expand our understanding of the process through inferring functional mechanisms that may generate these results. Our analysis is based on earlier models that predict the effects of focal contact formation on adhesion strength (Ward and Hammer, 1993). However, our analysis uses experimentally derived parameters, including adhesion strength, number of bonds, geometries of cell–substrate contacts, and FA areas, to allow direct comparisons between theoretical and experimental results. By modeling the density and distribution of bound integrins (both uncoupled and coupled to the cytoskeleton) within the contact area, we can calculate the overall adhesive forces that resist the applied hydrodynamic forces, in order to analyze systematically the contributions of different adhesive parameters to overall adhesion strength (Gallant and Garcia, 2007).

A. Resolving Forces for a Cell Under Hydrodynamic Shear

The force applied at the cell–ECM interface can be resolved using a force balance approach, which considers the static equilibrium of a cell attached to a micropatterned substrate under shear flow (Fig. 7A). On the basis of Newton's laws, forces applied to a stationary object must be balanced by counterforces applied by that object such that the sum of the forces and the sum of the moments must equal zero. Therefore, before a cell under shear flow detaches, a mechanical equilibrium exists where the applied hydrodynamic shear force (F_s) and torque (T_s) are balanced by horizontal (resultant F_{tan}) and vertical tensile (resultant F_T) bond forces and compressive forces (resultant F_c). Using the solution for F_s and T_s for a sphere in shear flow (Goldman *et al.*, 1967), F_T is determined to be the dominant force resisting the applied hydrodynamic loading (Fig. 7A). Assuming a peeling detachment mechanism in which bonds at the edge of attachment resist most of the detachment force relative to the center of the cell–ECM contact, F_T acts through a point at the periphery of the cell–substrate contact area. Therefore, the distance from the center of the contact area (C) to the point of application of F_T is approximately equal to the radius of the contact area. On the basis of these approximations, F_T is

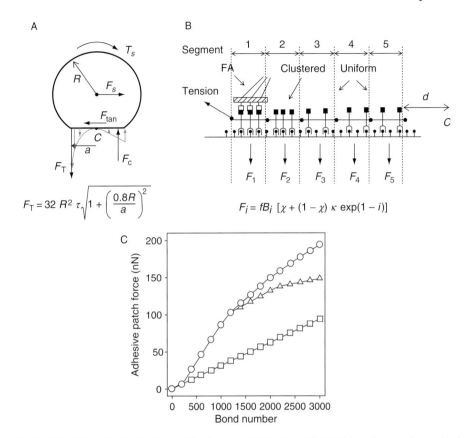

Fig. 7 Modeling of adhesive forces. The forces applied at the cell–ECM interface can be resolved using a free body diagram of a rounded cell under shear flow and simple mechanics. (A) On the basis of mechanical equilibrium, the applied shear stress is directly proportional to the dominant force resisting detachment, F_T. (B and C) this vertical tensile force can then be subject to a mathematical analysis of adhesion strengthening, to systematically dissect the contributions of different adhesive parameters to overall adhesion strength. As the bond number increases, adhesive patch force increases linearly for uniformly distributed bonds (□) and nonlinearly for clustered (△) and FA-associated (○) bonds, with FA association enhancing adhesive force by 30% over clustering. This 30% enhancement has been confirmed experimentally in adhesion strength measurements (Gallant *et al.*, 2005).

reduced to the expression in Eq. (4) and is directly proportional to the applied wall shear stress τ, which we use for the experimental measurement of adhesion strength.

$$F_T = 32R^2\tau\sqrt{1 + \left(\frac{0.8R}{a}\right)^2} \tag{4}$$

where R is the radius of the spherical adhesive object and a is the radius of the contact area.

B. Mathematical Analysis of Adhesion Strengthening Mechanics

The macroscopic force balance analysis yields values of the net force at the cell–ECM interface, which can then be integrated with a microscopic model to relate the applied force to the number, position, and state of the bound receptors that provide the forces to resist detachment. Due to the peeling detachment mechanism, the adhesive force resisting the applied hydrodynamic force, F_T, is generated by a small area of the total cell–ECM interface localized at the leading edge of the contact area. In our model, this "adhesive patch" is divided into segments (Fig. 7B), which contain the load-bearing, bound receptors. Three cases of bound state are considered:

1. Uniformly distributed bonds—bonds that are equally distributed among patch segments;
2. Clustered bonds—bonds localized to the outermost segment until a saturation number is reached (B_{max}) and then the next segment is filled;
3. FA-associated bonds—a fraction of clustered bonds (χ) that must fail simultaneously (as one rigid unit).

The force produced by each segment (F_i) is calculated using Eq. (4).

$$F_i = \phi B_i [\chi + (1 - \chi)\kappa \exp(1 - i)] \qquad (5)$$

where ϕ is the individual integrin–ligand bond strength, B_i is the number of bonds in segment i, κ modulates the exponential dependence of segment loading, and χ is the fraction of bonds associated with FA elements. The force and moment for all segments are added to calculate F_T, the adhesive patch force. Due to the exponential dependence of the adhesive patch force (ΣF_i) on the number of segments (i), the first five segments from the edge bear most of the load and the remaining segments contribute negligible resistant force. Therefore, the adhesive patch is located a distance d from the center of the contact area (C) (Fig. 2B).

Using experimentally determined values for the parameters in Eq. (4), we have derived quantitative relationships among adhesion strength, integrin binding, and FA size and position, as well as perform parametric simulations of the variables in the model, in order to understand the dependence of adhesion strength on these parameters (Gallant and Garcia, 2007). For instance, our analysis predicts that the coupling of FAs to bound integrins provides ~30% of the overall adhesion strength, with integrin binding over other segments providing the remaining balance (Gallant and Garcia, 2007; Gallant et al., 2005; Fig. 7C). Furthermore, this analysis shows that increasing the cell–ECM contact area (increasing d) increases adhesion strength proportionally and that the force required to detach an adhesive patch is consistently 200 nN, independent of the patch size (width of segments 1–5).

VII. Discussion

Incorporating experimental results with mechanical analyses provides unique insights into the adhesion strengthening process. This work provides a robust framework for the systematic and rigorous analysis of structure–function relationships in the regulation of cell adhesion strength. On the basis of this analysis, we have identified dominant mechanisms that regulate adhesion strength: (1) number of bound integrins, (2) position and distribution of bonds, (3) individual bond strength, and (4) coupling of integrins to cytoskeletal elements. We anticipate that this framework will be useful in elucidating mechanisms that regulate adhesive interactions in normal and pathological processes.

Acknowledgments

Funding was provided by National Institutes of Health (R01-GM065918) and an NSF Graduate Fellowship (KEM).

References

Bain, C. D., Troughton, E. B., Tao, Y.-T., Evall, J., Whitesides, G. M., and Nuzzo, R. G. (1989). Formation of monolayer films by the spontaneous assembly of organic thiols from solution onto gold. *J. Am. Chem. Soc.* **111,** 321–335.

Balaban, N. Q., Schwarz, U. S., Riveline, D., Goichberg, P., Tzur, G., Sabanay, I., Mahalu, D., Safran, S., Bershadsky, A., Addadi, L., and Geiger, B. (2001). Force and focal adhesion assembly: A close relationship studied using elastic micropatterned substrates. *Nat. Cell Biol.* **3,** 466–472.

Beningo, K. A., Dembo, M., Kaverina, I., Small, J. V., and Wang, Y. L. (2001). Nascent focal adhesions are responsible for the generation of strong propulsive forces in migrating fibroblasts. *J. Cell Biol.* **153,** 881–888.

Brands, R., and Feltkamp, C. A. (1988). Wet cleaving of cells: A method to introduce macromolecules into the cytoplasm. Application for immunolocalization of cytosol-exposed antigens. *Exp. Cell Res.* **176,** 309–318.

Chen, A., and Moy, V. T. (2000). Cross-linking of cell surface receptors enhances cooperativity of molecular adhesion. *Biophys. J.* **78,** 2814–2820.

Choquet, D., Felsenfeld, D. P., and Sheetz, M. P. (1997). Extracellular matrix rigidity causes strengthening of integrin-cytoskeleton linkages. *Cell* **88,** 39–48.

Danen, E. H. J., and Sonnenberg, A. (2003). Integrins in regulation of tissue development and function. *J. Pathol.* **201,** 632–641.

Dembo, M., and Wang, Y. L. (1999). Stresses at the cell-to-substrate interface during locomotion of fibroblasts. *Biophys. J.* **76,** 2307–2316.

DiMilla, P. A., Barbee, K., and Lauffenburger, D. A. (1991). Mathematical model for the effects of adhesion and mechanics on cell migration speed. *Biophys. J.* **60,** 15–37.

Doroszewski, J., Skierski, J., and Przadka, L. (1977). Interaction of neoplastic cells with glass surface under flow conditions. *Exp. Cell Res.* **104,** 335–343.

Engler, A. J., Griffin, M. A., Sen, S., Bonnemann, C. G., Sweeney, H. L., and Discher, D. E. (2004). Myotubes differentiate optimally on substrates with tissue-like stiffness: Pathological implications for soft or stiff microenvironments. *J. Cell Biol.* **166,** 877–887.

Furuta, Y., Ilic, D., Kanazawa, S., Takeda, N., Yamamoto, T., and Aizawa, S. (1995). Mesodermal defect in late phase of gastrulation by a targeted mutation of focal adhesion kinase, FAK. *Oncogene* **11,** 1989–1995.

Galbraith, C. G., Yamada, K. M., and Sheetz, M. P. (2002). The relationship between force and focal complex development. *J. Cell Biol.* **159,** 695–705.

Gallant, N. D., Capadona, J. R., Frazier, A. B., Collard, D. M., and Garcia, A. J. (2002). Micro-patterned surfaces to engineer focal adhesions for analysis of cell adhesion strengthening. *Langmuir* **18,** 5579–5584.

Gallant, N. D., and Garcia, A. J. (2007). Model of integrin-mediated cell adhesion strengthening. *J. Biomech.* **40,** 1301–1309.

Gallant, N. D., Michael, K. E., and Garcia, A. J. (2005). Cell adhesion strengthening: Contributions of adhesive area, integrin binding, and focal adhesion assembly. *Mol. Biol. Cell* **16,** 4329–4340.

Garcia, A. J. (2005). Get a grip: Integrins in cell-biomaterial interactions. *Biomaterials* **26,** 7525–7529.

Garcia, A. J., and Boettiger, D. (1999). Integrin-fibronectin interactions at the cell-material interface: Initial integrin binding and signaling. *Biomaterials* **20,** 2427–2433.

Garcia, A. J., Ducheyne, P., and Boettiger, D. (1997). Quantification of cell adhesion using a spinning disc device and application to surface-reactive materials. *Biomaterials* **18,** 1091–1098.

Garcia, A. J., and Gallant, N. D. (2003). Stick and grip: Measurement systems and quantitative analyses of integrin-mediated cell adhesion strength. *Cell Biochem. Biophys.* **39,** 61–73.

Garcia, A. J., Huber, F., and Boettiger, D. (1998a). Force required to break alpha5beta1 integrin-fibronectin bonds in intact adherent cells is sensitive to integrin activation state. *J. Biol. Chem.* **273,** 10988–10993.

Garcia, A. J., and Reyes, C. D. (2005). Bio-adhesive surfaces to promote osteoblast differentiation and bone formation. *J. Dent. Res.* **84,** 407–413.

Garcia, A. J., Takagi, J., and Boettiger, D. (1998b). Two-stage activation for alpha5beta1 integrin binding to surface-adsorbed fibronectin. *J. Biol. Chem.* **273,** 34710–34715.

George, E. L., Georges-Labouesse, E. N., Patel-King, R. S., Rayburn, H., and Hynes, R. O. (1993). Defects in mesoderm, neural tube and vascular development in mouse embryos lacking fibronectin. *Development* **119,** 1079–1091.

Giannone, G., Jiang, G., Sutton, D. H., Critchley, D. R., and Sheetz, M. P. (2003). Talin1 is critical for force-dependent reinforcement of initial integrin-cytoskeleton bonds but not tyrosine kinase activation. *J. Cell Biol.* **163,** 409–419.

Goldman, A. J., Cox, R. G., and Brenner, H. (1967). Slow viscous motion of a sphere parallel to a plane wall. 2. Couette flow. *Chem. Eng. Sci.* **22,** 653–660.

Hubbell, J. A. (1999). Bioactive biomaterials. *Curr. Opin. Biotechnol.* **10,** 123–129.

Hynes, R. O. (2002). Integrins: Bidirectional, allosteric signaling machines. *Cell* **110,** 673–687.

Jin, H., and Varner, J. (2004). Integrins: Roles in cancer development and as treatment targets. *Br. J. Cancer* **90,** 561–565.

Keselowsky, B. G., and Garcia, A. J. (2005). Quantitative methods for analysis of integrin binding and focal adhesion formation on biomaterial surfaces. *Biomaterials* **26,** 413–418.

Lauffenburger, D. A., and Horwitz, A. F. (1996). Cell migration: A physically integrated molecular process. *Cell* **84,** 359–369.

Lawrence, M. B., McIntire, L. V., and Eskin, S. G. (1987). Effect of flow on polymorphonuclear leukocyte/endothelial cell adhesion. *Blood* **70,** 1284–1290.

Lotz, M. M., Burdsal, C. A., Erickson, H. P., and McClay, D. R. (1989). Cell adhesion to fibronectin and tenascin: Quantitative measurements of initial binding and subsequent strengthening response. *J. Cell Biol.* **109,** 1795–1805.

Mammoto, A., Huang, S., Moore, K., Oh, P., and Ingber, D. E. (2004). Role of RhoA, mDia, and ROCK in cell shape-dependent control of the Skp2-p27kip1 pathway and the G1/S transition. *J. Biol. Chem.* **279,** 26323–26330.

McBeath, R., Pirone, D. M., Nelson, C. M., Bhadriraju, K., and Chen, C. S. (2004). Cell shape, cytoskeletal tension, and RhoA regulate stem cell lineage commitment. *Dev. Cell* **6,** 483–495.

McClay, D. R., Wessel, G. M., and Marchase, R. B. (1981). Intercellular recognition: Quantitation of initial binding events. *Proc. Natl. Acad. Sci. USA* **78,** 4975–4979.

Micoulet, A., Spatz, J. P., and Ott, A. (2005). Mechanical response analysis and power generation by single-cell stretching. *Chemphyschem.* **6**, 663–670.

Mohandas, N., Hochmuth, R. M., and Spaeth, E. E. (1974). Adhesion of red cells to foreign surfaces in the presence of flow. *J. Biomed. Mater. Res.* **8**, 119–136.

Munson, B. R., Young, D. F., and Okiishi, T. H. (2006). "Fundamentals of Fluid Mechanics." John Wiley and Sons, Inc., Hoboken, NJ.

Palecek, S. P., Horwitz, A. F., and Lauffenburger, D. A. (1999). Kinetic model for integrin-mediated adhesion release during cell migration. *Ann. Biomed. Eng.* **27**, 219–235.

Palecek, S. P., Loftus, J. C., Ginsberg, M. H., Lauffenburger, D. A., and Horwitz, A. F. (1997). Integrin-ligand binding properties govern cell migration speed through cell-substratum adhesiveness. *Nature* **385**, 537–540.

Palecek, S. P., Schmidt, C. E., Lauffenburger, D. A., and Horwitz, A. F. (1996). Integrin dynamics on the tail region of migrating fibroblasts. *J. Cell Sci.* **109**(Pt. 5), 941–952.

Polte, T. R., Eichler, G. S., Wang, N., and Ingber, D. E. (2004). Extracellular matrix controls myosin light chain phosphorylation and cell contractility through modulation of cell shape and cytoskeletal prestress. *Am. J. Physiol., Cell Physiol.* **286**, C518–C528.

Reinhart-King, C. A., Dembo, M., and Hammer, D. A. (2005). The dynamics and mechanics of endothelial cell spreading. *Biophys. J.* **89**, 676–689.

Sheetz, M. P., Felsenfeld, D. P., and Galbraith, C. G. (1998). Cell migration: Regulation of force on extracellular-matrix-integrin complexes. *Trends Cell Biol.* **8**, 51–54.

Stephens, L. E., Sutherland, A. E., Klimanskaya, I. V., Andrieux, A., Meneses, J., Pedersen, R. A., and Damsky, C. H. (1995). Deletion of beta 1 integrins in mice results in inner cell mass failure and peri-implantation lethality. *Genes Dev.* **9**, 1883–1895.

Tan, J. L., Tien, J., Pirone, D. M., Gray, D. S., Bhadriraju, K., and Chen, C. S. (2003). Cells lying on a bed of microneedles: An approach to isolate mechanical force. *Proc. Natl. Acad. Sci. USA* **100**, 1484–1489.

Ulman, A., Eilers, J. E., and Tillman, N. (1989). Packing and molecular-orientation of alkanethiol monolayers on gold surfaces. *Langmuir* **5**, 1147–1152.

Walker, J. L., Fournier, A. K., and Assoian, R. K. (2005). Regulation of growth factor signaling and cell cycle progression by cell adhesion and adhesion-dependent changes in cellular tension. *Cytokine Growth Factor Rev.* **16**, 395–405.

Ward, M. D., and Hammer, D. A. (1993). A theoretical analysis for the effect of focal contact formation on cell-substrate attachment strength. *Biophys. J.* **64**, 936–959.

Wehrle-Haller, B., and Imhof, B. A. (2003). Integrin-dependent pathologies. *J. Pathol.* **200**, 481–487.

Wilson, C. J., Clegg, R. E., Leavesley, D. I., and Pearcy, M. J. (2005). Mediation of biomaterial-cell interactions by adsorbed proteins: A review. *Tissue Eng.* **11**, 1–18.

Wozniak, M. A., Desai, R., Solski, P. A., Der, C. J., and Keely, P. J. (2003). ROCK-generated contractility regulates breast epithelial cell differentiation in response to the physical properties of a three-dimensional collagen matrix. *J. Cell Biol.* **163**, 583–595.

Xu, W., Baribault, H., and Adamson, E. D. (1998). Vinculin knockout results in heart and brain defects during embryonic development. *Development* **125**, 327–337.

Zamir, E., and Geiger, B. (2001). Molecular complexity and dynamics of cell-matrix adhesions. *J. Cell Sci.* **114**, 3583–3590.

Zimerman, B., Volberg, T., and Geiger, B. (2004). Early molecular events in the assembly of the focal adhesion-stress fiber complex during fibroblast spreading. *Cell Motil. Cytoskeleton* **58**, 143–159.

CHAPTER 15

Studying the Mechanics of Cellular Processes by Atomic Force Microscopy

Manfred Radmacher

Institute of Biophysics
University of Bremen
Bremen 28334, Germany

Abstract

The mechanical properties of cells are important for many cellular processes like cell migration, cell protrusion, cell division, and cell morphology. Depending on cell type, the mechanical properties of cells are determined mainly by the cell wall or the interior cytoskeleton. In eukaryotic cells, the stiffness is mainly determined by the cytoskeleton, which is made of several polymeric networks, including actin, microtubuli, and intermediate filaments. To study the mechanical properties of living cells at a subcellular resolution is of outmost importance to understanding the cellular processes mentioned above. One option is to use the atomic force microscopy (AFM) to measure the cell's elastic properties locally. By obtaining force curves, that is measuring the cantilever deflection while the tip is brought in contact and retracted cyclically, effectively the loading force indentation relation is measured. The elastic or Young's modulus can be calculated by applying simple models, like the Hertz model for spherical or parabolic indenters or Sneddon's modification for pyramidal indenters.

I. Introduction

The atomic force microscopy (AFM) is just 20-year old but has already become an important tool for studying biological structures and processes on cellular and molecular scales. AFM allows one to image live cells at a resolution significantly higher than that of light microscopy. Although its resolution is lower than the resolution of the electron microscope, the latter is limited to fixed samples. AFM can also provide unique and increasingly important information that is not available with other techniques, including true topographic information such as sample height, as well as mechanical and other physical properties of cells and cell–substrate adhesions.

The combination of positioning a very sharp tip at nanometer precision and applying small forces down to piconewtons allows experiments to be conducted that would otherwise be impossible or exceedingly difficult. In particular, the application of AFM to study mechanical properties of cells has opened a new route for addressing some long-standing cell biological questions. Here we present some of these applications and provide background information on current instrumentation and operating modes.

The AFM was invented in 1986 (Binnig et al., 1986) as an "offspring" of the scanning tunneling microscope (STM) (Binnig and Rohrer, 1982). The main difference between the two is that AFM allows the investigation of electrically nonconducting samples. AFM can, therefore, be operated in liquid environments (Drake et al., 1989), which makes it compatible with the physiological requirements of live cell specimen. Indeed, biological samples for AFM need not be chemically fixed or treated in any way. Some of the first examples of AFM imaging of native processes on a cellular scale included the activation of platelets (Fritz et al., 1993, 1994) and the protrusion of lamellipodia (Henderson et al., 1992). Investigations of molecular-scale processes followed soon afterward, and included observations of the activity of lysozyme (Radmacher et al., 1994b) and the degradation of DNA (Bezanilla et al., 1994). These examples also demonstrate that, besides pure topographic imaging, AFM can provide information on surface properties such as friction, elasticity, and adhesion. Indeed, force modulation provided an early basis for imaging or mapping mechanical properties of soft samples: by applying a modulation frequency of several tens of kilohertz and monitoring the response of the cantilever at this frequency, it is possible to infer various mechanical properties of the sample (Radmacher *et al.*, 1993). Despite the technical simplicity, however, it proved difficult to obtain reproducible and quantifiable data on cell samples using this approach. The currently preferred method is to generate pixel-by-pixel force curves by recording cantilever deflection while bringing the sample in and out of contact. This mode allows reproducible and quantitative measurements of elastic properties of cells and other soft samples. We will focus on this mode after a brief introduction of the instrument.

II. Instrumentation and Operation Modes

A. Principal Components

There are many different designs of AFMs, but the main parts of almost any AFM are similar: (1) a sharp tip mounted on a flexible cantilever that acts as a spring, (2) a laser diode, whose light is focused on the very end of the cantilever beam, (3) a position-sensitive photodiode that detects the laser beam reflected by the cantilever, and (4) a piezo device for positioning the sample relative to the tip in three orthogonal dimensions, x–y–z. The most critical component here is the scanning piezo (Fig. 1). Since cells are relatively large structures, several dozens

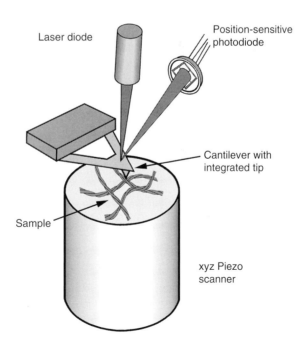

Fig. 1 Principal components of an AFM: cantilever with integrated tip, laser diode and photodiode for deflection sensing, and a piezo device for positioning the sample relative to the tip.

of micrometers in diameter and several micrometers in height, the scanning range must be appropriate for this size scale. Typical instruments used for cell imaging exhibit a lateral scan range of about 100 μm and a vertical scan range of about 10 μm or more. At these large extensions, piezos exhibit a variety of nonlinearities, creep, and hysteresis (dependence on history), and so on. Reproducible positioning is only possible when the position of the piezo is measured via a sensor, typically a capacitive sensor, and controlled with electronic feedback.

B. Cantilevers

Cantilevers with integrated tips are available in a large variety of materials and dimensions. Depending on the manufacturing process, cantilever materials may be amorphous silicon nitride or crystalline silicon. For the investigation of cells, the most important factor is the stiffness of the cantilever. Since cells are very soft and delicate, loading forces must be minimized, which implies the use of softest cantilevers available. Typical force (spring) constants of cantilevers used for imaging live cells range from 10 to 30 mN/m. Note that a force constant $k_c = 10$ mN/m equals 10 pN/nm, such that a deflection of 1 nm corresponds to a force of 10 pN, which is only about twice the force generated by a typical motor enzyme such as myosin. Importantly, k_c needs to be accurately known to make accurate force measurements, while AFM cantilevers are made on a large scale by silicon wafer methods that limit the accuracy of the nominal force constant reported by manufacturers. Therefore, k_c should be determined for each cantilever by methods such as those described below (Fig. 2).

Since soft cantilevers usually have long cantilever legs (typically 200–320 μm), they possess a low resonance frequency, which may be expressed as $\omega \sim (k_c/m_{\mathrm{eff}})^{1/2}$ where m_{eff} is the effective mass of the system. Whereas the resonance frequency in air can be deduced approximately from the geometrical dimensions (length, width,

Fig. 2 Scanning electron micrographs (SEM) of typical AFM cantilevers (left) and a close-up view of the tip (right). This chip offers a variety of cantilevers, which differ in geometrical dimensions. The longest triangular one, whose legs have a length of 320 μm, a width of 22 μm, and a thickness of 0.6 μm, has a force constant of only 10 mN/m.

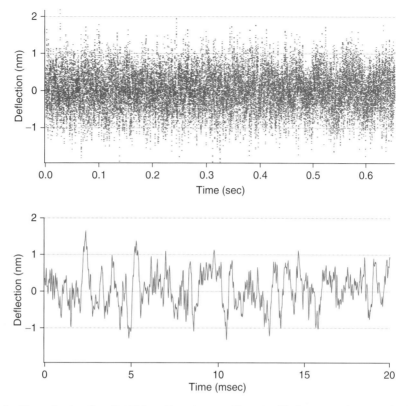

Fig. 3 Thermal noise of a soft AFM cantilever immersed in water. The lower trace is a zoom-in of the upper data set. Tip excursion shows a timescale of 1 msec or less, which matches nicely the resonance frequency as presented in Fig. 4. The amplitude of tip fluctuation in this case is 0.5 Å (rms). From Eq. (3), we can estimate a force constant of about 15 pN/nm.

thickness of the legs) and the physical properties of the material (density and Young's modulus) that combine to give k_c; in water, the resonance frequency is much lower. This is due to the fact that the cantilever drags some amount of water when being moved, drastically increasing its effective mass m_{eff}. Figure 3 shows the vibration of a soft cantilever water due to thermal noise, that is random motion driven by Brownian motion of the surrounding medium. From the amplitude of this random motion, the force constant of the cantilever can be determined. According to the equipartition theorem of statistical thermodynamics, each degree of freedom (or "mode") of any system possesses an average energy of:

$$\langle E \rangle = 0.5 k_B T \qquad (1)$$

where the brackets $\langle\rangle$ denote time average, k_B is Boltzmann's constant, and T is the absolute temperature. Assuming that the cantilever undergoes mainly bending vibrations at the resonance frequency (the principal mode), we can estimate the force constant from the fluctuations in tip position:

$$\langle E \rangle \sim 0.5 k_c \langle d^2 \rangle \qquad (2)$$

where $\langle d^2 \rangle$ is the mean squared displacement of the tip. By combining Eqs. (1) and (2), we obtain an approximate expression for the force constant k_c (Butt and Jaschke, 1995; Sader *et al.*, 1995):

$$k_c = \frac{k_B T}{\langle d^2 \rangle} \qquad (3)$$

However, a cantilever has more than one degree of freedom exhibiting bending and buckling modes at correspondingly higher frequencies. The ground frequency and higher harmonics are analogous to those of a guitar string and other oscillatory systems. Since the AFM detects contributions from all of the harmonics, the force constant from Eq. (3) represents an underestimate due to overestimation of displacements. To obtain a more accurate value of the force constant, we need to look at the power spectral density, or short power spectra, of the deflection signal (as a function of frequency, where the power of vibration is calculated as the square of the amplitude of vibration; Fig. 4). These plots show the intensity of vibration at the resonance frequency and at higher harmonics in both air and water. The energy of vibration at the resonance frequency is calculated either by integration of the peak or by fitting a model function. This procedure gives a force constant of about 20 pN/nm in our example, in comparison to 15 pN/nm obtained by the simple estimate from Eq. (3) and Fig. 3.

We can also see from the plots of Fig. 4 that the resonance frequency of a cantilever drops drastically when it is immersed in water (or other liquids), mainly due to the added effective mass of the water. This added mass results in a decrease of the resonance frequency, typically by a factor of about 3 (Walters *et al.*, 1996a). In our case, the resonance frequency drops from around 6.5 kHz down to 1.2 kHz. The corresponding increase in the characteristic response time of the cantilever limits the temporal resolution of the AFM; therefore, the acquisition time needed for imaging a cell or making other measurements is much longer in water than in air. Higher resonance frequencies can be achieved by reducing the dimensions of cantilevers that move less water and decrease m_{eff}, which has been shown with prototypes (Schäffer *et al.*, 1997; Walters *et al.*, 1996b). These short and soft cantilevers have become available just recently; however, due to its high price and dependence on suitable instrumentation, they have not been used widely yet.

C. Combination of Optical Microscopy and AFM

For the imaging of live cells, the combination of optical microscopy and AFM is very useful (Vesenka *et al.*, 1995). Optical microscopy allows positioning of the AFM tip onto a cell or even onto a particular region of a cell. The optical image

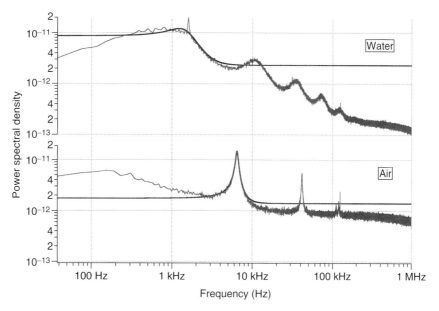

Fig. 4 Power spectral density of thermal vibration of the same cantilever in air and water. The resonance frequency in air is around 6.5 kHz, which can be clearly seen as a peak in the power spectral density of the deflection signal. Higher harmonic peaks can also be seen. When immersing this cantilever in water, the spring constant is not affected. However, since the cantilever drags water while fluctuating, the effective mass of the cantilever is largely increased. Consequently, the resonance frequency drops, in this case to 1.2 kHz. The blue curve is a fit of the data, for calculating the energy of vibration at these respective peaks. The resulting force constant is 20 pN/nm.

also allows the integrity and morphology of the cell to be monitored during the investigation by AFM. To increase contrast, it is very helpful to be able to use phase-contrast or differential interference-contrast optics (Lugmaier *et al.*, 2005), which has become available in commercial combined instruments (Figs. 5 and 6).

III. Operating Modes

A. Imaging Modes

In the simplest mode of operation, the contact mode, the AFM tip is brought into direct physical contact with the sample, while the cantilever deflection is monitored. Raster scanning the sample under the tip while holding the cantilever base height constant is called *constant height mode* and is generally only used for very flat and rigid samples, for example when looking at a glass surface or a crystal surface like mica. On soft samples with large height differences, such as with cells,

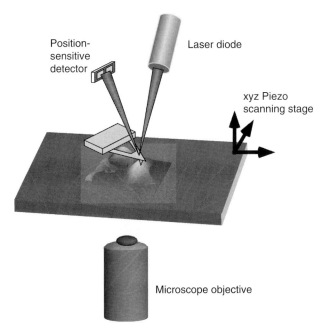

Fig. 5 Schematics of combined AFM and optical microscopy. The sample, a cell attached to a glass coverslip or a plastic Petri dish, is mounted on an xyz piezo transducer for positioning in AFM. Parallel optical images are obtained by a standard phase-contrast microscope.

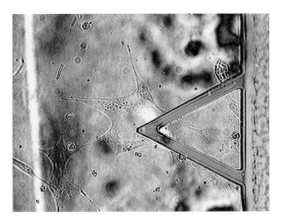

Fig. 6 Optical image through combined optical microscope and AFM. The cantilever has been positioned on a cluster of fibroblastic cells. The width of the cantilever legs is 32 μm.

the constant height mode will result in large changes of the cantilever deflection, corresponding to large changes in loading force. Such large forces often damage cells. To avoid this problem, a feedback is employed to adjust the sample height such that the cantilever deflection is held constant. This mode is then called *constant deflection mode*. However, due to the limited responsiveness of the feedback, especially in the mechanical transduction path (piezo transducers), some residual fluctuations in the deflection signal are still present.

Therefore, the true height information of the sample is separated in two image channels. The output of the feedback, which corresponds to the height signal, shows the overall topography of the sample. The fine details in topography that are not compensated by the height feedback signal generate residual deflections of the cantilever. This mode of operation is called *error signal mode* (Putman *et al.*, 1992). The separation of information in two channels can be seen in the parallel height and deflection images of Fig. 7. The overall height of the cell is 3–4 μm, whereas the residual fluctuations in the cantilever deflection correspond to about 100 nm.

By adding the height signal (Fig. 8A) and the simultaneously captured deflection signal (Figs. 8B and 9), we obtain an image reflecting the true height of the cell (Fig. 8C). A simulated deflection image may be constructed by calculating the spatial derivative of the true height in the fast scan direction (Fig. 8D), which strongly resembles the deflection data (Fig. 8B).

The modes described above are generally referred to as contact modes ("DC" modes). Alternative to contact modes are imaging modes where the sample height is modulated ("AC" modes), such that the tip is periodically retracted from the sample. This minimizes lateral forces and is very helpful when imaging single molecules (proteins, DNA, and so on) adsorbed on flat substrates. The simplest AC mode is called tapping mode. In tapping mode, the cantilever height is modulated at its resonance frequency, which is several hundred kilohertz for stiff

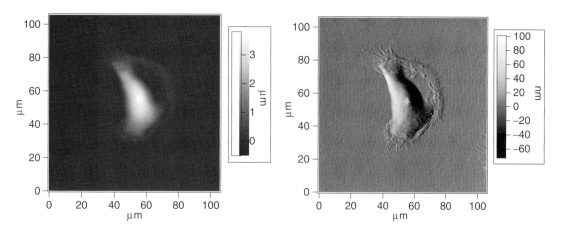

Fig. 7 AFM image of an NRK fibroblastic cell. The left image shows the height signal, which corresponds to the output of the feedback circuit that tries to keep the deflection signal constant. The right-hand side is the deflection signal, which corresponds to the residual error due to the finite response time of the feedback circuit. As a result, we obtain two images, one reflecting the overall topography and the other reflecting the fine corrugations on top of the height signal.

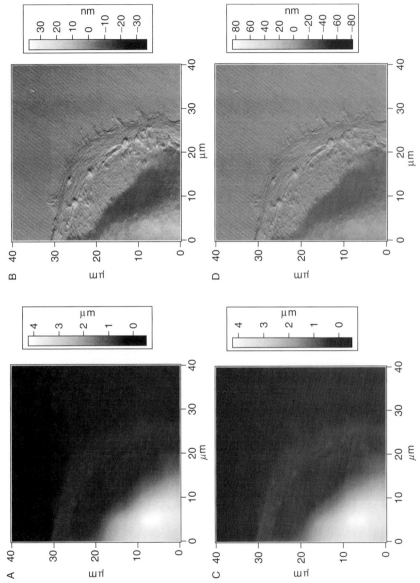

Fig. 8 The top row shows raw data as captured from the AFM. Panel (A) corresponds to the height image, panel (B) shows the deflection image. In panel (C), these two image channels have been added to create the true topography. However, this image is virtually indistinguishable from the raw height image as provided by the microscope. By differentiating this image in x-direction, which corresponds to the time axis during data capturing, we can create an image, which resembles strongly the deflection image (panel D).

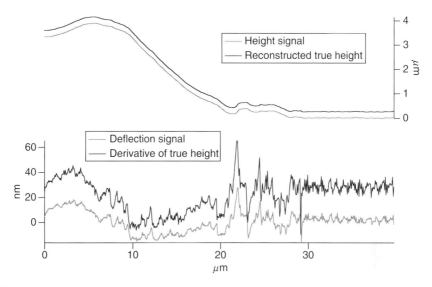

Fig. 9 Line sections through the images of Fig. 8. The height signal and the true height, created by adding height and deflection signals, are virtually indistinguishable, while the spatial derivative of the true height signal resembles to a large extent the deflection signal.

tapping cantilevers used for imaging in air. When the tip approaches the sample, the amplitude of cantilever motion is suppressed, which can be used as the input signal for distance feedback, just as the deflection signal is used in contact mode. In liquids, especially when using soft cantilevers, the resonance frequency of the cantilever is not very well defined and can only be detected when looking at the thermal motion of the cantilever as discussed earlier (Fig. 4). When actively driving the cantilever, as in tapping mode, the apparent resonances are dominated by the cantilever holder or the fluid in the fluid chamber. Nevertheless, the phrase "tapping mode in liquids" refers to driving a soft cantilever at a frequency of several kilohertz and using the amplitude response as the input of feedback (Hansma *et al.*, 1994; Putman *et al.*, 1994). Variants of this principle have been suggested like the pulsed force mode, which operates at a lower driving frequency of around 100 Hz. It can also be understood as running and analyzing fast force curves on the fly (see next paragraph). Pulsed force mode has been applied for imaging molecules or viruses, but not yet for cells.

B. Force Curve Mode

The measurement and quantification of interaction forces between an AFM tip and the sample is referred to as *force curve mode*. In a *force curve*, the lateral position of the tip is not changed, that is scanning is turned off, while the sample height is ramped such that the tip is brought in and out of contact and the deflection of the cantilever is monitored (Fig. 10). Since the cantilever deflection d and the loading force F are related via the force constant of the cantilever k_c, the force can be read directly from the plot showing deflection versus sample height (Fig. 11).

$$F = k_c d \qquad (4)$$

The force curve provides information on adhesion, molecular interactions (Florin et al., 1994), electrostatic interactions (Rotsch and Radmacher, 1997), stiffness (Radmacher et al., 1995), hardness (Almqvist et al., 2001), and other properties of the samples.

C. Force Volume Mode

By repositioning the tip to obtain force curves over different regions of the sample, it is possible to map interaction forces as a function of position. This mode, called force volume mode (Radmacher et al., 1994a), allows the mapping of electrostatic forces between the tip and sample (Rotsch and Radmacher, 1997) or the elastic properties of cells (Hofmann et al., 1997). It has also been used to distinguish between cells by their respective adhesion properties (Grandbois et al., 2000).

IV. Investigations of Live Cells

A. Imaging

The great potential of AFM to follow cellular dynamics at high resolution was recognized for over a decade (Fritz et al., 1993, 1994; Henderson et al., 1992; Schabert et al., 1994; Schoenenberger and Hoh, 1994), but advances have been hampered by the delicate nature of cells and damages due to forces applied by the AFM tip. Many studies have been performed after chemically fixing cells, which results in a rigidification of cells (Braet et al., 1997) and loss of the main advantages of AFM—the ability to follow the dynamics of cells such as protrusion of lamellipodia in response to chemical triggers (Rotsch et al., 2001) or the beating of myocyte cells (Domke et al., 1999). To overcome this limitation, an elegant study by Henderson et al. (Schaus and Henderson, 1997) showed that low forces and blunt tips minimize sample damage and allow imaging of live cells (Fig. 12).

B. Measuring Stiffness

The stiffness (perhaps better referred to as softness) of cells strongly influences topographic imaging by AFM, but it is also a physical characteristic of great intrinsic interest. The force curve mode of AFM, as explained above, may be employed to measure the stiffness of a cell.

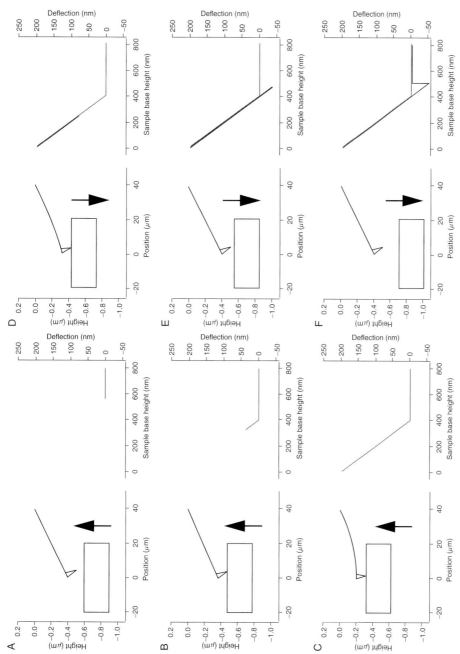

Fig. 10 Animation of data collection during a force curve. The tip first approaches the sample (A). Since the tip is not in contact with the sample yet, the cantilever deflection is constant. When the tip touches the sample (B), the cantilever starts to deflect and the deflection is proportional to the sample movement (C). The sample is then retracted, as indicated by the blue curve in (D), which shows the deflection during retraction. Ideally, the deflection during retraction remains identical to the deflection during approach at the same sample position. Here the blue curve has been offset by a tiny amount, to show both curves at the same time. However, if there is adhesion between sample and tip, the tip will stick to the sample beyond the point of contact (E), until it finally breaks free again and the deflection returns to zero (F).

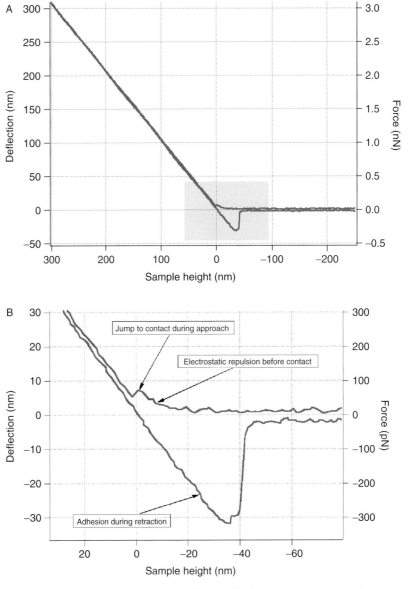

Fig. 11 Force curve on a glass slide. The cantilever deflection (left *y*-axis) is plotted versus sample height. The right *y*-axis gives the corresponding force, using a cantilever force constant of 10 mN/m. Panel (A) shows the three main parts of the force curve: deflection is constant before the tip makes contact (at negative sample heights), deflection is proportional to sample height (at positive sample heights), and a strong adhesion during retraction. More details can be seen when zooming into the region around the point of contact. In this case, the tip is deflected away from the sample before contact (due to electrostatic repulsion), but finally jumps into contact, due to strong van der Waals forces. These van der Waals forces cause also the large hysteresis during retract, which can be used to calculate the adhesion force (around 300 pN in this case).

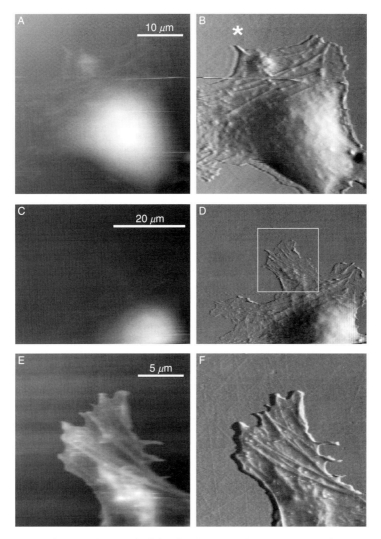

Fig. 12 Consecutive AFM images of a living fibroblast. Panels show three sets of data, each consisting of two simultaneously captured frames of height (A, C, E) and deflection images (B, D, F). The height images show the overall topography and the deflection images show the fine details of this topography. The first set of images show the extension of three leading lamellipodia (A, B). Several stress fibers extend perpendicular to the cell's extending edge and into the lamellipodia. These stress fibers can be clearly seen by zooming into the middle lamellipodium (C, D, corresponding to the boxed region in B). This lamellipodium retracted (F, asterisk), after scanning several times (four frames) and taking a force map (8000 force curves), as confirmed after zooming out of the region (E, F).

On a soft sample, the tip compresses the sample to some extent due to the application of the loading force. Therefore, the deflection d of the cantilever will not be proportional to the sample height z but deviates by the sample indentation δ

$$d = z - \delta \qquad (5)$$

If there is no sample compression, the indentation is 0 and Eq. (5) reduces to $d = z$: the proportionality of sample height and indentation as explained above for stiff samples. Intuitively, it is clear that the force needed to achieve a given indentation depends not only on the material's stiffness but also on the contact area between the tip and sample—the larger the contact area, the smaller the applied pressure and the indentation. In the case of a pyramidal tip, the contact area increases with increasing indentation and the relation between loading force and indentation is nonlinear.

The simplest model for describing the elastic response of a soft material to an indenter or "punch" is based on the work of Heinrich Hertz for spherical indenters (Hertz, 1882), which has been extended to the case of a conical punch by Sneddon (1965). A concise treatment can be found in books on contact mechanics (Johnson, 1994). In the case of a conical punch, the loading force can be calculated from the material properties (Young's modulus, E and Poisson ratio, v), the geometry of the punch (opening angle of the conical punch, α), and the indentation δ:

$$F = \frac{2}{\pi} \tan(\alpha) \frac{E}{1 - v^2} \delta^2 \qquad (6)$$

For a parabolic punch a similar relation can be derived, here R is the radius of curvature of the punch. This expression can also be used as an approximation for a spherical punch of radius R for small indentations ($\delta < R/10$):

$$F = \frac{4}{3} \frac{E}{1 - v^2} \delta^{3/2} \sqrt{R} \qquad (7)$$

The Poisson ratio v can have values between 0 and 0.5. A value of 0.5 corresponds to an incompressible material, which is a reasonable assumption for a well-hydrated soft polymeric gel such as the cytoskeleton.

These equations assume that the sample is homogeneous, isotropic, and infinitely thick. Strictly speaking, none of these assumptions necessarily applies to cells. The passive mechanical responses of cells are determined mainly by three cellular components: the cell membrane or cell wall, the cytoskeleton, and the cytosol. In the case of bacteria or plant cells, the cell wall is very stiff and is the determining component of cellular stiffness, while the contributions of cytoskeleton and cytosol can be neglected. The mechanical model of a bacteria, therefore, is a thin shell, as has been derived theoretically and shown experimentally (Arnoldi et al., 2000; Boulbitch, 1998). In the case of eukaryotic cells, the cell membrane is very soft and can be disregarded in most mechanical analyses. An exception is red blood cells, where elastic properties of the membrane have a significant influence on AFM measurements mainly due to volume constraints (Sen et al., 2005). For most other eukaryotic cells, the cytoskeleton, composed of various filamentous proteins—mainly actin, but also microtubules and intermediate filaments—and associated proteins, control the molecular architecture and mechanical properties. Especially in the case of actin cytoskeleton, numerous accessory proteins are known to control the lengths of filaments, cross-link the filaments, anchor the filaments to the membrane, induce the formation of filament bundles including stress fibers, and link the filaments to cell–substrate adhesion sites (Alberts and Lewis, 2002).

The cytoskeleton is a polymeric, fibrous network, which will not appear homogenous at a length scale comparable to or smaller than the mesh size (a few tens of nanometers). Additionally, the mechanical response of the cytoskeleton will not be isotropic in the presence of stress fibers or attachments to the membrane. Furthermore, sample thickness will be an issue for any estimate of mechanical properties, as a cell is never infinitely thick, and the use of equation for conical punch (6) introduces additional errors when applied to pyramidal AFM tips. Nevertheless, it proves instructive to examine the Hertzian-type relationships between loading force and indentation, as it turns out that the mechanical behavior of cells often follows closely the simple models of Eqs. (6) and (7). Figure 13 shows the prediction from the Hertz model for conical tips under typical conditions found in experiments with cells.

For soft samples, the Hertz model predicts a nonlinear force curve in the contact region as shown in Fig. 14. From this graph, the indentation and the loading force (deflection times force constant of cantilever) can be read directly. Typical data from cells are shown in Fig. 15. For small loading forces, the measured data fit closely the Hertz model, allowing the determination of a Young's modulus (Hofmann et al., 1997; Radmacher et al., 1996), despite the possible deviations from assumptions of the Hertz model as discussed above.

Because of the limitations of applicability of the Hertz model, other ways of quantifying mechanical responses of cells have been proposed such as calculating the area under the force curve (A-Hassan et al., 1998). However, there has been no simple extension of the Hertz model that describes more simply and appropriately the elastic response of cells, although there are ongoing efforts, both experimental and theoretical, to advance along this direction (Kasas et al., 2005; McElfresh et al., 2002; Rico et al., 2005).

A point that deserves further elucidation concerns the effect of sample thickness, which can be seen in the force data taken on a thin, flat lamellipodium (Fig. 16). As the indentation approaches the thickness of the region, no further compression is possible. Hence, the deflection becomes proportional to the sample height as on a stiff sample. Effectively, the cantilever senses only the stiffness of the underlying substrate. Not surprisingly, a Hertz fit only follows the data at very low loading forces, and it is questionable even under this condition how relevant the Hertz model fits can be. This effect has been characterized experimentally by taking force curves on thin and soft gelatin films (Domke and Radmacher, 1998), showing that the apparent stiffness indeed depends on sample thickness even if the material is homogeneous. Stiffness data on cells, particularly in flat regions, will therefore always reflect the sample thickness to some degree. Nonetheless, relative changes in

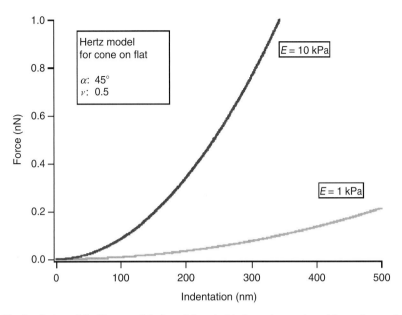

Fig. 13 Prediction of the Hertz model of a solid conical indenter interacting with a soft material of a Poisson ratio of 0.5 and a Young's modulus of 1 or 10 kPa, respectively. The half-opening angle α has been set to 45°, a typical value for AFM tips.

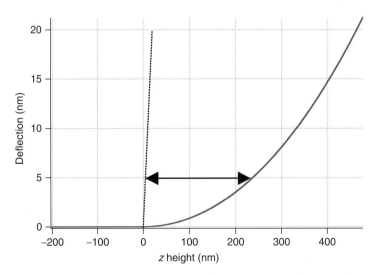

Fig. 14 A simulated force curve of a soft sample, using the Hertz model. On sample contact, the deflection follows a nonlinear relationship in comparison to the linear behavior of a stiff sample (dotted line). The difference between measured deflection on a soft sample and the corresponding data on a stiff sample is the indentation δ, which can be read directly from the graph.

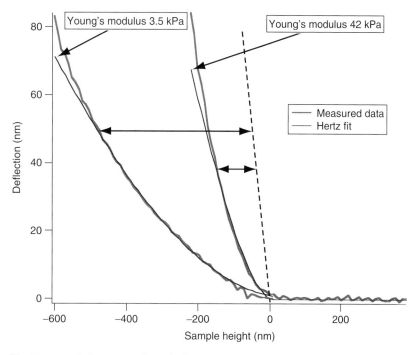

Fig. 15 Two sample force curves from the force map presented in Fig. 16. Each data curve (red) has been fit according to the Hertz model (blue), which yields an elastic modulus of 42 and 3.5 kPa, respectively. The dotted black line shows how a force curve on a stiff substrate would appear: the deflection is proportional to the sample height. The distance between the black line and the corresponding measured deflection (red lines) gives the sample indentation. The softer curve (left one) has been taken in a thick region of the cell, where the Hertz fit follows the data up to large indentation of 400 nm denoted by the double arrow line. The stiffer curve, recorded in the lamellipodial region, deviates from the Hertz fit at a smaller, indentation of about 100 nm, while the indentation becomes comparable to the thickness of the lamellipodial region at 200–300 nm.

stiffness can always be extracted from the data if a region undergoes limited changes in thickness, as in the lamellipodium (Schäfer and Radmacher, 2005).

There is also no simple model that takes into consideration potential interactions of a thin and soft material with the underlying stiff substrate, for example adhesion of the cell to the substrate. Two limiting cases are slip or nonslip boundaries, which determine if the cell is able to slide along the substrate. Several models have been proposed in the literature and applied to AFM data (Dimitriadis *et al.*, 2002; Mahaffy *et al.*, 2004; Ogilvy, 1993).

C. Dynamics of Cellular Mechanical Properties

The stiffness of eukaryotic cells as measured by AFM reflects mainly properties of the actin cytoskeleton. This can be proven by comparing fluorescence images after labeling specifically actin filaments with stiffness images or by monitoring

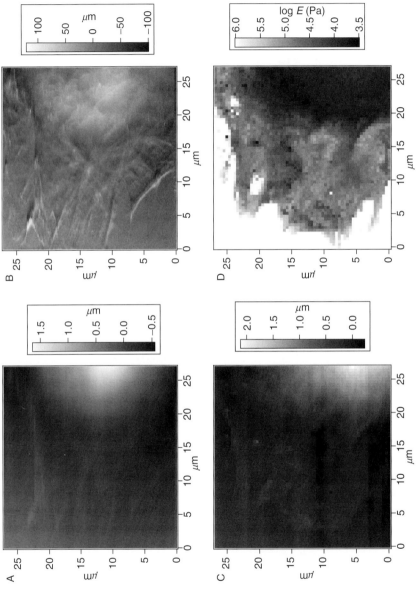

Fig. 16 Comparison of contact mode images and elasticity maps of the same area of a fibroblast. Panels (A) and (B) are the height channel and deflection channel, respectively, acquired during contact mode imaging. Panels (C) and (D) are the contact point and the elastic modulus, respectively, calculated from a force map.

changes in stiffness before and after degrading the actin cytoskeleton with drugs (Rotsch and Radmacher, 2000). Cellular processes that have been investigated while monitoring the mechanical properties include migration (Laurent *et al.*, 2005; Rotsch *et al.*, 1999), lamellipodial protrusion (Rotsch *et al.*, 2001), and division (Matzke *et al.*, 2001).

D. Effects of Cell Stiffness on AFM Imaging

The softness of cells affects AFM imaging in several ways. In *contact mode* imaging, the tip exerts a significant loading force (typically several hundreds of piconewtons) to facilitate tracking of the surface topography. This results in an indentation of 100 nm or more depending on the softness of the particular area of the cell. The contact radius will be on the same order, which generally limits the lateral resolution. However, if there are cytoskeletal structures at some depth under the membrane, the indentation might facilitate the visualization of such submembraneous structures, as depicted in Fig. 17.

The way that the contrast in contact mode imaging evolves can be understood best by analyzing the so-called force volume maps (Section III.C). A force volume map contains information on the effective sample height (including compression or indentation) for each applied loading force, and allows the construction of topographical images from the individual force curve at each location under a given loading force. In the example of Fig. 18, this has been done for the force volume data of a fibroblast. At low loading forces, the topography is smooth and featureless. At higher loading forces, the underlying cytoskeletal structures start to appear in the image.

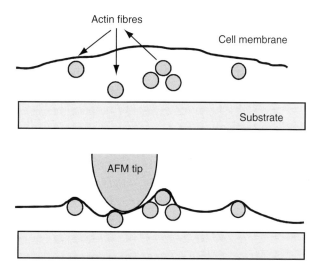

Fig. 17 Schematics of image formation in contact mode AFM of cells. The loading forces exerted by the tip on the sample cause cell indentation, allowing the tip to sense and image stiffer cytoskeletal structures, for example bundles of actin fibers, embedded deep in the cytoplasm.

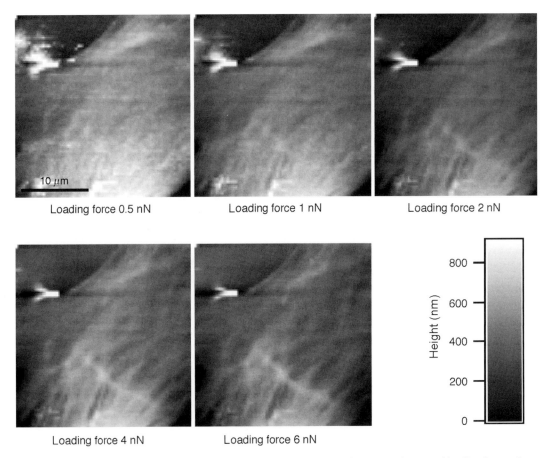

Loading force 0.5 nN   Loading force 1 nN   Loading force 2 nN

Loading force 4 nN   Loading force 6 nN

Fig. 18 Topographic images reconstructed from force volume maps for several loading force values. The image at low forces corresponds to the "true" topography of the cell surface, whereas the images at higher loading force mimic the appearance of cells in contact mode AFM.

As explained above, the softness of cells usually limits lateral resolution on the order of 100 nm or worse. Only when imaging flat areas of eukaryotic cells, where the underlying substrate supports the cell (Grimellec *et al.*, 1998; Pesen and Hoh, 2005), or when imaging noneukaryotic cells with a stiff cell wall, higher resolution can be expected. Such high-resolution imaging has been applied to examine cell growth and division of *Staphylococcus aureus* (Touhami *et al.*, 2004).

V. Outlook

An AFM can be used to not only image but also map the height and mechanical properties of living cells. The technique does not require fixation, allowing the investigator to follow changes during application of drugs or during cellular activities such as protrusion and division. Further improvements in method and data analysis are nonetheless needed. One promising extension of the method is in measuring cell mechanics as a function of driving frequency (Mahaffy *et al.*, 2000), which is equivalent to performing rheology by AFM (Chapter 1 by Janmey *et al.* and Chapter 2 by Kandow *et al.*, this volume). In addition, better mechanical models for cells and new experimental strategies are needed to address the issues of sample thickness, inhomogeneity, and anisotropy. It may be especially helpful to relate AFM data to data obtained from other techniques with a similarly high spatial resolution, including magnetic or optical tweezers and microbead rheology (Bausch *et al.*, 1998; Chapter 6 by Panorchan *et al.*, Chapter 7 by Crocker and Hoffman, Chapter 8 by Wang *et al.*, Chapter 19 by Lele *et al.*, and Chapter 20 by Tanase *et al.*).

Acknowledgments

I am very thankful for the SEM images by Fabian Heinemann and for the force volume data by Jan Christian Martens. This work has been supported by the University of Bremen under grant No. BFK 01/115/04.

References

A-Hassan, E., Heinz, W. F., Antonik, M. D., D'Costa, N. P., Nagaswaran, S., Schoenenberger, C.-A., and Hoh, J. H. (1998). Relative micro-elastic mapping of living cells by atomic force microscopy. *Biophys. J.* **74**(3), 1564–1578.

Alberts, V. B., and Lewis, J. (2002). "Molecular Biology of the Cell." Garland Publishing, Inc., New York.

Almqvist, N., Delamo, Y., Smith, B. L., Thomson, N. H., Bartholdson, Ä., Lal, R., Brzezinski, M., and Hansma, H. G. (2001). Micromechanical and structural properties of a pennate. *J. Microsc.* **202**(3), 518–532.

Arnoldi, M., Fritz, M., Bäuerlein, E., Radmacher, M., Sackmann, E., and Boulbitch, A. (2000). Bacterial turgor pressure can be measured by atomic force microscopy. *Phys. Rev. E* **62**(1), 1034–1044.

Bausch, A. R., Ziemann, F., Boulbitch, A. A., Jacobson, K., and Sackmann, E. (1998). Local measurements of viscoelastic parameters of adherent cell surfaces by magnetic bead microrheometry. *Biophys. J.* **75**, 2038–2049.

Bezanilla, M., Drake, B., Nudler, E., Kashlev, M., Hansma, P. K., and Hansma, H. G. (1994). Motion and enzymatic degradation of DNA in the atomic force microscope. *Biophys. J.* **67**, 1–6.

Binnig, G., Quate, C. F., and Gerber, C. (1986). Atomic force microscope. *Phys. Rev. Lett.* **56**(9), 930–933.

Binnig, G., and Rohrer, H. (1982). Scanning tunneling microscope. *Helv. Phys. Acta* **55**, 726–735.

Boulbitch, A. A. (1998). Deflection of a cell membrane under application of a local force. *Phys. Rev. Lett.* **57**(2), 2123–2128.

Braet, F., Rotsch, C., Wisse, E., and Radmacher, M. (1997). Comparison of fixed and living endothelial cells by atomic force microscopy. *Appl. Phys. A* **66**, S575–S578.

Butt, H.-J., and Jaschke, M. (1995). Calculation of thermal noise in atomic force microscopy. *Nanotechnology* **6**(1), 1–7.

Dimitriadis, E. K., Horkay, F., Maresca, J., Kachar, B., and Chadwick, R. S. (2002). Determination of elastic moduli of thin layers of soft material using the atomic force microscopy. *Biophys. J.* **82,** 2798–2810.

Domke, J., Parak, W. J., George, M., Gaub, H. E., and Radmacher, M. (1999). Mapping the mechanical pulse of single cardiomyocytes with the atomic force microscope. *Eur. Biophys. J.* **28,** 179–186.

Domke, J., and Radmacher, M. (1998). Measuring the elastic properties of thin polymer films with the AFM. *Langmuir* **14**(12), 3320–3325.

Drake, B., Prater, C. B., Weisenhorn, A. L., Gould, S. A. C., Albrecht, T. R., Quate, C. F., Cannell, D. S., Hansma, H. G., and Hansma, P. K. (1989). Imaging crystals, polymers and biological processes in water with AFM. *Science* **243,** 1586–1589.

Florin, E.-L., Moy, V. T., and Gaub, H. E. (1994). Adhesion forces between individual ligand-receptor pairs. *Science* **264,** 415–417.

Fritz, M., Radmacher, M., and Gaub, H. E. (1993). *In vitro* activation of human platelets triggered and probed by SFM. *Exp. Cell Res.* **205**(1), 187–190.

Fritz, M., Radmacher, M., and Gaub, H. E. (1994). Granula motion and membrane spreading during activation of human platelets imaged by atomic force microscopy. *Biophys. J.* **66**(5), 1328–1334.

Grandbois, M., Dettmann, W., Benoit, M., and Gaub, H. E. (2000). Affinity imaging of red blood cells with an atomic force microscope. *J. Histochem. Cytochem.* **48**(5), 719–724.

Grimellec, C. L., Lesniewska, E., Giocondi, M.-C., Finot, E., Vié, V., and Goudonnet, J.-P. (1998). Imaging of the surface of living cells by low-force contact-mode atomic force microscopy. *Biophys. J.* **75**(2), 695–703.

Hansma, P. K., Cleveland, J. P., Radmacher, M., Walters, D. A., Hillner, P. E., Bezanilla, M., Fritz, M., Vie, D., Hansma, H. G., Prater, C. B., Massie, J., Fukunaga, L., *et al.* (1994). Tapping mode atomic force microscopy in liquids. *Appl. Phys. Lett.* **64**(13), 1738–1740.

Henderson, E., Haydon, P. G., and Sakaguchi, D. S. (1992). Actin filament dynamics in living glial cells imaged by atomic force microscopy. *Science* **257,** 1944–1946.

Hertz, H. (1882). Über die Berührung fester elastischer Körper. *J. Reine Angew. Mathematik* **92,** 156–171.

Hofmann, U. G., Rotsch, C., Parak, W. J., and Radmacher, M. (1997). Investigating the cytoskeleton of chicken cardiocytes with the atomic force microscope. *J. Struct. Biol.* **119,** 84–91.

Johnson, K. L. (1994). "Contact Mechanics." Cambridge University Press, Cambridge.

Kasas, S., Wang, X., Hirling, H., Marsault, R., Huni, B., Yersin, A., Regazzi, R., Greeningloh, G., Riederer, B., Forrò, L., Dietler, G., and Catsicas, S. (2005). Superficial and deep changes of cellular mechanical properties following cytoskeleton disassembly. *Cell Motil. Cytoskeleton* **62,** 124–132.

Laurent, V., Kasas, S., Yersin, A., Schaffer, T., Catsicas, S., Dietler, G., Verkhovsky, A., and Meister, J. (2005). Gradient of rigidity in the lamelllipodia of migrating cells revealed by atomic force microscopy. *Biophys. J.* **89,** 667–675.

Lugmaier, R. A., Hugel, T., Benoit, M., and Gaub, H. E. (2005). Phase contrast and DIC illumination for AFM-Hybrids. *Ultramicroscopy* **104,** 255–260.

Mahaffy, R. E., Park, S., Gerde, E., Käs, J., and Shih, C. K. (2004). Quantitative analysis of the viscoelastic properties of thin regions of fibroblasts using atomic force microscopy. *Biophys. J.* **86,** 1777–1793.

Mahaffy, R. E., Shih, C. K., MacKintosh, F. C., and Käs, J. A. (2000). Scanning probe-based frequency-dependent microrheology of polymer gels and biological cells. *Phys. Rev. Lett.* **85**(4), 880–883.

Matzke, R., Jacobson, K., and Radmacher, M. (2001). Direct, high resolution measurement of furrow stiffening during the division of adherent cells. *Nat. Cell Biol.* **3,** 607–610.

McElfresh, M., Baesu, E., Balhorn, R., Belak, J., Allen, M. J., and Rudd, R. E. (2002). Combining constitutive materials modeling with atomic force microscopy to understand the mechanical properties of living cells. *Proc. Natl. Acad. Sci. USA* **99,** 6493–6497.

Ogilvy, J. A. (1993). A parametric elastic model for indentation testing of thin films. *J. Phys. D: Appl. Phys.* **26**, 2123–2131.

Pesen, D., and Hoh, J. (2005). Micromechanical architecture of endothelial cell cortex. *Biophys. J.* **88**, 670–679.

Putman, C. A. J., van der Werf, K. O., de Grooth, B. G., van Hulst, N. F., and Greve, J. (1994). Tapping mode atomic force microscopy in liquid. *Appl. Phys. Lett.* **64**(18), 2454–2456.

Putman, C. A. J., van der Werf, K. O., de Grooth, B. G., van Hulst, N. F., Greve, J., and Hansma, P. K. (1992). A new imaging mode in atomic force microscopy based on the error signal. *Proc. Soc. Photo-Opt. Instr. Eng.* **1639**, 198–204.

Radmacher, M., Cleveland, J. P., Fritz, M., Hansma, H. G., and Hansma, P. K. (1994a). Mapping interaction forces with the atomic force microscope. *Biophys. J.* **66**(6), 2159–2165.

Radmacher, M., Fritz, M., Hansma, H. G., and Hansma, P. K. (1994b). Direct observation of enzyme activity with the atomic force microscope. *Science* **265**, 1577–1579.

Radmacher, M., Fritz, M., and Hansma, P. K. (1995). Imaging soft samples with the atomic force microscope: Gelatin in water and propanol. *Biophys. J.* **69**(7), 264–270.

Radmacher, M., Fritz, M., Kacher, C. M., Cleveland, J. P., and Hansma, P. K. (1996). Measuring the elastic properties of human platelets with the atomic force microscope. *Biophys. J.* **70**(1), 556–567.

Radmacher, M., Tillman, R. W., and Gaub, H. E. (1993). Imaging viscoelasticity by force modulation with the atomic force microscope. *Biophys. J.* **64**, 735–742.

Rico, F., Roca-Cusachs, P., Gavara, N., Farrè, R., Rotger, M., and Navajas, D. (2005). Probing mechanical properties of living cells by atomic force microscopy with blunted pyramidal cantilever tips. *Phys. Rev. E Stat. Nonlin. Soft Matter Phys.* **72**, 021914–021924.

Rotsch, C., Jacobson, K., Condeelis, J., and Radmacher, M. (2001). EGF-stimulated lamellipod extension in mammary adenocarcinoma cells. *Ultramicroscopy* **86**, 97–106.

Rotsch, C., Jacobson, K., and Radmacher, M. (1999). Dimensional and mechanical dynamics of active and stable edges in motile fibroblasts investigated by atomic force microscopy. *Proc. Natl. Acad. Sci. USA* **96**, 921–926.

Rotsch, C., and Radmacher, M. (1997). Measuring electrostatic interactions with the atomic force microscope. *Langmuir* **13**(10), 2825–2832.

Rotsch, C., and Radmacher, M. (2000). Drug-induced changes of cytoskeletal structure and mechanics in fibroblasts: An atomic force microscopy study. *Biophys. J.* **78**, 520–535.

Sader, J. E., Larson, I., Mulvaney, P., and White, L. R. (1995). Method for the calibration of atomic force microscope cantilevers. *Rev. Sci. Instrum.* **66**(7), 3789–3797.

Schabert, F., Knapp, H., Karrasch, S., Häring, R., and Engel, A. (1994). Confocal scanning laser-scanning probe hybrid microscope for biological applications. *Ultramicroscopy* **53**(2), 147–157.

Schäfer, A., and Radmacher, M. (2005). Influence of myosin II activity on cellular stiffness. *Acta Biomater.* **1**, 273–280.

Schäffer, T. E., Viani, M., Walters, D. E., Drake, B., Runge, E. K., Cleveland, J. P., Wendman, M. A., and Hansma, P. K. (1997). An atomic force microscope for small cantilevers. *In* "Micromaching and Imaging," pp. 49–52. SPIE, San Jose, CA.

Schaus, S. S., and Henderson, E. R. (1997). Cell viability and probe-cell membrane interactions of XR1 glial cells imaged by atomic force microscopy. *Biophys. J.* **73**(9), 1205–1214.

Schoenenberger, C.-A., and Hoh, J. H. (1994). Slow cellular dynamics in MDCK and R5 cells monitored by time-lapse atomic force microscopy. *Biophys. J.* **67**, 929–936.

Sen, S., Subramaniam, S., and Discher, D. E. (2005). Indentation and adhesive probing of cell membrane with AFM: Theoretical model and experiments. *Biophys. J.* **89**, 3203–3213.

Sneddon, I. N. (1965). The relation between load and penetration in the axisymmetric Boussinesq problem for a punch of arbitrary profile. *Int. J. Eng. Sci.* **3**, 47–57.

Touhami, A., Jericho, M., and Beveridge, T. (2004). Atomic force microscopy of cell growth and division of *Staphylococcus aureus*. *J. Bacteriol.* **186**(1), 3286–3295.

Vesenka, J., Mosher, C., Schaus, S., Ambrosio, L., and Henderson, E. (1995). Combining optical and atomic force microscopy for life sciences research. *BioTechniques* **19**(2), 240–253.

Walters, D. A., Cleveland, J. P., Thomson, N. H., and Hansma, P. K. (1996a). Frequency shifts of cantilevers vibrating in various media. *Appl. Phys. Lett.* **69**(19), 2834–2836.

Walters, D. A., Cleveland, J. P., Thomson, N. H., Hansma, P. K., Wendman, M. A., Gurley, G., and Elings, V. (1996b). Short cantilevers for atomic force microscopy. *Rev. Sci. Instrum.* **67**(10), 3583–3590.

CHAPTER 16

Using Force to Probe Single-Molecule Receptor–Cytoskeletal Anchoring Beneath the Surface of a Living Cell

Evan Evans[*,†,‡] and Koji Kinoshita[*]

[*]Department of Biomedical Engineering
Boston University
Boston, Massachusetts 02215

[†]Department of Physics and Astronomy
University of British Columbia, Vancouver
British Columbia, Canada V6T 1Z1

[‡]Department of Pathology and Laboratory Medicine
University of British Columbia, Vancouver
British Columbia, Canada V6T 2B5

Abstract

The ligation of cell surface receptors often communicates a signal that initiates a cytoplasmic chemical cascade to implement an important cell function. Less well understood is how physical stress applied to a cell surface adhesive bond propagates throughout the cytostructure to catalyze or trigger important steps in these chemical processes. Probing the nanoscale impact of pulling on cell surface bonds, we discovered that receptors frequently detach prematurely from the interior cytostructure prior to failure of the exterior adhesive bond [Evans, E., Heinrich, V., Leung, A., and Kinoshita, K. (2005). Nano-to-micro scale dynamics of P-selectin detachment from leukocyte interfaces: I. Separation of PSGL-1 from the cell cytoskeleton. *Biophys. J.* 88, 2288–2298]. Retracting cells from receptor-surface attachments at many different speeds revealed that the kinetic rate for receptor–cytoskeletal unbinding increased exponentially with the level of force, suggesting disruption at a site of single-molecule interaction. Since many important enzymes and signaling molecules are closely associated with a membrane receptor–cytoskeletal linkage, pulling on a receptor could alter interactions among its constellation of associated proteins, perhaps switching some aspect of their function. Thus, if used in conjunction with cleverly engineered cell lines targeting receptor–cytoskeletal linkages, probing the kinetics of receptor–cytoskeletal unbinding with ultrasensitve force techniques can provide unique physical insight into the interactions involved in the chemical functions of a molecular adhesion complex. The aim of this chapter is to describe the nanomechanical methods needed to probe receptor–cytoskeletal anchoring beneath the surface of a living cell and to provide the analytical "thinking" needed to extract dissociation kinetics from the statistics of the various failure events observed in pulling. As demonstrations of the experimental approach and concepts, we will use examples taken from probing selectin and integrin receptors immobilized on glass microspheres and expressed on surfaces of leukocytes.

I. Generic Methods and Physical Foundations

A. Using Force to Probe Single-Molecule Interactions

A force probe is merely an ultrasensitive "spring" with a "sticky" tip that attaches to molecules on a target surface on contact, and that then stretches as the target is pulled away until the attachment breaks. Thus, the probe reports the history of stress experienced by an attachment until the moment of its failure signaled by recoil of the probe spring, establishing how long the molecular inter-action survives under stress (i.e., its lifetime). After taking into account the serial elasticity of the probe spring, changes of the force during displacement of the target both reveal the nanoscale compliance of the molecular attachment as well as abrupt transitions that may occur in conformational structure. Since biomolecular bonds can break at small force and small forces produce small spring deflections, probe instruments are usually designed with very soft springs as transducers and employ some type of high-resolution detection to quantify the spring deflection. The characteristic scale for force is the piconewton ($\sim 1 \times 10^{-10}$ gram weight), which follows from the thermal energy scale for activation of bond kinetics at room temperature ($k_B T \sim 4.08 \times 10^{-21}$, $J = 4.08$ pN nm) divided by a scale for molecular length gained in bond dissociation ($x_\beta \sim 1$ nm). Hence, with optical-imaging systems capable of tracking ~ 1-nm deflections of the probe tip, testing weak biomolecular interactions is best performed using a probe transducer with a spring constant $\kappa_f \sim$ pN/nm. Prominent examples of very sensitive force probes include the optical trap force probe (OTFP: Abbondanzieri *et al.*, 2005; Asbury *et al.*, 2003; Ashkin and Dziedzic, 1987; Kuo and Sheetz, 1993; Liphardt *et al.*, 2001, 2002; Schnitzer and Block, 1997; Simmons *et al.*, 1996; Smith *et al.*, 1996; Svoboda *et al.*, 1993) and the biomembrane force probe (BFP: Bayas *et al.*, 2006; Evans *et al.*, 1995, 2004, 2005; Merkel *et al.*, 1999; Perret *et al.*, 2004). Controlled by laser power and pipette suction pressure, the spring constants of an OTFP and a BFP can be tuned in a range of 10^{-3}–10^{-1} pN/nm and 0.1–1 pN/nm, respectively. Making these instruments especially well suited for probing molecular interactions at a cell surface, the OTFP and BFP also have fast response times (due to low hydrodynamic damping) and low force noise that enable soft touch to a surface as well as subpiconewton force precision over long time periods as demonstrated in Fig. 1 below.[1]

B. Testing Bonds on Solid Substrates

We begin by outlining the approach used to test mechanical strengths and dissociation kinetics of interactions between ligand and receptor molecules when immobilized on solid substrates (referred to as *in vitro* tests). As we show later, these measurements establish the essential experimental control needed to distin-guish failure kinetics of receptor–cytoskeletal connections inside a cell from the kinetics of the exterior ligand–receptor bond.

---

[1] Because of inadequate control of the surface impact force and drift, the conventional atomic force microscope (AFM, Binnig *et al.*, 1986) is less well suited at present for testing molecular interactions at cell surfaces with low forces. Employing a long-thin silicon nitride cantilever with relatively large "spring" constant (>10 pN/nm), the AFM measurement response is also encumbered to some extent by force noise and significant hydrodynamic damping. On the other hand, using a sensitive "optical lever" to achieve fine detection of tip deflection, the AFM has been used very effectively to probe the conformational stability and structure of protein domains (Carl *et al.*, 2001; Dietz and Rief, 2004; Law *et al.*, 2003; Li *et al.*, 2000; Müller *et al.*, 1999; Rief *et al.*, 1997, and so on).

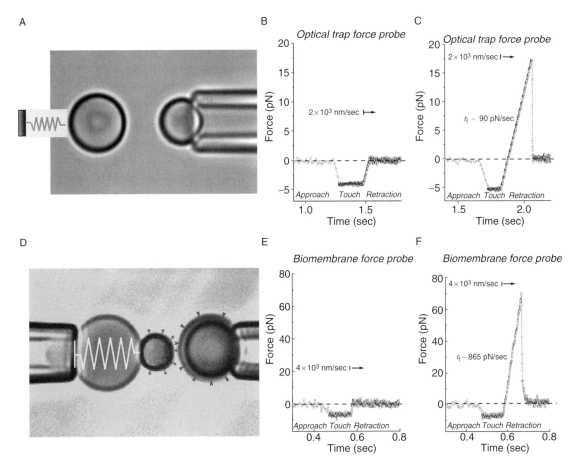

Fig. 1 Examples of testing protein–protein interactions between microspheres with an OTFP (A–C) and a BFP (D–F). Superposed on the images in (A and D), a "spring" identifies the force transducer. Taken from sequences of repeated cycles of approach/touch/retraction, force–time traces are shown with (C and F) and without (B and E) formation of a bond. Values of the force were derived from displacements of the probe particle multiplied by the "spring" constant (OTFP = 0.05 pN/nm and BFP = 0.25 pN/nm). As such, the precision in force (OTFP = ±0.2 pN SD and BFP = ±1.2 pN SD) was determined by the accuracy (±3–5 nm) of the high-speed video image software used to track the probe-tip movement. The OTFP in (A) was created by capturing a 3-μm glass microsphere in a sharply focussed beam of near-IR (1064 nm) laser light. Choosing the objective numerical aperture, the microsphere radius, and index of refraction, the OTFP spring constant was tuned by control of the laser beam intensity. Pressurized to a spherical shape by pipette suction, the BFP transducer in (D) was a PEG-biotinylated red blood cell; the BFP tip was added by bonding an avidinated 2-μm glass microsphere. Choosing the pipette radius R_p, the BFP spring constant was tuned using the pipette suction pressure to control the cell membrane tension. Section B contains additional details of the experimental methods.

In addition to the precision control and tracking of a force probe, the key to success in testing molecular interactions is to ensure a high probability of forming single-molecule attachments. Thus, surfaces of the micrometer-sized glass microspheres used for OTFP, BFP tips, and receptor targets are decorated with very low densities of recombinant proteins. Typically, this involves covalent coupling with a heterobifunctional (e.g., amine and thiol reactive) polymer linker (e.g., PEG), or noncovalent coupling through a previously covalently linked monoclonal antibody. [In the case of a BFP, the probe-tip microspheres are also bound with a large amount of PEG-biotin and coated with (multivalent) avidin, which enable strong adhesion to the PEG-biotinylated red blood cell.] After assembly in the microscope chamber, a target microsphere (selected and held by a micropipette, as shown to the right of Fig. 1A and D) is repeatedly brought to/from contact with the probe tip using computer-directed movements of a linear-piezo translator. By adequately reducing the density of reactive sites, a level of coverage is reached where repeated *soft* contact between the probe tip and target surfaces produces infrequent attachments, thereby indicating a high probability for forming single bond events (e.g., probability >0.9 when 1 attachment occurs out of 10 touches). Formation of an attachment is sensed when the probe "spring" shows a tensile (pulling) deflection as the target is retracted (cf. Fig. 1C and F). Failure of the attachment is signaled by recoil of the probe to its rest position. Complete histories of force over the course of approach—touch—separation with and without formation of a molecular attachment are demonstrated in Fig. 1. After numerous trials following the same course of approach—touch—and retraction, the subset of force–time traces showing bond events are identified and the events are analysed to establish the statistics (histograms) of breakage forces and dissociation times. As we show next, these histograms cumulate discrete outcomes of a random kinetic process, which changes throughout the time course (history) of force application.

C. Simple Physics of Breaking a Bond

Once a bond has formed, pulling a receptor target away from a ligand-decorated probe stretches the probe spring (cf. Fig. 1C and F) until the attachment breaks. The measurements of lifetime and maximum strength at the breakup represent one instance of a random kinetic process and one history of force application. In order to characterize the statistics expected for many such events, we need to consider how a probe interacts with the biomolecular complex and how pulling forces alter the process of attachment failure (Evans and Williams, 2002).

When coupled to an elastic probe, what actually determines the stability of a molecular bond is the "effective-spring" potential composed of the probe spring in series with the molecular tethers attaching the ligand and receptor to the probe tip and target. When the receptor is bound to a probe ligand, target retraction will stretch the "effective spring." The stretching force lowers both the energy minimum characterizing the dissociated state and the activation energy barrier that impedes dissociation of the bound complex as illustrated in Fig. 2A. We see that pulling on the bond drives the reaction away from its equilibrium state. Yet, if weakly stretched, the spring energy minimum will stay above the bound-state energy minimum (the deep valley to the left), and will promote quick return to the

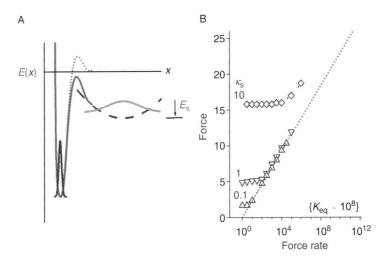

Fig. 2 (A) Schematic energy landscape (red-solid curve) of a chemical interaction along a "reaction coordinate" x defined by direction of pulling with a force probe. The dotted curve idealizes the stress-free interaction where a prominent energy barrier confines initial bound-state configurations of the ligand (dark-narrow distribution). Pulling on the bond with a probe adds a "spring-like" potential (dashed line) that lowers the energy barrier and captures dissociated states of the ligand (gray-broad distribution). Starting from the level set by binding affinity, $K_{eq} = \log_e(E_0/k_B T)$, the ratio of rebinding-to-dissociation rates k_{on}/k_{off} falls dramatically as the "spring" energy minimum deepens with an increase of tensile force, $E_s \approx f^2/2\kappa_s$, that is $k_{on}/k_{off} \sim \exp[(E_0 - E_s)/k_B T]$. Thus, as shown in (B), the most frequent force f^* for rupture begins under slow pulling above the threshold where the spring stretch eliminates rebinding, $E_s > E_0$, or $f^* \geq (2\kappa_s E_0)^{1/2}$. This threshold depends on the effective elastic constant κ_s of the probe and molecular linkages applying force to the bond (Evans, 2001). Pulling fast blocks rebinding and the kinetic frequency of bond rupture becomes the off rate, which rises effectively as an exponential of the force, $k_{off} \approx (1/t_{off}) \exp(f/f_\beta)$. The "kinetically limited" failure rate speeds up as the rate of force application is increased, predicting the most frequent failure force, $f^* \approx f_\beta \ln(r_f t_{off}/f_\beta)$. (Note that the force axis is normalized by the thermal activation scale f_β and force rate r_f by f_β/t_{off}. The values of effective elastic constant κ_s are defined by the scale, $f_\beta^2/k_B T$.)

bound state, should a transient dissociation event occur. Pulling further, the minimum of the "effective spring" potential will eventually drop below the energy minimum defining the bound state and the dissociation event will permanently rupture the attachment. Hence, in order to break a bond, the probe spring has to be stretched enough to assure that the forward rate k_{off} of escape from the receptor-bound state exceeds the reverse rate k_{on} of rebinding from the effective spring potential. As illustrated in Fig. 2B (Evans, 2001), the kinetic crossover from binding to dissociation establishes a threshold for "most frequent rupture force f^*" which is approximated by the harmonic mean of the "effective spring" constant κ_s and the free energy E_0 of ligand–receptor binding, that is $f^* \geq (2\kappa_s E_0)^{1/2}$. Even though free energies for ligand–receptor interactions can be significant ($\sim 10^{-19}$ J), the threshold force for near-equilibrium unbinding will be very small for soft polymer linkages ($\kappa_s \sim 0.1$ pN/nm or less). Consequently, instead of remaining constant, the forces measured at bond rupture now increase with increase in rate of force application which, in turn, is governed by the pulling speed V_{pull} and effective spring constant κ_s, that is $r_f = V_{pull}\kappa_s$, as illustrated in Fig. 2B.

D. Forcing Bonds to Dissociate Faster Than Their Spontaneous Off Rate

The simplest intuition that we can employ to describe the impact of force on dissociation kinetics has been idealized into a relation called the "Bell model" (Bell, 1978). In this view, off rate is postulated to increase exponentially with applied force starting from an *apparent*-unstressed rate $1/t_{off}^0$,

$$k_{off}(f) \approx \left(\frac{1}{t_{off}^0}\right)\exp\left(\frac{f}{f_\beta}\right) \qquad (1)$$

Examined in more detail (Evans and Ritchie, 1997), the model has been shown to characterize kinetic transitions traversing a very sharp energy barrier, which diminishes by one unit of thermal energy k_BT for each increment in force equal to the value f_β. In spite of the bold abstraction of kinetics implicit in the Bell approximation, Eq. (1) often agrees with the *bond failure rates* observed in tests of bonds over significant ranges of applied force as we show below. Yet, significant departures from this idealized model have revealed important-unexpected features of complex ligand–receptor interactions (Evans *et al.*, 2004; Marshall *et al.*, 2003; Perret *et al.*, 2004).

Given the force dependence $k_{off}(f)$ for dissociation rate and the history of force $f(t)$ applied to a bond, the likelihood $S(t)$ that an attachment will survive over time (in the absence of rebinding) is predicted by,

$$S(t) = \exp\{-\int_{0 \to t} k_{off}(f)dt'\} \qquad (2)$$

Thus, if held under a constant force (labeled *force clamp*), the probability of bond survival is expected to decay as a single exponential in time. The force dependence of the mean lifetime $t_{off}(f) = 1/k_{off}(f)$ can be assayed by collecting the statistics of bond failure at many different levels of force and plotting them on a logarithmic scale versus time,

$$\log_e\left[\frac{N(t)}{N_{tot}}\right] = -k_{off}(f)t \qquad (3)$$

N_{tot} is the total number of bonds that exist at $t = 0$ and $N(t)$ is the number of bonds that remain at time t. Demonstrating this force clamp assay, Fig. 3A shows an example of a mono-ICAM-1–$\alpha_L\beta_2$ integrin attachment held at constant force. Figure 3B shows the exponential-like decay in the statistics of survival obtained by testing a few hundred attachments at three levels of force. At first glance, this would seem to be the easiest method to establish the profile of lifetime under force. However, implementation and precise control of a "force clamp" can be difficult

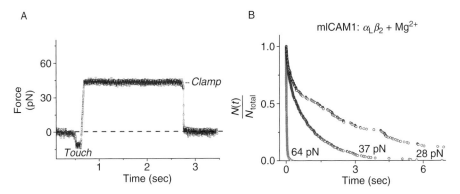

Fig. 3 (A) A force clamp test of an attachment between mono-ICAM-1 and $\alpha_L\beta_2$ integrin linked covalently to glass microspheres and activated in Mg^{2+}. (B) The statistics of survival from many tests of mono-ICAM-1–$\alpha_L\beta_2$ integrin attachments are shown plotted as a function of time, with more than 300 lifetimes measured at each force. The dramatic shortening of attachment survival with increase in the force shows the exponential-like dependence of the off rate on force. [Note that the $\alpha_L\beta_2$ heterodimer was a leucine-zipper construct produced by ICOS Corp. (Dr. Don Staunton) and made available to us through our collaboration with Dr. Scott Simon at University of California, Davis.]

when probing cell surface bonds. Moreover, the statistics of dissociation events include populations of spurious "nonspecific" interactions and multiple bonds. Thus, in the remainder of this chapter, we will focus on a method called the "force ramp," which enables discrimination of the failure events occurring at a cell surface from these spurious effects.

E. Force Ramp Method and Examples of Testing Protein–Protein Interactions *In Vitro*

Demonstrated by the examples in Fig. 1C and F, the "force ramp" method involves application of a steadily increasing force to attachments, that is $f = r_f t$. When the ramp rate r_f is varied from very slow to very fast, it is possible to quantify the profile of dissociation kinetics under force, obtaining the same information as with the "force clamp" but presented in a different functional form. Modeling the off rate of an interaction as exponentially dependent on force, the likelihood of survival over time is now found to depend on the force ramp rate r_f (Evans and Ritchie, 1997),

$$S(t) = \exp\left\{-\frac{1}{t_{off}^0} \int_{0 \to t} \left[\exp\left(\frac{r_f y}{f_\beta}\right) dy\right]\right\} = \exp\left\{-\left(\frac{r_f^0}{r_f}\right)\left[\exp\left(\frac{r_f t}{f_\beta}\right) - 1\right]\right\} \quad (4)$$

The kinetic parameters governing bond dissociation define the characteristic scale for ramp rate, $r_f^0 \equiv f_\beta/t_{off}^0$. Once ramp rate exceeds r_f^0, the period of survival diminishes at a rate much faster than $1/t_{off}^0$. As shown in Fig. 4A–C, measurements

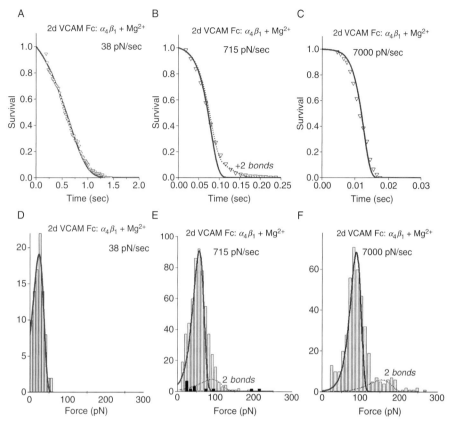

Fig. 4 Lifetimes (A–C) and rupture forces (D–F) from force ramp tests of interactions between a two domain construct of the vascular cell adhesion molecule (2d VCAM-1) and the integrin $\alpha_4\beta_1$ (VLA-4) linked covalently to glass microspheres and activated in Mg^{2+}. Modeling "kinetically limited bond failure," the superposed solid curves in (A–F) are survival probabilities [Eq. (4)] and force distributions [Eq. (5)] predicted by an exponential dependence of off rate on force: $k_{off}(f) \approx (0.6\text{–}0.8 \text{ per second}) \exp(f/14 \text{ pN})$. The dotted curves were added to show that the outliers at high force occur in the region expected for a small population of double-bonded attachments equally sharing the force. [Note that the $\alpha_4\beta_1$ integrin and domains 1–2 of the VCAM-1 ligand were Fc fusion proteins produced by CellTech Group R&D (now UCB-Group, UK) and kindly provided to us by Drs. Martyn Robinson and Henry Alistair.]

of survival times for interactions between VCAM (vascular cell adhesion molecule)-1 and $\alpha_4\beta_1$ integrin on microspheres closely follow the functional form predicted by Eq. (4). Because of the off rate exponentiation under force, the time frame for bond survival shortens dramatically with each order of magnitude increase in ramp rate.

Since force events are exactly correlated to bond failure times, $f = r_f t$, the probability distribution, $p(t) = k_{off}[f(t)]S(t)$, for survival events in time also predicts the statistics (histograms) of bond rupture forces, that is $p(f) = (1/r_f)k_{off}(f)S(f)$,

$$p(f) = \left(\frac{r_f^0}{r_f}\right) \exp\left\{\frac{f}{f_\beta} - \left(\frac{r_f^0}{r_f}\right)\left[\exp\left(\frac{f}{f_\beta}\right) - 1\right]\right\} \tag{5}$$

The distribution of force shifts progressively to higher force values as the ramp rate is increased, following the increase in most frequent force f^* (distribution peak) predicted by the proportionality to logarithm of ramp rate, that is $f^* = f_\beta \log_e(r_f/r_f^0)$. Thus, when measured over a wide range of ramp rates, linear correlations between the peak positions of force histograms and logarithms of ramp rate provide a direct assay of the kinetic parameters, that is f_β = slope and $r_f^0 = f_\beta/t_{off}^0$ = intercept at zero force. Again demonstrating off rate exponentiation under force, the probability distributions for force predicted by Eq. (5) are found to agree closely with the force histograms in Fig. 4D–F from the tests of VCAM-1–$\alpha_4\beta_1$ integrin bonds and move upward in proportion to the logarithm of ramp rate (an increment of $\sim 2.3f_\beta$ for each order of magnitude increase). Comparing the distributions of lifetimes and forces plotted in Fig. 4, we immediately appreciate the generic character of "kinetically limited bond failure": that is, bonds pulled slowly have long lifetimes but break at small forces, whereas the same bonds pulled fast have much shorter lifetimes and break at larger forces.

The steady ramp method provides a nanorheological spectrum for how the mechanical strength of an interaction changes with stress rate. Key to the analysis of probe tests at cell surfaces, the dynamic profile of bond strength versus log(ramp rate) establishes a critical assay for whether a bond will be "strong" or "weak" when present in a linkage with other bonds. Using the stress dependencies of bond strength for mono-ICAM-1–$\alpha_L\beta_2$ and VCAM-1–$\alpha_4\beta_1$ interactions obtained *in vitro*,[2] we demonstrate in Fig. 5 how the relative attribute of "strong" or "weak" can switch from one bond to another bond between slow and fast ramp rates. More importantly, the measurements of mono-ICAM-1–$\alpha_L\beta_2$ and VCAM-1–$\alpha_4\beta_1$ interactions *in vitro* predict the likelihoods that these bonds will survive or fail in tests of attachments to cell surfaces, which is critical to the analysis of membrane–cytoskeletal failure events. As discussed next, differences in the dynamical hierarchy of "strong" versus "weak" interactions expressed as functions of stress rate accounts for the diversity in force responses observed when probing cell surface bonds.

---

[2] Consistent with the role of "firm adhesion" in leukocyte biology performed by LFA–1 (Carman and Springer, 2003; Gahmberg et al., 1998; Krieglstein and Granger, 2001; Springer, 1994), the results in Fig. 5 show that ICAM-1–$\alpha_L\beta_2$ interactions are long lived and insensitive to stress rate. By comparison, the results in Fig. 4 show that VCAM-1–$\alpha_4\beta_1$ interactions are much shorter lived and require higher stress rates to reach the same level of strength, accounting for the "selectin–like" character attributed to VLA-4 in the trafficking and sorting of cell progenitors in the bone marrow (Frenette et al., 1998). Results for both interactions (in tests not shown here) also exhibit the expected hierarchical-allosteric activation by divalent cations ($Mn^{2+} > Mg^{2+} >> Ca^{2+}$).

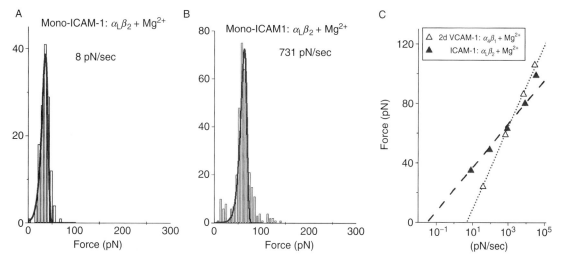

Fig. 5 (A and B) Rupture forces from ramp tests of interactions between mono-ICAM-1 and $\alpha_L\beta_2$ integrin linked covalently to glass microspheres and activated in Mg^{2+}. (C) The most frequent rupture forces plotted as functions of the logarithm of ramp rate show that mono-ICAM-1–$\alpha_L\beta_2$ integrin interactions are stronger than 2d VCAM-1–$\alpha_4\beta_1$ interactions under slow loading; but once the ramp rate exceeds 10^3 pN/sec, the mono-ICAM-1–$\alpha_L\beta_2$ integrin bonds are the statistically weaker interactions. The slopes of the dashed-line correlations in (C) provide the force scales ($f_\beta \sim 6$ and 14 pN) and apparent-unstressed lifetimes ($t_{off}^0 \sim 140$ sec and 1–2 sec) that characterize the exponential dependence of off rate on force for these interactions *in vitro*. Verifying that the "kinetically limited bond failure" model agrees with the statistics over all levels of force, the force scale and lifetime obtained for the mono-ICAM-1–$\alpha_L\beta_2$ integrin interaction were used to predict the force distributions [Eq. (5)] shown by the solid curves superposed on the histograms in (A and B). [Note that the $\alpha_L\beta_2$ heterodimer was a leucine-zipper construct produced by ICOS Corp. (Dr. Don Staunton) and made available to us through our collaboration with Dr. Scott Simon at University of California, Davis.]

II. Probing Bonds at Cell Surfaces

A. Phenomenology of Unbinding Events

The usual motivation for probing ligand–receptor interactions at cell surfaces is to establish the mechanical strength and dissociation kinetics of adhesive bonds, verifying when possible the origins of molecular activity through comparison to tests of recombinant forms of the molecules on solid substrates. Less obvious, we can surprisingly also test the mechanical strength and failure kinetics of the receptor–cytoskeletal linkage beneath the membrane. Again limiting probe-ligand coverage to achieve rare attachments, the experimental approach described above for testing bonds *in vitro* can be applied to interactions at cell surfaces. To minimize the hydrodynamic drag effects that accompany fast retraction (described in Section II.B), the best targets for probing receptor bonds at cell surfaces are small spherical cells captured from free suspension like the leukocytes seen in Fig. 6 taken from OTFP and BFP experiments. Held and maneuvered by a piezo-driven

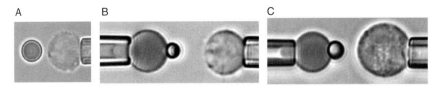

Fig. 6 (A) Image of a lymphocyte taken from a finger-prick blood sample, held by micropipette (at right) and tested against a 3-μm OTFP microsphere (at left). (B) Image of a B-lymphocyte taken from culture (obtained through a collaboration with Richard Larson, University of New Mexico), held by micropipette (at right) and tested against a 2-μm BFP microsphere (at left). (C) Image of a granulocyte (PMN) taken from a finger-prick blood sample, held by micropipette (at right) and tested against a 2-μm BFP microsphere (at left).

micropipette in the same way as a microsphere target, movement of the target away from the probe tip (retraction) at constant speed produces a variety of force responses after an attachment has formed, the simplest being a linear ramp of force as found when pulling on $\alpha_L\beta_2$ integrin (LFA-1) at PMN surfaces (cf. Fig. 7A). However, for other types of cells and receptors (including LFA-1 in the case of lymphocytes), the force–time trace begins with a steady linear increase in force but then deviates abruptly (as labeled by $\otimes$ in Fig. 7B and C), slowing progressively until detachment. The abrupt deviation in force marks the onset of viscous fluid-like extrusion of a membrane nanotube (called a *tether*). As the cell is pulled away from the site of receptor binding to the probe, the membrane and receptor separate from the underlying cytoskeletal structure. Recoil of the probe to zero force signals release of the cell membrane from the probe and the end of membrane extrusion. After subtracting out stretch of the probe, the initial linear rise in force shows that the cell has responded with a nanoscale ($\sim$200–500 nm) elastic-like deformation. Then released from the cytoskeleton, the membrane is pulled at a point by the probe to form a fast-growing nanotube often reaching several micrometer in length before detachment. More puzzling is that in some cases, pulling on a cell receptor at low speed causes membrane unbinding from the cytoskeleton but at fast speed, the membrane remains bound to the cytoskeleton until the receptor adhesive bond breaks. As we discuss below, the dynamics of receptor adhesive and cytoskeletal unbinding events can be explained by failure in a series of molecular bonds each with a different profile of "kinetically limited failure, that is strength versus applied force rate."[3]

Probing leukocyte adhesion receptors shows that attachments fail in some cases only at the cell surface connection to the probe (cf. Fig. 7A) whereas, in other cases,

[3] Note: force responses in tests of cell surface attachments in some cases reveal many events of membrane release and tether formation. However, cell tests with rare molecular attachments most often produce force responses like those in Fig. 7A–C with only one event of cell detachment preceded by one—or no—event of membrane release. Traces showing obvious multiple tethering events are discarded to increase the fraction of single–molecule attachments (Evans *et al.*, 2005).

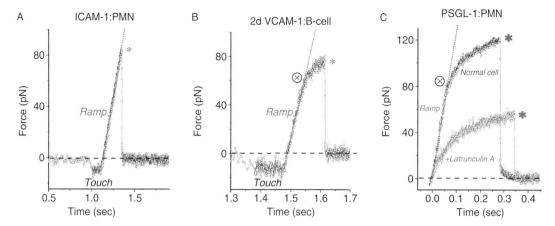

Fig. 7 Examples of leukocytes pulled away from receptor attachments to ligands on a BFP tip. (A) The force response during retraction of a PMN held by LFA-1 ($\alpha_L\beta_2$) to an ICAM-1 probe in Mn^{2+} shows only a precipitous event of adhesive detachment (pink star). By comparison, the force responses during retraction of a B-lymphocyte (B) held by VLA-4 ($\alpha_4\beta_1$) to a VCAM-1 probe in Mg^{2+} and during retraction of a PMN (C) held by PSGL-1 to a P-selectin probe in Ca^{2+} reveal membrane–cytoskeletal unbinding events ($\otimes$) prior to rupture of the probe-ligand bond (stars). Deviations from initial linear ramps and the slowing of force rates in (B and C) coincide with the periods of membrane tether extrusion after separation from the cytoskeleton. To demonstrate that receptor anchoring and the linear phase requires an intact cytoskeleton, the lower force trace in (C) shows the retraction of a PMN bound to a P-selectin probe in the presence of the actin-depolymerizing drug latrunculin A.

failure occurs first at the membrane–cytostructure interface (cf. Fig. 7B and C) and then at the cell connection to the probe. However, failure events in single traces are insufficient to establish if the outcome is the typical response to be observed with a high probability, or if the same outcome would occur at all pulling speeds. Likewise, it remains unclear as to whether the phenomenological events of cell detachment and membrane separation from the cytoskeleton originate from breakup at a specific site—or diverse sites—of molecular interactions. To address these issues and quantitatively analyze the kinetics of bond failure events, large numbers of cell–probe attachments must be tested and the tests repeated over a wide range of pulling speeds. More precisely, bond failure events need to be examined over a very broad span in the actual rate of force applied to the cell–probe attachments. Because the material structure of the cell interface is soft, the initial rate of force application to an attachment is typically 20–30% of the rate expected from the product of pipette retraction speed and probe spring constant.[4]

---

[4] The actual force rate $r_f = \Delta f/\Delta t$ (slope of the initial linear regime) depends on the elastic stiffness of the cell k_{cell}. The cell stiffness parameter can be derived from the ratio of the measured force rate r_f to the nominal loading rate $k_f v_{pull}$ (probe spring constant × cell retraction speed) using the expression $k_{cell} = k_f/(k_f v_{pull}/r_f - 1)$ based on stretch of two springs in series.

The force rate then drops rapidly after membrane release from the cytoskeleton. Thus, precise analysis of the process leading to each bond failure event ideally involves treatment of the entire history of force $f(t)$ up to the moment of failure.

In the case of membrane–cytoskeletal unbinding events, the force history is easily quantified by measuring the slope of the initial linear ramp and the transition force $f_\otimes$ that signifies onset of membrane tether flow (Evans *et al.*, 2005). By comparison, as apparent in Fig. 7B and C, the force history seen thereafter up to the time of cell detachment is very nonlinear and flattens as if approaching a plateau. Although it is possible to model the force flattening that defines the second phase of force history prior to cell detachment, this requires a detailed study of the tether growth process at different levels of force (Heinrich *et al.*, 2005), which is tedious and often difficult to express in a simple mathematical form. To avoid dealing with the force history following membrane–cytoskeletal separation, and at the same time reliably treat the kinetics of adhesive bond dissociation, the practical approach is to test the ligand–receptor interaction *in vitro* and establish a kinetic profile for the ligand–receptor off rate versus force. By quantifying this profile, the force and time statistics can be predicted for cell detachments under any history of pulling force [as used for example in an analysis of PMN tethering under flow by King *et al.* (2005)]. As we demonstrate in later examples, this *a priori* definition of the probability for adhesive bond survival under force is essential in the analysis of membrane–cytoskeletal failure.

B. Experimental Frustrations

Before discussing methods of analyzing results from tests of cell surface bonds, it is important to describe the difficulty that must be recognized when probing attachments to cell surfaces. As already noted, cells are soft structures. This means that initial touch of a cell with a small impingement force often produces a large contact area and, concomitantly, that the subsequent pulling at a point on the cell surface can result in a deformation much larger than the probe spring deflection. The former increases the level of nonspecific binding between the cell surface and the probe whereas the latter necessitates increasing retraction speeds by three- to four-fold to reach loading rates comparable to those applied *in vitro*. Perhaps most complicating in cell tests, the combination of large contact area and fast retraction speed can result in a large hydrodynamic "suction" between the surfaces, which appears as a spurious attachment force. The nonspecific suction force arises from the well-known "Reynolds lubrication" effect, where viscous dissipation strongly impedes flow of surrounding fluid into a narrow gap between the cell surface and probe tip. Along with the separation speed, the magnitude of the nonspecific force depends on the area and closeness of cell–probe contact, which are governed in turn by impingement force, cell surface deformability, and surface roughness. Even for very small impingement forces (<10 pN), touching a cell to a 1-μm radius probe tip can result in a contact area of ~ 0.3 μm^2 given a typical cell-interface stiffness of ~ 0.2 pN/nm. For this reason, when pulling leukocytes away from soft contact to a nonspecific BFP, hydrodynamic suction forces become prominent at retraction speeds >25 μm/sec, significantly increasing the frequency of attachments and masking the presence of small specific forces. The emergence of strong hydrodynamic coupling at fast pulling speed is illustrated in the force histograms (cf. Fig. 9A and B) taken from tests of PMN attachments to ICAM-1 probes. By testing small spherical cells and controlling initial cell contact to limit the impingement force, it is possible to minimize the nonspecific interaction and better detect attachments held by weak specific interactions. However, it is still often necessary to keep force rates low (<10^4 pN/sec) in order to avoid hydrodynamic coupling when testing weak specific interactions.

C. Most Likely Site of Molecular Unbinding When Probing Bonds at Cell Surfaces

The precision measurements of cell attachment forces in Fig. 7 demonstrate that membrane–cytoskeletal unbinding events can be readily identified when they occur prior to failure of the cell–probe adhesive bond. The question is how do we analyze these tests in such a way that we can attribute membrane–cytoskeletal unbinding to a specific site of bond failure? To address this question, we need to recognize that a receptor adhesion complex is held together by a number of molecular bonds, all stressed to some extent by the pulling force, and any one of which can be the first to fail. Modeled at the most simple level, the force applied in pulling on a cell surface receptor is viewed to propagate through a serial linkage of molecular connections inside the cell (cf. sketch in Fig. 8A) where each component interaction is characterized by an exponential dependence of off rate on applied force [Eq. (1)].

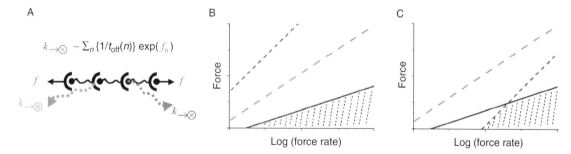

Fig. 8 (A) Schematic of a serial linkage in which each bond experiences the same pulling force and dissociation of any molecular component will cause failure. Each bond is modeled by an exponential dependence of off rate on force [Eq. (1)], defined by its force scale [i.e., $f_n = f/f_\beta(n)$] and unstressed off rate $1/t_{off}(n)$, where n identifies a particular bond. (B and C) Graphical hierarchy of "kinetically limited failure" in a linkage of three bonds under increasing force. Plotted versus the logarithm of force rate, the "most likely" force for failure of a bond maps to a particular linear regime (dark-short dash, gray-long dash, solid) as defined by its force scale f_β and force rate intercept f_β/t_{off}. The shaded area shows the region of mechanical stability where the linkage is unlikely to fail. The upper boundary for stability at any force rate is the linear regime identifying the "weakest" bond. The weakest bond can switch from one site to another as seen in panel (C) (the dark-short dash interaction being weakest at low force rate then gray-long dash at high force rate).

On the basis of this idealized model and assuming a steadily increasing force, we can construct a simple graph that expresses the hierarchy of mechanical strengths predicting the most likely molecular site for bond failure (Evans, 1998). Illustrated in Fig. 8B and C for a linkage of three bonds, the shaded areas represent the regions of mechanical stability where no bonds are expected to fail. The limit for stability at a particular force rate is set by the molecular component most vulnerable to failure (i.e., the "weakest" bond). Figure 8B demonstrates the only hierarchy of "kinetically limited failure" where a molecular site for "most likely" failure remains the same bond at all force rates. Here, the zero force intercepts of the force rate curves, $r_f^0 = f_\beta / t_{off}$, for the bonds follow a numerical order reciprocal to that of their force scales (slopes), that is $f_\beta(1) > f_\beta(2) > f_\beta(3)$ and $r_f^0(1) < r_f^0(2) < r_f^0(3)$. With special exceptions, this numerical order implies that the bond with the fastest off rate remains the most likely site of failure at all force rates. On the other hand, if the force rate intercepts and slopes do not array in reciprocal numerical order, crossovers will occur between regimes as the force rate is increased (cf. Fig. 8C). In this case, the relative attribute of "strong" versus "weak" switches from one bond to another at a higher force rate. Again with special exceptions, the scenario in Fig. 8C implies that the bond with the slowest spontaneous off rate usually remains the strongest connection. But of the other two bonds, the one with the slower off rate will be the stronger when the linkage is pulled slowly whereas the faster-dissociating bond will be the stronger when the linkage is pulled rapidly because its off rate is less sensitive to force (steeper slope). There is a third possible scenario (not shown), where the site of failure switches through all of the bonds if the intercepts follow the same numerical order as the slopes. In the next sections, we present specific examples of membrane–cytoskeletal unbinding and adhesive detachment events that illustrate the behavior seen in Fig. 8B and C.

D. Adhesive Failure Without Membrane–Cytoskeletal Separation

Consistent with the persistent hierarchy of bond strengths schematized in Fig. 8B, retracting PMNs following attachment of LFA-1 integrin to a di-ICAM-1 probe nearly always produces a linear ramp of force that ends by precipitous detachment (cf. Fig. 7A). Even though observed at all pulling speeds, a consistent event of abrupt detachment could arise from either failure of the ligand bond or extraction of the receptor from the lipid bilayer. For this reason, a key step in analyzing results in this case as well as the more complicated cases to follow will be to compare the force rate dependence of the probe detachment forces to that found for the ligand–receptor unbinding *in vitro*. Hence, to establish that the ICAM-1–LFA-1 interaction was indeed the most frequent site of failure, we predicted distributions for the probe detachment force using the parameters [Eq. (1)] derived for the off rate kinetics of ICAM-1–$\alpha_L \beta_2$ interactions under force *in vitro* and compared the computed distributions to the statistics of cell detachment forces.[5] Summarizing the outcome, Fig. 9A–C shows that the most likely rupture of di-ICAM-1–LFA-1 attachments to PMNs in Mn^{2+} occurs at nearly the same forces for all force rates as rupture of mono-ICAM-1–$\alpha_L\beta_2$ bonds between microspheres. Since nearly all of the adhesive detachments occurred in

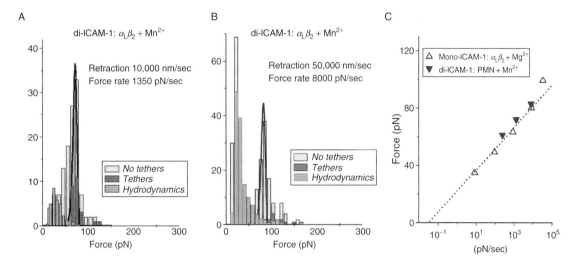

Fig. 9 (A and B) Histograms of cell detachment forces obtained from tests of di-ICAM-1–LFA-1 bonds at PMN surfaces in Mn^{2+} under pulling speeds that produced initial force ramps of 1350 and 8000 pN/sec. Light gray bins identify the majority of cell detachments that showed no prior event of membrane–cytoskeletal unbinding; dark gray bins identify a few cell detachments preceded by membrane–cytoskeletal unbinding and tether formation. [Also plotted are the nonspecific forces (gray bins) obtained using a probe decorated with an irrelevant protein, which demonstrate the emergence of strong hydrodynamic coupling at fast retraction speeds.] Modeled by random "unzipping" of a dimer of monomeric interactions,[5] the exponential dependence of bond failure rate on force [Eq. (1)] found for mono-ICAM-1–$\alpha_L\beta_2$ bonds *in vitro* (cf. Fig. 5) was used to predict the force distributions for di-ICAM-1–LFA-1 bonds (solid curves) superposed on each histogram. (C) The most frequent rupture forces plotted as functions of the logarithm of ramp rate show that the strengths of di-ICAM-1–LFA-1 bonds at PMN surfaces were nearly the same as strengths of mono-ICAM-1–$\alpha_L\beta_2$ bonds between microspheres over the experimental range of force rate.

---

[5] When probing LFA-1 on leukocytes like PMNs, we have only been able to obtain viable specific attachments if the probe tips are decorated with a Fc dimer of ICAM-1 (di-ICAM-1). Activating the LFA-1 in a Mn^{2+} buffer (or in a Mg^{2+} buffer that also contained the activating monoclonal 240Q), we found the peaks in force distributions (Fig. 9A and B) to be uniformly narrower and slightly stronger (2–3 pN) at all stress rates than in tests of mono-ICAM-1:$\alpha_L\beta_2$ bonds *in vitro*. This outcome implies that the ICAM-1 Fc dimer interaction acted like a *zipper* of two monomeric interactions. To verify the agreement between kinetics found *in vitro* and the force distributions from the cell tests, we modeled the distributions of forces with a probability density for random "unzipping" of two bonds, $p_2(f) = p_1(f)$ $[C_{frq}(f)]$, where $p_1(f)$ is the distribution for single bond failure and $C_{frq}(f) = (f_\beta/r_f)\int_{1\to f} k_{off}(y)dy = (f_\beta/r_f)$ $[\exp(f/f_\beta)-1]$ describes the cumulated rate of prior bond failure at a particular force (Williams and Evans, 2002).

traces showing no prior event of membrane–cytoskeletal unbinding, we conclude that anchoring of the membrane and LFA-1 to the PMN cytoskeleton is statistically stronger than ICAM-1–LFA-1 adhesive bond.

E. Membrane–Cytoskeletal Unbinding Followed by Adhesive Detachment

Also consistent with the persistent hierarchy of bond strengths schematized in Fig. 8B, retracting PMNs following attachment of platelet sialo-glycoprotein-1 (PSGL-1) to a P-selectin probe nearly always results in membrane unbinding from the PMN cytostructure prior to probe detachment, when the force rates exceed 200 pN/sec (cf. Fig. 7C; Evans *et al.*, 2005). Considering only single tethering events, we hypothesized that the abrupt termination of the elastic-like regimes (marked by the symbol $\otimes$ in Fig. 7C) originated from some type of discrete molecular unbinding event within the membrane–receptor linkage to the PMN cytostructure allowing a membrane tether to form. To examine this hypothesis, the forces $f_\otimes$ at membrane separation and force rates r_f leading up to each event were measured in thousands of force traces, at pulling speeds between 400 and 150,000 nm/sec. Illustrated by the examples in Fig. 10A and B, these data provided

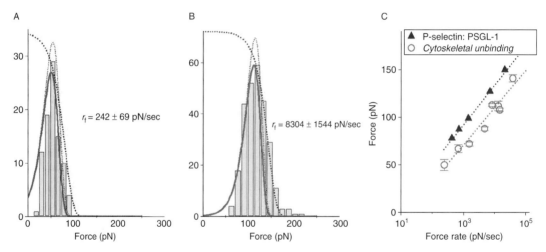

Fig. 10 (A and B) Histograms of rupture forces for membrane–cytoskeletal unbinding events found in tests of PSGL-1–P-selectin bonds at PMN surfaces in Ca^{2+} with initial force ramps of 242 and 8304 pN/sec. (C) The most frequent forces for membrane–cytoskeletal unbinding (circles) are plotted as functions of the logarithm of applied force rates and compared with the strengths of P-selectin–PSGL-1 bonds (triangles) measured *in vitro* (Evans *et al.*, 2004). Modeling the rate of membrane–cytoskeletal unbinding as an exponential function of force [Eq. (1)] and using the linear correlation (light gray-dotted line) in (C) to provide the thermal activation scale for force (slope = 17 pN), the distributions for "kinetically limited failure" [Eq. (5)] were computed and fit to each histogram by varying the unstressed off rate $1/t_{off}^0$ within a very narrow range (0.7–1.4 sec^{-1}). To illustrate the corrections to distributions for preemptive adhesive failure, the initial fitting distributions (cf. light gray-dotted curves in A and B) are shown compared to the probabilities predicted for survival of the PSGL-1–P-selectin adhesive bond (illustrated by the black-dotted curves in A and B).

histograms of membrane separation events at eight values of the initial force rate r_f between 200 and 40,000 pN/sec. Plotting the most frequent forces for membrane separation in each histogram as a function of the logarithm of the mean force rates r_f ($\pm$SD) produced a linear correlation consistent with the "kinetically limited failure" of a molecular complex. To critically test the implied outcome, the slope was used to define a force scale, $f_\beta = 17$ pN, for off rate exponentiation [Eq. (1)] that could be used to predict the distributions of force at each force rate and aligned to the peak in each histogram using the unstressed off rate $1/t^0_{\mathrm{off}}$ as fitting parameter. As shown by the solid curves in Fig. 10A and B, a very narrow range of the unstressed off rate, $1/t^0_{\mathrm{off}} = 0.7$–$1.7$ sec^{-1}, provided good agreement between the predicted distributions and histograms at all force rates. This outcome suggests that the rate of membrane–PSGL-1 unbinding from the PMN cytostructure is regulated by a weak complex whose rate of failure increases exponentially under stress as described by $k_{\mathrm{off}}(f) \approx (0.7$–$1.7$ sec$^{-1})\exp(f/17$ pN$)$. Summarized in Fig. 10C, and consistent with the tandem sequence of events, the membrane–PSGL-1 linkage to the PMN cytoskeleton fails most often at a force below the level needed to rupture P-selectin–PSGL-1 adhesive bonds *in vitro* (Evans *et al.*, 2004), when force rates exceed 200 pN/sec. Confirmed by analysis of PMN detachment times (Heinrich *et al.*, 2005), the membrane separation from the cytoskeleton seems to act like a mechanical fuse that limits the force to enable longer survival of P-selectin attachments to PMNs.

F. Overlapping Bond Failure Processes

Significantly more complicated to analyze, retraction of B-lymphocytes following attachment of VLA-4 to a VCAM-1 probe (cf. Fig. 11) produces a mixture of force responses at a fixed force rate, with some traces showing membrane–cytoskeletal unbinding followed by probe detachment, the other traces only probe detachment. Illustrated by the force histograms in Fig. 11A and B, the populations for only probe detachment and those for membrane–cytoskeletal unbinding plus probe detachment are seen to overlap to an extent that depends on the force rate. In particular, the relative proportion of membrane–cytoskeletal unbinding events is seen to increase significantly at the higher force rate. To analyze this more complex situation, the first step was to predict the probability [Eq. (4)] for survival of the adhesive bond, $S_{\mathrm{ab}}(f)$, and force distributions, $p_{\mathrm{abf}}(f) = k_{\mathrm{abf}}(f)S_{\mathrm{ibid}}(f)$, expected for adhesive bond failure without membrane–cytoskeletal unbinding, based on the exponential dependence of failure rate $k_{\mathrm{abf}}(f)$ on force [Eq. (1)] found for 2d VCAM-1–$\alpha_4\beta_1$ bonds *in vitro* (cf. Fig. 4). Then, also based on exponential amplification under force, a trial function was postulated for the rate $k_{\mathrm{mcf}}(f)$ of membrane–cytoskeletal unbinding and used to compute the probability for survival $S_{\mathrm{mc}}(f)$ of the membrane–receptor anchoring. Treated as independent random events, the trial function for the joint distribution of B-cell detachment

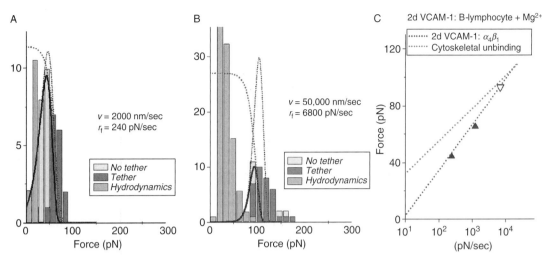

Fig. 11 (A and B) Detachment forces in tests of bonds between a 2d VCAM-1 probe and the VLA-4 receptor on B-lymphocytes in Mg^{2+}, shown here for initial force ramps of 240 and 6800 pN/sec. Light gray bins identify cell detachments without a prior event of membrane–cytoskeletal unbinding; dark gray bins identify cell detachments preceded by a membrane–cytoskeletal unbinding event and tether formation. [Again, the nonspecific forces found using a probe decorated with an irrelevant protein (gray bins) have been added to emphasize the strong hydrodynamic coupling at fast retraction speeds.] Modeling failure rates for membrane–cytoskeletal unbinding and adhesive detachment as exponentials of force [Eq. (1)], the changes in fractions of cell detachment events with—and without—membrane tethering under increase in pulling speed were found to be matched using the rate, $k_{mcf}(f) \approx (0.04 \text{ sec}^{-1})$ $\exp(f/10 \text{ pN})$, to characterize membrane–cytoskeletal unbinding and the off rate described in Fig. 4 for 2d VCAM-1–$\alpha_4\beta_1$ interactions *in vitro* for adhesive bond failure. The distributions predicted for adhesive bond failure without membrane tethering are shown as solid curves. The survival probabilities for best-fit of the membrane–cytoskeletal unbinding rate are shown as dotted curves, demonstrating the emergence of tethering events in force traces prior to adhesive bond failure at fast force rates. (C) The most frequent forces for adhesive bond failure without membrane tethering (triangles) are shown to correlate with the "kinetically limited failure" found for 2d VCAM-1–$\alpha_4\beta_1$ bonds *in vitro* (dark-dotted line, cf. Fig. 4) and to merge with the "kinetically limited failure" derived for membrane–cytoskeletal anchoring (light-dotted line) at fast force rates.

without membrane–receptor unbinding, $p_{abf}(f)S_{mc}(f)$,[6] was fit to the statistics of B-cell detachment without membrane tethering (cf. Fig. 11A and B) through variation of the parameters f_β and t_{off}^0 defining the kinetic rate $k_{mcf}(f)$. As further verification, the corresponding force distributions, $k_{mcf}(f)S_{mc}(f)S_{ab}(f)$, were computed for membrane–cytoskeletal unbinding prior to B-cell detachment and compared to the statistics of forces at onset of tether formation. The dynamic

---

[6] For "kinetically limited failure" of two bonds in series, the failure rates (Eq. [1]) for the bonds under force $k_1(f), k_2(f)$, and Eq. (4) describe the joint survival probability $S_1(f)S_2(f)$. The probability distribution [Eq. (5)] then specifies the force statistics for the two failure events, $p_1(f) = k_1(f)S_1(f)S_2(f)$ and $p_2(f) = k_2(f)S_1(f)S_2(f)$.

profiles obtained for each "kinetically limited failure" are plotted in Fig. 11C and demonstrate that the increasing overlap in the events reveal approach to a switch between the sites of "strong" and "weak" failure at fast force rates like that schematized in Fig. 8C. In this case, the adhesive bond is weaker than the complex anchoring the membrane to the cytoskeleton under slow force rates but then exceeds the strength of membrane–cytoskeletal anchoring at fast force rates.

Again demonstrating a switch in sites of "strong" and "weak" failure under fast force loading, retraction of lymphocytes bound with the activating mab 240Q following attachment of LFA-1 to a di-ICAM-1 probe results in overlapping subpopulations of membrane–cytoskeletal unbinding and cell detachment forces. However, the outcome is exactly opposite to the example of VLA-4 attachments to a VCAM-1 probe described above. In this case, the subpopulations of events were found to switch from events of cell detachment nearly always preceded by membrane tethering at slow force rates to events of cell detachments without tethering at fast force rates (cf. Fig. 12A and B). As above, the first step in the analysis was to predict the probability for survival of the adhesive bond $S_{ab}(f)$ and force distributions $p_{abf}(f)$ expected for adhesive bond failure without membrane–cytoskeletal unbinding. In this case, the failure rate $k_{abf}(f)$ for di-ICAM-1–LFA-1 was based on "unzipping" of two mono-ICAM-1–$\alpha_L\beta_2$ bonds *in vitro* (cf. Fig. 5 and footnote 5). Next, using Eq. (1) to describe the rate $k_{mcf}(f)$ of membrane–cytoskeletal unbinding and compute the probability for survival $S_{mc}(f)$ of the membrane–receptor anchoring, trial distributions for forces of membrane–cytoskeletal unbinding and cell detachment without tethering were fit to the populations of these events measured at all force rates as illustrated in Fig. 12A and B, again varying the parameters f_β and t^0_{off} defining the rate $k_{mcf}(f)$. The dynamic profiles for each "kinetically limited failure" are plotted in Fig. 12C, demonstrating the emergence of preemptive adhesive bond failure at fast pulling speeds and the durability of membrane–cytoskeletal anchoring.

III. Future Challenge and Opportunity

Adhesive bonding between a cell and its substrate plays a much larger role in biology than simply arresting a cell at a surface. The mechanical strength of an adhesive attachment is usually attributed to the receptor–ligand interaction, but this view clearly overlooks the impact of the adhesive force on molecular connections linking the receptor to the cytostructure below the membrane surface. Probing single bonds at cell surfaces with nanoscale resolution, we find that the response provides a quantitative assay for how ligand–receptor and membrane–cytostructural interactions regulate the mechanical strength of a cell adhesion complex. In doing this, we have demonstrated that failure local to a subset (or complex) of membrane receptor–cytoskeletal interactions appears to limit adhesion strength in many cases. By quantifying abrupt transitions in force response leading up to—and including—adhesive detachment, we have described

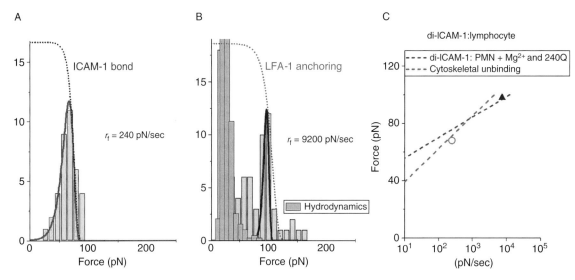

Fig. 12 (A) Membrane–cytoskeletal unbinding forces obtained in tests of bonds between a di-ICAM-1 probe and the LFA-1 receptor on lymphocytes bound with mab 240Q in Mg^{2+} under a slow initial force ramp of 240 pN/sec. (B) Cell detachment forces in tests of bonds between the same di-ICAM-1 probe and LFA-1 on lymphocytes bound with mab 240Q in Mg^{2+} under a fast initial force ramp of 9200 pN/sec. [The nonspecific forces found using a probe decorated with an irrelevant protein (gray bins) emphasize the strong hydrodynamic coupling at fast retraction speeds.] Modeling failure for membrane–cytoskeletal unbinding and adhesive detachment as exponentials of force [Eq. (1)], the distributions of cell detachment forces with—and without—membrane tethering under increase in pulling speed were found to be matched using the rate, $k_{mcf}(f) \approx (0.02\ sec^{-1}) \exp(f/10\ pN)$, to characterize membrane–cytoskeletal unbinding and using the off rate in Fig. 5 for mono-ICAM-1–$\alpha_L\beta_2$ interactions *in vitro* to describe "unzipping" of a di-ICAM-1 adhesive interaction (see text). The force distribution for membrane tethering at slow force rate is shown by the solid curve in (A). The force distribution predicted for adhesive bond failure without membrane tethering at fast force rate is shown by the solid curve in (B). The corresponding probabilities for survival of membrane–cytoskeletal anchoring and survival of the adhesive bond are superposed as dotted curves in panels (A) and (B), respectively. (C) The most frequent forces for adhesive bond failure without membrane tethering at fast force rate (triangle) and for membrane–cytoskeletal unbinding at slow force rate (circle) are plotted adjacent to the "kinetically limited failure" characteristics describing di-ICAM-1–LFA-1 bonds and membrane–cytoskeletal anchoring, respectively, which cross at an intermediate force rate of $\sim 10^3$ pN/sec.

a nanomechanical method to quantify the failure kinetics of the bonds governing both adhesive detachment and receptor anchoring to the cytostructure. The challenge is to relate these kinetic properties to the components that make up the putative structural complex. Using the methods described here, the obvious approach to accomplish this goal is to probe cells engineered with strategic alterations in receptor–cytoskeletal interactions. Although a significant task, many of these lines have already been established for integrin receptors. In this way, we have the opportunity to "move inside the cell" and produce a molecular map of the "mechanical circuitry" that anchors receptor tail domains to the cytostructure. Much more bold and exciting, the future objective should be to then explore the role that this "circuitry" plays in signaling and regulation of important chemical pathways.

Acknowledgments

The experimental examples described in this chapter are from work supported by National Institutes of Health grants HL65333 and HL31579. The authors greatly appreciate the collaborative support of Dr. Scott Simon, University of California Davis, in the studies of the recombinant $\alpha_L\beta_2$ integrin as well as Dr. Richard Larson and his colleagues at the University of New Mexico School of Medicine in the studies of B-lymphocyte interactions with recombinant 2d VCAM-1.

References

Abbondanzieri, E. A., Greenleaf, W. J., Shaevitz, J. W., Landick, R., and Block, S. M. (2005). Direct observation of base-pair stepping by RNA polymerase. *Nature* **438,** 460–465.

Asbury, C. L., Fehr, A. N., and Block, S. M. (2003). Kinesin moves by an asymmetric hand-over-hand mechanism. *Science* **302,** 2130–2134.

Ashkin, A., and Dziedzic, J. M. (1987). Optical trapping and manipulation of viruses and bacteria. *Science* **235,** 1517–1520.

Bayas, M. V., Leung, A., Evans, E., and Leckband, D. (2006). Lifetime measurements reveal kinetic differences between homophilic cadherin bonds. *Biophys. J.* **90,** 1385–1395.

Bell, G. I. (1978). Models for the specific adhesion of cells to cells. *Science* **200,** 618–627.

Binnig, G., Quate, C. F., and Gerber, Ch. (1986). Atomic force microscope. *Phys. Rev. Lett.* **56,** 930–933.

Carl, P., Kwok, C. H., Manderson, G., Speicher, D. W., and Discher, D. E. (2001). Forced unfolding modulated by disulfide bonds in the Ig domains of a cell adhesion molecule. *Proc. Natl. Acad. Sci. USA* **98,** 1565–1570.

Carman, C. V., and Springer, T. A. (2003). Integrin avidity regulation: Are changes in affinity and conformation underemphasized? *Curr. Opin. Cell Biol.* **15,** 547–556.

Dietz, H., and Rief, M. (2004). Exploring the energy landscape of GFP by single-molecule mechanical experiments. *Proc. Natl. Acad. Sci. USA* **101,** 16192–16197.

Evans, E. (1998). Energy landscapes of biomolecular adhesion and receptor anchoring at interfaces explored with dynamic force spectroscopy. *Faraday Discuss Chem. Soc.* **111,** 1–15.

Evans, E. (2001). Probing the relation between force—lifetime—and chemistry in single molecular bonds. *Annu. Rev. Biophys. Biomol. Struct* **30,** 105–128.

Evans, E., Heinrich, V., Leung, A., and Kinoshita, K. (2005). Nano-to-micro scale dynamics of P-selectin detachment from leukocyte interfaces: I. Separation of PSGL-1 from the cell cytoskeleton. *Biophys. J.* **88,** 2288–2298.

Evans, E., Leung, A., Heinrich, V., and Zhu, C. (2004). Mechanical switching and coupling between two pathways for dissociation in the PSGL-1:P-selectin adhesion bond. *Proc. Natl. Acad. Sci. USA* **101,** 11282–11286.

Evans, E., and Ritchie, K. (1997). Dynamic strength of molecular adhesion bonds. *Biophys. J.* **72,** 1541–1555.

Evans, E., Ritchie, K., and Merkel, R. (1995). Sensitive force technique to probe molecular adhesion and structural linkages at biological interfaces. *Biophys. J.* **68,** 2580–2587.

Evans, E., and Williams, P. (2002). Dynamic force spectroscopy: I. Single bonds. Dynamic force spectroscopy: II. Multiple bonds. *In* "Physics of Bio-Molecules and Cells, *Ecoles des HOUCHES d'Ete* LXXV," pp. 145–203. EDP Sciences, Springer-Verlag.

Frenette, P. S., Subbarao, S., Mazo, I. B., and von Andrian, U. H. (1998). Endothelial selectins and vascular cell adhesion molecule-1 promote hematopoietic progenitor homing to bone marrow. *Proc. Natl. Acad. Sci. USA* **95**, 14423–14428.

Gahmberg, C. G., Valmu, L., Fagerholm, S., Kotovuori, P., Ihanus, E., Tian, L., and Pessa-Morikawa, T. (1998). Leukocyte integrins and inflammation. *Cell. Mol. Life Sci.* **54**, 549–555.

Heinrich, V., Leung, A., and Evans, E. (2005). Nano-to-micro scale dynamics of P-selectin detachment from leukocyte interfaces: II. Tether flow terminated by P-selectin dissociation from PSGL-1. *Biophys. J.* **88**, 2299–2308.

King, M., Heinrich, V., Evans, E., and Hammer, D. (2005). Nano-to-micro scale dynamics of P-selectin detachment from leukocyte interfaces: III. Numerical simulation of tethering under flow. *Biophys. J* **88**, 1676–1683.

Krieglstein, C. F., and Granger, D. N. (2001). Adhesion molecules and their role in vascular disease. *Am. J. Hematol.* **14**, 44S–54S.

Kuo, S. C., and Sheetz, M. P. (1993). Force of single kinesin molecules measured with optical tweezers. *Science* **260**, 232–234.

Law, R., Carl, P., Harper, S., Dalhaimer, P., Speicher, D. W., and Discher, D. E. (2003). Cooperativity in forced unfolding of tandem spectrin repeats. *Biophys. J.* **84**, 533–544.

Li, H.-B., Oberhauser, A. F., Fowler, S. B., Clark, J., and Fernandez, J. M. (2000). Atomic force microscopy reveals the mechanical design of a modular protein. *Proc. Natl. Acad. Sci. USA* **97**, 6527–6531.

Liphardt, J., Dumont, S., Smith, S. B., Tinoco, I., Jr., and Bustamante, C. (2002). Equilibrium information from nonequilibrium measurements in an experimental test of Jarzynski's equality. *Science* **296**, 1832–1835.

Liphardt, J., Onoa, B., Smith, S. B., Tinoco, I., Jr., and Bustamante, C. (2001). Reversible unfolding of single RNA molecules by mechanical force. *Science* **292**, 733–737.

Marshall, B. T., Long, M., Piper, J. W., Yago, T., McEver, R. P., and Zhu, C. (2003). Direct observation of catch bonds involving cell-adhesion molecules. *Nature* **423**, 190–193.

Merkel, R., Nassoy, P., Leung, A., Ritchie, K., and Evans, E. (1999). Energy landscapes of receptor-ligand bonds explored with dynamic force spectroscopy. *Nature* **397**, 50–53.

Müller, D. J., Baumeister, W., and Engel, A. (1999). Controlled unzipping of a bacterial surface layer with atomic force microscopy. *Proc. Natl. Acad. Sci. USA* **96**, 13170–13174.

Perret, E., Leung, A., Feracci, H., and Evans, E. (2004). *Trans*-bonded pairs of E-cadherin exhibit a remarkable hierarchy of mechanical strengths. *Proc. Natl. Acad. Sci. USA* **101**, 16472–16477.

Rief, M., Gautel, M., Oesterhelt, F., Fernandez, J. M., and Gaub, H. E. (1997). Reversible unfolding of individual titin immunoglobulin domains by AFM. *Science* **276**, 1109–1111.

Smith, S. B., Cui, Y., and Bustamante, C. (1996). Overstretching B-DNA: The elastic response of individual double-stranded and single-stranded DNA molecules. *Science* **271**, 795–799.

Schnitzer, M. J., and Block, S. M. (1997). Kinesin hydrolyses one ATP per 8-nm step. *Nature* **388**, 386–390.

Simmons, R. M., Finer, J. T., Chu, S., and Spudich, J. A. (1996). Quantitative measurements of force and displacement using an optical trap. *Biophys. J.* **70**, 1813–1822.

Springer, T. A. (1994). Traffic signals for lymphocyte recirculation and leukocyte emigration: The multistep paradigm. *Cell* **76**, 301–314.

Svoboda, K., Schmidt, C. F., Schnapp, B. J., and Block, S. M. (1993). Direct observation of kinesin stepping by optical trapping interferometry. *Nature* **365**, 721–727.

Williams, P., and Evans, E. (2002). Dynamic force spectroscopy: II. Multiple bonds. *In* "Physics of Bio-Molecules and Cells, *Ecoles des HOUCHES d'Ete* LXXV," pp. 186–203. EDP Sciences, Springer-Verlag.

CHAPTER 17

High-Throughput Rheological Measurements with an Optical Stretcher

Bryan Lincoln, Falk Wottawah, Stefan Schinkinger, Susanne Ebert, and Jochen Guck

Institut für Experimentelle Physik I
Universität Leipzig
Linnéstrasse 5
04103 Leipzig, Germany

METHODS IN CELL BIOLOGY, VOL. 83
Copyright 2007, Elsevier Inc. All rights reserved.

0091-679X/07 $35.00
DOI: 10.1016/S0091-679X(07)83017-2

Abstract

   The cytoskeleton is a major determinant of the mechanical strength and morphology of most cells. The composition and assembly state of this intracellular polymer network evolve during the differentiation of cells, and the structure is involved in many cellular functions and is characteristically altered in many diseases, including cancer. Here we exploit the deformability of the cytoskeleton as a link between molecular structure and biological function, to distinguish between cells in different states by using a laser-based optical stretcher (OS) coupled with microfluidic handling of cells. An OS is a cell-sized, dual-beam laser trap designed to nondestructively test the deformability of single suspended cells. Combined with microfluidic delivery, many cells can be measured serially in a short amount of time. With this tool it could be shown that optical deformability is sensitive enough to monitor subtle changes during the progression of cells from normal to cancerous and even a metastatic state. Stem cells can also be distinguished from more differentiated cells. The surprisingly low number of cells required for this assay reflects the tight regulation of the cytoskeleton by the cell. This suggests the possibility of using optical deformability as an inherent cell marker for basic cell biological investigation, diagnosis of disease, and sorting of stem cells from heterogeneous populations, obviating the need for external markers or special preparation. Many additional biological assays can be easily adapted to utilize this innovative physical method. This chapter details the setup and use of the microfluidic OS, the analysis and interpretation of data, and the results of a typical experiment.

I. Introduction

   Physical properties of cells are in many ways as important to cell biology as signaling pathways and gene expression levels. While at first sight, the mechanics of cells might appear to have little to do with the everyday work of biologists, cell mechanical properties are largely determined by the cytoskeleton, a polymeric network of various filaments, which plays an important role in many cellular processes (Alberts *et al.*, 1994). There are three different types of filaments that, together with their accessory proteins, collectively form the cytoskeleton. Actin filaments are semiflexible polymers that form the actin cortex at the perimeter of the cell or span the cell interior and are often bundled to form stress fibers in strongly adhesive cells. Actin is often thought to govern cell mechanics at small deformations. Microtubules, the second type of filaments, are fairly rigid, rod-like polymers that act as highways for motor protein transport and are especially important during mitosis. Both actin and microtubules are very dynamic and can rapidly assemble and disassemble. Intermediate filaments are flexible polymers and are more static than the other two types of filaments. They protect the integrity of the cell at large deformations ($>10\%$). While cells also contain nuclei and other organelles and are surrounded by the cell membrane, these structures do not seem to contribute as much to a cell's resistance to external forces.

The cytoskeleton is not only the main determinant of cell mechanics, it is also involved in many vital cellular processes. Cells expend energy to regulate their biochemical environment and actively control the conditions that lead to filament polymerization, severing, bundling, cross-linking, and sliding. In this way, the cytoskeleton is always changing and adapting to its environment. The dynamic nature of this system is critical for processes such as differentiation, mitosis, motility, intracellular transport, phagocytosis, and mechanotransduction (Elson, 1988). The connection between cell function and the cytoskeleton is also evident from multiple G-protein-mediated signaling pathways that affect the cytoskeleton (Bloch et al., 2001). This link between the processes mediated by a well-regulated cytoskeleton and cellular mechanical properties can be exploited to study these processes. Whenever a cell alters its cytoskeleton, its mechanical properties change, which can be monitored by appropriate techniques. Changes can be brought about by physiological processes, by pathological perturbations, or in response to manipulations of a researcher. While any cellular process that involves the cytoskeleton can be the target of such a study, there are some examples that seem most promising.

Among physiological processes involving the cytoskeleton, the effects of mitosis on cell mechanics can be used to discriminate proliferating cells from postmitotic cells. Differentiated cells will also likely have a distinct cytoskeleton from progenitor cells, which may be identified in a heterogeneous population. Likewise, motile cells such as activated macrophages can be discriminated from stationary or nonactivated cells.

The cytoskeleton can also be modified by the addition of certain drugs and chemicals or by specific genetic modifications. Toxins such as cytochalasins, latrunculins, phalloidin, nocodazole, taxol, or bradykinin, which disrupt or stabilize specific targets in the cytoskeleton, lead to measurable changes in the physical properties of cells. Similarly, the influence of unknown chemicals, drugs, or molecules (e.g., siRNA) and the effect of overexpression or knockout of certain genes on the cytoskeleton can be tested by monitoring the mechanical resistance of cells. Viability tests also fall into this category, as dead cells certainly will exhibit different mechanical properties from live cells. This could be useful for drug-screening applications or for assessing transfection efficiencies.

There are also many well-known examples of pathological changes that affect the mechanical properties of cells. These include cytoskeletal alterations of blood cells that cause capillary obstructions and circulatory problems (Worthen et al., 1989); genetic disorders of intermediate filaments that lead to problems with skin, hair, liver, colon, and motor neuron diseases such as amyotrophic lateral sclerosis (ALS) (Fuchs and Cleveland, 1998), and various blood diseases including malaria (Suresh et al., 2005), sickle-cell anemia, hereditary spherocytosis, or immune hemolytic anemia (Bosch et al., 1994). Especially well investigated is the progression of cancer where the changes include a reduction in the amounts of constituent polymers and accessory proteins, and restructuring of the cytoskeletal network (Ben-Ze'ev, 1985; Cunningham et al., 1992; Moustakas and Stournaras, 1999; Rao and Cohen, 1991), with a corresponding change in cellular mechanical properties (Guck et al., 2005; Lekka et al., 1999; Park et al., 2005; Thoumine and Ott, 1997;

Wachsstock *et al.*, 1994; Ward *et al.*, 1991; Wottawah *et al.*, 2005a,b). All of these examples suggest that mechanical properties can serve as a cell marker to investigate cellular processes, to characterize cells, and to diagnose diseases.

How does an approach that uses mechanical properties as a physical "marker" of cell state compare in sensitivity to molecular biological approaches? From polymer physics we know that the mechanical strength of a network of filaments does not depend linearly on the constituent proteins: for example, twice as much actin makes a cell more than twice as stiff. Indeed, the elastic modulus of a polymer network scales with the concentration of filaments and the concentration of cross-linking molecules raised to a power between 2.2 and 6, depending on network structure (Gardel *et al.*, 2004; Janmey, 1991; Wilhelm and Frey, 2003). This means that even small changes in molecular composition of the cytoskeleton and its accessory proteins are dramatically amplified in cell mechanical properties. Thus, unlike many other techniques such as Western blots, gel electrophoresis, microarrays, or FACS analysis, the measured parameter contains a built-in amplification mechanism. This benefit is accompanied by the ability to determine this parameter for single cells—not on cell populations. A few altered cells can be identified in principle against the background of many unaltered cells, leading to an excellent signal-to-noise ratio. This is especially important when only few cells are available in the first place. In addition, the intrinsic nature of the mechanical properties renders any sort of tagging preparation (radioactive or fluorescent labeling, and so on) unnecessary, saving time and cost, while leaving the cells alive, intact, and ready for further analysis or use.

While there are many techniques covered in this volume that can sensitively detect changes in mechanical properties, here we present a particular type of laser trap called an optical stretcher (OS), for the controlled and nondestructive deformation of cells (Fig. 1).

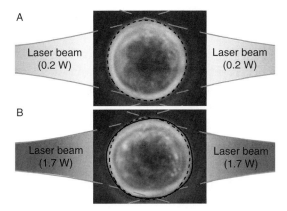

Fig. 1 Working principle of the optical stretcher. (A) A cell is trapped between two counterpropagating divergent laser beams at low power. The contour of the cell is indicated by the dashed line. (B) When the laser power is increased, the cell remains at the same position but is stretched out. The solid line indicates the new contour of the cell compared to the nondeformed state (dashed).

By its nature, an optical trap will draw a cell into the peak of laser intensity from anywhere in the vicinity of the trap. This has the added advantage that the measurement process can be combined with microfluidic delivery of a cell suspension from which individual cells are serially centered and analyzed. In other words, the cell comes to the measurement probe rather than the other way around, opening the potential for high-throughput measurements of many individual cells. Since the cells are in suspension, they can be easily sorted by standard microfluidic means. The deformability of cells can thus be assessed and exploited as a very sensitive, built-in marker for identification, characterization, and sorting of single living cells. This could complement other flow cytometry techniques except that this method does not require external labeling, is gentler, can be done under sterile, microfluidic conditions, and leaves the cells in an uncontaminated state for further investigations.

II. Rationale

The working principle of the microfluidic OS relies on some counterintuitive physical effects. The laser trap is composed of two identical counterpropagating laser beams that have a measurable impact on microscopic objects such as cells. The photons have momentum and can exert forces on contact, much like air molecules on sails. Since air molecules cannot penetrate the sail, the sail is largely propelled in the direction of the wind. Unlike air, however, light can penetrate transparent objects, and due to fundamental physical laws—conservation of energy and momentum—the forces can act in the direction opposite to the propagation of light (Ashkin and Dziedzic, 1973; Casner and Delville, 2001). This happens everywhere on the surface where light enters or exits the object. A detailed description of the underlying physics can be found in Guck *et al.* (2000, 2001) and Schinkinger (2006). The result is forces everywhere on the surface of an object (Fig. 2).

Force distributions over the surface depend on various parameters including the refractive indices of the cell and the surrounding medium and the size and position of the laser beam with respect to the size of the cell. The cumulative effect of the forces on the surface with one-laser beam present is asymmetric so that the particle moves in the direction of the resulting net force, toward the laser beam axis and away from the light source. However, if there is a second identical laser beam propagating in the opposite direction, the forces from the two beams will simply add up, the object will move toward the center point between the two laser beams and remain there. This effect is the key to the trapping and centering of suspended cells when flowed into the proximity of the laser beams. At this point there ensues a tug-of-war between two equally strong opponents. If the forces applied on both sides are strong enough, an object in the middle can be stretched out (Figs. 1B and 2B). As the two "opponents" stretching biological cells in the center are two intense light beams, this arrangement is called an optical stretcher.

Deforming cells in this way can be used to gather information about cell mechanical properties and the underlying structures. A cell that is deformed more is

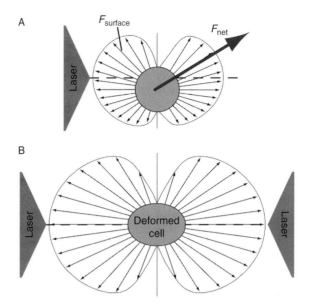

Fig. 2 Illustration of the forces encountered by a cell in the vicinity of laser beams. (A) One laser beam incident from the left induces forces all over the surface. Their distribution is anisotropic so that the sum of all surface forces results in a net force that moves the cell toward the laser axis (dashed line) and away from the light source. (B) With two identical laser beams from left and right and the cell being on axis, the distribution of surface forces will be symmetric so that their sum results in no net force at all. The cell is trapped as any displacement from this position will distort the distribution of surface forces and will lead to a restoring net force. In this trapping position and with sufficient laser power, the cell will be deformed by the surface forces, which are always present.

softer than a cell that is deformed less—assuming identical optical properties. Consequently, differences in cytoskeleton (and its softness) may be detected by comparing the deformability.

III. Methods

A. Basic Experimental Setup

We first describe the most current optical setup with specific source information. The basic requirements are two axially aligned, counterpropagating, divergent laser beams, emanating from optical fibers aligned within a fluid medium and mounted on a microscope for observation (Fig. 3).

The laser is an ytterbium fiber laser (YLD-10-1064, IPG Photonics, Germany), operating at a wavelength of $\lambda = 1064$ nm. The output of the laser is an optical fiber (HI1060, Corning, Germany), which is first spliced to a 2×2 fiber coupler (Gould Fiber Optics, Maryland) that splits the light into a separate fiber with a

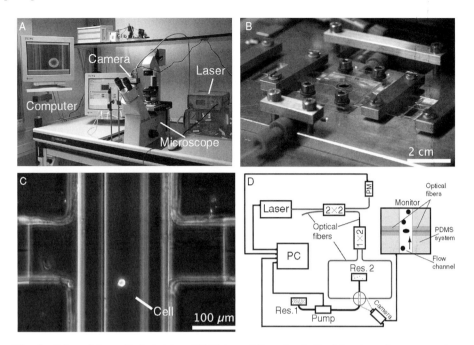

Fig. 3 Setup of the optical stretcher. (A) Picture of the setup in the laboratory. Laser, computer, microscope, and camera are the basic components. The entire setup is modular and can be combined with any microscope. (B) Close-up of the microfluidic chamber mounted on the microscope stage. The details are described in Section IV.C. (C) Phase-contrast image as seen through the microscope. The ends of the optical fibers are visible on the left and right edges. Vertically through the middle runs the glass capillary containing the cell suspension. (D) Schematic of the setup as described in the text.

1:99 ratio. The fiber carrying 1% of the light can be used as a power reference during the experiment. The remaining 99% of the laser light is subsequently split by a 1 × 2 fiber coupler (Gould Fiber Optics) in a precise 50:50 ratio into each of two fibers that provide the laser light used for the dual beam optical trap. A good antiparallel alignment of the laser beams is essential. Several effects may arise from incorrect alignment: cells might rotate around the laser beam axis or circulate between the laser beams. If the alignment becomes even poorer, a cell may not be trapped at all. The procedure of alignment will be described in Section IV.C.1.

Cells are delivered serially into the OS through a microfluidic flow channel (Gast *et al.*, 2006; Lincoln *et al.*, 2004, 2007), which is described in detail in Section IV.C. A cell that comes close enough to the laser beams is pulled toward the center of the beams and trapped at low laser power (50–200 mW per beam). If the power of the laser light is then increased to 0.6–1.5 W per beam, depending on the cell type, the cell will stretch out along the laser beam axis.

The microfluidic chamber is mounted onto an inverted phase-contrast microscope (DM IRB, Leica, Germany) where stretching is generally observed through a 63×, Ph2, CORR, 0.65-NA air objective (Leica, Germany). A long working distance of the objective, up to 2 mm, is important because the microfluidic chamber is thicker than a regular coverslip. An 8-bit digital camera (A202k, Basler, Germany), operating at a frame rate of 10–100 frames/sec, is attached to the microscope for recording the experiment. The entire setup is computer controlled with a LabVIEW program and appropriate A/D cards as described in detail in Section III.D.

B. Preparation of Cells

The OS measures suspended cells, lacking any surface attachments. Cells that grow and divide naturally in a suspended state are the ideal targets for this technique because they do not have to be detached prior to the measurement. Various blood cells and blood cell lines grown in suspension can be measured either in their growth medium or in PBS with EDTA to prevent coagulation.

Cells that grow on a substrate can be detached and suspended before measurement with the OS. Detachment is usually achieved using 0.25% trypsin/EDTA, depending on the cell type, following the same approach used for passaging cells. Detached, rounded cells are then resuspended in fresh medium or PBS with 5-mM glucose. Care should be taken to minimize perturbation to the cells. Also, adding a pH buffer to the medium to maintain a physiological pH level will help to minimize effects on the mechanical properties of the cells due to pH changes. For very sensitive cells such as stem cell cultures, EDTA alone or Accutase (PAA Laboratories, Pasching, Austria) can be substituted for trypsin/EDTA.

Resuspended cells can be diluted to the desired concentration, which will depend on the method of delivery of cells into the OS. As an example, using the capillary channel as described in Section IV.C, the preferred cell density is in the range of $(2–5) \times 10^5$ cells/ml. A lower density results in the infrequent appearance of cells and thus a prolonged waiting period between measurements. A higher density risks trapping of multiple cells at once, and clogging of the capillary if enough cells clump together. The total volume of cell suspension can be as low as 0.2 ml, although a volume of about 1–2 ml is more commonly used. Most cell types can be handled at room temperature for at least 2 h without interfering with results. For cells that remain highly active, this temperature can be lowered in an attempt to reduce cell activity, as spontaneous protrusive activities may be confused with laser-induced stretching.

C. Measurement Process

The OS is used to deform cells one at a time. In order for this to happen, each cell must be delivered to the vicinity of the trap while the power of the two laser beams are low enough to trap cells without significantly deforming them. The cells are injected into the microfluidic system through a T-junction (available, for example, from Upchurch Scientific, Oak Harbor, Washington) with a small

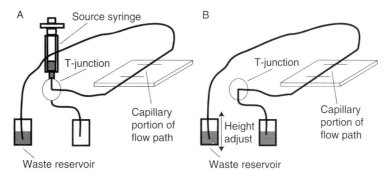

Fig. 4   Drawing illustrating cell injection and transport through the microfluidic system. (A) The cells are injected with a syringe. (B) The syringe is removed and the T-junction rotated to connect the two reservoirs through a continuous channel. Flow is controlled by adjusting the height of the waste reservoir.

syringe. The T-junction is then switched to remove the syringe and to connect the collection reservoir to a source reservoir containing the medium (Fig. 4). This procedure generates a single-flow channel containing the trap region, with a reservoir on each end. The cells are transported within the flow, which can be controlled by adjusting the relative heights of two reservoirs clamped to vertical poles. This method allows the experimenter quick, hands-on control of the flow. Automated options involving pumps are discussed in Section IV.C.4.

To start the experiment, the waste reservoir is lowered to initiate forward flow. When a cell appears in the field of view of the microscope, the waste reservoir is raised to slow down, stop, or even reverse the flow briefly, until the cell becomes trapped by the laser beams. The height of the reservoir is then fine adjusted with a micromanipulator until the background flow is stopped. This is important because any background flow can cause a rotation of the cell about the optical axis, which can lead to measurement artifacts as discussed in Section III.F. Once a cell is trapped and remains stationary, a computer program takes over which controls the deformation process. After the measurement, the cell is released by turning off the laser and the entire process is repeated with the next cell.

D. Computer Control of the Measurement

The entire experiment is controlled by a computer. It contains a frame grabber card (e.g., NI 1428, National Instruments, Austin, Texas) to which the digital camera is connected. The laser is connected via a BNC cable to an analog output card (e.g., NI 6713, National Instruments), which controls the current of the pump diode in the fiber laser and therefore the laser power.

The software necessary for the coordination of laser power application, image acquisition, and data reporting is written in LabVIEW (National Instruments; available on request). The program is able to generate a time-dependent power pattern of many forms such as step-, ramp-, sine waves, or frequency sweep functions. Most often a constant power, which generates constant stretching forces, is applied for several seconds. For image acquisition, we use the Vision-stream module for LabVIEW (Alliance Vision, Montelimar, France), which allows high frame rates by storing the data stream directly as a binary image stack file on the hard drives of a RAID array.

E. Image Analysis

The raw data of each cell's deformation comes as a series, or stack, of phase-contrast images taken via video microscopy. The next step is to extract temporal changes of the cells' contour from these images. An algorithm was specifically developed for the detection of cell edges in phase-contrast images (Fig. 5; available on request). Phase-contrast images of living cells are characterized by a bright halo

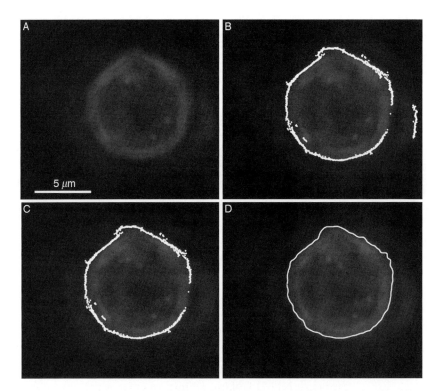

Fig. 5 Contour detection of a cell. (A) Phase-contrast image of cell. A particularly difficult image is shown to demonstrate the power of the algorithm. (B) Raw edge data determined by contour detection. (C) Edge data after the application of a standard deviation filter. Outliers beyond a certain distance of the calculated center of the cell are neglected. (D) Fourier fit of raw data points used for subsequent analysis.

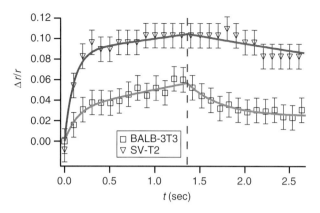

Fig. 6 Typical deformation behavior of two different fibroblast cell lines. Shown is the relative radial deformation as a function of time for a normal (BALB-3T3) and a malignantly transformed (SV-T2) fibroblast cell line. The deforming stress is being applied at $t = 0$ for about 1.4 sec (end of stress indicated by the dashed line) after which the cells are allowed to relax. The difference between the two different cell types is evident. Compared to the malignant cell, the normal cell does not get deformed as much and also relaxes faster after release of the stress.

surrounding the cell. The algorithm detects this halo based on contrast, width, and steepness, and the outer edge of the halo is taken to represent the edge of the cell. A defined image region is scanned from top to bottom and from left to right, and an edge point is detected for each line by means of the defined parameters. Subsequent analysis and connection of all the edge points then gives the entire contour of the cell.

A detailed description of the algorithm can be found in Schinkinger (2006). The speed of the algorithm allows real-time analysis. An alternative algorithm that is slower but more thorough has been described in Lincoln *et al.* (2007). From the contour information, the relative radial deformation along the laser axis is extracted as a function of time (Fig. 6).

F. Interpretation of the Data

For further analysis, the relative radial extension, or strain (Chapter 1 by Janmey *et al.* and Chapter 2 by Kandow *et al.*, this volume), $\gamma(t) = \Delta r/r$ along the laser beam axis is considered. The extension of the cells in general can be classified as viscoelastic, which implies that these materials respond in an intermediate way between purely elastic and viscous behavior.

Viscoelastic behavior can often be fit by the following equation as a function of the time of stress application:

$$\gamma(t) = \sigma_0 \left(\frac{b_1}{a_1} - \frac{a_2}{a_1^2} \right) \left(1 - \exp\left(-\frac{a_1}{a_2} t \right) \right) + \frac{\sigma_0}{a_1} t \tag{1}$$

Here a_1, a_2, and b_1 are the parameters for fitting the curve. σ_0 is the maximum stress applied to the cell along the laser axis. It can be calculated from

$$\sigma_0 = \frac{n_{med} I_0}{c} (2 - R + R^2) \left(\frac{n_{cell}}{n_{med}} - 1 \right) \qquad (2)$$

where c is the speed of light in vacuum ($c \approx 2.99 \times 10^8$ m/sec), n_{med} is the refractive index of the medium (usually $n_{med} \approx 1.335$), and n_{cell} is the refractive index of the cell. A table of typical values of n_{cell} for various cell types can be found below (Table I). R is the amount of reflection of light at the interface between the medium and the cell for normal incidence, which can be calculated with

$$R = \left(\frac{n_{cell} - n_{med}}{n_{cell} + n_{med}} \right)^2 \qquad (3)$$

Finally, I_0 is the intensity of the laser light on the laser axis at the position of the cell, which can be calculated from the total power P and the radius of the laser beam at the position of the cell ω with the following formula

$$I_0 = \frac{2P}{\omega^2 \pi} \qquad (4)$$

The value of ω depends on various parameters as discussed in Section IV.C.

The behavior after switching off the stress, when the cell relaxes back to an equilibrium extension, can be fit by:

$$\gamma(t) = \sigma_0 \left(\frac{b_1}{a_1} - \frac{a_2}{a_1^2} \right) \left(1 - \exp\left(-\frac{a_1}{a_2} t \right) \right) \exp\left(-\frac{a_1}{a_2} (t - t_1) \right) + \frac{\sigma_0}{a_1} t \qquad (5)$$

Using this functional dependence, the fit parameters a_1, a_2, and b_1 can be determined using programs such as MatLab (The Mathworks, Natick, MA). Typical rheological parameters such as the shear modulus G, the steady state viscosity η, or

Table I

Refractive Indices of Various Cell Types Determined in Analogy to the Method Described by Barer and Joseph (1954, 1955a,b), Along with Standard Error of the Mean Within the Cell Population

Cell type	n_{cells}
Mouse fibroblasts (BALB/3T3)	1.3722 ± 0.0036
Malignantly transformed mouse fibroblasts (SV-T2)	1.3711 ± 0.0039
Human leukemia cells (HL 60)	1.3735 ± 0.0019
Nonmetastatic human breast cancer cells (adenocarcinoma) (MCF-7)	1.3640 ± 0.0011
Metastatic human breast cancer cells (modMCF-7)	1.3654 ± 0.0016

ModMCF-7 cells were generated by treating MCF-7 cells with 100-nM 12-*O*-tetradecanoylphorbol-13-acetate (TPA) according to Johnson *et al.* (1999).

the relaxation time τ (Chapter 1 by Janmey *et al.*, this volume) can be calculated via:

$$G = \frac{1}{2(1 + \mu)} \frac{a_1^2}{a_1 b_1 - a_2} \qquad (6)$$

$$\eta = \frac{1}{2(1 + \mu)} a_1 \qquad (7)$$

$$\tau = b_1 \qquad (8)$$

where μ is Poisson's ratio, which may be assumed to be $\mu \approx 0.45$–0.50 for biological material (Mahaffy *et al.*, 2000).

Furthermore, the shear relaxation modulus $G(t)$ or complex shear modulus as a function of angular frequency ω, $G(\omega) = G'(\omega) + iG''(\omega)$ can be calculated.

$$G(t) = \frac{1}{2(1 + \mu)} \left(\frac{a_1 b_1 - a_2}{b_1^2} \exp\left(-\frac{t}{b_1}\right) + \frac{a_2}{b_1} \delta(t) \right) \qquad (9)$$

$$G'(\omega) = \omega \int_0^\infty G(t)\sin(\omega t)\mathrm{d}t = \frac{1}{2(1 + \mu)} \left(\frac{\omega^2(a_1 b_1 - a_2)}{1 + \omega^2 b_1^2} \right) \qquad (10)$$

$$G''(\omega) = \omega \int_0^\infty G(t)\cos(\omega t)\mathrm{d}t = \frac{1}{2(1 + \mu)} \left(\frac{\omega a_1 - \omega^3 a_2 b_1}{1 + \omega^2 b_1^2} \right) \qquad (11)$$

In this manner, the OS can be used to determine the usual rheological parameters for characterizing reconstituted cytoskeletal networks or cells as addressed throughout this volume.

However, such detailed analysis is often unnecessary when comparing different populations of cells with the objective of high-throughput mechanical characterization of the cells, which is the particular strength and unique utility of the microfluidic OS. Instead, it is often sufficient to examine the strain at one or more discrete time points (Fig. 7), expressed as the optical deformability $\delta_0(t)$

$$\delta_0(t) = \frac{\gamma(t) - \gamma(0)}{\gamma(0)} \qquad (12)$$

This simplified comparison is reliable in situations where the stress applied does not change between cells. However, it will not be applicable if the cell types being compared have significantly different optical properties, which is quantified via a *t*-test of their measured values (see Table I for a list of refractive indices measured for some cell types).

For example, a cell with the same mechanical properties but a higher refractive index would feel more stress, deform more, and falsely appear softer. To allow comparison of results between cell types with very different refractive indices, or

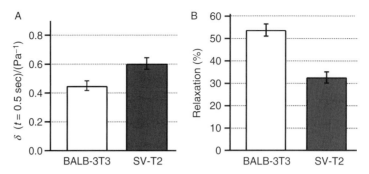

Fig. 7 Deformability analysis of two different fibroblast cell lines also shown in Fig. 6. (A) Normalized deformability of malignantly transformed (SV-T2) fibroblasts after 0.5 sec of stretching is significantly larger than that of normal (BALB-3T3) fibroblasts. (B) Also the extent of relaxation 1 sec after the release of the stress is significantly different. Normal cells have relaxed further toward their original shape than malignant cells. This distinction is possible with only 30 cells measured for each type. Error bars represent standard error of the mean.

obtained under different experimental conditions (different laser power, different fiber distance), the deformability is normalized by the peak stress actually acting on the cell surface.

$$\delta(t) = \frac{\delta_0(t)}{\sigma_0} \tag{13}$$

Figure 7 shows some results obtained with this approach.

In order to achieve this data quality, it is important to detect and eliminate artifacts, which may be caused by cells that rotate when the stress is applied (Fig. 8A–C), or by cells that move vertically and change their focal plane as they become stretched (Fig. 8D). This slight change of focus can cause an apparent shift in the measured axial length and is often avoidable by increasing the trap power slightly. While rotations can usually be eliminated by good flow control, their effect on the strain curves is indistinguishable from an active behavior of the cells. Conversely, apparent active behavior is often explained by a rotation of a passive but nonspherical cell.

In all of the problematic cases shown in Fig. 8, the measured deformation, as obtained by tracking the shape of the outer contour of the suspended cell over time, is affected by combining the true deformation with an unknown shape change that cannot be separated from the desired deformation information. Fortunately, many of the artifacts are obvious by observation of raw images. However, the remaining, more subtle cases may still artificially increase the spread in distribution, usually toward zero or negative values, and a high enough percentage of artifacts may result in an apparent normal distribution about a false value.

Data can be improved further by removing cells whose response does not conform to a true passive viscoelastic deformation. Figure 9 shows deformation

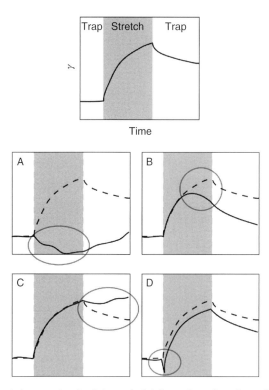

Fig. 8 The top panel shows a sketch of the typical deformation of a cell as a function of time. The power settings *trap* and *stretch* refer to low and high laser power, that is, stress applied to the cell, respectively. (A–D) Four examples of unusual apparent cell responses (solid curves), which can be identified as artifacts. Circles emphasize the difference from the normal response (dashed lines).

curves at various levels of selectivity. The first level contains the entire set of data with all obvious rotations removed and will be denoted as *stationary*.

This set still includes cells with artifacts as shown in Fig. 8. By removing cells that exhibit the behavior illustrated in Fig. 8A, where the cell does not respond passively to the applied stress, we end up with a subgroup of 41 cells in this example, denoted as *deforming*. The histogram in Fig. 9B illustrates that removing these cells reveals a normal distribution centered at a higher average deformation. Further selectivity can be achieved by removing cells that show a deformation curve as in Fig. 8B, where the rate of deformation becomes negative while the stress is still being applied, and in Fig. 8C, where the cell does not relax back after the stress has been reduced. These abnormal responses are likely caused by reorientation of the cell relative to the trap in addition to the deformation itself. Removing these cells from the analysis lead to a reduced data set denoted *viscoelastically deforming* which includes 10 of 62 cells in this example. These cells are the ideal candidates to be individually fit in order to obtain physical values, such as G, η, and τ, see

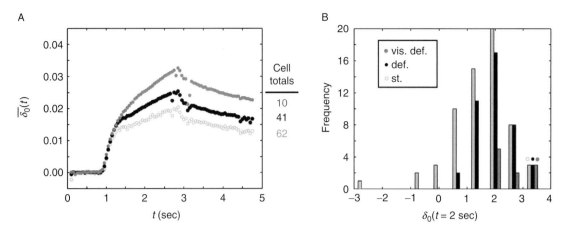

Fig. 9 Illustration of the improvement of data quality by the removal of cells that produce certain artifacts. (A) Deformation curves averaged over all cells fulfilling certain criteria as a function of time. Different levels of selectivity are denoted as *stationary* (*st.*), *deforming* (*def.*), and *viscoelastically deforming* (*vis. def.*), as described in the text. The stress is applied for 2 sec starting at $t = 1$ sec. (B) Histogram showing the actual distribution of deformability measured at $t = 2$ sec for the three different levels of selectivity.

Eqs. (6)–(8). However, looking at the histogram in Fig. 9B, it can be seen that this group is undersampled and, thus, nonnormal. A normal population is a good indicator of a low standard error and a reliable mean value for the deformability.

Figure 10 illustrates the reproducibility of this procedure and the power of using deformability as a cell marker. Here, cells were allowed to grow in four separate but identically prepared dishes. One dish was measured each day for 4 days in order to determine the change in deformability as a function of cell density on the dish. Using the *deforming* category just described, it can be seen that the average strain as a function of time remained unchanged during the first 3 days. On the fourth day, when the final dish had reached confluency, the average strain was reduced. This indicates that either confluent conditions can lead to cell stiffening, or higher proliferation rates on the first 3 days are correlated with an increased deformability. This also serves to illustrate the sensitivity of optical deformability for characterizing cells.

IV. Additional Notes on Equipment

A. Optical Fibers

In the most practical setup, light is delivered to the trapping region with single-mode optical fibers. They have the advantage of providing an ideal Gaussian intensity distribution, are easy to handle, and simpler to align than free-space laser beams.

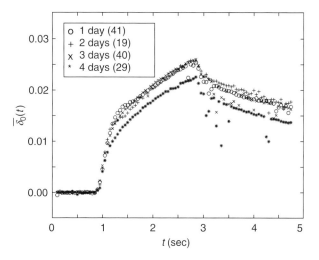

Fig. 10 Averaged deformation curves of BALB-3T3 fiborblasts cultured for several consecutive days. The figures in parentheses indicate the number of cells in the *deforming* category measured on that particular day. While the cells behave identically on the first 3 days, the deformability decreases when they reach confluency.

Single-mode optical fibers have an outer diameter of 125 μm and an inner core diameter of $d \approx 6$ μm. This is significantly smaller than the core size of multimode fibers that are in the range of at least 50 μm. Although trapping and stretching have been demonstrated also with multimode fibers, the forces cannot be calculated in an easy manner, as the intensity distribution is more difficult to determine than for single-mode beams. Light can either be coupled into the bare ends of the fiber (usually impractical unless sufficient expertise is available) or created directly inside the fiber in a fiber laser.

B. Laser Source

The laser should be operating in the near-infrared at wavelengths $\lambda > 750$ nm. In this range, the absorption of proteins and chromophores in most eukaryotic cells is negligible. However, the absorption of water increases with increasing wavelength, which would favor a laser with shorter wavelengths. A laser source at ≈ 800 nm would be ideal, while lasers at 1064 nm still offer a good compromise between these two concerns.

Driven by the rapid development in telecommunications, fiber lasers have become simple and cost-effective laser sources. In this type of laser, the light is generated directly inside the fiber, offering an easy to operate, stable, and reliable source of light for the OS application. Although fiber lasers are currently not yet available at ≈ 800 nm, their development is foreseeable in the near future. Fiber lasers are available between 500 mW and 150 W. In the present description, an ytterbium-doped single-mode 10-W fiber laser at 1064-nm wavelength (IPG Photonics) is used. A sensible alternative is to use two less powerful lasers (e.g., each with 2-W power capacity) that can be controlled separately. This enables individual control of the two beams, which offers further flexibility in the type of laser pattern applied or in the fine control needed to correct for any slight differences in the laser power between the two fiber ends.

Ti:Sapphire lasers, whose emission can be chosen in the wavelength range from ~650 to 1100 nm, appear to be ideal light sources for optical trapping and stretching applications of cells. However, they are very expensive compared to fiber lasers (~$150,000 compared to ~$10,000) and are difficult to couple into single-mode optical fibers at sufficient power (>1 W) (Section IV.A). There are also fiber-coupled diodes available (<$3000) with sufficient power (up to 5 W) and at an appropriate wavelength (e.g., 808 nm). However, due to the nonideal beam divergence properties of high-power diode lasers, they are only available with multimode fibers (Section IV.A).

A further requirement of the laser is continuous wave (CW) operation. Even though high-power-pulsed lasers exist, their application in the OS is not recommended due to multiphoton effects potentially causing unwanted photochemistry and cell damage.

A word of caution seems to be appropriate at this point. Due to the nature of the infrared light and relatively high laser powers required, an appropriate protection of the sensitive CCD chip in the camera should be ensured by the use of infrared filters permanently installed in the optical path of the microscope. Furthermore, appropriate laser goggles should be worn during setup and when looking directly through the eyepieces of the microscope.

C. Microfluidic Chip

1. Open Setup

The simplest setup, without microfluidic delivery of the cells, is achieved by aligning the optical fibers against a straight edge (backstop) attached to a glass slide (Fig. 11) (Constable *et al.*, 1993; Guck *et al.*, 2001, 2002). The backstop can be a number of materials, for example a larger multimode fiber glued down on the glass. The optical fibers delivering the laser beam are guided into place by miniature three-axis translation stages.

This simple arrangement is sufficient for the quick measurement of a few cells, which are brought to the trap by pipetting of suspended cells onto the space between the fibers. As the cells settle, the fibers can each be translated along the backstop in order to move the trap into their path. Pulling the fibers away from each other actually helps to create a little flow directed toward the trap and to pull in cells settling off to one side. Once a cell is trapped, the distance between the fibers can still be adjusted with the translation stages. It is important to record the

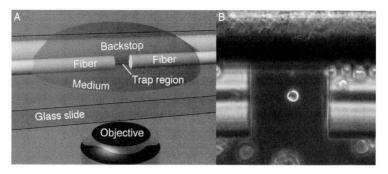

Fig. 11 The optical stretcher without microfluidic delivery (open setup). (A) Drawing of the open setup with all relevant parts labelled. The miniature three-axis aligners used to hold and move the fibers are external to the image. (B) Phase-contrast image (bottom view) of the open setup with a cell trapped at the center. Other cells that have settled to the floor are visible but out of focus. From this image the distance between the end of the optical fiber and the trapped cell can be determined, which is used for the calculation of the deforming stress.

distance between the fibers for each cell, which affects both the trap conditions and the peak stress σ_0 (Fig. 11B).

In this open setup, the laser beams propagate through a single medium (PBS or cell culture medium) and the beams diverge according to the equation

$$\omega(d) = \omega_0 \sqrt{1 + \left(\frac{\lambda d}{n_{\text{med}} \pi \omega_0^2}\right)} \tag{14}$$

where d is the distance between the fiber end and the cell, ω_0 is the radius of the laser beam where it exits the fiber, a distance defined as one-half the mode field diameter (for a single-mode fiber at 1064 nm this is usually about 3.1 μm), and λ is the wavelength of the laser in vacuum (here $\lambda = 1064$ nm). The value of ω, calculated for the distance d measured (Fig. 12), is then used for calculation of the peak stress σ_0, Eq. (2), which in turn is used to normalize the deformation according to Eq. (13).

Despite its successful use for the necessary proof of principle, the open setup has a number of limitations. It is difficult to maintain a sterile condition and experiments can become progressively dirtier due to the open nature of the system. This is especially problematic if any debris becomes stuck to the end of a fiber, which distorts the beam shape and necessitates the time-consuming process of removing, cleaning, and realigning the fibers against the backstop. Also the volume of the drop of medium is constantly changing due to the addition and removal as well as steady evaporation. This not only increases the osmolarity but also has the adverse effect of changing the optical condition for each measurement, which affects the quality and contrast of the phase-contrast images and impairs image analysis with the algorithm described. Nevertheless, this is a simple and cheap implementation of the OS when only a few cells are to be measured.

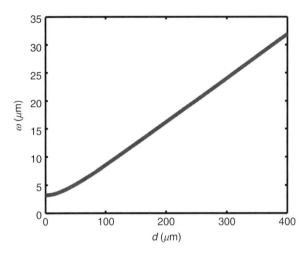

Fig. 12 Plot showing the divergence of the laser beam as a function of distance d from the fiber end. The beam width ω was calculated for $\omega_0 = 3.1\ \mu$m and $\lambda = 1064$ nm according to Eq. (14).

2. Single-Channel Design

If many cells are to be measured in a single experiment, some microfluidic delivery needs to be implemented. This is slightly more involved than the open setup, but a viable solution has recently been described (Lincoln *et al.*, 2007). A square glass microcapillary ($80 \times 80\ \mu$m^2 inside dimension, Vitrocom, Mountain Lakes, New Jersey) serves as the flow channel and the laser beams are sent through the capillary walls to trap and stretch the cells inside (Fig. 13). This type of setup can be made rather easily without access to sophisticated microfabrication facilities, and has been used reliably to measure 50–100 cells/h. The separation of the optical fibers from the cell suspension has the obvious advantage that the ends of fiber do not collect debris, which could affect beam properties. Also, the glass capillary can be cleaned rather vigorously with any kind of solvent, although rinsing with ethanol and deionized water is usually sufficient. Once assembled, this setup can be reused for many experiments for up to several months.

While in the open setup the beam passes through a single medium, here the beam passes through regions of index-matching gel, the glass wall of the capillary, and the growth medium. Index-matching immersion oil or gel (Thorlabs, Newton, New Jersey) is used to reduce reflections of the laser light as it enters the capillary wall. This necessitates a different formula for the calculation of the beam width at the position of the cell.

$$\omega = \omega_0 \sqrt{1 + \frac{\lambda B^2}{n_{\text{gel}} \pi \omega_0^2}} \qquad (15)$$

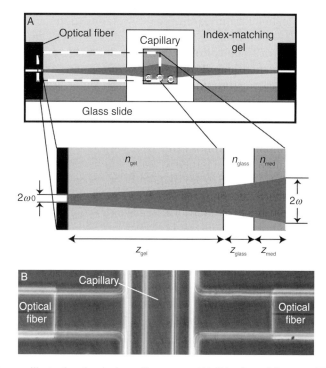

Fig. 13 Diagram illustrating the single capillary setup. (A) Side view of the setup. The laser beams emanating from the optical fibers pass through the capillary walls to trap and stretch the cells inside. The inset shows various materials, with the relevant distances and refractive indexes, that the beam has to propagate through. (B) Phase-contrast microscopy image of the experimental setup (bottom view).

with

$$B = \frac{z_{\text{med}} n_{\text{gel}}}{n_{\text{med}}} + \frac{z_{\text{glass}} n_{\text{gel}}}{n_{\text{glass}}} + z_{\text{gel}} \qquad (16)$$

where n_x denotes the refractive index of the material and z_x the distance as depicted in Fig. 13. Usually, $n_{\text{gel}} = 1.449$, $n_{\text{glass}} = 1.474$, $z_{\text{med}} = z_{\text{glass}} = 40\ \mu\text{m}$, and only z_{gel} is adjustable (Fig. 13). An advantage with this setup is that once the fibers are installed and z_{gel} becomes fixed, ω has to be calculated only once. Thus, the peak stress can only vary if the refractive index is different among the cell types examined (Table I).

3. Arbitrary-Channel Design

By making the channels from the elastomer poly(dimethylsiloxane) (PDMS) instead of using a glass capillary, one can gain the ability to design multichannel flow systems with the potential for sorting cells after measurement, either by flow or by some sort of downstream selection mechanism. However, reliable production of these channels requires microfabrication facilities and experience. An example of a commercial chip created in this way can be found in Gast *et al.* (2006). In addition to allowing sorting, multiple input channels could be used to exchange the medium around a cell while it is trapped and repeatedly measured, for example to deliver drugs into its environment. The ability to easily vary the channel design in such systems allows for the development toward higher measurement rates as well as automated flow control as described below.

4. Flow Control

Single- and arbitrary-channel designs have successfully used a height-regulated pressure differential system, as described in Section II.C (Fig. 4), to start and stop the flow of cells manually. Peristaltic and syringe pumps can be used to control the flow velocity more exactly, which would be necessary for any multichannel design that controls the timing of separate inflows or positions cells for sorting. Such a pump system also has the potential for complete computer-controlled automation.

D. Microscopy Technique

1. Phase Contrast

Phase-contrast microscopy is the technique predominantly used with the OS. Inverted phase-contrast microscopes are typically found in cell biology laboratories. High-NA (0.6–0.7; air) phase-contrast objectives with magnifications of $40\times$ to $100\times$ and large working distances (2 mm) are needed for both imaging into the flow channel and providing the resolution for analysis. The high contrast of phase-contrast images facilitates visual observation, but makes it difficult to identify the exact edge of a suspended cell. However, as long as the edge is determined consistently, analysis of relative deformation is sufficient for the comparison of different cell lines or for monitoring changes in deformability. In this case, the presence of a symmetric halo provides an easy target for automated image analysis. Alternatively, Nomarski (or differential interference contrast; DIC) imaging can also be used, although the advantage of DIC to provide a more reliable image of the cell edge has to be weighed against the increased cost and asymmetric contrast along the cell edge (shadow effect). This latter aspect would require modifications of the contour analysis algorithm described in Section III.E.

2. Fluorescence Microscopy

The OS can easily be combined with fluorescence microscopy. However, due to the large working distance of the objective required to image into the microfluidic channel, it is not possible to use oil immersion objectives, which are ideally suited for fluorescence applications. The necessity to use an air objective limits the resolution and the use of bright samples. However, high-NA air objectives were found to be sufficient for measuring the deformation of the cells and for checking the presence fluorescent cell markers for correlation studies.

3. Confocal Laser-Scanning Microscopy

The OS has also been used on confocal laser-scanning microscopes. While the use of non-oil objectives is less critical here, the resolution of the image is limited by the slight motion of cells in suspension and the relatively slow scanning process. Nevertheless, large structures such as the membrane, the actin cortex, or the nucleus can clearly be imaged. Also, when only the deformation is to be monitored, a reduction of the scanned area down to a narrow rectangle or even a line along the laser axis is sufficient, which can be done at very high frame rates.

E. Camera

Almost any standard CCD camera operating at a rate of >10 frames/sec is suitable for video microscopic recording of the OS. Digital CCD cameras, as opposed to analog video cameras, offer a greater adjustability in resolution, frame rates, precision (10–12 bit instead of 8 bit for analog), and simplicity (no frame grabber required), while analog video cameras are usually cheaper. For the current setup, we use a Basler A202k digital camera (Basler Vision AG, Ahrensburg, Germany) with a CameraLink connection. A favorable development is the use of cameras with a Firewire or fast USB connection to the computer, which offer a greater flexibility than conventional video cameras since frame grabbers are not required.

V. Discussion

Using the OS, we have successfully studied the microscopic origin of global cell deformation behavior (Wottawah et al., 2005a,b) and have seen statistically relevant differences in deformability between normal and cancerous fibroblasts (Guck et al., 2005; Lincoln et al., 2004). The cancer cells were consistently softer than their normal counterparts, likely caused by a thinner actin cortex and a comparable elastic modulus (Ananthakrishnan et al., 2006). The softness of cancer cells is further correlated with their increased proliferation rate and motility; both appear consistent with a more flexible cytoskeleton. This finding has since been confirmed on leukemia cells, oral mucosa cells, and breast epithelial cells. Even metastatic cells can be distinguished from less aggressive cancer cells, which is not possible with other approaches. The idea of using the deformability of individual cells as a diagnosis marker for cancer has already entered the clinical arena and is being tested in trials. Preliminary results are promising.

This use of optical deformability to characterize cells is not just limited to pathological cases. Normal processes such as differentiation, which may be viewed as an opposite process of transformation, are also being investigated. For example, we have found that stem cells are softer than differentiated cells. The high proliferation rate of unspecialized master cells is again correlated with high deformability, consistent with the notion that the differentiation into mature cells requires a build up of cytoskeleton. This functional approach offers a sensible way of defining the nature of stem cells, and opens novel possibilities in identifying and sorting stem cells from heterogeneous populations. An obvious feature of the OS, compared to all other methods in this volume, is its ability to measure cells in suspension without any mechanical contact. However, this environment is different from most cells' natural environment, which regulates cell function through interactions with the extracellular matrix and neighboring cells. The natural, 3D environment is also typically softer than 2D substrates used in cell culture (Li *et al.*, 2003), which are known to induce some nonphysiological cell properties. While suspended cells are also in a nonphysiological natural environment (an exception being blood cells), they tend to be more homogenously structured and more amenable to theoretical modeling (Ananthakrishnan *et al.*, 2005, 2006; Wottawah *et al.*, 2005a,b). In addition, although the absolute values of elasticity may differ between cells embedded in a 3D matrix and in suspension, similar differences between cell types in tissue have been detected with cells in a suspended state using the OS (Guck *et al.*, 2005; Mahaffy *et al.*, 2004). The suspended state further facilitates integration with microfluidic devices and high-throughput applications.

For future development, the image analysis algorithms developed still require high-quality images, which limit the range of cell types to be investigated quantitatively. Improved approaches to analyzing the deformation are currently being developed. In addition, cells that are grossly nonspherical are less accessible than spherical cells to theoretical modeling both optically, for determining the forces acting on the cell, and mechanically, for extracting meaningful material constants. Fortunately, most cells encountered lend themselves easily to modeling.

An additional uncertainty on a single-cell level is the refractive index of cells. Variation in the refractive index may cause an apparent spreading of mechanical properties. The measurement of refractive indexes at the single/subcellular level is currently not feasible in a practical fashion. Future improvement in the measurement will further improve the discriminating power of the method, while the refractive index itself could serve as an additional inherent marker.

Finally, there are cell types and experimental conditions where the cells cannot be passively deformed, while the cells remain actively responsive to applied forces. Such active responses can be easily detected and viewed as a useful characteristic feature, particularly when the objective of the study is to determine the differences between different cell types or states, which is the major strength of this technology.

VI. Summary

In summary, the OS has matured from a novel physical phenomenon to a reliable method for many relevant cell biological questions. As presented in this section, the technique can be used to measure the deformability of 50–100 cells/h with little setup time beyond its initial construction, which requires only several hours. The technique enables mechanical investigation of a significant number of suspended cells with ease and accuracy and adds a new dimension to the investigation and sorting of cells. With appropriate technological development, the microfluidic OS might one day reach the same prevalence in biological laboratories for the high-throughput cell screening and sorting as flow cytometers today.

Acknowledgments

The authors would like to thank Christian Dietrich, Martin Kühner, Karin Schütze, Dennis Discher, Josef Käs, Cornelia Deckert, Julia Dietrich, and Kort Travis for stimulating discussions. This work was supported financially by the Humboldt Foundation, the European Fund for Regional Development (EFRE) 2000–2006, and the state of Saxony.

References

Alberts, B., Johnson, A., Lewis, J., Raff, M., Roberts, K., and Walter, P. (1994). "Molecular Biology of the Cell." Garland Publishing, New York.

Ananthakrishnan, R., Guck, J., Wottawah, F., Schinkinger, S., Lincoln, B., Romeyke, M., Moon, T. J., and Käs, J. (2005). Modelling the structural response of a eukaryotic cell in the optical stretcher. *Curr. Sci.* **88,** 1434–1440.

Ananthakrishnan, R., Guck, J., Wottawah, F., Schinkinger, S., Lincoln, B., Romeyke, M., Moon, T. J., and Käs, J. (2006). Quantifying the contribution of actin networks to the elastic strength of fibroblasts. *J. Theor. Biol.* **242,** 502–516.

Ashkin, A., and Dziedzic, J. M. (1973). Radiation pressure on a free liquid surface. *Phys. Rev. Lett.* **30,** 139–142.

Barer, R., and Joseph, S. (1954). Refractometry of living cells. Part I: Basic principles. *Q. J. Microsc. Sci.* **95,** 399–423.

Barer, R., and Joseph, S. (1955a). Refractometry of living cells. Part II: The immersion medium. *Q. J. Microsc. Sci.* **96,** 1–26.

Barer, R., and Joseph, S. (1955b). Refractometry of living cells. Part III: Technical and optical methods. *Q. J. Microsc. Sci.* **96,** 423–447.

Ben-Ze'ev, A. (1985). The cytoskeleton in cancer cells. *Biochim. Biophys. Acta* **780,** 197–212.

Bloch, W., Fan, Y., Han, J., Xue, S., Schöneberg, T., Ji, G., Lu, Z. J., Walther, M., Fässler, R., Hescheler, J., Addicks, K., and Fleischmann, B. K. (2001). Disruption of cytoskeletal integrity impairs G_i-mediated signaling due to displacement of G_i proteins. *J. Cell Biol.* **154**(4), 753–762.

Bosch, F. H., Werre, J. M., Schipper, L., Roerdinkholder-Stoelwinder, B., Huls, T., Willekens, F. L., Wichers, G., and Halie, M. R. (1994). Determinants of red blood cell deformability in relation to cell age. *Eur. J. Haematol.* **52,** 35–41.

Casner, A., and Delville, J.-P. (2001). Giant deformations of a liquid-liquid interface induced by the optical radiation pressure. *Phys. Rev. Lett.* **87,** 054503.1–054503.4.

Constable, A., Kim, J., Mervis, J., Zarinetchi, F., and Prentiss, M. (1993). Demonstration of a fiber-optical light-force trap. *Opt. Lett.* **18,** 1867–1869.

Cunningham, C. C., Gorlin, J. B., Kwiatkowski, D. J., Hartwig, J. H., Janmey, P. A., Byers, H. R., and Stossel, T. P. (1992). Actin-binding protein requirement for cortical stability and efficient locomotion. *Science* **255**, 325–327.

Elson, E. L. (1988). Cellular mechanics as an indicator of cytoskeletal structure and function. *Annu. Rev. Biophys. Biophys. Chem.* **17**, 397–430.

Fuchs, E., and Cleveland, D. W. (1998). A structural scaffolding of intermediate filaments in health and disease. *Science* **279**, 514–519.

Gardel, M. L., Shin, J. H., MacKintosh, F. C., Mahadevan, L., Matsudaira, P., and Weitz, D. A. (2004). Elastic behavior of cross-linked and bundled actin networks. *Science* **304**, 1301–1305.

Gast, F. U., Dittrich, P. S., Schwille, P., Weigel, M., Mertig, M., Opitz, J., Queitsch, U., Diez, S., Lincoln, B., Wottawah, F., Schinkinger, S., Guck, J., *et al.* (2006). The microscopy cell (MicCell), a versatile modular flow through system for cell biology, biomaterial research, and nanotechnology. *Microfluidics Nanofluidics* **2**, 21–36.

Guck, J., Ananthakrishnan, R., Cunningham, C. C., and Käs, J. (2002). Stretching biological cells with light. *J. Phys. Condens. Matter* **14**, 4843–4856.

Guck, J., Ananthakrishnan, R., Mahmood, H., Moon, T. J., Cunningham, C. C., and Käs, J. (2001). The optical stretcher: A novel laser tool to micromanipulate cells. *Biophys. J.* **81**, 767–784.

Guck, J., Ananthakrishnan, R., Moon, T. J., Cunningham, C. C., and Käs, J. (2000). Optical deformability of soft biological dielectrics. *Phys. Rev. Lett.* **84**, 5451–5454.

Guck, J., Schinkinger, S., Lincoln, B., Wottawah, F., Ebert, S., Romeyke, M., Lenz, D., Erickson, H. M., Ananthakrishnan, R., Mitchell, D., Käs, J., Ulvick, S., *et al.* (2005). Optical deformability as an inherent cell marker for testing malignant transformation and metastatic competence. *Biophys. J.* **88**, 3689–3698.

Janmey, P. A. (1991). Mechanical properties of cytoskeletal polymers. *Curr. Opin. Cell Biol.* **3**, 4–11.

Johnson, M. D., Torri, J. A., Lippman, M. E., and Dickson, R. B. (1999). Regulation of motility and protease expression in PKC-mediated induction of MCF-7 breast cancer cell invasiveness. *Exp. Cell Res.* **247**, 105–113.

Lekka, M., Laidler, P., Gil, D., Lekki, J., Stachura, Z., and Hrynkiewicz, A. Z. (1999). Elasticity of normal and cancerous human bladder cells studied by scanning force microscopy. *Eur. Biophys. J.* **28**, 312–316.

Li, N., Tourovskaia, A., and Folch, A. (2003). Biology on a chip: Microfabrication for studying the behavior of cultured cells. *Crit. Rev. Biomed. Eng.* **31**(5–6), 1–66.

Lincoln, B., Erickson, H. M., Schinkinger, S., Wottawah, F., Mitchell, D., Ulvick, S., Bilby, C., and Guck, J. (2004). Deformability-based flow cytometry. *Cytometry* **59A**, 203–209.

Lincoln, B., Schinkinger, S., Travis, K., Wottawah, F., Ebert, S., Sauer, F., and Guck, J. (2007). Reconfigurable microfluidic integration of a dual-beam laser trap with biomedical applications. *Biomed. Microdev.* (accepted for publication).

Mahaffy, R. E., Park, S., Gerde, E., Käs, J., and Shih, C. K. (2004). Quantitative analysis of the viscoelastic properties of thin regions of fibroblasts using atomic force microscopy. *Biophys. J.* **86**, 1777–1793.

Mahaffy, R. E., Shih, C. K., MacKintosh, F. C., and Käs, J. (2000). Scanning probe-based frequency-dependent microrheology of polymer gels and biological cells. *Phys. Rev. Lett.* **85**, 880–883.

Moustakas, A., and Stournaras, C. (1999). Regulation of actin organisation by TGF-β in H-*ras*-transformed fibroblasts. *J. Cell Sci.* **112**, 1169–1179.

Park, S., Koch, D., Cardenas, R., Käs, J., and Shih, C. K. (2005). Cell motility and local viscoelasticity of fibroblasts. *Biophys. J.* **89**, 4330–4342.

Rao, K. M., and Cohen, H. J. (1991). Actin cytoskeletal network in aging and cancer. *Mutat. Res.* **256**, 139–148.

Schinkinger, S. (2006). Optical Deformability as a Sensitive Physical Parameter in Cell Characterization. Ph.D. Dissertation. Universität Leipzig, Germany.

Suresh, S., Spatz, J., Mills, J. P., Micoulet, A., Dao, M., Lim, C. T., Beil, M., and Seuerlein, T. (2005). Connections between single-cell biomechanics and human disease states: Gastrointestinal cancer and malaria. *Acta Biomater.* **1**, 15–30.

Thoumine, O., and Ott, A. (1997). Comparison of the mechanical properties of normal and transformed fibroblasts. *Biorheology* **34,** 309–326.

Wachsstock, D. H., Schwarz, W. H., and Pollard, T. D. (1994). Cross-linker dynamics determine the mechanical properties of actin gels. *Biophys. J.* **66,** 801–809.

Ward, K. A., Li, W. I., Zimmer, S., and Davis, T. (1991). Viscoelastic properties of transformed cells: Role in tumor cell progression and metastasis formation. *Biorheology* **28,** 301–313.

Wilhelm, J., and Frey, E. (2003). Elasticity of stiff polymer networks. *Phys. Rev. Lett.* **91,** 108103.1–108103.4.

Worthen, G. S., Schwab, B., III, Elson, E. L., and Downey, G. P. (1989). Mechanics of stimulated neutrophils: Cell stiffening induces retention in capillaries. *Science* **245,** 183–186.

Wottawah, F., Schinkinger, S., Lincoln, B., Ananthakrishnan, R., Romeyke, M., Guck, J., and Käs, J. (2005a). Optical rheology of biological cells. *Phys. Rev. Lett.* **94,** 098103.1–098103.4.

Wottawah, F., Schinkinger, S., Lincoln, B., Ebert, S., Müller, K., Sauer, F., Travis, K., and Guck, J. (2005b). Characterizing single suspended cells by optorheology. *Acta Biomater.* **1,** 263–271.

CHAPTER 18

Measuring Mechanical Properties of Embryos and Embryonic Tissues

Lance Davidson* and Ray Keller†

*Department of Bioengineering
University of Pittsburgh
Pittsburgh, Pennsylvania 15260

†Department of Biology and Morphogenesis and Regenerative
Medicine Institute, University of Virginia
Charlottesville, Virginia 22904

Abstract

Cell- and tissue-scale mechanics integrates numerous processes within the developing embryo to bring about both local cell movements and global tissue sculpting during morphogenesis. Rapid advances in molecular genetic approaches have outpaced quantitative approaches to study mechanics in early embryos. In this chapter, we present a device that is capable of carrying out a stress–relaxation test to measure the time-dependent elastic modulus of embryonic tissue explants a few hundreds of micrometers on a side. The device can detect forces from 25 nN to several micronewtons and allows the determination of extremely low modulii, as low as 5 Pa, found in frog embryonic tissues during gastrulation. We describe this device along with the methodology of its use and discuss the general challenges of working with embryonic tissues.

I. Introduction

A. Biomechanics and Developmental Biology

How diverse body plans, tissues, and organs are formed through a conserved set of molecular genetic processes is one of the great paradoxes of modern biology. The processes of morphogenesis and organogenesis that produce embryonic animal forms and tissues involve complex integration of molecular-, cellular-, and tissue-scale processes operating through biochemical, biophysical, and biomechanical pathways. Recent molecular genetic analysis has elaborated many of the biochemical details of embryogenesis yet little is known about the complementary biophysical and biomechanical pathways and how these pathways are integrated in developing tissues. For instance, pathways leading to the differentiation of endoderm (Levine and Davidson, 2005; Loose and Patient, 2004) are conserved throughout the chordates yet the early morphogenesis of those tissues and subsequent organogenesis produces organs of various forms and functions (Gilbert, 2006; Stern, 2004). In contrast to rapid advances in molecular genetics, only a few quantitative analyses of biomechanics have been carried out (Adams et al., 1990; Davidson et al., 1995, 1999; Gustafson and Wolpert, 1963; Hutson et al., 2003; Moore et al., 1995; Selman, 1958; Wiebe and Brodland, 2005).

Understanding morphogenesis and development will require a much deeper insight into the intrinsic forces and deformations that help shape an organism. The classical school of developmental mechanics—"Entwicklungsmechaniks"—put considerable emphasis on biomechanical mechanisms and physical models (Holtfreter, 1939, 1943, 1944; Lewis, 1947; Spek, 1918), and many of their proposed mechanisms were both plausible and supported by circumstantial evidence. Measurements of forces and tissue mechanical properties were clearly being motivated, but how cells generated and transmitted forces was comparatively unclear. A few attempts to measure forces during embryogenesis included the force developed in closing neural folds (Selman, 1955, 1958; Waddington, 1939), the internal blastocoel pressure required to stop invagination in echinoderms (Moore, 1941), and the tension developed by the cleavage furrow in echinoderms (Rappaport, 1977). In the last several decades, molecular biology and developmental genetics have come to the forefront, and a dominant paradigm for understanding mechanism has been based on genetic or molecular interdiction of the expression or the function of a molecule to see what fails. These approaches have excelled in identifying many components essential for morphogenic processes and also in providing many tools to manipulate experimentally the presence or function of these components. The challenge presented by these advances is to learn how gene products actually generate or modulate the forces, pattern these forces, and provide the cells and tissues with mechanical properties to transmit these forces in appropriate patterns. A new focus on biomechanics has also been encouraged by cell biological advances in understanding the molecular basis of cell adhesion, cell movement, cell polarity, force generation, and other aspects of cell behavior relevant to morphogenesis. There are now many methods for measuring forces at the molecular (Howard, 2001) and the cellular levels (see the section on "Measuring Mechanical Properties and Force Output of Cells"), and the time is ripe for bringing a biomechanical analysis to bear on the problem of morphogenesis. Here, we will focus on methods of measuring and analyzing forces and mechanical properties of embryonic tissues.

B. Why Do We Want to Understand the Mechanical Basis of Morphogenesis?

Biomechanics is one of the few universal principles of biology. The physical processes of solid and fluid mechanics illuminate the physical function of organisms at a variety of spatial and temporal scales. Physical processes within molecules, cells, tissues, and whole organisms can be quantitatively understood from fundamental principles. The application of classical physics and engineering principles to embryonic tissues is not dependent on the species, the size or shape of the tissue, or the temperature at which the embryos are raised. In this regard, biomechanics can be applied universally in the same manner that molecular genetic approaches can be applied across species, extending results and insights to the process of human development.

Koehl (1990) laid the framework for a biomechanical analysis of a morphogenetic process:

> "(1) a qualitative description of the phenomena and a qualitative statement of the theories about the physical mechanisms underlying it; (2) qualitative experiments where components of the system are hypothesized to be involved in the process are removed or altered, and the consequences to the process are observed; (3) quantitative description of the process, including (a) morphometric analysis of the structures involved and (b) kinematic analysis of their motions, as well as (c) measurement of the forces exerted during the process and (d) measurement of the mechanical properties of the tissues subjected to the forces; (4) quantitative statement of theories of the mechanisms responsible for the phenomena (i.e. mathematical models); and (5) empirical tests (both qualitative and quantitative) of the predictions of the models." (p. 368)

Such a biomechanical analysis is not necessarily a sequential investigation based on this list, but can also include synergistic investigations applying complementary theoretical and experimental approaches.

There are many questions concerning the mechanical basis of morphogenesis and many opportunities to better define the important biomechanical features of the embryo, ranging from mechanical properties to tissue architecture and forces generated. These are all thought to play important roles in developmental processes and are essential to complement ongoing cell, molecular, and genetic efforts toward understanding the molecular basis of morphogenesis.

By necessity we use the term "mechanical properties" since these are empirically determined values of elasticity and viscosity collected under defined experimental protocols (Chapter 1 by Janmey *et al.* and Chapter 2 by Kandow *et al.*, this volume). Although a measured mechanical property of a sample may differ when different experimental protocols are used, this does not indicate that one value or the other is incorrect. Differences may simply reflect the different assumptions made during the establishment of the protocol or in subsequent analyses. We will discuss here two basic experimental protocols, the stress-relaxation protocol and the isometric strain protocol. Mechanical properties of intact embryonic tissues can be measured with a variety of other techniques from cone-and-plate rheometers to atomic force microscopy (AFM)-based indenters that we will not discuss. Differences in the results from these techniques can arise from differences in the physical principles used in the experiment as well as differences in the biophysical principles used to analyze the results.

II. Applying and Measuring Forces of 10 nN to 10 μN

The regime of forces acting within embryos, either generated autonomously or due to forces applied from surrounding tissues, dictates the physical techniques used to measure them. For instance, forces on the order of 1 μN are considerably beyond the capacity of laser-based optical trapping methods and are far below the sensitivity of thin film-based strain sensors. Even though the AFM is capable of measuring these forces, this method has not proven to be practical in our hands due to geometry of the force sensors and the cost of the instruments. Instead, we use fine glass needles or optic fibers to apply and measure forces.

To measure or apply forces in a range useful for embryonic tissues, the needles must be of an appropriate diameter and length. Such needles are made from capillary tubes (1.0-mm outer diameter, 0.58-mm inner diameter; Harvard Apparatus, Holliston, MA) that have been pulled out with a microprocessor-controlled needle puller (Sutter Instrument Company, Novato, California). With practice long fine glass needles can be pulled out to 2–3 cm. Needles whose ends "feather out" should be trimmed or discarded. Needles can be produced in quantity with a variety of lengths and stored for future use in a large Petri dish on rails of modeling clay.

Optical fibers are a good alternative to glass needles. Conventional glass optical fiber is typically produced in 125-μm diameter and is available as either raw cable or "patch" cables with a variety of fiber optic connectors preassembled onto the ends (Digikey, Thief River Falls, Minnesota). Before fibers can be used, they must be stripped of three layers of protective material. The outermost layer is a sheath that is easily removed by a razor blade. The next layer consists of strands of Kevlar fibers that protect the fibers from overextension. These can be cut down with scissors. The innermost layer of plastic cladding is the most difficult to remove and requires a special fiber-stripping tool designed for the diameter of the inner glass core (Newport Corporation, Irvine, California). With practice 100-mm lengths of optical fibers can be stripped within 10 min.

Needles or fibers must be mounted stably to micromanipulators. Glass needles can be mounted in conventional needle holders (Harvard Apparatus). Optical fibers can be mounted with epoxy to glass capillary tubes such as those used to fabricate glass needles. Since epoxy can swell when exposed to aqueous media, the connection between the optical fiber and the capillary tube can be sealed with thin coat of nail polish. Once mounted, optical fibers can also be stored on clay rails in Petri dishes.

Our experimental protocols require that needles or glass fibers distribute forces evenly across a flat surface. One approach is to mount a flat "plate" directly onto the end of the fiber. We have found a variety of materials for this purpose. Sheets of mica (Ted Pella, Redding, California) can be cut with scalpels into fine rectangles $100 \times 200 \ \mu m^2$ in dimension. Flecks of glitter (Jewel Glitter, gold ultrafine; Mark Enterprises, Newport Beach, California) are regular hexagons 300 μm in diameter. Plastic or polyester shims, 12 μm and thicker (Small Parts, Inc., Miami Lakes, Florida), can be cut with scissors to fine rectangles with dimensions of 200 μm.

These plates are attached to the side of a fiber near its tip with a cyanoacrylate-based glue (Loctite Superglue). Precut plates and a small drop of superglue are placed in proximity on a piece of parafilm (American Can Company, Fisher Scientific, Pittsburgh, Pennsylvania) in the same field of view under a stereomicroscope. The tip of the fiber is then dragged through the glue until a small bulb of glue forms. The fiber is then quickly moved to the plate and brought into contact with the top surface of the plate. The small droplet of glue will spread onto the plate. The fiber is held onto the plate until the glue dries, which takes only a few minutes. Fibers with attached plates can be stored on clay rails in Petri dishes.

The spring stiffness constant, k, of each fiber is determined directly by hanging small wire weights or indirectly against the deflection of a previously calibrated fiber. The uncalibrated fiber mounted in a microcapillary tube is held horizontally with a micromanipulator (Marzhauser M33, Wetzlar-Steindorf, Germany). For direct calibration, we use fine copper wire (California Fine Wire Products, Corona, California). The mass per millimeter is determined by weighing a 1-m length of wire. Small standard lengths of wire 4, 8, and 12 mm in length are cut and folded in half. Forceps are used under a stereomicroscope to hang individual wires from the end of the fiber (Fig. 1A). Spatially calibrated images of the deflection of the tip (Fig. 1B–D) are captured using a horizontally placed microscope with a $20\times$ objective and a mounted charge-coupled device (CCD) camera. The deflections are measured using image analysis software (ImageJ; written by Wayne Rasband, NIMH, NIH, available for download at http://rsb.info.nih.gov/ij/). The spring stiffness constant k is determined by plotting the displacement of the fiber tip per force applied, where k is obtained from the inverse of the slope m for the displacements observed with different weights (Fig. 1E; $k = 1/m$). The force–displacement curve is generally linear and should intersect the origin.

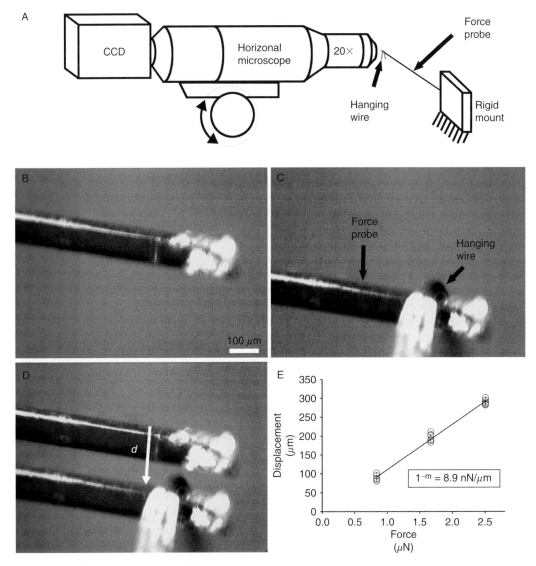

Fig. 1 Measuring the spring constant of an optical fiber. (A) A simple projection microscope positioned horizontally collects images of the end of the optical fiber force probe. The force probe is mounted rigidly so that weights hung on the end of the probe generate vertically directed forces. (B) A close-up of the end of the optical fiber with no weight using a 20× objective and (C) with a short length of wire. (D) Analysis of the images before and after the placement of the wire provides a measure of the displacement (*d*) caused by the weight of wire. (E) Displacements for a set of wires of different lengths produce different applied forces. The slope of the displacement as a function of force is the spring constant for that optical fiber.

Alternatively, the spring stiffness constant k for a fiber can be determined by deflecting the tip of the uncalibrated fiber against the tip of a previously calibrated fiber. We carry out this calibration on the stage of an inverted compound microscope (IX70, Olympus America, Melville, New York) with the uncalibrated fiber held fixed in a micromanipulator (Marzhauser M33, Wetzlar-Steindorf, Germany) mounted onto the microscope stand. The calibrated fiber mounted in a capillary tube is affixed to an automated X-Y stage positioner (Prior Scientific, Rockland, Massachusetts) and moved in discrete steps against the uncalibrated probe. The displacement of the calibrated fiber is used to determine the applied force, and the spring stiffness constant is determined by the slope of displacements of the uncalibrated fiber in response to the force applied by the calibrated fiber. Again, the calibration curve should be linear and should intersect the origin. At this stage, it is useful to uniquely label each fiber since each will have its own spring stiffness constant.

Following this protocol, probes of glass fibers or glass needles 30–40 mm in length can be constructed with spring stiffness in the range of 10 nN/μm. From engineering beam theory (Gere and Timoshenko, 1984), the spring stiffness of an end-loaded cantilevered beam is $(3EI)/L^3$, where EI is the flexural rigidity of the beam determined by the elastic modulus and the cross-sectional geometry of the beam (E is elastic modulus and I is the second moment of the cross-sectional area) and L is the length of the beam. For a beam with a solid circular cross section like an optic fiber, the spring constant is proportional to (d^4/L^3), where d is the diameter of the beam. Thus, it is relatively easy to increase the stiffness of a fiber by cutting off a segment from the end. Longer fibers can be prepared or smaller diameter fibers can be obtained (Polymicro Technologies, Phoenix, Arizona) to reduce the spring constant of the fiber.

III. Nanonewton Force Apparatus: Parts, Function, and Operation

The nanonewton force apparatus (nNFA) is a device to measure forces in tissue explants. These explants are typically solid cellular "bricks" made of cells and extracellular matrix of roughly $800 \times 300 \times 300$ μm^3. Forces exerted by the tissue as the result of morphogenetic movements or of compression range from 50 nN to 10 μN. The nNFA can be used in one of two modes: (1) measuring forces generated by tissue movements or (2) determining the time-varying elastic modulus of tissue explants. The first requires measuring the forces generated by an explant over the time period during which normal morphogenetic movement is occurring, typically several hours. The second use involves a more temporally limited experiment. The tissue under investigation is subject to a stress-relaxation materials testing protocol. Briefly, the tissue is compressed until a predetermined strain is attained. The viscoelastic properties of the tissue are then determined from the forces that the tissue generates in resistance to this compression, and from the change of these forces as components of the tissue rearrange to minimize the effects of compression. A stress-relaxation experiment typically takes less than 5 min. Tests over longer timescales may well be important to morphogenesis, but greater attention must be paid to active force generation by the tissue.

The nNFA consists of four major components (Fig. 2A). At the heart of the nNFA is a closed-loop DC servo micropositioner with encoder (US Eurotek, Lakeview, California), which carries the tissue sample into contact with the fiber optic probe. The positioner reports the base position of the fiber optic probe. Laser-generated light (820 nm, 15 mW; Newport Corporation) is fed into the fiber at one end and exits at the other end near the face of a quadrant detector (model 1242, UDT Instruments, Baltimore, MD), which detects light emitted by the probe tip and reports the tip position. The third component of the nNFA is the fiber optic force probe, consisting of the same optic fiber and a plate glued to the fiber near the detector (see above). The probe and the quadrant detector are aligned using manually controlled miniature XYZ micropositioners (Optosigma, Santa Ana, California). The nNFA is completed by a personal computer with data acquisition and hardware control software (LabVIEW, National Instruments Corporation, Austin, Texas). The computer controls the position of the tissue sample and records the position of the tip of the probe reported by the quadrant detector.

IV. Preparation of Tissue Samples

For the study of amphibian embryos, tissue samples are isolated at the desired stages using microsurgical techniques (Keller *et al.*, 1999; Moore *et al.*, 1995). Tissues from embryos of other species can be isolated by similar means, with the limitation that some embryos are more suited to microsurgery than others. For example, we find that avian or rodent embryos are particularly difficult to dissect compared to *Xenopus laevis*. In any case, the preparation of the tissue presents certain challenges. Mechanical measurements in fields of materials science and engineering generally deal with samples of consistent, well-defined geometric features, and these parameters often prove stable over time, although there are exceptions (e.g., granular solids and polymers). In contrast, the geometry of isolated embryonic tissues varies from sample to sample, depending on the microsurgeon's skill, the stage of development of the embryo, and the type of tissue and its behavior. For example, in a uniaxial compression stress-relaxation test (see following paragraph), the protocol requires compression by a specified, repeatable amount. However, the edges of explanted tissues are never perfectly straight or parallel, making determination of the start point of compression difficult. One solution is to cut the explant as cleanly as possible, with parallel edges, then apply a "preload" of compression that is a small fraction of the force applied on compression but enough to flatten the edges of the explant (Moore *et al.*, 1995).

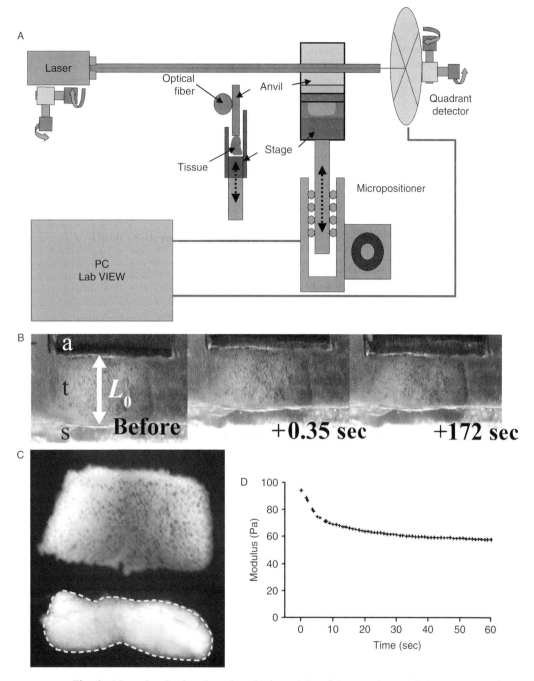

Fig. 2 Measuring the time-dependent elastic modulus of tissue explants with the nanonewton force apparatus (nNFA). (A) A schematic of the nNFA. (B) Three frames from a stress-relaxation experiment. Before a 10–15% compression, the tissue explant is placed on the stage and the anvil is brought

V. Measurement of the Time-Dependent Elasticity of Embryos or Tissue Explants

There are numerous standard engineering measures of the mechanical properties of materials. However, due to the limited quantities of embryonic tissues (both in size and number of samples), we will focus only on the stress-relaxation measure of the time-dependent elastic modulus (Moore *et al.*, 1995). A more general discussion of the stress-relaxation approaches and various representations of viscoelastic material can be found in Chapter 1 by Janmey *et al.* and Chapter 2 by Kandow *et al.*, this volume and in Findley *et al.* (1989). In brief, the test involves compressing a piece of tissue along a single axis (a uniaxial test) to a fixed strain while simultaneously measuring the force generated as the sample resists the compression. A time-lapse sequence of the entire course of the experiment is recorded with a CCD camera mounted on a stereoscope. After the tissue sample has been tested, it is immediately removed from the apparatus and fixed (overnight in 4% formaldehyde). The sample is then bisected with a scalpel to expose the cross-sectional area perpendicular to the axis of compression through which the force of compression was measured. The cross-sectioned tissue is then imaged with a CCD camera mounted on a stereoscope and the area is measured with image analysis software (ImageJ). We calculate the time-dependent modulus during the time course of the stress-relaxation experiment by dividing the time-dependent resistance force by the cross-sectional area of the sample.

The time-dependent modulus represents the bulk viscoelastic property of the tissue. It is an empirical measure of material properties under the specific conditions of the experimental test. For instance, the rate at which strain is initially applied can alter the modulus in the first few seconds, but will not alter the long-term elastic response of the tissue unless the tissue rips or tears. Alternatively, application of a larger strain may change the magnitude of the long-term elastic response. In order to compare the modulus from sets of different samples under the same test conditions, two different types of parameters can be compared, either the elastic modulus at a specific time (e.g., empirical) or the calculated coefficients of a specific material model (e.g., spring constants and coefficients of viscosity from a fitted spring and dashpot model). Since sample sizes are generally small, from 10 to 20, statistical significance of each of these parameter values can be obtained using nonparametric statistical tests such as the Mann–Whitney U test (Sokal and Rohlf, 1981) or the Jonckheere test (Siegel and Castellan, 1988). We typically compare empirical parameters within a study (e.g., comparing changes during development

---

into contact. The displacement of the tip of the optical fiber is measured on the quadrant detector and a video of the stress-relaxation test is recorded. The instantaneous strain at the start of the stress-relaxation test is calculated from the video and the force exerted from the explant is determined from the spring constant and the displacement reported by the quadrant detector. (C) Explant compressed in (B) is bisected and its cross-sectional area (white dashed line) measured. (D) Time-dependent elastic modulus of explant shown in (B) and (C).

or after experimental manipulations), but it is generally desirable to use standard material model parameters when comparing results with those obtained from different approaches or materials (e.g., comparing tissue properties to the bulk viscoelastic properties of purified gels from cone-and-plate rheometry).

VI. Spring and Dashpot Models of Viscoelasticity Represent More Complex Structural Sources

The viscoelastic response of a tissue can be represented by physically analogous structures composed of springs and dashpots organized in parallel or series. The structures are chosen because they mimic mathematically the response of the tissue to applied forces (Findley *et al.*, 1989). These spring and dashpot "constructs" are helpful aids to extend our intuition to otherwise abstract mathematical representations. The cell or tissue biology responsible for either the "spring" or "dashpot" must be worked out experimentally. For instance, the viscous response of a tissue might be due to one or more biophysical processes such as interstitial fluid flow, cytoplasmic fluid flow, cytoskeletal rearrangement, or cell rearrangement. Similarly, the elastic response of a tissue might be due to the elastic response of individual parts of the cell, such as the nucleus (Chapter 11 by Lammerding *et al.*, this volume) or cytoskeleton, the elastic response of the tissue as a cellular foam, or the response of tissue elements like the extracellular matrix.

Spring and dashpot models do not capture cell-patterned or architectural contributions to the mechanical properties of the tissues. Engineering studies on the mechanics of cellular foams identify several possible sources of material anisotropy (i.e., stiffer in one direction than the other; Gibson and Ashby, 1997). The material itself may be anisotropic with aligned cytoskeletal, extracellular elements, or directionally aligned cells. Simple viscoelastic models also cannot capture tissue architectural features such as the notochord-somite boundary (i.e., interfaces), the neural floorplate, or multiple layers of different tissue materials. To accommodate these architectural features requires more elaborate structural-based models or more complex mathematical-material models.

VII. Challenges of Working with Embryonic Tissues

Some of the problems of preparing embryonic tissues for measurement were discussed above, but there are additional problems that arise in the course of making measurements.

One such problem arises when tissue geometry and mechanical properties change due to wound healing. Once tissue is cut out of the embryo, it will begin to heal, and the shape of the tissue can change rapidly, depending on the organization of the tissue and its inherent properties. Often the samples consist of deep mesenchymal cells and an epithelial sheet on one or more sides. The free edges of the sheets will begin to migrate over the deeper, mesenchymal tissue, often with surprising speed, and the geometry of the excised tissue may change very quickly, potentially in ways that will alter the measurements. For example, if the tissue has a tendency to curl and is already bent when placed in the measuring apparatus, the tissue may buckle under much lower loads than if it were straight when the test begins. In addition, if the tissue becomes completely covered by the epithelium, it could form a closed hydraulic system of very different mechanical response compared with an explant with mesenchymal cells exposed on one or more sides. Explanted epithelial and mesenchymal tissues, or combinations of both, usually begin to round up in a fashion similar to cell aggregates (Foty and Steinberg, 2005; Holtfreter, 1939), and thus the tissue will be changing shape during a test that is designed for static tissues of a rectangular shape.

Another problem is choosing a saline medium in which to do the tests. Mechanical tests are no different than many other assays and experimental protocols done on living embryonic tissues in culture; what you get may depend on the medium used. Once tissue is excised from the internal, controlled environment of the embryo, many of its properties are affected by the physiological medium in which it is cultured. Calcium levels and pH affect cell adhesion and, therefore, potentially affect the mechanical properties of the tissue. The outer surface of the epithelial sheet covering many free-living aquatic embryos is normally exposed to very low ionic strength solutions, for example pond water, whereas the internal (deep) ends of these cells and the internal (deep) mesenchymal cells are normally exposed to a higher ionic strength environment. Most media are not ideal for one or the other tissue. The best choice of mediums is to use one that supports normal morphogenesis. For many of our assays, we use Danilchik's for Amy (DFA; 53-mM NaCl, 5-mM Na_2CO_3, 4.5-mM K gluconate, 32-mM Na gluconate, 1-mM $CaCl_2$, 1-mM $MgSO_4$, 1 g of bovine serum albumin (BSA) per liter, and buffered to pH 8.3 with 1-M bicine), a physiological saline similar in ionic composition, and pH to blastocoel fluid of the frog (Gillespie, 1983; Sater et al., 1993).

Finally, embryonic tissues are dynamic and will change shape during the test. If the movement is fast and the test long, the measurements will reflect a complex behavior. For example, the dorsal mesoderm of the frog gastrula actively converges (becomes narrower) in the mediolateral dimension and extends (becomes longer) in the anterior–posterior direction, a behavior that drives a major part of gastrulation and also elongates the body axis of vertebrates (Keller, 2002; Keller et al., 2000). As the tissue extends, it becomes stiffer and also actively pushes with a force equivalent to about 0.6 μN for a single explant of the tissue in an isometric assay (Moore, 1994; Moore et al., 1995). During uniaxial stress-relaxation tests to measure stiffness along the anterior–posterior axis, the tissue is actively narrowing in the mediolateral direction and actively pushing or elongating along the anterior–posterior direction (Moore et al., 1995). Therefore, to make meaningful measurements, the relaxation period must be relatively short, 3 min in this case (Moore et al., 1995). In any experiment, tissue shape changes due to healing or active, normal movements should be taken into account.

VIII. Use of Standard Engineering Terms and Units

As in other areas of biomechanics, mechanical analysis of embryonic tissues should use standard definitions, protocols, and units of measure as typically developed in engineering and the physical sciences. Definitions of important terms, such as force, stress, strain, tension, compression, shear, and others, are given in a number of primers on the subject (Chapter 1 by Janmey *et al.* and Chapter 2 by Kandow *et al.*, this volume; Koehl, 1990; Koehl and Wainwright, 1985; Wainwright *et al.*, 1976). Uniform standards for testing methods and units of measure should be encouraged as this allows comparison of results from different studies, something that is not always possible particularly when "creative" biomechanical tests that do not actually measure the same parameters are used in various studies (Koehl, 1990).

IX. Future Prospects

Application of biomechanical methods to embryonic tissues is in its infancy, and the current technology and methods of application can be developed much further. The above examples describe compressive tests, but basically the same machines can be adapted to tensile tests by allowing test tissues to attach to adhesive substrates applied to the mounting plates, thus allowing the instrument to pull rather than push on the tissue. In addition to stress-relaxation tests, isometric force measurements can be done (Moore, 1994) and in some cases, it might be desirable to do force velocity measurements. For example, the dorsal axial and paraxial mesoderm of *Xenopus* can extend with a pushing force of about $0.6\ \mu N$ in an isometric test, but a more relevant parameter might be the force that it can exert at its normal rate of extension in the embryo, which would indicate the force it is contributing to the stretching of the surrounding tissues. This test would involve using the computer-controlled measuring probe to determine the retarding load necessary to limit the extension to a particular rate. To our knowledge, only two instruments appropriate for tissue level measurements of this type have been built and used, the one described above and the one developed by Moore and coworkers (Morre, 1994; Moore *et al.*, 1995). As it becomes clear what parameters should be measured for various embryonic tissue architectures, new designs and performance parameters can be implemented.

References

Adams, D. S., Keller, R., and Koehl, M. A. (1990). The mechanics of notochord elongation, straightening and stiffening in the embryo of *Xenopus laevis*. *Development* **110**(1), 115–130.

Davidson, L. A., Koehl, M. A., Keller, R., and Oster, G. F. (1995). How do sea urchins invaginate? Using biomechanics to distinguish between mechanisms of primary invagination. *Development* **121**, 2005–2018.

Davidson, L. A., Oster, G. F., Keller, R. E., and Koehl, M. A. (1999). Measurements of mechanical properties of the blastula wall reveal which hypothesized mechanisms of primary invagination are physically plausible in the sea urchin Strongylocentrotus purpuratus. *Dev. Biol.* **209,** 221–238.

Findley, W. N., Lai, J. S., and Onaran, K. (1989). "Creep and Relaxation of Nonlinear Viscoelastic Materials." Dover Publications, Inc, New York.

Foty, R. A., and Steinberg, M. S. (2005). The differential adhesion hypothesis: A direct evaluation. *Dev. Biol.* **278,** 255–263.

Gere, J. M., and Timoshenko, S. P. (1984). "Mechanics of Materials." Brooks/Cole Engineering Division, Monterey, California.

Gibson, L. J., and Ashby, M. F. (1997). "Cellular solids: Structure and properties." Cambridge University Press, New York.

Gilbert, S. F. (2006). "Developmental Biology," Vol. xviii, 817 p. Sinauer Associates, Inc., Publishers, Sunderland, MA.

Gillespie, J. I. (1983). The distribution of small ions during the early development of *Xenopus laevis* and *Ambystoma mexicanum* embryos. *J. Physiol. (Lond.)* **344,** 359–377.

Gustafson, T., and Wolpert, L. (1963). The cellular basis of morphogenesis and sea urchin development. *Int. Rev. Cytol.* **15,** 139–214.

Holtfreter, J. (1939). Gewebeaffinitat, ein Mittel der embryonalen formbildung. *Arch. Exp. Zellforsch.* **23,** 169–209.

Holtfreter, J. (1943). A study of the mechanics of gastrulation (part 1). *J. Exp. Zool.* **94,** 261–318.

Holtfreter, J. (1944). A study of the mechanics of gastrulation (part 2). *J. Exp. Zool.* **95,** 171–212.

Howard, J. (2001). "Mechanics of Motor Proteins and the Cytoskeleton," 367 pp. Sinauer Associates, Inc., Publishers, Sunderland, MA.

Hutson, M. S., Tokutake, Y., Chang, M. S., Bloor, J. W., Venakides, S., Kiehart, D. P., and Edwards, G. S. (2003). Forces for morphogenesis investigated with laser microsurgery and quantitative modeling. *Science* **300,** 145–149.

Keller, R. (2002). Shaping the vertebrate body plan by polarized embryonic cell movements. *Science* **298,** 1950–1954.

Keller, R., Davidson, L., Edlund, A., Elul, T., Ezin, M., Shook, D., and Skoglund, P. (2000). Mechanisms of convergence and extension by cell intercalation. *Philos. Trans. R. Soc. Lond. B* **355,** 897–922.

Keller, R., Poznanski, A., and Elul, T. (1999). Experimental embryological methods for analysis of neural induction in the amphibian. *In* "Methods in Molecular Biology: Molecular Embryology: Methods and Protocols" (P. Sharpe and I. Mason, eds.). Humana Press, Inc., Totowa, NJ.

Koehl, M. A. R. (1990). Biomechanical approaches to morphogenesis. *Semin. Dev. Biol.* **1,** 367–378.

Koehl, M. A., and Wainwright, S. A. (1985). Biomechanics. *In* "Handbook of Phycological Methods: Ecological Field Methods—Macroalgae" (M. M. Littler and D. S. Littler, eds.), pp. 291–313. Cambridge University Press, Cambridge.

Levine, M., and Davidson, E. H. (2005). Gene regulatory networks for development. *Proc. Natl. Acad. Sci. USA* **102,** 4936–4942.

Lewis, W. H. (1947). Mechanics of invagination. *Anat. Rec.* **97,** 139–156.

Loose, M., and Patient, R. (2004). A genetic regulatory network for *Xenopus* mesendoderm formation. *Dev. Biol.* **271,** 467–478.

Moore, A. R. (1941). On the mechanics of gastrulation in Dendraster excentricus. *J. Exp. Zool.* **87,** 101–111.

Moore, S. W. (1994). A fiber optic system for measuring dynamic mechanical properties of embryonic tissues. *IEEE Trans. Biomed. Eng.* **41,** 45–50.

Moore, S. W., Keller, R. E., and Koehl, M. A. R. (1995). The dorsal involuting marginal zone stiffens anisotropically during its convergent extension in the gastrula of *Xenopus leavis*. *Development* **121,** 3130–3140.

Rappaport, R. (1977). Tensiometric studies of cytokinesis in cleaving sand dollar eggs. *J. Exp. Zool.* **201** (3), 375–378.

Sater, A. K., Steinhardt, R. A., and Keller, R. (1993). Induction of neuronal differentiation by planar signals in *Xenopus* embryos. *Dev. Dyn.* **197,** 268–280.

Selman, G. G. (1955). Studies on the forces producing neural closure in amphibia. *Proc. R. Phys. Soc. Edinb.* **24,** 24–27.

Selman, G. G. (1958). The forces producing neural closure in amphibia. *J. Embryol. Exp. Morphol.* **6,** 448–465.

Siegel, S., and Castellan, J. N. J. (1988). "Nonparametric Statistics for the Behavioral Sciences," 399 p. McGraw-Hill Book Company, New York.

Sokal, R. R., and Rohlf, F. J. (1981). "Biometry." W. H. Freeman and Company, New York.

Spek, J. (1918). Differenzen im quellungszustand der plasmakolloide als eine ursache der gastrulainvagination, sowie der einstulpungen und faltungen von zellplatten uberhaupt. *Kolloidchemische Beihefte* **9,** 259–399.

Stern, C. D. (2004). "Gastrulation: from Cells to Embryo," xvi, 731 p. Cold Spring Harbor Laboratory Press, Cold Spring Harbor, NY.

Waddington, C. H. (1939). Order of magnitude of morphogenetic forces. *Nature* **144**(3649), 637.

Wainwright, S. A., Biggs, W. D., Currey, J. D., and Gosline, J. M. (1976). "Mechanical Design in Organisms." John Wiley and Sons, New York.

Wiebe, C., and Brodland, G. W. (2005). Tensile properties of embryonic epithelia measured using a novel instrument. *J. Biomech.* **38,** 2087–2094.

PART IV

Mechanical Stimuli to Cells

CHAPTER 19

Tools to Study Cell Mechanics and Mechanotransduction

Tanmay P. Lele,★,[1] **Julia E. Sero,**★ **Benjamin D. Matthews,**★,†
Sanjay Kumar,★,[2] **Shannon Xia,**★ **Martin Montoya-Zavala,**★
Thomas Polte,★ **Darryl Overby,**★,[3] **Ning Wang,**‡
and Donald E. Ingber★

★Vascular Biology Program
Departments of Pathology and Surgery
Children's Hospital and Harvard Medical School
Boston, Massachusetts 02115

†Department of Pediatrics
Massachusetts General Hospital and Harvard Medical School
Boston, Massachusetts 02114

‡Department of Mechanical Science and Engineering
University of Illinois at Urbana-Champaign
Urbana, Illinois 61801

Abstract
 I. Introduction
 II. Control of Cell Shape, Cytoskeletal Organization, and Cell Fate Switching
     A. Microcontact Printing of Micropatterned Substrates for Cell Culture
     B. Application Notes on Microcontact Printing
     C. Extension and Future Development of Microcontact Printing
 III. Probing Cell Mechanics, Cytoskeletal Structure, and Mechanotransduction
     A. Magnetic Twisting Cytometry (MTC)
     B. Applications of MTC
     C. Extension and Future Development of MTC

[1] Department of Chemical Engineering, University of Florida, Gainesville, Florida 32611.
[2] Department of Bioengineering, University of California, Berkeley, Berkeley, California 94720.
[3] Department of Biomedical Engineering, Tulane University, New Orleans, Louisiana 70118.

METHODS IN CELL BIOLOGY, VOL. 83
Copyright 2007, Elsevier Inc. All rights reserved.
0091-679X/07 $35.00
DOI: 10.1016/S0091-679X(07)83019-6

Abstract

Analysis of how cells sense and respond to mechanical stress has been limited by the availability of techniques that can apply controlled mechanical forces to living cells while simultaneously measuring changes in cell and molecular distortion, as well as alterations of intracellular biochemistry. We have confronted this challenge by developing new engineering methods to measure and manipulate the mechanical properties of cells and their internal cytoskeletal and nuclear frameworks, and by combining them with molecular cell biological techniques that rely on microscopic analysis and real-time optical readouts of biochemical signaling. In this chapter, we describe techniques like microcontact printing, magnetic twisting cytometry, and magnetic pulling cytometry that can be systematically used to study the molecular basis of cellular mechanotransduction.

I. Introduction

Cellular mechanotransduction refers to the processes by which cells convert physical forces into changes in intracellular biochemistry. These processes are critical for control of cell growth, migration, differentiation, and apoptosis during organogenesis and wound repair. Destabilization of cell and tissue structure, or dysfunctional mechanotransduction, can lead to the development of numerous diseases and debilitating conditions, including atherosclerosis, hypertension, asthma, osteoporosis, and cancer (Ingber, 2003a). Analysis of how cells sense and respond to mechanical stress has been limited by the availability of techniques that can apply controlled mechanical forces to living cells while simultaneously measuring changes in cell and molecular distortion, as well as alterations of intracellular biochemistry. We have confronted this challenge by developing new engineering methods to measure and manipulate the mechanical properties of cells and their internal cytoskeletal and nuclear frameworks, and by combining them with molecular cell biological techniques that rely on microscopic analysis and real-time optical readouts of biochemical signaling.

The methods we describe here emerged from systematic testing of an underlying hypothesis relating to cell structure, matrix mechanics, and mechanotransduction that has driven the work in our laboratory for over 25 years. Early views of the cell prior to the mid-1970s postulated a mechanical structure consisting primarily of an elastic membrane surrounding a viscoelastic cytoplasm (e.g., like a balloon filled with molasses or jello), and many engineering models of the cell are still based on this vision (Dong and Sung, 1991). In contrast, we proposed that a cell uses a specific form of architecture known as "tensegrity" to structure its cytoskeleton, and that cells are composed of an interconnected network of tensed cables and membranes stabilized by compressed struts and substrate anchors (i.e., more like a tent than a water balloon) (Ingber, 1993a, 2003b; Ingber and Jamieson, 1985). This model predicts that contractile microfilaments and intermediate filaments in anchored cells function as cables which distribute tensile forces throughout the cytoplasm and nucleus, whereas compressive forces are resisted through cell adhesions by underlying extracellular matrix (ECM) and by neighboring cells, as well as by internal cytoskeletal struts such as microtubules or cross-linked actin bundles.

This view of cell structure led us to propose that transmembrane receptors such as integrins that physically integrate the cytoskeleton into the ECM may function as mechanoreceptors that provide a preferred path for mechanical force transfer across the cell surface (Ingber, 1991; Ingber and Jamieson, 1985). The cellular tensegrity model assumes that forces channeled through these discrete linkages become focused locally at the site of ECM anchorage within specialized cytoskeletal anchoring complexes known as "focal adhesions," and at distant sites throughout the cytoplasm and nucleus due to mechanical connectivity across discrete molecular linkages throughout the ECM–cytoskeleton–nuclear matrix lattice.

Importantly, many of the enzymes and substrates that mediate most of the cell's metabolic machinery perform their functions when physically immobilized on these macromolecular scaffolds (Ingber, 1993b). Thus, an important corollary to the tensegrity model of cell mechanics is that forces applied to cells and transmitted to the cytoskeleton through transmembrane integrin receptors may be converted into changes in intracellular biochemistry and gene expression at the nanometer scale through stress-dependent distortion of cytoskeletal-associated molecules (Ingber, 1997, 2006; Ingber and Jamieson, 1985). Force-induced changes in the shape of these load-bearing proteins will alter their chemical potential, thereby changing their kinetic and thermodynamic behavior, and hence, altering their biochemical activities.

If cells are structured using tensegrity, then cell shape stability will depend on a mechanical force balance between cytoskeletal traction forces and the ability of the ECM substrate to resist these stresses which will create a state of isometric tension, or a tensional "prestress" in the cytoskeleton and linked ECM. The tensegrity model therefore predicts that at the whole tissue level, changes in ECM mechanics may alter cell shape, cytoskeletal organization, and the steady state of mechanical balance, or prestress, inside the cytoskeleton, and that this, in turn, may alter intracellular biochemistry. This is important because it suggests that local variations in ECM structure and mechanics that are observed in developing tissues may contribute to the regional differentials in growth and motility that drive morphogenetic changes of tissue form in the embryo, as well as in certain disease processes (Huang and Ingber, 1999; Ingber and Jamieson, 1985).

Traditional tools in molecular cell biology cannot be used to test these biophysical hypotheses. Instead, we have had to miniaturize different types of engineering analysis methods, to develop new ones, and to combine them with novel molecular cell biological approaches to test the tensegrity theory and gain greater insight into cellular biophysics. Initially, we devised methods to control how cells physically interact with ECM substrates in order to test directly whether cell shape distortion influences intracellular biochemistry and cell behavior. Results of these studies confirmed that mechanical forces functioned as bioregulators.

We then shifted our focus to explore how cells sense and respond to mechanical forces at the molecular level. Due to the small size of living cells and our desire to determine the contributions of discrete cytoskeletal filament networks, integrins, ECM, and cytoskeletal prestress to cell mechanics and mechanotransduction, these methods needed to provide the ability to apply controlled forces to specific molecules while simultaneously measuring changes in molecular displacement and biochemical activities inside the cell at the nanometer scale. Additional insight into cell structure and mechanotransduction came from the development of different methods that provided ways to physically disrupt discrete cytoskeletal elements with nanometer resolution in living cells, without interfering with the function of other cellular components or compromising cell viability.

By applying these methods to systematically analyze the molecular basis of cellular mechanotransduction, and combining these techniques with methods developed by other laboratories, such as traction force microscopy, we confirmed that mechanochemical conversion is mediated by forces channeling through integrins and the cytoskeleton. Our results also provide direct experimental evidence that cells use tensegrity to mechanically stabilize themselves (Brangwynne *et al.*, 2006; Kumar *et al.*, 2006; Lele *et al.*, 2006; Maniotis *et al.*, 1997; Wang *et al.*, 1993). In addition to providing new analytical approaches to study cellular biophysics and mechanotransduction, some of these methods are beginning to be adapted to provide novel biomaterial control interfaces, and they may facilitate integration of living cells into machines (e.g., biochips, medical devices, biodetectors, computers) in the future.

II. Control of Cell Shape, Cytoskeletal Organization, and Cell Fate Switching

On the basis of past work from cell and developmental biology, and the concept that cells and tissues might use tensegrity architecture, we proposed over 20 years ago that the spatial differentials of cell growth and function that drive tissue morphogenesis might be controlled mechanically through local variations in physical interactions between cells and their ECM (Huang and Ingber, 1999; Ingber and Jamieson, 1985). Changes in the cellular force balance would produce both local and global cytoskeletal rearrangements inside the cell and thereby, drive changes in intracellular biochemistry that influence cell fate decisions such as whether cells will grow, differentiate, or die. In early studies, we developed a method to control cell shape by varying the density of ECM molecules coated on otherwise nonadhesive dishes; cells spread on surfaces with high ECM-coating densities, but retracted and rounded on low coating concentrations (Ingber and Folkman, 1989b). In addition, adherent cells, such as capillary endothelial cells (Ingber, 1990; Ingber and Folkman, 1989a) and primary hepatocytes (Mooney *et al.*, 1992), respond to growth factors and proliferate when spread, whereas they differentiate in the same medium under conditions that prevent cell extension (e.g., when adherent to low ECM densities). However, these results were difficult to interpret because, in addition to differences in cell shape, altering ECM-coating densities also may change the degree of integrin receptor clustering on the cell surface, which regulates the ability of these ECM receptors to activate intracellular signal transduction (Schwartz *et al.*, 1991).

We therefore set out to design an experimental system in which cell shape distortion could be varied independently of either the concentration of soluble hormones or the local ECM ligand-binding density. The approach we took was to microfabricate ECM islands of defined size, shape, and position on the micrometer scale, surrounded by nonadhesive barrier regions. Living cells exert traction forces on their ECM adhesions, and thereby spread and flatten themselves against standard culture substrates. Our approach was therefore based on the concept that cells would adhere to small ECM islands and extend themselves until they reached the nonadhesive barrier region where they would stop, and effectively take on the bounding shape of their "container." The degree of extension and flattening of single cells could therefore be controlled by plating the cells on different sized islands coated with the same high density of ECM molecule, in the same growth factor-containing medium; only the degree of cell distortion would vary. We accomplished this by adapting a microcontact printing technique that was initially developed as an alternative way to manufacture microchips for the computer industry by the laboratory of our collaborator, George Whitesides (Department of Chemistry and Chemical Biology, Harvard University) (Singhvi *et al.*, 1994; Whitesides *et al.*, 2001; Xia and Whitesides, 1998).

A. Microcontact Printing of Micropatterned Substrates for Cell Culture

The microcontact printing technique is a form of "soft lithography" which uses the elastomeric material, poly(dimethylsiloxane) (PDMS), to create flexible molds that retain the surface topography of silicon masters. The surfaces of silicon masters are, in turn, patterned using standard photolithography techniques which involves coating with photoresist polymer layers, exposure to UV light, removal of noncrosslinked polymers, and surface etching (Xia and Whitesides, 1998). These stamps can be used to transfer pattern elements and create multiple replica substrates when combined with chemical inks that form self-assembled monolayers (SAMs) (Prime and Whitesides, 1991).

We adapted this method to create patterns of ECM molecules that support cell adhesion in order to microengineer culture substrates with defined shapes and sizes on the micrometer scale (Fig. 1). Typically for micropatterned substrates for cell-sized adhesive islands, the desired pattern is drawn using a computer-aided design software (e.g., AutoCAD), and printed on a transparent plastic sheet that functions as the photomask. The printing is done typically by commercial services with high-resolution laser printers, although masks with features in the order of hundreds of micrometers may be generated using a consumer laser printer. Standard photolithographic techniques are used to etch this pattern into a photoresist polymer layer (e.g., 2-μm thick coating of polymethylmethacrylate) that is coated over a planar silicon wafer to generate a "master" with topographic surface features corresponding to that of the photomask (Chen *et al.*, 1998; Chapter 13 by Sniadecki and Chen, this volume for more details). Liquid PDMS prepolymer (1:10 elastomer base:curing agent; Sylgard 184, Dow Corning) is poured over the master, cured at 60 °C for 2 h and then peeled away by hand. This procedure yields clear and flexible PDMS stamps with raised surface features that correspond precisely to the original photomask pattern (Fig. 1). To engineer cell culture substrates on the micrometer scale for experimental studies, the surface of the flexible stamp is inked with a cotton applicator stick saturated with an ethanolic solution of methyl-terminated long-chain alkanethiols (e.g., hexadecanethiol, Sigma, Missouri), dried with compressed nitrogen or argon, and brought into conformal contact for 30 sec with a clean gold substrate. The substrate is previously prepared by depositing a 40-nm layer of gold on a titanium-primed glass slide using an e-beam metal evaporator (note that this step involves the same equipment that is used for preparing samples for electron microscopy; Whitesides *et al.*, 2001). The PDMS stamp is applied to the gold-coated slide by placing the stamp onto the slide, and then pressed down lightly with a forceps or finger (Fig. 1) until the pattern is visibly in contact with the entire surface. This can sometimes be seen by a

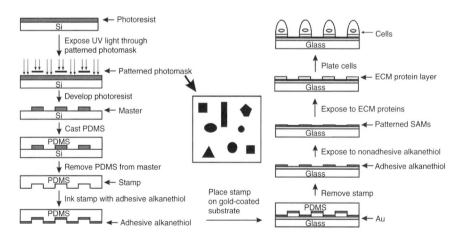

Fig. 1 Microcontact printing using soft lithography and SAMs. Method for microfabrication of stamps and microcontact printing SAMs on gold-coated glass substrates.

refraction of light on the substrate. Only the raised features of the stamp touch the gold and transfer the methyl-terminated alkanethiols within those geometrically defined regions to the surface of the substrate, where they self-assemble into a semicrystalline planar lattice (Xia and Whitesides, 1998). The PDMS stamps can be cleaned by sonicating in ethanol for 30 min and used repeatedly for several hundred times. The silicon master itself can be used to make hundreds of PDMS stamps if kept in a dust-free environment.

After removing the stamp, the glass substrate is covered with a solution of alkanethiol that is terminated with oligo(ethylene glycol) [OEG, e.g., $(EG)_3OH$]; this backfills all remaining (unstamped) areas of the exposed gold and thereby creates one single-planar SAM that covers the entire surface of the substrate, but exposes different terminal groups in different patterned regions. The substrate is then rinsed sequentially with ethanol, water, phosphate-buffered saline (PBS) and finally incubated in PBS containing ECM proteins (e.g., fibronectin, laminin, different collagen types) at a concentration of 50 μg/ml for 3 h at room temperature. After washing off excess protein (taking care not to allow the substrate to dry out) and transferring the substrates to culture medium, the microcontact-printed substrate is ready for cell culture.

As the terminal OEG groups of the alkanethiols within the barrier regions prevent protein adsorption, the soluble ECM proteins do not adhere to these regions. Instead, proteins preferentially adsorb to the patterned regions of the substrate that were stamped with methyl-terminated alkanethiols. Thus, this procedure results in creation of a microarray of adhesive ECM islands with the precise size, shape, and distribution depicted in the original photomask, surrounded by nonadhesive, OEG-coated barrier regions (Fig. 2A).

B. Application Notes on Microcontact Printing

For applications that require single cells to be confined within single adhesive islands, cell spreading and shape can be controlled with micrometer resolution by the size and shape of the islands, as long as the area of the island is equal to or less than the maximum area of spreading for that particular cell type (e.g., ~3000 and 4000 μm^2 for bovine and human endothelial cells, respectively; Chen et al., 1998), and the nonadhesive spacing between adjacent islands is large enough to prevent cell spreading across multiple islands. For example, the spacing thresholds at which bovine and human endothelial cells can bridge across neighboring islands are 10 and 20 μm, respectively, and this bridging ability appears to be directly related to cell size (i.e., human endothelial cells are about twice as large as bovine) (Chen et al., 1998). On the other hand, the interisland spacing cannot be made too large because this can lead to "sagging" of the PDMS stamp between neighboring islands and a loss of pattern fidelity. In short, both design features (i.e., island size and interisland distances) need to be determined empirically for each new cell type that is introduced into this experimental system. These studies can be carried out in

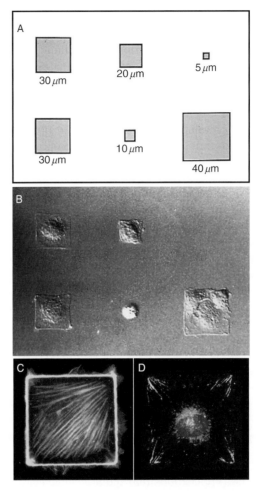

Fig. 2 Control of cell shape and cytoskeletal organization on micropatterned adhesive islands. (A) Schematic design of an array of different shaped square adhesive islands (top) and Nomarski microscopic images of bovine capillary endothelial cells plated on fibronectin-coated adhesive islands patterned according to this design using the microcontact printing method. Note that cells change their size and shape to match that of their adhesive islands. (Scale bar is 30 μm.) [Modified from Chen *et al.* (1997) with permission]. (C) Higher magnification fluorescence microscopic view of a cell cultured on a $40 \times 40 \ \mu m^2$ square ECM island and stained for actin fibers using Alexa488-phalloidin. Note that stress fibers orient preferentially along the diagonals (central nuclei are stained with DAPI). (D) A similar magnification view of another cell cultured on a similar island and immunostained for vinculin. Vinculin-containing focal adhesions concentrate within the corner regions at the points where the ends of stress fibers insert on the underlying ECM substrate. [Modified from Parker *et al.* (2002) with permission.]

medium with or without serum; however, the best shape control is generally observed under serum-free conditions.

The microcontact printing technique is extremely useful for holding single cells in particular configurations, as various types of adherent cells spread to take on the size and geometric form (e.g., square, triangle, circle, and so on) of the micro-engineered islands (Fig. 2A and B). Thus, it can be used to analyze directly the relation between cell shape and biological control. This method allowed us, for example, to demonstrate unequivocally that cell spreading produced through physical interactions between cells and their ECM adhesions can control cell fate decisions. Capillary endothelial cells, hepatocytes, and other cell types proliferated when allowed to spread on large (2500 μm^2) islands in the presence of soluble growth factors (Fig. 3A); however, they shut off growth and differentiated (liver cells produced blood proteins, capillary cells formed hollow tubes) when cultured on smaller islands that promote moderate spreading ($\sim$1000–2000 μm^2) in the same medium (Chen *et al.*, 1997; Dike *et al.*, 1999; Singhvi *et al.*, 1994). Moreover, capillary cells switched on the cellular suicide (apoptosis) program when grown on single tiny (<500 μm^2) ECM islands that induced almost complete cell rounding (Chen *et al.*, 1997; Fig. 3A).

As described above, when the interisland spacing is decreased, cells can bridge across multiple islands. We took advantage of this observation to culture capillary cells on multiple islands that had the size of individual focal adhesions (5 μm in diameter) and were closely spaced (10-μm apart). Under these conditions, each cell spread extensively (Fig. 3A), even though it contacted the same total amount of ECM area as when it was confined to single small ECM islands (10-μm diameter circles) that promoted apoptosis. In contrast to cells that were constrained in a round form on single 10-μm islands, cells that spread over many 5-μm islands proliferated, hence demonstrating that cell shape distortion is critical in governing cell life and death (Chen *et al.*, 1997).

Substrates created with microcontact printing are also extremely useful for analysis of the effects of physical interactions between cells and their ECM adhesions on intracellular cytoskeletal organization and focal adhesion formation. For example, immunofluorescence staining for vinculin within cells spread over a dense microarray of multiple small (3- to 5-μm diameter) ECM islands revealed that focal adhesion assembly is governed by the degree to which the entire cell can spread: cells held in a round form by attachment to single 10-μm ECM islands could not form well-developed focal adhesions (Chen *et al.*, 2003), even though they adhered to the same high ECM density. Moreover, the organization of the focal adhesion itself varied depending on the size of the island and its position beneath the cell, as well as on the level of tension in the cytoskeleton.

Analysis of cells cultured on larger (single-cell sized) square ECM islands revealed that intracellular stress fibers align predominantly along the diagonals of the square (Fig. 2C), while focal adhesions preferentially form at corners (Brock *et al.*, 2003; Parker *et al.*, 2002; Fig. 2D). Importantly, when the square cells were stimulated with motility factors, they preferentially extended lamellipodia,

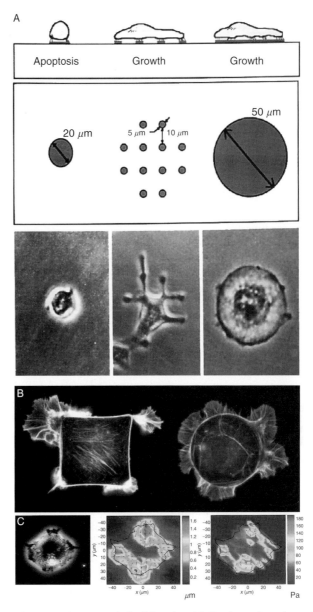

Fig. 3 Cell distortion-dependent control of cell function. (A) Top image is a schematic of a design for a micropatterned substrate containing single sparse small or large circular ECM islands, or a dense array of tiny (5 μm) focal adhesion-sized islands. The bottom phase-contrast microscopic images show that capillary endothelial cells take on the shape of the single small (left panel) or large (right panel) islands, but spread over multiple closely spaced tiny islands (middle panel). Spread cells proliferate whereas those that remain spherical undergo apoptosis even though the total amount of cell–ECM contact area is similar in cells on the single small island and multiple tiny islands. [Modified from Chen *et al.* (1997) with permission.] (B) Fluorescence staining of the actin cytoskeleton reveals that cells

filopodia, and microspikes from their corners, whereas there was no bias in round cells on circular islands (Parker *et al.*, 2002; Fig. 3B). Use of polygonal ECM islands with different shapes and numbers of corners, but similar size (900 μm^2), revealed that cells extend new motile processes more frequently from corners with acute, rather than obtuse angles (Brock *et al.*, 2003). By using the microcontact printing technique to microfabricate square ECM islands on the surface of flexible polyacrylamide gels used for traction force microscopy (Pelham and Wang, 1997; Wang and Pelham, 1998; Chapter 2 by Kandow *et al.*, this volume), we were able to demonstrate that cells exert greatest traction forces in these same corner regions (Wang *et al.*, 2002a; Fig. 3C). Separate studies revealed that lamellipodia extension can be inhibited by dissipating cytoskeletal tension (Brock *et al.*, 2003; Parker *et al.*, 2002). Thus, use of the microcontact printing method allowed us to demonstrate that global control of cell shape can result in spatial patterning of cytoskeletal prestress which, in turn, influences local molecular biochemical responses inside the cell, including the positioning of focal adhesions and lamellipodia that are central to directional cell motility. The microcontact printing technique also has been adapted for analysis of signal transduction and other biochemical studies (Chen *et al.*, 1998; Polte *et al.*, 2004). However, because cells must be plated sparsely to maintain single cells on individual islands, it is sometimes difficult to obtain large amounts of protein or RNA for biochemical analysis (i.e., many similar dishes must be cultured in parallel).

C. Extension and Future Development of Microcontact Printing

Taken together, these studies show that microfabricated ECM islands allow one to discriminate clearly between signals conveyed by soluble factors, direct binding of immobilized ECM molecules, and physical cues associated with cell shape distortion. For this reason, this method or modified versions of this technique (Tan *et al.*, 2004) are now a mainstay in studies on the mechanisms by which cell shape and behavior are controlled by physical interactions between cells and ECM. The microcontact printing method could also be modified by directly conjugating peptides that mediate integrin receptor binding and cell adhesion (e.g., RGD cell-binding site from fibronectin) to the terminal (EG)OH group of the OEG-alkanethiol (Kato and Mrksich, 2004; Roberts *et al.*, 1998). Use of these SAMs eliminates the need to add ECM protein to the substrate, as they are themselves sufficient to promote cell attachment, spreading, and growth (Roberts *et al.*, 1998).

---

stimulated with motility factors (e.g., FGF, PDGF) preferentially extend new motile processes (e.g., lamellipodia, filopodia) from their corner regions when cultured on square ECM islands (left), whereas cells on circular islands do not display this preference (right). (C) Phase-contrast (left) and traction force microscopic images of substrate displacements (middle) and traction fields (right) beneath cells cultured on square ECM islands that are fabricated on the surface of flexible polyacrylamide gels. Traction forces concentrate in the corner regions where new lamellipodia preferentially form. [Modified from Parker *et al.* (2002) with permission.] (See Plate 27 in the color insert section.)

New ECM proteins secreted by cells cannot adsorb to these RGD-coated substrates because of the dense, underlying layer of (EG)OH groups, thus these substrates may be particularly useful for studies that are designed to discriminate between direct "outside-in" signaling from the ECM substrate versus signaling mediated by *de novo* deposition of new proteins by the adherent cells.

Microcontact printing is convenient and suitable for patterning large areas of a surface (up to ~100 cm^2) with multiple adhesive islands in a single printing. It can routinely transfer patterns with island features having dimensions on the size of 1 μm with an edge roughness of $\leq$100 nm (Xia and Whitesides, 1998). This resolution is more than sufficient for control of cell shape and position; however, it can be enhanced with some additional steps to create patterns with 40-nm features if nanoscale control is desired (Rogers *et al.*, 1998). Multiple methodological improvements also have been made. For example, one modification allows direct printing of proteins onto PDMS-coated substrates, thereby removing the need for special metal-coating equipment (Tan *et al.*, 2004; Chapter 13 by Sniadecki and Chen, this volume). This method overcomes some major limitations of the original microcontact printing technique, namely the unsuitability of gold substrates for microfluorimetry (e.g., calcium imaging) and the high cost of materials. Direct stamping of PDMS also allows visualization of live cells at high magnification, by printing proteins onto PDMS-coated coverslip-bottomed dishes (MatTek).

Finally, multiple electrochemical (Jiang *et al.*, 2005; Yeo *et al.*, 2001, 2003; Yousaf *et al.*, 2001) and optochemical (Ryan *et al.*, 2004) methods also have been incorporated into the microcontact printing technique. These modifications allow dynamic and localized alteration of the chemical composition of micropatterned substrates, and enable reversible or dynamic control over cell behaviors. For example, some of these substrates permit rapid release of an adherent cell from the physical constraints of its island, by converting surrounding nonadhesive barrier areas into adhesive regions, and thus may be particularly useful for studies on directional cell motility. The power of this method lies in the creativity of the investigator.

III. Probing Cell Mechanics, Cytoskeletal Structure, and Mechanotransduction

A number of models predict that mechanical stresses are not transmitted equally across all points on the cell surface. For example, the cellular tensegrity model was one of the first to suggest that forces will be preferentially transferred across transmembrane adhesion receptors, such as integrins, that physically anchor the cytoskeleton to the ECM (Ingber, 1991; Ingber and Jamieson, 1985). Moreover, if the cytoskeleton behaves like other tensegrity structures, then its mechanical stiffness should vary as a function of cytoskeletal prestress. To test these hypotheses, we needed to develop a method to apply controlled mechanical stresses to integrins and other surface proteins on living cells, and simultaneously measure changes in cell and cytoskeletal mechanics. We first accomplished this by developing a magnetic twisting cytometry (MTC) technique in which controlled shear forces are applied to surface receptors via bound ligand-coated ferromagnetic microbeads (Wang *et al.*, 1993). We later developed related magnetic pulling cytometry (MPC) methods that apply tensional forces to superparamagnetic beads coated and bound in a similar manner (Alenghat *et al.*, 2000; Matthews *et al.*, 2004b; Overby *et al.*, 2005).

Magnetic techniques offer a number of advantages over other methods that are commonly used to probe cell mechanics such as optical tweezers (Dai and Sheetz, 1995) and atomic force microscopy (Shroff *et al.*, 1995). While all these methods can apply localized forces to molecules on the cell surface, the advantages of micromagnetic techniques include: (1) a much wider range of stress (from piconewtons to many nanonewtons) can be applied to specific cell surface receptors, (2) a much larger frequency range (0–1000 Hz) of forces can be applied, (3) hundreds to thousands of cells and bound beads can be analyzed simultaneously, (4) cells may be mechanically probed continuously for hours or even days without potential heating problems, (5) forces can be applied inside cells by allowing cells to engulf the beads, and (6) magnetic systems are more robust, easier to use, and cheaper to build.

A. Magnetic Twisting Cytometry (MTC)

MTC uses ferromagnetic microbeads (1- to 10-μm diameter) to apply twisting forces (shear stresses) to specific receptors on the surface membrane of living cells. Carboxylated ferromagnetic beads (4.4-μm diameter; Spherotek) are coated with ligand (e.g., ECM molecules, RGD peptide from the cell-binding region of fibronectin, receptor-specific antibodies) by incubating the beads (5 mg) with 1 mg/ml of ligand in 0.1-M sodium phosphate buffer (pH 5.5) containing 1-mg/ml 1-ethyl-3 (-3-dimethylaminopropyl)carbodiimide (EDAC) for 2 h with gentle agitation to prevent settling. Coated beads are rinsed twice and stored in sterile PBS containing 0.1% BSA for up to 1 week at 4 °C. For experiments, coated ferromagnetic beads are added to the cells ($\sim$20 beads/cell) in serum-free, chemically defined medium and incubated for 10 min at 37 °C followed by gentle washing three times with PBS to remove unbound beads prior to force application. To apply twisting forces to beads and bound surface receptors, a strong (1000 G) but very brief (10 μsec) magnetic pulse is applied to the receptor bound ferromagnetic beads using a horizontal Helmholtz coil (Fig. 4A). This induces and aligns the magnetic dipoles of the beads in the horizontal direction. Within a few seconds, a weaker (0–80 G), but sustained, magnetic field is applied in the perpendicular direction using a second vertically oriented Helmholtz coil. As a result, the beads are twisted, thus applying shear forces directly to the bound receptors. The average bead rotation and angular strain induced by the twisting field is measured using an in-line magnetometer (Valberg and Butler, 1987; Fig. 4B), and rotational shear stress is computed (see next paragraph) based on knowledge of the twisting field and angular strain of the bead (mechanical anisotropy of the cell may affect the degree

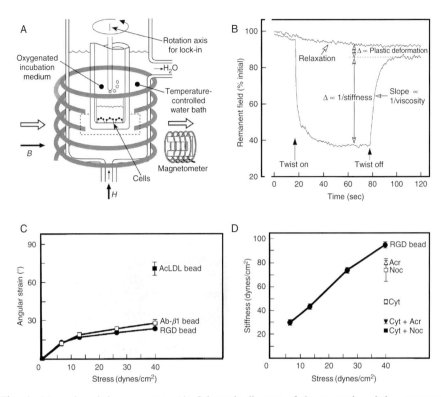

Fig. 4 Magnetic twisting cytometry. (A) Schematic diagram of the magnetic twisting cytometer [modified from Wang *et al.* (1993) with permission]. Small tubes (single wells of detachable 96-well plates) containing cells and bound ferromagnetic beads are placed in medium in a central tube and gassed with 95% O_2/5% CO_2 in the temperature-controlled water-jacketed chamber. A 1000-G horizontal magnetic field (B) is used to magnetize the beads with a pair of horizontal magnetic coils (open arrows). The magnetic field generated by the magnetized beads (horizontal closed arrow) is measured by an in-line magnetometer. Next, a perpendicular 30-G magnetic field (*H*) applied by a vertical coil outside the chamber (vertical closed arrow) is used to twist the beads. Ambient magnetic noise is minimized by appropriate orientation of the four magnetometer probes, an external superalloy shield, and rotating the entire chamber around the vertical axis at 10 Hz. (B) An example of data obtained from MTC analysis of normal cardiocytes. The relaxation curve (relaxation) represents spontaneous remanent field decay in the absence of a twisting field. Cytoskeletal stiffness is inversely related to the extent of the decrease in the remanent field after the twisting field is applied, and cytoskeletal apparent viscosity is inversely related to the slope of remanent field recovery after the twisting field is removed. The residual angular strain representing permanent cytoskeletal deformation after twisting field removal is also indicated (plastic deformation). (C) Stress–strain relation in living capillary endothelial cells measured with bound ferromagnetic beads coated with ligands for transmembrane integrin receptors (anti-β1 integrin antibodies, synthetic RGD peptide) or control transmembrane metabolic acetylated low-density lipoprotein (AcLDL) receptors [modified from Tagawa *et al.* (1997) with permission]. Cells preferentially stiffen and exhibit decreased bead rotation (as evident from the flattening of the rotational stress–angular strain curve), when twisting forces are applied to integrins. Angular strain was calculated as the arc cosine of the ratio of remanent field after 1 min of twist to the field at time 0. Applied stress was determined as described in the text. (D) Cell stiffness (ratio of stress to strain in radians after 1-min twist) measured through integrin bound ferromagnetic beads in the absence or presence of cytochalasin

of angular rotation of the bead). In this manner, the mechanical behavior of the integrins linked to the cytoskeleton can be characterized in living cells by measuring stress–strain responses (Fig. 4C and D), as well as analyzing creep and elastic recoil behavior (Wang and Ingber, 1995; Fig. 4B). Here we describe how several biomechanical parameters (stiffness, permanent deformation, and apparent viscosity) are measured.

The applied torque T_{mag} is given by $T_{mag} = \mu_0 \mathbf{M} \times \mathbf{H_a}$, where μ_0 is permeability of free space, vector $\mathbf{M}$ is the microbead magnetic moment, and $\mathbf{H_a}$ is the external twisting field. Since the specific gravity of magnetic beads is much lower than 10^4 (Wang and Ingber, 1994), we can neglect inertial effects; thus the net torque on the microbead is assumed to be zero. If we assume that elastic and viscous contributions are mechanically in parallel, then their torques are additive. Therefore, $T_{mag} = T_{elas} + T_{vis}$ where T_{elas} and T_{vis} are elastic and viscous (or frictional) torques, respectively (assuming that cells generate negligible torques to resist the applied magnetic torque). This gives $\mu_0 \mathbf{M} \mathbf{H_a} \sin\theta = -\kappa V \eta \omega + \kappa V E (90° - \theta)$ where θ is the angle between $\mathbf{M}$ and $\mathbf{H_a}$, κ is the particle shape factor and equals 6 for a spherical microbead, V is the microbead volume, η is the viscosity, and ω is the angular velocity of the microbead rotation and equals $d\theta/dt$.

The value θ above is also the angular strain of the microbead induced by the twisting field. To correct for the effects of relaxation (i.e., force-induced cell remodeling), we assume that twisting-induced rotation and relaxation are two independent processes so that $\theta = 90° - \cos^{-1}[B(t)_{twist}/B(t)_{relax}]$. $B(t)_{twist}$ is the magnitude of the remanent field (i.e., the magnetic induction that remains in the material after removal of the applied twisting field that is measured by the in-line magnetometer in the magnetizing direction). $B(t)_{relax}$ is the remanent field resulting from the relaxation by the cell and is measured by the in-line magnetometer at the steady state after the twisting field is turned off. The applied stress $\sigma = cH_a[B(t)_{twist}/B(t)_{relax}]$, where H_a is the applied external field and c is the bead calibration constant that is determined by placing the bead in a viscous standard (Wang and Ingber, 1994). Thus, the applied shear stress σ and angular strain θ can be calculated, from which the cell stiffness E can be computed as $E = \sigma/\theta$, the ratio of stress to angular strain. Permanent deformation is defined as the nonrecovered bead rotation after the twisting field is turned off (i.e., when applied stress is removed; Fig. 4B) relative to the bead rotation when the stress is applied, and is given by $(\theta_2/\theta_1) \times 100\%$ where 2 and 1 represent states when the applied stress is on and off, respectively).

D (CytoD) which disrupts the actin cytoskeleton in living cells. Cellular stiffness increases linearly with stress. The stiffness dramatically decreases when cells are treated with CytoD, nocodazole (Noc), or acrylamide (Acr) to disrupt microfilament, microtubule, or intermediate filament integrity, respectively [modified from Wang *et al.* (1993) with permission]. Each agent alone only produced partial inhibition, whereas combination of multiple agents completely inhibited the cell-stiffening response. These results show that the mechanical resistance measured through integrins is due to an integrated cytoskeletal strengthening response that involves all three filament systems.

Cellular viscosity is obtained by measuring the time course of the disappearance of angular strain θ when the applied field is turned off. The equation above gives $\kappa V \eta \omega = \kappa V E (90° - \theta)$, solving this differential equation yields $\phi = \phi_0 e^{-t/(\eta/c)} = \phi_0 e^{-t/(\tau)}$ where $\phi = 90° - \theta$, $\phi = \phi_0$ at $t = 0$ and $\tau = \eta/E$; the later relation allows the estimation of cell viscosity through knowledge of τ and E. τ is defined as the time when $\phi = \phi_0(e^{-1}) = 0.34\phi_0$ and is measured experimentally. Thus, MTC can be used to obtain cell stiffness, cell apparent viscosity (frictional contribution), and permanent deformation (remodeling associated changes) of the cell and cytoskeleton when controlled shear stresses are applied to different cell surface receptors.

B. Applications of MTC

The MTC technique was used to demonstrate directly that forces applied to integrins and nonadhesion receptors on the same cell produce different responses. For example, beads bound to integrins via RGD ligands or anti-β1 integrin antibodies exhibited significantly less angular rotation in response to applied magnetic field, compared to beads bound to transmembrane receptors that do not physically couple to the internal cytoskeleton such as metabolic receptors, growth factor receptors, or histocompatibility antigens (Wang *et al.*, 1993; Yoshida *et al.*, 1996; Fig. 4C). The mechanical stiffness of many different types of cells also was found to increase in direct proportion to the level of applied shear stress (linear strain-hardening behavior) (Fig. 4D). Use of specific pharmacological modifiers or genetic knockout techniques further revealed that all three cytoskeletal filament systems—actin microfilaments, microtubules, and intermediate filaments—contribute to this mechanical response (Fig. 4D). Moreover, cell stiffness varied with cytoskeletal prestress: increasing basal levels of tension in the cytoskeleton using stimulators of actomyosin interactions made the cells more rigid, whereas dissipating tension immediately increased cell flexibility (Wang *et al.*, 2002b). A mathematical formulation of the tensegrity model predicts this behavior (Coughlin and Stamenovic, 1998, 2003; Laurent *et al.*, 2002; Stamenovic *et al.*, 1996).

The MTC method also has been used to determine how different transmembrane adhesion receptors, including various types of integrins, cadherins, selectins, and urokinase receptors (Potard *et al.*, 1997; Wang and Ingber, 1995; Yoshida *et al.*, 1996), differ in their ability to support transmembrane mechanical coupling to the cytoskeleton. Additionally, when the same stress was applied to different-sized magnetic beads (e.g., 4.5-μm vs 1.4-μm diameter), larger beads appeared to be stiffer or more resistant to rotation than smaller beads (Wang and Ingber, 1994). This result is consistent with the idea that, when applied to beads ligated to integrins, MTC probes the underlying 3D structure of the cytoskeleton rather than the 2D structure of the cell's surface membrane (Stamenovic and Coughlin, 2000).

C. Extension and Future Development of MTC

MTC has been combined with signal transduction experiments to analyze the molecular basis of cellular mechanotransduction. For example, MTC was used to show that shear stress applied to integrin receptors activates the cAMP signaling pathway and leads to mechanical activation of gene transcription driven by cAMP response elements (Meyer *et al.*, 2000). Importantly, application of the same stress to beads bound to transmembrane metabolic receptors or to nonactivated integrin receptors (i.e., that do not have the RGD-binding site occupied) did not activate cAMP signaling in these cells, demonstrating that mechanical deformation of the plasma membrane alone is not sufficient to activate this mechanotransduction pathway.

MTC also has been combined with high resolution *in situ* hybridization to reveal that mRNA and ribosomes are simultaneously recruited to focal adhesions that form on the cytoplasmic face of surface-bound magnetic microbeads coated with integrin ligands, but not with ligands for other receptors (Chicurel *et al.*, 1998). When shear stress was applied to integrins through these beads using MTC, mRNA and ribosome recruitment was significantly augmented, and this response was suppressed by inhibiting cytoskeletal tension generation. These findings suggest that rapid posttranscriptional changes in gene expression may be mediated by repositioning of translational components to sites of signal reception and that the level of prestress in the cytoskeleton governs this response. In addition, force application via MTC was shown to regulate endothelin-1 gene expression in a prestress-dependent manner (Chen *et al.*, 2001).

Finally, MTC can be used both with large populations of cells to obtain averaged rheological properties and at the single-cell level with simultaneous microscopic visualization. The latter is achieved by fabricating a device containing orthogonal Helmholtz coils that can be placed on a microscope stage. When placed on a microscope, twisting forces are applied to many beads on a cell while using optical techniques to analyze local structural and biochemical responses of beads located in particular positions on the cell surface, or inside the cytoplasm (when engulfed, the degree of engulfment can be determined by cytoskeletal immunostaining after fixation or using GFP-tagged cytoskeletal proteins in live cells). Modified versions of the MTC method also have been developed to characterize dynamic cell mechanical behavior (oscillatory MTC; Fabry *et al.*, 2001), to apply forces in multiple directions (3D MTC; Hu *et al.*, 2004), and to explore force-induced displacements of molecular elements inside the cytoplasm and the nucleus (3D magnetic tomography; Hu *et al.*, 2003, 2004). Because these methods combine MTC with high-resolution microscopic imaging methods, nanoscale displacements of the bead can be measured in conjunction with measurements of changes in cytoskeletal elements in the cytoplasm and nucleus to explore how mechanical stresses are distributed throughout the cell (Chapter 8 by Wang *et al.*, this volume for more details).

D. Magnetic Pulling Cytometry (MPC)

MPC, also known as magnetic tweezers, is a related magnetic micromanipulation technology that can also apply forces to specific surface receptors through bound ligand-coated microbeads and measure local cell rheology. However, the technique differs from MTC in that superparamagnetic beads (4.5 μm; Dynal) are utilized instead of ferromagnetic beads, and a magnetic needle is used to apply tensional forces locally to individual surface receptor-bound beads on single cells (i.e., rather than global shear stresses to large populations of cells). Unlike ferromagnetic beads that maintain their magnetic moment, the induced magnetic moment disappears from superparamagnetic beads on the removal of external magnetic field. One advantage of this method over MTC is that large-scale distortion of the cell can be produced, whereas MTC only twists the beads in place. This method is especially well suited for studying force-induced changes of the mechanical properties and biochemical signaling functions at focal adhesions or other receptor-mediated anchoring complexes in single living cells.

Our group has fabricated magnetic needles that utilize either a stationary permanent magnet (Fig. 5) or an electromagnet to induce bead magnetization (Alenghat *et al.*, 2000; Matthews *et al.*, 2004a,b, 2006; Overby *et al.*, 2004, 2005; Figs. 6 and 7). The advantage of the permanent magnetic needle is that virtually any investigator can assemble one of these devices in a short time at minimal cost. Fabricating the electromagnetic needle is more involved and expensive, but it offers a wider range of dynamic control (Fig. 7) as well as higher levels of force application (up to 50 nN). Thus, a permanent magnetic needle is ideal for quick, preliminary experiments to test hypotheses, whereas the electromagnetic needle is preferred for studies involving more rigorous characterization.

The size and shape of the tips of the magnetic needles used in both devices are designed to maximize magnetic field gradient intensity while minimizing their size, so that they can be positioned as close to the cell membrane as possible (Fig. 6D). This is important because the magnetic field increases exponentially with decreasing distance to the needle tip (Fig. 5B and C). All magnetic needles can be calibrated by pulling the magnetic beads through a glycerol solution with a known viscosity (1 kg/m/sec or 1000 cP) (Fig. 5B). After recording the beads' velocities through the fluid, Stokes' formula for low Reynolds number flow, force $= 3\pi\eta Dv$, is used to deduce the forces on the beads, where η is the viscosity of the fluid, D is the bead diameter, and v is the velocity of the bead through the fluid. Due to the strong dependence of magnetic forces on distance, it is important that the magnetic needle be placed at the identical position during calibration as used during experimental manipulation.

The permanent magnetic microneedle system (Matthews *et al.*, 2004b) consists of a standard stainless steel needle attached to a permanent neodymium iron boron disk magnet (Edmund Industrial Optics, New Jersey) attached to an aluminum rod that is mounted on a microscope micromanipulator (Eppendorf, Germany) (Fig. 5A). To measure the local mechanical properties of bound receptors and

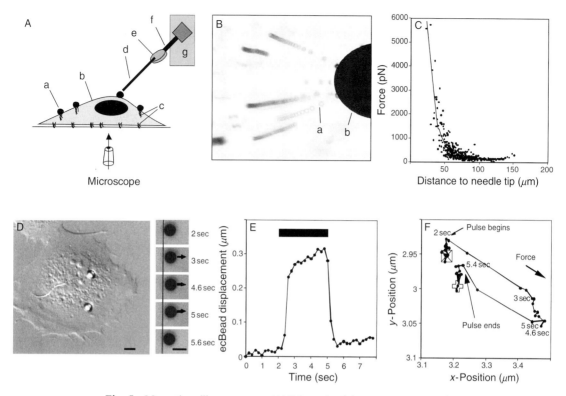

Fig. 5 Magnetic pulling cytometry. (A) Schematic of the permanent magnetic needle device. Ligand-coated superparamagnetic microbeads (a) bind and cluster receptors on the surface of cultured cells (b). When integrins are clustered, focal adhesions (c) form at the site of bead binding. A magnet microneedle consisting of a stainless steel needle (d) attached to a permanent magnet (e) fastened to an aluminum rod (f) that is mounted on a micromanipulator (g) is used to apply force to the receptor-bound beads, while viewing the cell and measuring bead displacements using an optical inverted microscope. (B) Forces applied to the beads are estimated by carrying out control studies as visualized in this composite of time-lapse images of bright-field views showing magnetic microbeads being pulled through glycerol in response to magnetic stress application by the microneedle. The beads (a) are attracted to the magnet along magnetic field lines perpendicular to the needle tip (b); forces exerted on individual beads are calculated using the Stokes equation as described in the text. (C) Force exerted on individual 4.5-μm diameter superparamagnetic Dynal beads as a function of the distance from the tip of the magnetic needle (calculated from data shown in B). (D) A Nomarski microscopic view of an adherent endothelial cell with two 4.5-μm RGD beads bound to integrins on its apical surface. A series of higher magnification bright-field images recorded over ~6 sec showing bead displacement to the right in response to application of a similarly oriented force (130 pN) pulse between 2 and 5 sec of the recording period (arrows) is shown at the right. (Scale bar is 5 μm.) (E) Bead displacement as a function of time before, during, and after the 3 sec force pulse (solid rectangle). (F) Map of changes in the positions of the bead measured in E during the same time course. Note the bead does not return to its original position after the force pulse ceases. [Modified from Matthews *et al.* (2004b) with permission.]

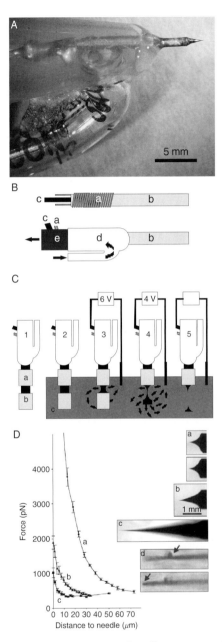

Fig. 6   The electromagnetic needle device for magnetic pulling cytometry. (A) Photograph of the pole tip of the electromagnetic needle device and integrated cooling system. (B) Diagram of the internal (top) and external (bottom) views of the magnetic needle system. A 50-μm diameter insulated copper electromagnet wire (a) is wound around a 1-mm diameter soft permalloy magnet core (b) and a 1.5-mm insulated copper wire (c) is soldered to the proximal end of the magnet core for electropolishing.

cytoskeletal linkages at bead-binding sites, cells with bound beads are maintained at 37 °C using a heated stage (Omega Engineering, Inc., Connecticut) and visualized on an inverted microscope (Nikon Diaphot 300, Japan). The magnetic needle is then used to apply force pulses to the bead, by moving the tip momentarily to the vicinity of the cell, while bead displacement is measured optically (Fig. 5D). To accomplish this, the manipulator speed is set to 1000 μm/sec, and the magnet tip is oriented at 45° relative to the substrate and positioned initially in the culture medium 600-μm away from the cell-bound beads to be tested. The needle tip is moved rapidly within 70–125 μm from the bead, held in position for any time desired (generally between 1 sec and 5 min), and then quickly returned to its original position using the micromanipulator. Time-lapse imaging (4 Hz) with a CCD camera (Hamamatsu, Japan) attached to the microscope is used to record bead motion. The centroid position of each bead is then determined frame by frame using IPLab (version 3.2.4, Scanalytics, Inc., Virginia), and the maximum bead displacement computed (Fig. 5E and F). To minimize cumulative effects from multiple force pulses, subsequent force pulses are applied to cells in the same dish that are >2-mm away from previously stimulated cells.

The electromagnetic needle device (Matthews *et al.*, 2004a) is powered by a simple voltage source and both the magnetic field and micromanipulator are controlled by a computer (Fig. 7). The electromagnetic needle is fabricated by winding multiple (>1000) loops of insulated electromagnet copper wire (25- to 50-μm diameter; Matthews *et al.*, 2004a) around a magnetic permalloy core in one or more layers (a 121-mm diameter composed of 81% nickel/19% iron can be obtained from Fine Metals Corporation, Ashland, Virginia; the permalloy core is annealed separately by Amuneal Manufacturing Corporation, Philadelphia, Pennsylvania) (Fig. 6B). The proximal portion of the permalloy core and the entire

---

The core with wound wire is placed within a thermoregulating water jacket (d) fashioned from a 1.5-ml Eppendorf tube with the exposed tip of the core extending through its distal surface. Arrows indicate direction of water flow through jacket and into a plastic outflow tube (e). (C) Electropolishing method for modification of pole tip geometry. (1) Two protective plastic cylindrical masks (a, b) are placed over the surface of the permalloy core tip so that the tip is completely covered and a controllable region of the core between the masks is exposed. (2) The tip with masks is then lowered into an acid solution (c). (3) An electric current (solid arrows) applied with a power supply set at 6-V DC is passed through the permalloy core thereby electrochemically polishing the exposed surface of the permalloy core. Once the core narrows by 50%, the distal plastic cap is removed and electropolishing is continued at 4 V (4) until the distal end of the permalloy core breaks off; the current is then shut down (5). The initial surface area of exposed core in step 1 determines final tip geometry. (D) Control of the magnetic field gradient by altering pole tip geometry. Needle pole tips with different tapered shapes of increasing lengths were created by exposing different areas of the core using different initial separations between the two plastic masks of 1.5 mm (a), 3 mm (b), 6 mm (c), and 15 mm (d) during the electropolishing procedure. In (d), 100× magnification images focused at the arrow indicates a 250-nm magnetic bead on shaft of needle tip (top), and 100-nm radius needle tip (below). Repeated fabrication protocol produced similar tip geometries (a). The lines indicate the force–distance relationship for respective pole tip geometries measured using 4.5-μm magnetic beads in glycerol as described in Fig. 5. [Modified from Matthews *et al.* (2004a) with permission.]

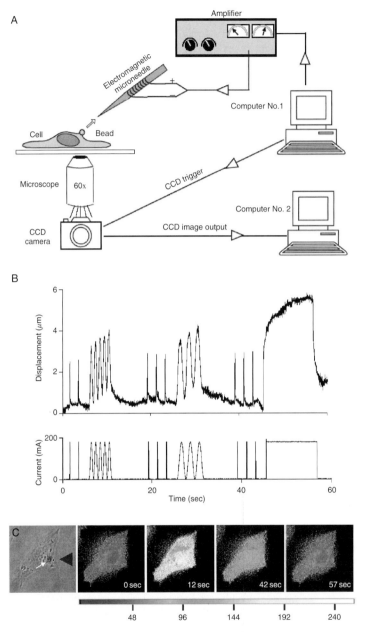

Fig. 7 Electromagnetic pulling cytometry for analysis of cell mechanics and mechanotransduction. (A) Experimental system. A ligand-coated superparamagnetic bead is bound to receptors on the surface of a living cell, while a magnetic force (arrow) is applied to the bead using the electromagnetic microneedle. An amplifier supplies current to the electromagnetic microneedle and is controlled using LabVIEW on computer No. 1 to generate any arbitrary user-defined force regimen. The bead

electromagnet coil are housed within a temperature-regulated water chamber (Fig. 6A).

The tip of the permalloy core of the electromagnet can be electrochemically sharpened (Fig. 6C) to between 200-nm and 20-μm diameter depending on the desired application, as magnetic field gradient intensity increases with tip curvature (Fig. 6D). To generate conical-shaped shafts with fine tips, two cylindrical plastic shields (1-mm internal diameter) cut from the ends of 200-μl Eppendorf pipette tips are fit over the ends of the core, leaving an exposed section of core between them (Fig. 6C). The pole tip is then immersed into a solution containing 8:7:5 phosphoric acid, sulfuric acid, and water, and a 6-V potential is applied through the permalloy and the acid solution to electropolish (etch) the exposed surface of the rod. When the diameter of the exposed material reaches $\sim$40–50% of its original size, the potential is stopped, the distal plastic shield removed, and the electropolishing is resumed at 0–4 V (for better control) until the distal portion of the permalloy tip falls off. Progressively increasing the spacing between the shields results in a gradual increase in the taper. Continued electropolishing after the distal portion of the permalloy tip has fallen off results in a progressively duller tip and shorter neck.

One major problem associated with the use of small-scale electromagnets is the resistive heating of the electromagnet which can locally denature biomolecules and injure living cells, while also causing thermal expansion and movements of the electromagnet core. This movement, which can be 15–20 μm, eliminates precise control over the distance between the magnetic particle and the tip of the electromagnet, thereby hindering accurate control of the forces applied. We therefore incorporated a temperature-regulating water flow chamber (Fig. 6A) to ensure that the tip temperature remains within the design range of less than 2 °C variation and eliminates the possibility of the coil melting during application of currents beyond a brief pulse ($>$500 msec).

We use tosyl-activated superparamagnetic beads (4.5-μm diameter; Dynabeads M-450, Dynal) that are coated with receptor ligands by incubating the beads and proteins in pH 9.4 carbonate buffer overnight in the cold. The beads are then washed and stored in medium containing 1% BSA for up to 3 weeks at 4 °C

---

displacement is optically recorded using a microscope and an externally triggered CCD camera, and the images are stored on a second computer (No. 2) using IPLab. Connections and arrowheads indicate the direction of information flow. (B) Examples of bead displacements (top) measured during a dynamic force regimen driven by electric current waveforms (bottom) consisting of multiple ($n = 8$) subsecond (100 msec) force pulses interspersed with two periods of sinusoidal oscillations (1.0 Hz, 0.5 Hz, from left to right) and ending with a single period of prolonged force (10 sec). Two hundred milliampere corresponds to $\sim$300-pN force. (C) Analysis of cellular mechanotransduction. Application of a high stress ($>$1 nN) to cell surface integrins via a bound RGD-coated magnetic bead (white arrow) using MPC (left, phase contrast image; black arrowhead indicates position of electromagnet tip) increases intracellular calcium within seconds, as detected using microfluorimetric ratio imaging with the calcium-sensitive dye, FURA-2, and shown in the time series of pseudocolored fluorescence images at the right (color bar indicates intracellular calcium concentrations in nM). [Modified from Overby et al. (2005) and Matthews et al. (2006) with permission.] (See Plate 28 in the color insert section.)

(Plopper and Ingber, 1993). Immediately before an experiment, cells are incubated with beads (~20 beads/cell) for 10 min and then washed three times with PBS to remove unbound beads. An electronic (Eppendorf) micromanipulator is then used to position the tip of the microneedle near surface receptor-bound magnetic beads, while viewing through an inverted microscope with a 20–100× objective lens (Eclipse TE2000E, Nikon) (Fig. 7A). Images of the beads are recorded using an externally triggered CCD camera (CoolSnap HQ, Roper Scientific) and imaging software (IPLab, Scanalytics, Fairfax, Virginia). A computer-controlled amplifier (Model SRL40-6, Sorensen, San Diego, California) is used to provide electric current to the electromagnetic microneedle. A desired voltage waveform is generated with the LabVIEW software (ver. 5.0, National Instruments, Austin, Texas) and sent to the control terminals of the amplifier. IPLab or MATLAB software (version 6, The MathWorks, Natick, Massachusetts) can be used for analyzing the image sequence, quantifying bead displacements, and temporally aligning bead displacements against force waveforms.

In most experiments, the needle tip is positioned to a predetermined height (~5–100 μm) and horizontal distance (~5–60 μm) from the magnetic bead prior to force application. These distances and the electrical current determine the magnitude of forces applied to the magnetic beads. The LabVIEW program simultaneously initiates the waveform to the electromagnet, actuates image acquisition by the CCD camera, and provides digital data acquisition. Centroids of beads (Fig. 7B) are analyzed to determine the bead displacement or viscoelastic creep response as a function of time.

Compared to an optical laser tweezer which may be used to only briefly (10–15 sec) trap single integrin-bound beads on the surface of cells (Choquet *et al.*, 1997), the magnetic needle can apply forces to beads for much longer periods of time. This is because the laser trap generates a very steep force gradient at its edges. When a cell adaptively reinforces its adhesions to the bead, it moves to the perimeter of the trap. If the trap is not displaced to accommodate this, the resistive force drops rapidly, allowing the bead to escape and the cell to readjust itself to a zero-force state (Choquet *et al.*, 1997). In contrast, forces can be applied to multiple beads for longer periods of time (minutes to days) using a magnetic needle, which generates a broad magnetic field gradient near the needle tip. Mechanical forces also can be maintained on the bead during and after adaptive cellular strengthening (Matthews *et al.*, 2006). In addition, the magnitude of force that can be applied using the optical trap is limited due to the risk of thermal injury to the cells by the laser at higher energy levels—even for laser wavelengths that might seem far from the absorption band of water or biomolecules. The magnetic needle, on the other hand, causes no damage to the cell even at proximity, allowing strong forces (up to 10 nN on 4.5-μm beads) to be applied. Finally, unlike the optical trap, MPC can easily apply brief pulses or cyclical force regimens (Fig. 7B) which can be used to estimate the local static and dynamic mechanical properties of the cell, and in particular, of bead-associated focal adhesions or other surface receptor–cytoskeleton linkages.

E. Applications of MPC

Using the permanent magnetic needle, we have shown that cells exhibit different types of mechanical adaptation responses when tensional forces are applied through beads bound to integrin receptors. The cells display an immediate viscoelastic response due to the local passive material properties of associated integrin–cytoskeleton linkages and recruited focal adhesion proteins (Matthews *et al.*, 2004b, 2006) that results in almost complete elastic recovery. They also exhibit an early adaptive behavior characterized by pulse-to-pulse attenuation in bead displacement in response to an oscillatory force regimen composed of short (3 sec) force pulses. A later form of adaptive cell stiffening is observed in response to sustained (>15 sec) static stresses, and a fourth type of adaptation is a large-scale repositioning response in which beads exposed to prolonged (>1 min) static stresses are retracted back by the cell against the applied force (Fig. 5F).

MPC also can be used to analyze the molecular biochemical basis of cell mechanical responses. For example, we found that the immediate viscoelastic response and early adaptation behavior are affected by dissipating cytoskeletal prestress whereas the later adaptive response to longer stresses is not (Matthews *et al.*, 2006). The large-scale repositioning of beads in response to prolonged stress is prevented by inhibition of myosin-based tension generation, and by blocking mechanosensitive ion channels. In addition, the large-scale repositioning response requires that integrins be chemically activated through occupancy of the RGD-binding site. Thus, MPC enabled us to demonstrate that cells use multiple distinct mechanisms to sense and respond to static and dynamic changes in the level of mechanical stress applied to cell surface integrin receptors.

MPC also was used in combination with calcium ratio imaging in cells loaded with the calcium-sensitive dye, FURA-2, to show that application of >1-nN forces to integrin-bound beads elicits a wave of calcium release within 2–5 sec after force application which can be blocked by treating cells with the stress-sensitive ion channel blocker gadolinium chloride (Matthews *et al.*, 2006). In more recent studies, we have been able to detect increases in calcium within <100 msec after force application. These findings obtained with MPC indicate that force application to integrins can activate stress-sensitive ion channels directly at the site of force application within the same focal adhesion. This method may be particularly useful for dissecting out the earliest steps involved in mechanotransduction across integrins.

IV. Discussion and Future Implications

The interplay between mechanics and chemistry that occurs inside the living cytoplasm has profound effects on cell behavior, and understanding the molecular biophysical basis of mechanotransduction represents a critically important challenge in biology. New insight into mechanotransduction has resulted from interdisciplinary approaches that creatively and seamlessly combine tools and techniques across disciplines. As described in this chapter, we have developed various hybrid techniques that borrow approaches from magnetics and materials engineering, and integrated them with more conventional cell and molecular biological tools to meet this challenge.

Our methods for microfabricating culture substrates have led to new insights into the physical and molecular basis of cell fate switching, as well as potentially offering new approaches for engineering artificial tissues and creating novel cell–biomaterial interfaces. The magnetic cytometry methods have provided new insight into how cells behave mechanically, as well as which molecules are used by cells to sense mechanical signals and to convert them into a biochemical response.

A major challenge for the future of this field is to expand our knowledge of cellular mechanotransduction within the context of living tissues and organisms as opposed to studying isolated cells in Petri dishes (Ingber, 2006). For example, establishment of regional variations of cell and ECM mechanics over micrometer distances may be critical for control of normal morphogenesis, and when dysregulated, may lead to tissue disorganization and tumor formation (Huang and Ingber, 1999, 2005; Ingber, 2005; Ingber and Jamieson, 1985; Moore *et al.*, 2005; Nelson *et al.*, 2005; Paszek *et al.*, 2005). Thus, entirely new methods are needed to measure the mechanics of individual cells and molecules *in situ* within tissue, organs and whole living organisms. Our magnetic methods are particularly useful in this regard because they are not limited by optical transmission. MTC, for example, has already been adapted to measure mechanical properties of macrophages in the lungs of living human patients (Stahlhofen and Moller, 1992). However, we and others will still need to develop entirely new methods (or combine old) to develop ways to image, manipulate and probe ECM, cell and subcellular structures in real-time *in vivo* to fully understand how structure and function are fully integrated in living tissues. A first step in this process might involve refinements of some of the methods we described here so that they can be used in studies with organ explants, embryonic rudiments, or 3D cell cultures. This may be facilitated by the development of new microfabrication approaches that allow fine spatial control over ECM ligands, cytokines and living cells in 3D, while also providing mechanical, optical and electrochemical inputs and readouts of cell behavior at multiple size scales.

Development of new methods often opens entirely novel avenues of investigation, and this certainly has been true in the field of cellular mechanotransduction. However, the methods described here may also have uses beyond study of mechanobiology. The magnetic cytometry techniques could, for example, provide a way to create real-time cellular sensors that act as optical readouts of mechanical force in the future. In fact, we have already created living cellular switches that can be actuated magnetically using MPC and read out optically by creating gene reporter constructs driven by cAMP signals (Overby *et al.*, 2004) that are elicited by force application (Meyer *et al.*, 2000). The magnetic microbeads we used for MTC and MPC also can be coated with enzymes and placed in different magnetic field configurations to create artificial ECMs (e.g., fibrin gels) with defined structure on the nanometer scale that might be useful for tissue-engineering applications (Alsberg *et al.*, 2006). Thus, the development of methods for analysis of cell mechanics and mechanotransduction may impact science and medicine in ways one might have never imagined.

References

Alenghat, F. J., Fabry, B., Tsai, K. Y., Goldmann, W. H., and Ingber, D. E. (2000). Analysis of cell mechanics in single vinculin-deficient cells using a magnetic tweezer. *Biochem. Biophys. Res. Commun.* **277,** 93–99.

Alsberg, E., Feinstein, E., Joy, M. P., Prentiss, M., and Ingber, D. E. (2006). Magnetically-guided self-assembly of fibrin matrices with ordered nanoscale structure for tissue engineering. *Tissue Eng.* **12,** 3247–3256.

Brangwynne, C. P., MacKintosh, F. C., Kumar, S., Geisse, N. A., Talbot, J., Mahadevan, L., Parker, K. K., Ingber, D. E., and Weitz, D. A. (2006). Microtubules can bear enhanced compressive loads in living cells because of lateral reinforcement. *J. Cell Biol.* **173,** 733–741.

Brock, A., Chang, E., Ho, C. C., LeDuc, P., Jiang, X., Whitesides, G. M., and Ingber, D. E. (2003). Geometric determinants of directional cell motility revealed using microcontact printing. *Langmuir* **19,** 1611–1617.

Chen, C. S., Alonso, J. L., Ostuni, E., Whitesides, G. M., and Ingber, D. E. (2003). Cell shape provides global control of focal adhesion assembly. *Biochem. Biophys. Res. Commun.* **307,** 355–361.

Chen, C. S., Mrksich, M., Huang, S., Whitesides, G. M., and Ingber, D. E. (1997). Geometric control of cell life and death. *Science* **276,** 1425–1428.

Chen, C. S., Mrksich, M., Huang, S., Whitesides, G. M., and Ingber, D. E. (1998). Micropatterned surfaces for control of cell shape, position, and function. *Biotechnol. Prog.* **14,** 356–363.

Chen, J., Fabry, B., Schiffrin, E. L., and Wang, N. (2001). Twisting integrin receptors increases endothelin-1 gene expression in endothelial cells. *Am. J. Physiol., Cell Physiol.* **280,** C1475–C1484.

Chicurel, M. E., Singer, R. H., Meyer, C. J., and Ingber, D. E. (1998). Integrin binding and mechanical tension induce movement of mRNA and ribosomes to focal adhesions. *Nature* **392,** 730–733.

Choquet, D., Felsenfeld, D. P., and Sheetz, M. P. (1997). Extracellular matrix rigidity causes strengthening of integrin-cytoskeleton linkages. *Cell* **88,** 39–48.

Coughlin, M. F., and Stamenovic, D. (1998). A tensegrity model of the cytoskeleton in spread and round cells. *J. Biomech. Eng.* **120,** 770–777.

Coughlin, M. F., and Stamenovic, D. (2003). A prestressed cable network model of the adherent cell cytoskeleton. *Biophys. J.* **84,** 1328–1336.

Dai, J., and Sheetz, M. P. (1995). Mechanical properties of neuronal growth cone membranes studied by tether formation with laser optical tweezers. *Biophys. J.* **68,** 988–996.

Dike, L. E., Chen, C. S., Mrksich, M., Tien, J., Whitesides, G. M., and Ingber, D. E. (1999). Geometric control of switching between growth, apoptosis, and differentiation during angiogenesis using micro-patterned substrates. *In Vitro Cell. Dev. Biol. Anim.* **35,** 441–448.

Dong, C. S. R., and Sung, K. L. (1991). Cytoplasmic rheology of passive neutrophils. *Biorheology* **28,** 557–567.

Fabry, B., Maksym, G. N., Shore, S. A., Moore, P. E., Panettieri, R. A., Jr, Butler, J. P., and Fredberg, J. J. (2001). Selected contribution: Time course and heterogeneity of contractile responses in cultured human airway smooth muscle cells. *J. Appl. Physiol.* **91,** 986–994.

Hu, S., Eberhard, L., Chen, J., Love, J. C., Butler, J. P., Fredberg, J. J., Whitesides, G. M., and Wang, N. (2004). Mechanical anisotropy of adherent cells probed by a three-dimensional magnetic twisting device. *Am. J. Physiol., Cell Physiol.* **287,** C1184–C1191.

Hu, S. H., Chen, J., Fabry, B., Numaguchi, Y., Gouldstone, A., Ingber, D. E., Fredberg, J. J., Butler, J. P., and Wang, N. (2003). Intracellular stress tomography reveals stress focusing and structural anisotropy in cytoskeleton of living cells. *Am. J. Physiol., Cell Physiol.* **285,** C1082–C1090.

Huang, S., and Ingber, D. E. (1999). The structural and mechanical complexity of cell-growth control. *Nat. Cell Biol.* **1**, E131–E138.

Huang, S., and Ingber, D. E. (2005). Cell tension, matrix mechanics, and cancer development. *Cancer Cell* **8**, 175–176.

Ingber, D. E. (1990). Fibronectin controls capillary endothelial cell growth by modulating cell shape. *Proc. Natl. Acad. Sci. USA* **87**, 3579–3583.

Ingber, D. E. (1991). Integrins as mechanochemical transducers. *Curr. Opin. Cell Biol.* **3**, 841–848.

Ingber, D. E. (1993a). Cellular tensegrity: Defining new rules of biological design that govern the cytoskeleton. *J. Cell Sci.* **104**(Pt. 3), 613–627.

Ingber, D. E. (1993b). The riddle of morphogenesis: A question of solution chemistry or molecular cell engineering? *Cell* **75**, 1249–1252.

Ingber, D. E. (1997). Integrins, tensegrity, and mechanotransduction. *Gravit. Space Biol. Bull.* **10**, 49–55.

Ingber, D. E. (2003a). Mechanobiology and diseases of mechanotransduction. *Ann. Med.* **35**, 564–577.

Ingber, D. E. (2003b). Tensegrity II. How structural networks influence cellular information processing networks. *J. Cell Sci.* **116**, 1397–1408.

Ingber, D. E. (2005). Mechanical control of tissue growth: Function follows form. *Proc. Natl. Acad. Sci. USA* **102**, 11571–11572.

Ingber, D. E. (2006). Cellular mechanotransduction: Putting all the pieces together again. *FASEB J.* **20**, 811–827.

Ingber, D. E., and Folkman, J. (1989a). How does extracellular matrix control capillary morphogenesis? *Cell* **58**, 803–805.

Ingber, D. E., and Folkman, J. (1989b). Mechanochemical switching between growth and differentiation during fibroblast growth factor-stimulated angiogenesis *in vitro*: Role of extracellular matrix. *J. Cell Biol.* **109**, 317–330.

Ingber, D. E., and Jamieson, J. D. (1985). Cells as tensegrity structures: Architectural regulation of histodifferentiation by physical forces transduced over basement membrane. *In* "Gene Expression During Normal and Malignant Differentiation" (L. C. Andersson, C. G. Gahmberg, and P. Ekblom, eds.), pp. 13–32. Academic Press, Orlando.

Jiang, X., Bruzewicz, D. A., Wong, A. P., Piel, M., and Whitesides, G. M. (2005). Directing cell migration with asymmetric micropatterns. *Proc. Natl. Acad. Sci. USA* **102**, 975–978.

Kato, M., and Mrksich, M. (2004). Using model substrates to study the dependence of focal adhesion formation on the affinity of integrin-ligand complexes. *Biochemistry* **43**, 2699–2707.

Kumar, S., Maxwell, I. Z., Heisterkamp, A., Polte, T. R., Lele, T. P., Salanga, M., Mazur, E., and Ingber, D. E. (2006). Viscoelastic retraction of single living stress fibers and its impact on cell shape, cytoskeletal organization, and extracellular matrix mechanics. *Biophys. J.* **90**, 3762–3773.

Laurent, V. M., Canadas, P., Fodil, R., Planus, E., Asnacios, A., Wendling, S., and Isabey, D. (2002). Tensegrity behaviour of cortical and cytosolic cytoskeletal components in twisted living adherent cells. *Acta Biotheor.* **50**, 331–356.

Lele, T. P., Pendse, J., Kumar, S., Salanga, M., Karavitis, J., and Ingber, D. E. (2006). Mechanical forces alter zyxin unbinding kinetics within focal adhesions of living cells. *J. Cell. Physiol.* **207**, 187–194.

Maniotis, A. J., Chen, C. S., and Ingber, D. E. (1997). Demonstration of mechanical connections between integrins, cytoskeletal filaments, and nucleoplasm that stabilize nuclear structure. *Proc. Natl. Acad. Sci. USA* **94**, 849–854.

Matthews, B. D., LaVan, D. A., Overby, D. R., Karavitis, J., and Ingber, D. E. (2004a). Electromagnetic needles with submicron pole tip radii for nanomanipulation of biomolecules and living cells. *Appl. Phys. Lett.* **85**, 2968–2970.

Matthews, B. D., Overby, D. R., Alenghat, F. J., Karavitis, J., Numaguchi, Y., Allen, P. G., and Ingber, D. E. (2004b). Mechanical properties of individual focal adhesions probed with a magnetic microneedle. *Biochem. Biophys. Res. Commun.* **313**, 758–764.

Matthews, B. D., Overby, D. R., Mannix, R., and Ingber, D. E. (2006). Cellular adaptation to mechanical stress: Role of integrins, Rho, cytoskeletal tension, and mechanosensitive ion channels. *J. Cell. Sci.* **119**(Pt. 3), 508–518.

Meyer, C. J., Alenghat, F. J., Rim, P., Fong, J. H., Fabry, B., and Ingber, D. E. (2000). Mechanical control of cyclic AMP signalling and gene transcription through integrins. *Nat. Cell Biol.* **2,** 666–668.

Mooney, D., Hansen, L., Vacanti, J., Langer, R., Farmer, S., and Ingber, D. (1992). Switching from differentiation to growth in hepatocytes: Control by extracellular matrix. *J. Cell. Physiol.* **151,** 497–505.

Moore, K. A., Polte, T., Huang, S., Shi, B., Alsberg, E., Sunday, M. E., and Ingber, D. E. (2005). Control of basement membrane remodeling and epithelial branching morphogenesis in embryonic lung by Rho and cytoskeletal tension. *Dev. Dyn.* **232,** 268–281.

Nelson, C. M., Jean, R. P., Tan, J. L., Liu, W. F., Sniadecki, N. J., Spector, A. A., and Chen, C. S. (2005). Emergent patterns of growth controlled by multicellular form and mechanics. *Proc. Natl. Acad. Sci. USA* **102,** 11594–11599.

Overby, D. R., Alenghat, F. J., Montoya-Zavala, M., Bei, H. C., Oh, P., Karavitis, J., and Ingber, D. E. (2004). Magnetic cellular switches. *IEEE Trans. Magn.* **40,** 2958–2960.

Overby, D. R., Matthews, B. D., Alsberg, E., and Ingber, D. E. (2005). Novel dynamic rheological behavior of focal adhesions measured within single cells using electromagnetic pulling cytometry. *Acta Biomater.* **1,** 295–303.

Parker, K. K., Brock, A. L., Brangwynne, C., Mannix, R. J., Wang, N., Ostuni, E., Geisse, N. A., Adams, J. C., Whitesides, G. M., and Ingber, D. E. (2002). Directional control of lamellipodia extension by constraining cell shape and orienting cell tractional forces. *FASEB J.* **16,** 1195–1204.

Paszek, M. J., Zahir, N., Johnson, K. R., Lakins, J. N., Rozenberg, G. I., Gefen, A., Reinhart-King, C. A., Margulies, S. S., Dembo, M., Boettiger, D., Hammer, D. A., and Weaver, V. M. (2005). Tensional homeostasis and the malignant phenotype. *Cancer Cell* **8,** 241–254.

Pelham, R. J., Jr., and Wang, Y. (1997). Cell locomotion and focal adhesions are regulated by substrate flexibility. *Proc. Natl. Acad. Sci. USA* **94,** 13661–13665.

Plopper, G., and Ingber, D. E. (1993). Rapid induction and isolation of focal adhesion complexes. *Biochem. Biophys. Res. Commun.* **193,** 571–578.

Polte, T. R., Eichler, G. S., Wang, N., and Ingber, D. E. (2004). Extracellular matrix controls myosin light chain phosphorylation and cell contractility through modulation of cell shape and cytoskeletal prestress. *Am. J. Physiol., Cell Physiol.* **286,** C518–C528.

Potard, U. S., Butler, J. P., and Wang, N. (1997). Cytoskeletal mechanics in confluent epithelial cells probed through integrins and E-cadherins. *Am. J. Physiol.* **272,** C1654–C1663.

Prime, K. L., and Whitesides, G. M. (1991). Self-assembled organic monolayers: Model systems for studying adsorption of proteins at surfaces. *Science* **252,** 1164–1167.

Roberts, C., Chen, C., Mrksich, M., Martichonok, V., Ingber, D., and Whitesides, G. M. (1998). Using mixed self-assembled monolayers presenting RGD and (EG)(3)OH groups to characterize long-term attachment of bovine capillary endothelial cells to surfaces. *J. Am. Chem. Soc.* **120,** 6548–6555.

Rogers, J. A., Paul, K. E., and Whitesides, G. M. (1998). Quantifying distortions in soft lithography. *J. Vac. Sci. Technol. B* **16,** 88–97.

Ryan, D., Parviz, B. A., Linder, V., Semetey, V., Sia, S. K., Su, J., Mrksich, M., and Whitesides, G. M. (2004). Patterning multiple aligned self-assembled monolayers using light. *Langmuir* **20,** 9080–9088.

Schwartz, M. A., Lechene, C., and Ingber, D. E. (1991). Insoluble fibronectin activates the Na/H antiporter by clustering and immobilizing integrin alpha 5 beta 1, independent of cell shape. *Proc. Natl. Acad. Sci. USA* **88,** 7849–7853.

Shroff, S. G., Saner, D. R., and Lal, R. (1995). Dynamic micromechanical properties of cultured rat atrial myocytes measured by atomic force microscopy. *Am. J. Physiol.* **269,** C286–C292.

Singhvi, R., Kumar, A., Lopez, G. P., Stephanopoulos, G. N., Wang, D. I., Whitesides, G. M., and Ingber, D. E. (1994). Engineering cell shape and function. *Science* **264,** 696–698.

Stahlhofen, W., and Moller, W. (1992). Investigation of the defense system of the human lungs with ferrimagnetic particles. *J. Aerosol Med.* **5,** 221–228.

Stamenovic, D., and Coughlin, M. F. (2000). A quantitative model of cellular elasticity based on tensegrity. *J. Biomech. Eng.* **122,** 39–43.

Stamenovic, D., Fredberg, J. J., Wang, N., Butler, J. P., and Ingber, D. E. (1996). A microstructural approach to cytoskeletal mechanics based on tensegrity. *J. Theor. Biol.* **181,** 125–136.

Tagawa, H., Wang, N., Narishige, T., Ingber, D. E., Zile, M. R., and Cooper, G., 4th. (1997). Cytoskeletal mechanics in pressure-overload cardiac hypertropy. *Circ. Res.* **80**(2), 295–296.

Tan, J. L., Liu, W., Nelson, C. M., Raghavan, S., and Chen, C. S. (2004). Simple approach to micropattern cells on common culture substrates by tuning substrate wettability. *Tissue Eng.* **10,** 865–872.

Valberg, P. A., and Butler, J. P. (1987). Magnetic particle motions within living cells. Physical theory and techniques. *Biophys. J.* **52,** 537–550.

Wang, N., Butler, J. P., and Ingber, D. E. (1993). Mechanotransduction across the cell surface and through the cytoskeleton. *Science* **260,** 1124–1127.

Wang, N., and Ingber, D. E. (1994). Control of cytoskeletal mechanics by extracellular matrix, cell shape, and mechanical tension. *Biophys. J.* **66,** 2181–2189.

Wang, N., and Ingber, D. E. (1995). Probing transmembrane mechanical coupling and cytomechanics using magnetic twisting cytometry. *Biochem. Cell Biol.* **73,** 327–335.

Wang, N., Ostuni, E., Whitesides, G. M., and Ingber, D. E. (2002a). Micropatterning tractional forces in living cells. *Cell Motil. Cytoskeleton* **52,** 97–106.

Wang, N., Tolic-Norrelykke, I. M., Chen, J., Mijailovich, S. M., Butler, J. P., Fredberg, J. J., and Stamenovic, D. (2002b). Cell prestress. I. Stiffness and prestress are closely associated in adherent contractile cells. *Am. J. Physiol., Cell Physiol.* **282,** C606–C616.

Wang, Y. L., and Pelham, R. J., Jr (1998). Preparation of a flexible, porous polyacrylamide substrate for mechanical studies of cultured cells. *Methods Enzymol.* **298,** 489–496.

Whitesides, G. M., Ostuni, E., Takayama, S., Jiang, X., and Ingber, D. E. (2001). Soft lithography in biology and biochemistry. *Annu. Rev. Biomed. Eng.* **3,** 335–373.

Xia, Y., and Whitesides, G. M. (1998). Soft lithography. *Angew. Chem. Int. Ed. Engl.* **37,** 550–575.

Yeo, W. S., Hodneland, C. D., and Mrksich, M. (2001). Electroactive monolayer substrates that selectively release adherent cells. *Chembiochem* **2,** 590–593.

Yeo, W. S., Yousaf, M. N., and Mrksich, M. (2003). Dynamic interfaces between cells and surfaces: Electroactive substrates that sequentially release and attach cells. *J. Am. Chem. Soc.* **125,** 14994–14995.

Yoshida, M., Westlin, W. F., Wang, N., Ingber, D. E., Rosenzweig, A., Resnick, N., and Gimbrone, M. A., Jr. (1996). Leukocyte adhesion to vascular endothelium induces E-selectin linkage to the actin cytoskeleton. *J. Cell Biol.* **133,** 445–455.

Yousaf, M. N., Houseman, B. T., and Mrksich, M. (2001). Using electroactive substrates to pattern the attachment of two different cell populations. *Proc. Natl. Acad. Sci. USA* **98,** 5992–5996.

CHAPTER 20

Magnetic Tweezers in Cell Biology

Monica Tanase, Nicolas Biais, and Michael Sheetz

Department of Biological Sciences
Columbia University
New York, New York 10027

Abstract

We discuss herein the theory as well as some design considerations of magnetic tweezers. This method of generating force on magnetic particles bound to biological entities is shown to have a number of advantages over other techniques: forces are exerted in noncontact mode, they can be large in magnitude (order of 10 nanonewtons), and adjustable in direction, static or oscillatory. One apparatus built in our laboratory is described in detail, along with examples of experimental applications and results.

METHODS IN CELL BIOLOGY, VOL. 83
Copyright 2007, Elsevier Inc. All rights reserved.

0091-679X/07 $35.00
DOI: 10.1016/S0091-679X(07)83020-2

I. Introduction

Form in biology, from cells to tissues and ultimately whole organisms, relies heavily on the sensing and generation of appropriate forces. The integrated response of the cell to forces controls cell growth and differentiation as well as extracellular matrix remodeling (Chen *et al.*, 2004; Chiquet *et al.*, 2003; Tamada *et al.*, 2004; Vogel and Sheetz, 2006). In the living organism, motile activity modulates the force over time and static forces are a rarity (Ito *et al.*, 2006; Maksym *et al.*, 2000; Meshel *et al.*, 2005; Murfee *et al.*, 2005). Thus, it is necessary to understand the dynamics of cellular force responses in order to understand cell growth and differentiation. An important example is the requirement for a substantive rather than overly soft substrate for normal cell growth since cancerous cells can often grow on soft agar (Discher *et al.*, 2005; Engler *et al.*, 2004; Georges and Janmey, 2005; Jiang *et al.*, 2006; Kostic and Sheetz, 2006). Changes in oncogenes are involved in enabling growth on soft agar (Giannone and Sheetz, 2006). We have relatively few tools for local modulation of forces at the nanonewton/ micrometer levels that are typically observed in cells. Magnetic tweezers offer such capabilities and have many advantages over other force-generating systems.

A wide variety of methods have been developed to generate and measure cellular forces: micromechanical devices (Desprat *et al.*, 2005; Galbraith and Sheetz, 1997; Thoumine and Ott, 1997), fluid flow-based systems (Thomas *et al.*, 2004), atomic force microscopes (AFM) (Felix *et al.*, 2005; Lal and John, 1994), and optical (Sheetz, 1998) and magnetic tweezers (Bausch *et al.*, 1999; Crick and Hughes, 1950). They all have advantages as well as limitations in their applicability. This chapter focuses on magnetic tweezers. The term is generally used to describe an apparatus that applies a force to magnetic particles through magnetic field gradients. This method is noninvasive, as it allows micromanipulation without direct contact of particles bound to the biological entity under investigation: molecule, organelle, and cell. Such systems come in various designs and levels of complexity depending on the application pursued, and generally consist of an arrangement of permanent magnets or electromagnets mounted on an optical microscope stage. The aim of this chapter is to present the underlying principles of magnetic tweezers and to provide information to assist in optimally choosing and designing a system (see also complementary information in Chapter 19 by Lele *et al.*, this volume). The main performance parameters that need to be considered in building a magnetic force generation apparatus are:

1. Amplitude and direction of the force;
2. Timescale over which the force needs to be maintained or modulated— signal frequency;
3. Size of the assay—spatial range at which the force profile has the desired characteristics.

Advances in technologies and materials are continually expanding the available range of the parameters above, and thus the spectrum of capabilities of magnetic tweezer systems. Amplitudes can vary from a few piconewtons (pN) (Strick *et al.*, 1996) to tens of nanonewtons (nN) (Bausch *et al.*, 1998; Strick *et al.*, 1996). While many of the reported magnetic systems apply force in one direction only, a growing number of groups have been reporting designs that provide spatial flexibility in the direction of the force, including full 3D systems (Fisher *et al.*, 2005). The timescale for force generation ranges from milliseconds (ms) to days, and frequencies can be as high as 5–10 kilohertz (kHz). The size of the assay can span from micrometers (Barbic *et al.*, 2001; Jie *et al.*, 2004) to centimeters (Haber and Wirtz, 2000). In designing a magnetic tweezers system, it is important to understand that these three categories of parameters are never independent of each other. Having an apparatus that operate in the high bandwidth of any of these parameters will significantly limit the range in the other two. Some of these limitations may be addressed partially by increasing the complexity of the system. Herein, we survey different possible magnetic tweezer assemblies and make note of some of the challenges involved in building systems that push the limits on the performance parameters.

II. Physics of Magnetic Tweezers

The working principle is that a magnetic particle placed in a magnetic field gradient will be subject to a force directed toward the source of the field. Two main components interact to create magnetic force: the profile of the external field and the magnetic properties of the particles used.

All materials are influenced to some degree by the presence of a magnetic field **B**, and their response is quantified by the magnetic moment **μ** of the material. If this is nonzero, the material is said to be *magnetic*. The intensity and properties of the moment dictate the response of the material to externally applied fields. Most materials used for the construction of tweezer systems and the particles used to transduce the force to biological entities come in two major flavors: *paramagnetic/ superparamagnetic* and *ferromagnetic*. *Paramagnetic* materials acquire a magnetic moment only when an external magnetic field is applied and are entirely nonmagnetic in zero field. *Ferromagnetic* materials are different in the sense that once exposed to a magnetic field, they become magnetized and will retain a certain fraction of their magnetization even after the field has been removed. Most commercially available beads (Fig. 1A) are *superparamagnetic*, a behavior similar to paramagnetism. Superparamagnetism occurs in materials containing ferromagnetic components (crystallites, nanoparticles) of dimensions small enough to cause the loss of their magnetic cohesion, that is their permanent magnetism, on removal of the external magnetic field. This is the case of most magnetic beads consisting of ferrite nanoparticles embedded in a spherical latex matrix. In contrast, ferromagnetic materials, such as nickel or cobalt nanowires (Fig. 1B), exhibit magnetic properties even when no external field is present (Fert and Piraux, 1999; Wernsdorfer *et al.*, 1996). The magnetic moment of a ferromagnet depends not only on the value of the external field but on history of the magnetization of the

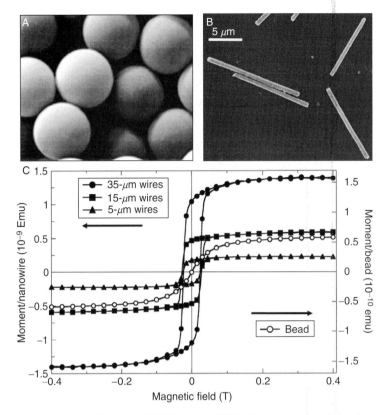

Fig. 1 Scanning electron microscope (SEM) images of (A) 4.5-μm paramagnetic beads (courtesy of Invitrogen Corporation, Dynal bead-based separations) and (B) electrodeposited nickel nanowires (Tanase *et al.*, 2005). (C) Hysteresis curves showing the dependence of the magnetic moment per particle μ versus external magnetic field **B** for paramagnetic beads (1.5-μm diameter) and ferromagnetic nickel nanowires (100-nm diameter, 5-, 15- and 35-μm long); figure courtesy of Daniel H. Reich. Saturation occurs when the field is large enough to align all constituent magnetic moments, while remanence is the residual moment after the "relaxation" of alignment when the field is zero. Note that the hysteresis of the superparamagnetic beads is zero, as they exhibit no magnetic lag or memory effect. Emu on the *y*-axis stands for electromagnetic unit, the CGS unit for electromagnetic moment.

material, and this dependence is called *hysteresis* (Fig. 1C). The amount of magnetization that is retained when the driving field is removed is called *remanence*, and is an important parameter in the design of a magnetic tweezers system. We note here that it is possible to *demagnetize* a ferromagnetic material (such as the core of an electromagnet, see Section III.A) by rapidly cycling the external field between the positive and negative values necessary for saturation, and gradually dampening the amplitude of oscillations down to zero field. Such cycling will zero out the unwanted remanent magnetism of the ferromagnetic material.

A magnetic particle in a field is subject to mechanical forces of magnetic origin due to the interplay between its magnetic moment μ and the external field **B**.

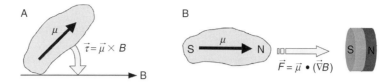

Fig. 2 Magnetic torque τ and force **F**. Magnetic particles can be (A) rotated and (B) displaced via external magnetic fields.

A torque $\tau = \mu \times \mathbf{B}$ will tend to rotate and align the particle's moment with the external field, and in many cases the particle rotates itself to accommodate (Fig. 2A). If the field is not constant but presents a gradient $\nabla\mathbf{B}$, the particle is subject to a force proportional to the local field gradient $\mathbf{F} = (\mu \cdot \nabla)\mathbf{B}$ directed toward the regions of higher magnetic field (Fig. 2B).

In summary, a magnetic particle will rotate to align its magnetic moment parallel to the magnetic field direction, and translate toward the regions of higher field. While magnetic particles can consequently be pulled in the direction of increasing fields, they can never be pushed away. If flexibility is required in the directionality of the force, multiple magnetic poles, as well as the use of nonpermanent field, need to be implemented in the system (de Vries *et al.*, 2005; Drndic *et al.*, 2001; Fisher *et al.*, 2005; Huang *et al.*, 2002).

III. Magnetic Field Considerations

We discuss here methods of generating magnetic fields and various aspects of the field gradient profile. Unless noted, the observations here refer to the field of a single magnetic pole. In the case of multiple poles, magnetic circuitry principles need to be considered.

A. Sources of Magnetic Field

In a magnetic tweezers apparatus, permanent magnets (Matthews *et al.*, 2004) or electromagnets can be used as sources of field. While a permanent magnet generates a static permanent field, electromagnets convert electrical currents into magnetic field and allow control of the field through control of the current.

Permanent magnets are the most accessible method for producing a magnetic field (Fig. 3A). When made of rare earth materials, they can generate fields as large as 0.8 T (for comparison, Earth's magnetic field is $\sim 0.5 \times 10^{-4}$ T). Their field is very steady but in order to modulate it the magnets need to be physically displaced—action that limits the time response of the system and can potentially add mechanical noise. In contrast, electromagnets typically generate fields that are orders of magnitude lower than those of equal or equivalent sized permanent magnets, but they have the great advantage that the amplitude of the magnetic field can be controllably and rapidly modulated. The simplest version of an

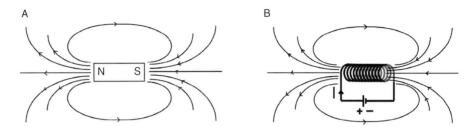

Fig. 3 Schematic of magnetic field profiles for (A) permanent magnet and (B) and an electromagnetic solenoid.

electromagnet is a coiled conductive wire. If multiple coils form a cylindrical geometry, the structure is called a *solenoid* (Fig. 3B). When electrical current runs through the wire, a magnetic field is established in the direction of the solenoid's axis. We note here that cylindrical permanent magnets and solenoids produce magnetic fields with similar profiles (Fig. 3), albeit different magnitudes.

The strength of the field produced by an electromagnet is set by the value of the current in the wire and the coil geometry. To amplify the magnetic field, a magnetic core can be positioned inside the solenoid. Cores are typically made from soft ferromagnetic materials with high saturation and low remanence. While such designs allow stronger local fields, they introduce another problem: the existence of a remnant magnetic field. Even when the current is turned off, the core remains magnetized to some extent, and therefore a remnant force is exerted on the targeted particle. As previously discussed, this effect can be reduced by running a cycle of demagnetization on the core or by using a more elaborate design with multiple coils. To produce large forces using an electromagnet, either large-diameter coils or high electrical currents are needed. The size of the solenoid (length and diameter) will be limited by the physical specifications of the apparatus and the ease of use. Large currents may generate heating in the coils, detrimental to the magnetic core and ultimately to the biological sample. To circumvent this problem, heat sinks or cooling systems are typically used (Haber and Wirtz, 2000).

B. Magnet Pole Design

While the amplitude of the field is important, it is the gradient of the magnetic field $\nabla \mathbf{B}$ that directly factors into the magnitude of the force. The gradient is the strongest close to the magnet, as the field drops in magnitude faster nearer the magnet than further away from it. A convenient rule of thumb in approximating the gradient one can obtain with a magnet is that most of the field will vanish within a distance that is roughly the size of the pole diameter. The size of the magnet is therefore directly coupled to the size of the experimental assay. If, for example, the force needs to be constant over an extended range such as the field of view of the microscope, then $\nabla \mathbf{B}$ needs to also be constant over the same area. This can be obtained either by using a blunt large magnet that will generate a slowly decreasing magnetic field, that is small gradient, or by doing the experiment further away from any type of pole, in the region where the variance of the field tapers off. Elaborate systems with multiple poles can produce constant gradients but at the expense of the force amplitude. Generation of large forces require large field gradients. This is typically obtained by tapering off the magnet or the electromagnetic core at the pole on the sample side to concentrate the magnetic flux. At the same time, the specimen should be positioned close to the magnetic tip. A sharpened pole face gives a larger gradient and force. For example, if a magnet generates a 0.5-T field and the cross section is 1 cm, the average gradient near the magnet is on the order of 50 T/m and the force is $\mu_{\text{particle}} \times 50$ T/m. If the magnet is sharpened (e.g., magnetized sewing needle) such that the pole face is small, the gradient can be much larger but over a much smaller distance. The cross section could be made as small as 10 μm, and for a face field of 0.5 T the local magnetic gradient is on the order of 5×10^4 T/m. The force in this case would be three orders of magnitude larger than that with the blunt geometry, provided that the moment of the particles is the same.

While magnetic tweezers are force clamps by the nature of the interaction, they can be designed to serve as position clamps. Such a design cannot be passive as it requires active control of the position of the particle and entails at the very least a position feedback system (Gosse and Croquette, 2002). Overall, electromagnets offer more control on the spatial and temporal profile of the magnetic field generated but at the price of increasing the technical challenges.

IV. Magnetic Particle Selection

In order to apply a magnetic force to a biological sample, a magnetic particle needs to be bound to the sample. A variety of particles, typically superparamagnetic beads, are currently used for physical manipulation of cells and biomolecules (Hafeli *et al.*, 1997). For most of the magnetic tweezer applications, consistency in the magnetic and geometric characteristics of the particles is very important. A large selection of magnetic beads is available with very low standard deviations in both diameter and magnetic content (Bangs Laboratories Fishers, Indiana; Polysciences Inc., Warrington, Pennsylvania; Invitrogen Corporation, Carlsbad, California). The most common type of paramagnetic bead consists of a spherical latex matrix containing dispersed magnetic nanoparticles. As previously mentioned, the force scales with the magnetic moment of the particle and, consequently, to the volume of magnetic material present in the particle. The overall volume of the bead is not the only relevant factor for the magnitude of force; another determining factor is the magnetic content. For example, 1-μm-diameter Dynal beads (MyOne, Invitrogen Corporation, Carlsbad, California) have 34% magnetic content, while the 2.7-μm beads (M270, Invitrogen Corporation, Carlsbad, California) contain 20% ferrite material.

Another type of magnetic particles available from Polysciences, Inc., 1- to 2-μm-diameter BioMagPlus, has been specifically engineered for separation applications.

They consist of a solid ferrite core amounting to 90% magnetic content, and are coated with a thin latex layer. The resulting magnetic force per particle is larger than other types of beads of similar diameter. One other difference is that unlike the spherical beads, these are irregularly shaped in order to produce a greater surface area, 20–30 times that of a spherical particle of the same size (Polysciences, Inc., technical data sheet No. 618). While the greatly increased area results in greater molecular binding, the volume and therefore magnetic content are different from particle to particle, resulting in a large variance (~100%) in the magnetic force across the population. The advantage of using these beads with magnetic tweezers is that much larger forces can be applied through small particles, but the force calibration has very large error margins. However, if the accuracy of the force measurement is not important but high binding efficiency is, such particles are preferred.

Magnetic nanowires (Fig. 1B) constitute another class of nanostructures suitable for magnetic tweezer applications (Reich *et al.*, 2003; Tanase *et al.*, 2005). These are quasi-one-dimensional cylindrical structures with large aspect ratios, with diameters in the 1–1000 nm range and lengths from tens of nanometers to tens of micrometers. While there are various methods for fabricating nanowires, one particularly attractive approach is electrodeposition into nanoporous templates (Whitney *et al.*, 1993). Due to the fabrication process, the composition along the length of electrodeposited nanowires can be precisely modulated, which in turn enables precise control of the architecture of the magnetic properties (Blondel *et al.*, 1994; Chen *et al.*, 2003). In addition, by using ligands that bind selectively to different segments of a multicomponent wire, it is possible to introduce spatially modulated multiple functionalization in the wires (Bauer *et al.*, 2003; Tanase *et al.*, 2001). Also, their strong shape anisotropy gives rise to preferential direction of the magnetic moment and therefore new properties (Hultgren *et al.*, 2005). The unique features of magnetic nanowires greatly expand the range of functions performed by magnetic particles. These include:

- A large remnant magnetic moment offering the prospect of low-field manipulation, whereas larger fields are required for the beads to become magnetic enough to be effective;
- Large forces and torques that can be applied to cells and biomolecules; forces can be up to 1000 times larger than the forces on beads of comparable volumes;
- Larger surface area providing increased adhesion surface;
- Multifunctional surfaces, as multiple bioactive ligands can be selectively bound to the different segments of multicomponent nanowires.

Magnetic tweezers are also used for studies of the inside of living cells (Basarab *et al.*, 2003; Francois *et al.*, 1996; Marion *et al.*, 2005). In some cases, the magnetic beads or nanowires are too large and a different category of particles is needed. One option is the use of ferrofluids which are magnetic nanoparticles in aqueous suspensions. The ferrofluids can be loaded into the cells via endocytosis, allowing the study of the structure of cytoplasm and organelles in which the particles are concentrated (Marion *et al.*, 2005; Valberg and Albertini, 1985; Wilhelm *et al.*, 2003).

One other factor that needs to be considered in choosing a suitable type of particle is whether fluorescence imaging is involved. As latex is an autofluorescent material, most of the magnetic beads on the market exhibit this effect. Examples of particles that are not autofluorescent are the irregularly shaped BioMagPlus and ferromagnetic nanowires.

As can be concluded, particles come in different sizes, shapes, surfaces, and optical and magnetic properties. Depending on the application intended, one or another type of particle may be more desirable, and the choice of use depends among others on the type of forces and torques needed as well as on the available surface chemistry.

V. Basic Solenoid Apparatus

In this section, we describe the design and construction of a magnetic tweezers apparatus currently used in our laboratory (Fig. 4), built to generate forces as large as 10 nN with frequencies from 0 to 1 kHz, values that match and exceed those found in tissues. One use of this system is the application of local forces at the position of interest on cells: lamellipodium, lamella, and perinuclear region, via magnetic beads attached to specific receptors on the cellular dorsal surface. The trajectory of the beads is the result of the interplay between the magnetic force and the force exerted by the cell on the bead. As the force applied on the cell via the beads can be modulated as desired within the system's specifications, the cellular response to a specific mechanostimulus can be investigated.

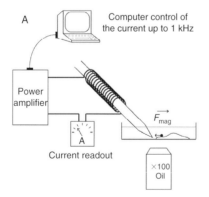 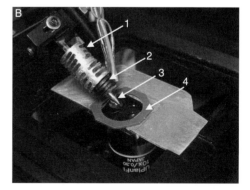

Fig. 4 (A) Schematic representation of the magnetic tweezers and imaging system. The current in the solenoid is controlled via computer through a power amplifier. (B) Electromagnetic system used in our laboratory, composed of (1) a heat-dissipating aluminium sheath, (2) three copper coils, (3) ferromagnetic core, and (4) the experimental chamber.

An electromagnet with a ferromagnetic core sharpened at the sample side was built to concentrate the magnetic field and allow generation of high-amplitude forces on micrometer-sized particles (Fig. 4A). The core of the electromagnet is 2.5 mm in diameter and is made of a soft magnetic alloy with very high saturation and low remanence (Hyperco50, The MuShield Company, Manchester, New Hampshire) in order to allow high magnetic flux without saturation at the tip, while minimizing the magnitude of the remnant fields inside the core after applying large currents. To further assure the repeatability of the measurement, a demagnetizing cycle is used to zero the field inside the core prior to each experiment. We note here one practical drawback of using magnetic materials with small-remnant field that they are not easily amenable to machining. Typically, magnetic materials with small remnant fields are quite brittle and great care must be taken to maintain an intact sharpened tip once the machining is completed.

A key component for this electromagnetic system is the current in the coils (fabricated in-house, 20 turns per coil, copper wire). In order to meet our system design needs, a power transconductance amplifier was generously provided by the Center for Computer Integrated Systems for Microscopy and Manipulation at University of North Carolina (Fisher *et al.*, 2005). This can independently supply three separate coils (Fig. 4A) with currents proportional to their input voltages. The drive amplifier is powered by the output of a National Instruments Acquisition Board, controlled through a LabVIEW designed computer user interface (Fig. 4B). This system permits the current in each coil to reach up to 5 A at frequencies up to 1 kHz. A three-axis micromanipulator allows tip positioning with submicron precision.

Another factor that can negatively influence the experiment is the heat generation due to the use of large currents. A significant increase in temperature would adversely affect the magnetic properties of the core and decrease the amplitude and gradient of the generated magnetic field. Additionally, if the heat transfer results in even a few degrees rise in the biological sample, the experimental results could be significantly affected. The easiest way to address the heating due to high currents is to introduce even the thinnest gap between the core and the coils to serve as a thermally insulating buffer, and to add aluminum heat sinks directly onto the core of the electromagnet. In our system, heat dissipation is adequate to prevent a temperature increase in the biological sample while producing high local field gradients. Currents up to 2 A can be maintained indefinitely without detectable heating of the coils. Higher currents may be used, but only intermittently in short pulses to allow time for heat dissipation.

VI. Force Calibration

The output of a *force calibration procedure* is a graph of the force versus the distance from the magnetic tip, and is a function of the magnetic particle used, the current in the coils, and in some cases the angle between the core axis and

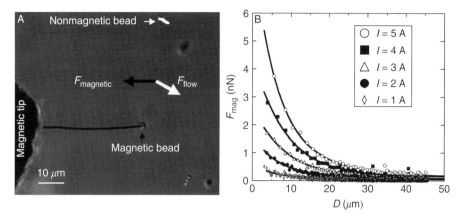

Fig. 5 Force calibration of magnetic tweezers. (A) Trajectories of a 2.7-μm magnetic bead (black trace) and a 1-μm polystyrene bead (white trace) suspended in 1000-cp standard viscosity oil. (B) Graph of force versus distance from the edge of the magnetic tip for coil currents of 1, 2, 3, 4, and 5 A.

the particle's position vector (Fig. 5B). Once this *force calibration curve* has been generated, it can be used to determine the value of the magnetic force on a bead attached to a biological entity, by recording its location relative to the magnetic tip.

It is theoretically possible to calculate the force exerted on a magnetic particle in a known field profile. However, inherent inaccuracies in the physical and geometrical properties of the system components make this approach rather impractical. It is common practice to determine the forces empirically, by tracking the displacement of the particles through stationary fluids of calibrated viscosity. An important dimensionless number in fluid dynamics is the Reynolds number, used for determining whether a flow is laminar or turbulent. It is defined as the ratio of inertial to viscous forces $Re = v_s \rho L / \eta$, where v_s is the mean fluid velocity, ρ is the fluid density, η is the dynamic fluid viscosity, and L is the characteristic particle length (particle diameter in the case of magnetic beads). For microparticles, Reynolds number is typically very small ($Re < 10^{-5}$), so viscous drag dominates over all other hydrodynamic effects and the flow is laminar (Lifshitz and Landau, 1959). In this regime, in response to an applied force F, a particle will move with terminal velocity $v = F/D$, where D is the appropriate drag coefficient. For a spherical particle, Stokes' law gives $D = 6\pi\eta r$, where r is the radius of the bead. In the case of nanowires, the drag coefficient for a cylinder has the same functional form as for a sphere, but r in this case is the effective radius and may be determined by approximating the wire as a prolate ellipsoid (Lamb, 1945). Consequently, by tracking the displacement of an unbound microparticle under magnetic force, the velocity versus position dependence can be obtained. The force can then be calculated as $F = 6\pi\eta r v$.

A. Calibration Sample Protocol

We found that the displacement of the magnetic beads under force induces motion of the viscous fluid surrounding them, and that after a certain period a concerted flow is established in the direction of the magnetic field gradient. In order to be able to use the described Stokes calibration, the drift of the fluid needs to be either avoided or taken into account. To monitor the displacement of the fluid itself, nonmagnetic beads are added into the fluid (Fig. 5A). These are nonresponsive to the magnetic force and are used as fluid flow markers. For optical identification, the two types of beads are chosen to be of different diameter, with the nonmagnetic beads smaller than the magnetic ones. The density of the magnetic beads is kept very low in order to delay the onset of fluid flow during calibration. The final density of beads in suspension depends on how well the beads are dispersed in the mix. On visual inspection, the optimum density of magnetic beads is such that they are spaced at least 20 diameters from each other, and the density of polystyrene beads is approximately three to four times larger. The volumes in the following procedure are to be used as a starting point and one should visually inspect the sample to determine the optimal dilution.

Calibration chamber: Small volume silicone chamber (9-mm diameter; 1-mm height; press-to-seal silicone isolators—Grace Bio-Labs, Inc., Bend, Oregon, Cat. No. JTR8R-1.0) fitted with cover glass bottom.

Materials: Magnetic carboxylate spheres (Dynal, M270), polystyrene beads (Bangs Laboratories, Fishers, Indiana, Cat. No. PS03N), calibrated viscosity silicone oil (dimethylpolysiloxane, Sigma-Aldrich, St. Louis, Missouri, Cat. No. DMPS1C-1000G), rare earth magnets (NdFeB discs, Amazing Magnets, Irvine, CA, Cat. No. D250C).

Stock solutions of both types of beads have to be homogenized prior to use, by gentle vortexing.

Silica beads

- Make a 1:10 dilution in milliQ water, briefly vortex.
- Add 5 μl of the diluted bead solution to 95 μl of milliQ water, centrifuge the beads down (4 min, 10,000 rpm) in a tabletop centrifuge, and carefully remove as much supernatant as possible.
- Add 1 ml of silicone oil and mix thoroughly. (Note: beads stored in aqueous suspensions are difficult to disperse in oils as they often close pack in micellar structures.)

Magnetic beads

- Make a 1:10 dilution of magnetic beads in milliQ water, briefly sonicate.
- Place 5 μl of the diluted magnetic bead solution into an Eppendorf tube (10^4 beads in the case of M270 beads from a stock concentration of 2×10^7 beads/ml).
- Using the magnet, collect all the beads onto one side of the water droplet, and with the magnet in place, use a small pipette tip to remove as much of the water supernatant as possible.

- Add 200 μl of oil suspension of silica beads, mix thoroughly with a 200-μl pipette tip to disperse and incorporate the magnetic beads.
- Place 65 μl of the resulting mixture in the calibration chamber.

Potential problems

1. Bead aggregation: carboxylate beads in water do not mix well into oil, micelles form. Remove as much water as possible, mix thoroughly.
2. If the suspension of polystyrene and magnetic beads has incorporated air bubbles, place the tube in a vacuum chamber until clear.
3. Glass binding: some of the beads are near the glass substrate and their mobility is reduced due to glass binding; these beads should not be used for calibration. Note that beads in suspension settle; therefore, gentle mixing with a pipette tip to resuspend them is required after 1–2 h (settling time is proportional to the oil viscosity).

B. Calibration Procedure

Place the calibration chamber on the optical microscope, search for a suitable single magnetic bead in the neighborhood of a few polystyrene ones, all in the same plane of focus. Insert the magnetic tip into the silicone oil to the depth of the chosen beads, as judged under the microscope. For calibration, a magnetic bead is considered "suitable" if it is:

- not surrounded by water
- positioned at least 20 diameters away from another magnetic bead
- more than 5 diameters away from the glass bottom of the chamber, as judged using the focusing mechanism of the microscope

Allow few minutes for the solution to settle and the beads to become mostly still. Start the image acquisition and apply the desired magnetic force. Allow the bead to reach the magnetic tip. Zero the current in the coil and demagnetize the core. Wait until the flow has ceased, then look for the next bead. Repeat as many times as needed at different angles from the needle for statistical significance.

C. Direction of Magnetic Force

As the solenoid and the core are not horizontal, the magnetic force will not be horizontal either. To minimize the vertical component, we flatten the bottom of the magnetic tip [method also reported in Bausch *et al.* (1998)]. With this geometry the force on the beads is mostly horizontal, with a slight angle that increases in the proximity of the tip. We estimate that 3 to 5 μm away from the tip, the force vector forms a 10° angle with the horizontal plane. In this case, the vertical component of the force accounts for 15% of the total force giving a proportional

15% error in the calibration curve, which is within the system's overall errors. We note here that a number of magnetic tweezer systems have been designed to generate vertical force (Assi *et al.*, 2002; Gosse and Croquette, 2002). The force versus distance calibration procedure often involves correlation of the diffraction rings of an ascending bead (Gosse and Croquette, 2002).

The force produced by any magnetized cylinder such as the tweezers core varies as a function of the angle between the bead's position vector and the axial direction of the cylindrical magnet core. The greatest force at equal distance from the tip is in the axial direction, and the force decreases as the angle increases. However, the dependence on the angle of approach becomes insignificant in the proximity of the tip, at a length scale on the order of the radius of the magnetic tip. In our case, the radius was on the order of 50 μm, and no difference was observed in the force profile across 50 μm from the tip, for angles up to 45° from the core axis.

D. Bead-Tracking System

Optical images are acquired by either a high-resolution CCD camera or through a video acquisition system allowing 30 frames/sec. The location of the bead is determined frame-by-frame by a position-tracking algorithm based on a cross-correlation image analysis (Gelles *et al.*, 1988) giving subpixel resolution (0.1–0.2 pixel error). The code is implemented as a plug-in in ImageJ (NIH, available in public domain). The subsequent data interpretation and modeling is done in IGOR Pro (WaveMetrics Inc., Lake Oswego, Oregon).

E. Data Interpretation

In an image sequence of a magnetic bead approaching the tip under magnetic forces, multiple targets should be tracked for accuracy in addition to the calibrating bead. The motion of at least one nonmagnetic bead reveals the displacement of the fluid itself and serves as a background displacement vector. The position of the magnetic tip should also be tracked, as even a few micrometers of positional drift would result in an erroneous calibration of the force, especially in the regions closest to the tip.

The error in the calibration of forces includes standard deviations for the bead's magnetic content and diameter, actual temperature of the oil (optical imaging can locally heat the observed volume of fluid and locally change the viscosity of these standardized oils—dimethylpolysiloxane), unaccounted vertical component of magnetic forces, and errors in position tracking. We estimate that the overall error is less than 15%.

VII. Experimental Procedures

In the *in vivo* cellular host environments such as the ECM, tissues, or organs, oscillations occur due to the contractile activity of cells or cyclical activities of the host organism. The magnetic tweezers assay can mimic the manner by which extracellular forces are applied to cells in tissues by allowing well-controlled generation of pulsatory mechanical signals of large magnitudes. We detail here example applications where constant and oscillatory forces were applied to beads bound to mammalian cells, and we describe the data interpretation along with cautionary notes on potential problems and sources of errors.

Spherical 2.7-μm-diameter Dynal beads (M270, Polysciences, Inc.) were functionalized with a pentamer of fibronectin's integrin-binding domain FNIII7–10 (Jiang *et al.*, 2003) according to the protocol included in the product data sheet (Dynabeads, M270, carboxylic acid technical sheet, Rev. No. 002). The beads were placed on laminin-coated glass substrates and the cells were subsequently allowed to spread. In a typical "spreading assay," when the protruding edge of a cell makes contact with a bead, integrins are ligated and activated and initial adhesive contacts form. The beads are therefore bound to the cell membrane and displaced by the rearward actin flow in a radial trajectory toward the nucleus. The velocity of the beads depends on a number of factors, including the level of motile cell activity, the type of ligands present on the beads, and the region of the cell the beads traverse. Bead displacement is, for example, faster on the lamellipodium, slower and less directional in the perinuclear region, and fairly constant in speed and direction across the lamella. Beads functionalized with FN pentamer traverse the lamellar region of a spreading cell at an average speed of 70 nm/sec (Jiang *et al.*, 2003).

The magnetic tweezers allow the application of localized stress to the dorsal surface of the cells via beads bound to specific receptors. The assay enables the study of cellular response to spatially localized mechanostimulus through observation of the trajectory resulting from the interplay between the magnetic and cellular forces. We describe here one representative case where large magnetic force was used (0.5–0.6 nN) (Fig. 6). Once the bead was observed to engage in rearward motion, a constant level of force was applied in the direction opposite to the cell force (Fig. 6A). The graph in Fig. 6B shows the distance between the bead and the cell edge D_{edge} versus time (upper panel), as well as the corresponding magnitude of the applied magnetic force (lower panel). As can be seen in Fig. 6A when the bead is moving away from the edge of the cell and toward the nucleus, D_{edge} increases indicating that $F_{cell} > F_{mag}$. When the magnetic force prevails over the cell force, the bead moves toward the edge of the cell and D_{edge} decreases. When the force is initially applied, a rapid displacement of the bead occurs in the direction of the magnetic tip (markers 2 to 3 in Fig. 6B and C) and can be attributed partly to rolling and partly to a viscoelastic response of the cell. This is followed typically by fluctuations in the bead's velocity and direction of movement even when the level

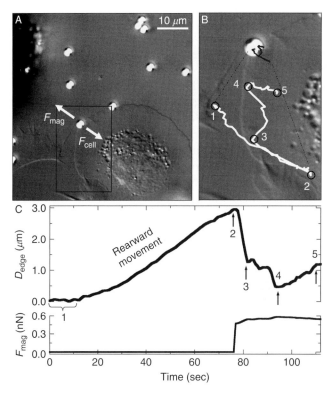

Fig. 6 Constant force assay. (A) Magnetic tweezers (tip located at top left corner) exert force on 2.7-μm magnetic beads bound to a spreading fibroblast. (B) Close-up of the rectangular region marked in panel (A). The actual trajectory of the bead is marked in black, and shown enlarged as the white trace. Prior to application of force, the rearward flow of actin displaces the bead from the cell edge (position 1–2), toward the nucleus. When the external force is applied (position 2), the bead moves under the competing action of the two forces, cellular and magnetic. (C) D_{edge} versus time graph. In response to application of force, the cell responds by pulling in a contractile manner (positive- and negative-rearward velocities) and reinforces the cytoskeletal adhesion to the stress site (position 3–4). The end of the adaptive period is marked by the recovery of the bead's constant rate of displacement (position 4–5).

of force is sustained (e.g., by doing experiments in a far region from the tip where the variation of the force with position is slow), then by the restoration of constant rearward bead velocity (markers 4 to 5). Since the magnetic force is constant, the variations in bead velocity (seen between markers 3 to 4) indicate fluctuations in the cell force and/or breakage of the association. As the cell adapts to the local stress, it generates pulsatory traction forces at the site of the mechanical signal, that is the bead. This is reminiscent of the periodic lamellipodial contractions in spreading and migrating cells (Dobereiner *et al.*, 2005; Dubin-Thaler *et al.*, 2004; Giannone and Sheetz, 2006; Giannone *et al.*, 2004). By using the magnetic tweezers assay, controlled activation of mechanosignaling can be induced at levels of force (up to about 10 nN) and with large numbers of beads not possible with other systems. In contrast, the maximum force of the optical tweezers is only about 100 pN and nanometer precision tracking systems are needed to keep the force constant for a single bead.

Cells show adaptive strengthening in response to large modulated stresses (Matthews *et al.*, 2006). Application of oscillatory local stimuli via the magnetic tweezers permits observation of these changes through the tracking of the beads' displacements. In the experiment illustrated in Fig. 7, a square wave force of 1.4- to 1.6-nN amplitude and 4-sec period (2-sec pulse width) is applied to a bead bound to the lamellipodium of a spreading cell. After the initial forward thrust with the force application, the bead shows oscillation movement with a fairly constant amplitude for a few cycles, then the amplitude begins to decrease indicating

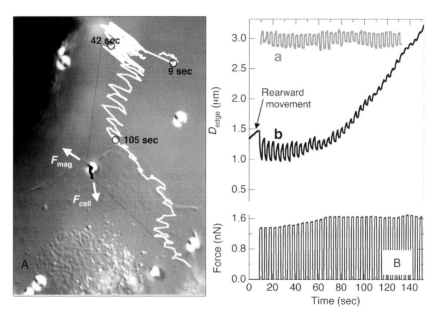

Fig. 7 Modulated force assay. (A) Oscillatory force is applied to a 2.7-μm bead bound to the lamellipodium of a spreading fibroblast. The trajectory of the bead is shown in black for the actual size, and white for the close-up. For 9 sec preceding the force application, the bead undergoes rearward movement. The amplitude of oscillations begins decreasing at the 42-sec time point, and rearward movement is restored at 105 sec. (B) Upper panel: D_{edge} versus time graph for the bead on cell (black trace b), and one bead on the substrate (gray trace a) for comparison. Lower panel: magnetic force versus time. The amplitude of oscillation can be seen to decrease and the velocity of the bead (curve slope) to increase to values comparable to that before stress.

reinforcement of the integrin–cytoskeleton adhesion. Only afterward (96 sec after application of force) the cell reengages the bead in the rearward movement while maintaining a small amplitude of bead oscillation. Such an assay can be used on cells expressing fluorescent adhesion proteins to further elucidate the dynamics of cellular response to mechanical stress. We note here that modulated force wave-forms have been extensively used to study the rheology of the cytoplasm, the actomyosin gel, organelles (Basarab *et al.*, 2003; Fabry *et al.*, 2001; Hu *et al.*, 2004; Keller *et al.*, 2001; Mack *et al.*, 2004; Matthews *et al.*, 2006; Valberg and Albertini, 1985; Wilhelm *et al.*, 2003; Ziemann *et al.*, 1994), and whole-cell assays (Thoumine and Ott, 1997).

In the experiments described here, the following targets were tracked in the image analysis stage:

- The bead under observation.
- Beads on the substrate that are bound to the glass through FN-laminin linkage. These beads roll under magnetic force and return to their initial position when the force is turned off, unless breakages occur in the protein structure anchoring them to the substrate. Their motion under force constitutes a qualitative *in situ* control of the magnetic force.
- The location of magnetic tip.
- Fiduciary structures on the cover glass to enable the tracking of substrate drift.

Observation of the motion of all these targets allows one to check and account for mechanical coupling, drift, noise, and system integrity. It also serves to verify the repeatability of the experimental conditions. The pattern of bead motion on the substrate should, for example, be similar if the experimental conditions are not modified. Observed changes may indicate changes in the calibrated force, substrate, or bead coating.

In this chapter, we have discussed a variety of magnetic tweezer designs and different types of applications of magnetic particles to the study of mechanotrans-duction and cell mechanical properties. Because of the different goals in each application, the optimal magnetic particles and optimal magnetic field configura-tion will differ. The particular application that we have discussed is to produce high forces on beads specifically bound to matrix receptors. For those experiments, a single sharpened magnetic probe is optimal. Important considerations in the design include thermal isolation of the tip from the coils to prevent sample heating and the choice of the beads. Although the measurement of absolute force is relatively inaccurate, at these large forces relative values provide useful measures of the mechanisms controlling force. Further, *in vivo* there are large oscillations in the level of mechanical forces that are applied to the cells. Thus, the ability to produce high and variable forces so rapidly is important for probing the mechanisms of cell mechanical functions.

References

Assi, F., Jenks, R., Yang, J., Love, C., and Prentiss, M. (2002). Massively parallel adhesion and reactivity measurements using simple and inexpensive magnetic tweezers. *J. Appl. Phys.* **92,** 5584–5586.

Barbic, M., Mock, J. J., Gray, A. P., and Schultz, S. (2001). Scanning probe electromagnetic tweezers. *Appl. Phys. Lett.* **79,** 1897–1899.

Basarab, G. H., Karoly, J., Peter, B., Ferenc, I. T., and Gabor, F. (2003). Magnetic tweezers for intracellular applications. *Rev. Sci. Instrum.* **74,** 4158–4163.

Bauer, L. A., Reich, D. H., and Meyer, G. J. (2003). Selective functionalization of two-component magnetic nanowires. *Langmuir* **19,** 7043–7048.

Bausch, A. R., Moller, W., and Sackmann, E. (1999). Measurement of local viscoelasticity and forces in living cells by magnetic tweezers. *Biophys. J.* **76,** 573–579.

Bausch, A. R., Ziemann, F., Boulbitch, A. A., Jacobson, K., and Sackmann, E. (1998). Local measurements of viscoelastic parameters of adherent cell surfaces by magnetic bead microrheometry. *Biophys. J.* **75,** 2038–2049.

Blondel, A., Meier, J. P., Doudin, B., and Ansermet, J.-P. (1994). Giant magnetoresistance of nanowires of multilayers. *App. Phys. Lett.* **65,** 3019–3021.

Chen, C. S., Tan, J., and Tien, J. (2004). Mechanotransduction at cell–matrix and cell–cell contacts. *Annu. Rev. Biomed. Eng.* **6,** 275–302.

Chen, M., Sun, L., Bonevich, J. E., Reich, D. H., Chien, C. L., and Searson, P. C. (2003). Tuning the response of magnetic suspensions. *Appl. Phys. Lett.* **82,** 3310–3312.

Chiquet, M., Renedo, A. S., Huber, F., and Fluck, M. (2003). How do fibroblasts translate mechanical signals into changes in extracellular matrix production? *Matrix Biol.* **22,** 73–80.

Crick, F. H. C., and Hughes, A. F. W. (1950). The physical properties of cytoplasm: A study by means of the magnetic particle method Part I. Experimental. *Exp. Cell Res.* **1,** 37–80.

de Vries, A. H. B., Krenn, B. E., van Driel, R., and Kanger, J. S. (2005). Micro magnetic tweezers for nanomanipulation inside live cells. *Biophys. J.* **88,** 2137–2144.

Desprat, N., Richert, A., Simeon, J., and Asnacios, A. (2005). Creep function of a single living cell. *Biophys. J.* **88,** 2224–2233.

Discher, D. E., Janmey, P., and Wang, Y. L. (2005). Tissue cells feel and respond to the stiffness of their substrate. *Science* **310,** 1139–1143.

Dobereiner, H. G., Dubin-Thaler, B. J., Giannone, G., and Sheetz, M. P. (2005). Force sensing and generation in cell phases: Analyses of complex functions. *J. Appl. Physiol.* **98,** 1542–1546.

Drndic, M., Lee, C. S., and Westervelt, R. M. (2001). Three-dimensional microelectromagnet traps for neutral and charged particles. *Phys. Rev. B* **63,** 085321-1–085321-4.

Dubin-Thaler, B. J., Giannone, G., Dobereiner, H. G., and Sheetz, M. P. (2004). Nanometer analysis of cell spreading on matrix-coated surfaces reveals two distinct cell states and STEPs. *Biophys. J.* **86,** 1794–1806.

Engler, A. J., Griffin, M. A., Sen, S., Bonnemann, C. G., Sweeney, H. L., and Discher, D. E. (2004). Myotubes differentiate optimally on substrates with tissue-like stiffness: Pathological implications for soft or stiff microenvironments. *J. Cell Biol.* **166,** 877–887.

Fabry, B., Maksym, G. N., Shore, S. A., Moore, P. E., Panettieri, R. A., Jr., Butler, J. P., and Fredberg, J. J. (2001). Signal transduction in smooth muscle: Selected contribution: Time course and heterogeneity of contractile responses in cultured human airway smooth muscle cells. *J. Appl. Physiol.* **91,** 986–994.

Felix, R., Pere, R.-C., Nuria, G., Ramon, F., Mar, R., and Daniel, N. (2005). Probing mechanical properties of living cells by atomic force microscopy with blunted pyramidal cantilever tips. *Phys. Rev. E* **72,** 021914-1–021914-10.

Fert, A., and Piraux, L. (1999). Magnetic nanowires. *J. Mag. Mag. Mater.* **200,** 338–358.

Fisher, J. K., Cummings, J. R., Desai, K. V., Vicci, L., Wilde, B., Keller, K., Weigle, C., Bishop, G., Taylor, R. M., II, Davis, C. W., Boucher, R. C., O'Brien, T. E., *et al.* (2005). Three-dimensional force microscope: A nanometric optical tracking and magnetic manipulation system for the biomedical sciences. *Rev. Sci. Instrum.* **76,** 053711-1–053711-11.

Francois, A., Bernard, Y., Andrew, P., and Stanislas, L. (1996). A magnetic manipulator for studying local rheology and micromechanical properties of biological systems. *Rev. Sci. Instrum.* **67,** 818–827.

Galbraith, C. G., and Sheetz, M. P. (1997). A micromachined device provides a new bend on fibroblast traction forces. *Proc. Natl. Acad. Sci. USA* **94,** 9114–9118.

Gelles, J., Schnapp, B. J., and Sheetz, M. P. (1988). Tracking kinesin-driven movements with nanometre-scale precision. *Nature* **331,** 450–453.

Georges, P. C., and Janmey, P. A. (2005). Cell type-specific response to growth on soft materials. *J. Appl. Physiol.* **98,** 1547–1553.

Giannone, G., Dubin-Thaler, B. J., Dobereiner, H. G., Kieffer, N., Bresnick, A. R., and Sheetz, M. P. (2004). Periodic lamellipodial contractions correlate with rearward actin waves. *Cell* **116,** 431–443.

Giannone, G., and Sheetz, M. P. (2006). Substrate rigidity and force define form through tyrosine phosphatase and kinase pathways. *Trends Cell Biol.* **16,** 213–223.

Gosse, C., and Croquette, V. (2002). Magnetic tweezers: Micromanipulation and force measurement at the molecular level. *Biophys. J.* **82,** 3314–3329.

Haber, C., and Wirtz, D. (2000). Magnetic tweezers for DNA micromanipulation. *Rev. Sci. Instrum.* **71,** 4561–4570.

Hafeli, U., Schutt, W., and Joachim, T. (1997). "Scientific and Clinical Applications of Magnetic Carriers." Plenum, New York.

Hu, S., Eberhard, L., Chen, J., Love, J. C., Butler, J. P., Fredberg, J. J., Whitesides, G. M., and Wang, N. (2004). Mechanical anisotropy of adherent cells probed by a three-dimensional magnetic twisting device. *Am. J. Physiol. Cell Physiol.* **287,** C1184–C1191.

Huang, H., Dong, C. Y., Kwon, H.-S., Sutin, J. D., Kamm, R. D., and So, P. T. C. (2002). Three-dimensional cellular deformation analysis with a two-photon magnetic manipulator workstation. *Biophys. J.* **82,** 2211–2223.

Hultgren, A., Tanase, M., Felton, E. J., Bhadriraju, K., Salem, A. K., Chen, C. S., and Reich, D. H. (2005). Optimization of yield in magnetic cell separations using nickel nanowires of different lengths. *Biotechnol. Prog.* **21,** 509–515.

Ito, S., Majumdar, A., Kume, H., Shimokata, K., Naruse, K., Lutchen, K. R., Stamenovic, D., and Suki, B. (2006). Viscoelastic and dynamic nonlinear properties of airway smooth muscle tissue: Roles of mechanical force and the cytoskeleton. *Am. J. Physiol. Lung Cell. Mol. Physiol.* **290,** L1227–L1237.

Jiang, G., Giannone, G., Critchley, D. R., Fukumoto, E., and Sheetz, M. P. (2003). Two-piconewton slip bond between fibronectin and the cytoskeleton depends on talin. *Nature* **424,** 334–337.

Jiang, G., Huang, A. H., Cai, Y., Tanase, M., and Sheetz, M. P. (2006). Rigidity sensing at the leading edge through alphavbeta3 integrins and RPTPalpha. *Biophys. J.* **90,** 1804–1809.

Jie, Y., Dunja, S., and John, F. M. (2004). Near-field-magnetic-tweezer manipulation of single DNA molecules. *Phys. Rev. E* **70,** 011905-1–011905-5.

Keller, M., Schilling, J., and Sackmann, E. (2001). Oscillatory magnetic bead rheometer for complex fluid microrheometry. *Rev. Sci. Instrum.* **72,** 3626–3634.

Kostic, A., and Sheetz, M. P. (2006). Fibronectin rigidity sensing through Fyn and p130Cas recruitment at the leading edge. *Mol. Biol. Cell.* **17,** 2684–2695.

Lal, R., and John, S. A. (1994). Biological applications of atomic force microscopy. *Am. J. Physiol. Cell Physiol.* **266,** C1–C21.

Lamb, H. (1945). "Hydrodynamics," 6th edn. Dover, New York.

Lifshitz, E., and Landau, L. (1959). "Fluid Mechanics." Pergamon Press, Oxford.

Mack, P. J., Kaazempur-Mofrad, M. R., Karcher, H., Lee, R. T., and Kamm, R. D. (2004). Force-induced focal adhesion translocation: Effects of force amplitude and frequency. *Am. J. Physiol. Cell Physiol.* **287,** C954–C962.

Maksym, G. N., Fabry, B., Butler, J. P., Navajas, D., Tschumperlin, D. J., Laporte, J. D., and Fredberg, J. J. (2000). Mechanical properties of cultured human airway smooth muscle cells from 0.05 to 0.4 Hz. *J. Appl. Physiol.* **89,** 1619–1632.

Marion, S., Guillen, N., Bacri, J.-C., and Wilhelm, C. (2005). Acto-myosin cytoskeleton dependent viscosity and shear-thinning behavior of the amoeba cytoplasm. *Eur. Biophys. J.* **34,** 262–272.

Matthews, B. D., Overby, D. R., Alenghat, F. J., Karavitis, J., Numaguchi, Y., Allen, P. G., and Ingber, D. E. (2004). Mechanical properties of individual focal adhesions probed with a magnetic microneedle. *Biochem. Biophys. Res. Commun.* **313,** 758–764.

Matthews, B. D., Overby, D. R., Mannix, R., and Ingber, D. E. (2006). Cellular adaptation to mechanical stress: Role of integrins, Rho, cytoskeletal tension and mechanosensitive ion channels. *J. Cell Sci.* **119,** 508–518.

Meshel, A. S., Wei, Q., Adelstein, R. S., and Sheetz, M. P. (2005). Basic mechanism of three-dimensional collagen fibre transport by fibroblasts. *Nat. Cell Biol.* **7,** 157–164.

Murfee, W. L., Hammett, L. A., Evans, C., Xie, L., Squire, M., Rubin, C., Judex, S., and Skalak, T. C. (2005). High-frequency, low-magnitude vibrations suppress the number of blood vessels per muscle fiber in mouse soleus muscle. *J. Appl. Physiol.* **98,** 2376–2380.

Reich, D. H., Tanase, M., Hultgren, A., Bauer, L. A., Chen, C. S., and Meyer, G. J. (2003). Biological applications of multifunctional magnetic nanowires (invited). *J. Appl. Phys.* **93,** 7275–7280.

Sheetz, M. P. (1998). "Laser Tweezers in Cell Biology," Vol. 55. Academic Press, San Diego, California.

Strick, T. R., Allemand, J.-F., Bensimon, D., Bensimon, A., and Croquette, V. (1996). The elasticity of a single supercoiled DNA molecule. *Science* **271,** 1835–1837.

Tamada, M., Sheetz, M. P., and Sawada, Y. (2004). Activation of a signaling cascade by cytoskeleton stretch. *Dev. Cell* **7,** 709–718.

Tanase, M., Bauer, L. A., Hultgren, A., Silevitch, D. M., Sun, L., Reich, D. H., Searson, P. C., and Meyer, G. J. (2001). Magnetic alignment of fluorescent nanowires. *Nano Lett.* **1,** 155–158.

Tanase, M., Felton, E. J., Gray, D. S., Hultgren, A., Chen, C. S., and Reich, D. H. (2005). Assembly of multicellular constructs and microarrays of cells using magnetic nanowires. *Lab Chip* **5,** 598–605.

Thomas, W. E., Nilsson, L. M., Forero, M., Sokurenko, E. V., and Vogel, V. (2004). Shear-dependent 'stick-and-roll' adhesion of type 1 fimbriated Escherichia coli. *Mol. Microbiol.* **53,** 1545–1557.

Thoumine, O., and Ott, A. (1997). Time scale dependent viscoelastic and contractile regimes in fibroblasts probed by microplate manipulation. *J. Cell Sci.* **110,** 2109–2116.

Valberg, P. A., and Albertini, D. F. (1985). Cytoplasmic motions, rheology, and structure probed by a novel magnetic particle method. *J. Cell Biol.* **101,** 130–140.

Vogel, V., and Sheetz, M. (2006). Local force and geometry sensing regulate cell functions. *Nat. Rev. Mol. Cell Biol.* **7,** 265–275.

Wernsdorfer, W., Doudin, B., Mailly, D., Hasselbach, K., Benoit, A., Meier, J., Ansermet, J. P., and Barbara, B. (1996). Nucleation of magnetization reversal in individual nanosized nickel wires. *Phys. Rev. Lett.* **77,** 1873.

Whitney, T. M., Jiang, J. S., Searson, P. C., and Chien, C. L. (1993). Fabrication and magnetic properties of arrays of metallic nanowires. *Science* **261,** 1316–1319.

Wilhelm, C., Cebers, A., Bacri, J. C., and Gazeau, F. (2003). Deformation of intracellular endosomes under a magnetic field. *Eur. Biophys. J.* **32,** 655–660.

Ziemann, F., Radler, J., and Sackmann, E. (1994). Local measurements of viscoelastic moduli of entangled actin networks using an oscillating magnetic bead micro-rheometer. *Biophys. J.* **66,** 2210–2216.

CHAPTER 21

Optical Neuronal Guidance

Allen Ehrlicher, Timo Betz, Björn Stuhrmann, Michael Gögler, Daniel Koch, Kristian Franze, Yunbi Lu, and Josef Käs

Lehrstuhl für die Physik Weicher Materie
Fakultät für Physik und Geowissenschaften
Universität Leipzig, Linnéstr. 5, Leipzig D-04103, Germany

Abstract

We present a novel technique to noninvasively control the growth and turning behavior of an extending neurite. A highly focused infrared laser, positioned at the leading edge of a neurite, has been found to induce extension/turning toward the beam's center. This technique has been used successfully to guide NG108–15 and PC12 cell lines [Ehrlicher, A., Betz, T., Stuhrmann, B., Koch, D. Milner, V. Raizen, M. G., and Kas, J. (2002). Guiding neuronal growth with light. *Proc. Natl. Acad. Sci. USA* 99, 16024–16028], as well as primary rat and mouse cortical neurons [Stuhrmann, B., Goegler, M., Betz, T., Ehrlicher, A., Koch, D., and Kas, J. (2005). Automated tracking and laser micromanipulation of cells. *Rev. Sci. Instr.* 76, 035105]. Optical guidance may eventually be used alone or with other methods for controlling neurite extension in both research and clinical applications.

I. Introduction

A. Neuron Structure

In the central nervous system, there are two principal kinds of neuronal cells: glial cells and neurons. Glial cells vastly outnumber neurons in vertebrates, and they are much softer than neurons, suggesting that they may function as padding to protect neurons from mechanical trauma (Lu *et al.*, 2006); however, glial cells perform other unique functions. For example, Müller cells, a special kind of glial cells in vertebrate retinas, have been found to act as living optical fibers, guiding light from the vitreous body to the photoreceptor cells (Franze *et al.*, 2007).

Unlike glial cells, neurons are the information processing/transmitting neuronal cells, and are generally composed of two main regions: neurites and the soma. Neurites are further subdivided into multiple dendrites and a single axon. The dendrites (Greek: dendron = tree) are highly branched neurites, or extensions, which receive input stimuli and direct the excitation to the soma. The soma contains the cell nucleus and other organelles, and is the most rigid part of the cell (Lu *et al.*, 2006). Typically, the axon is the longest neurite, ranging from tens of micrometers to meters in length, depending on the type of neuron and the species of animals. Axons create the "forward-active" efferent structure of the network, meaning that electrochemical activities integrated from other neurons and sources are transmitted from dendrites, through the soma, and outward along the axon to the next cell in the path through a synapse.

B. Neuronal Cells in Development

During the development of the nervous system, neurons are rapidly inter-connected via migrating neurites, which must be guided through a chemically noisy and physically crowded microenvironment. The highly dynamic structure at the tip of neurites that interprets guidance cues into a connection path is known as the growth cone (Fig. 1). As the growth cone is the focus of our study, its

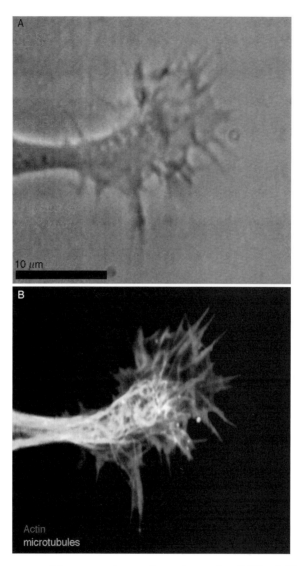

Fig. 1 Phase-contrast and fluorescence image of growth cone in a PC12 cell. The growth cone is imaged in phase-contrast (top panel) and fluorescence microscopy (bottom panel). Actin filaments are stained with fluorescent rhodamine–phalloidin and shown in red, microtubules are stained with indirect immunofluorescence and shown in green. One can see that while most microtubules are concentrated near the central stump, a few individuals are able to penetrate far out into the lamellipodium. The actin-rich projections emanating from the outer lamellipodial edge are filopodia. Additionally, one can see a strong correlation between the darker areas of the phase-contrast image and actin distribution in the fluorescence image. Scale bar is 10 μm. (See Plate 29 in the color insert section.)

properties are described in some detail. Actin filaments (red in Fig. 1) fill the peripheral region of the growth cone, known as the lamellipodium. The spike-like protrusions of actin bundles that protrude outward beyond the lamellipodium edge are known as filopodia (Fig. 1). The central region of the growth cone and the axon are filled with microtubules (green in Fig. 1), which extend into the growth cone and are largely restricted by the actin-rich lamellipodium (Dent and Kalil, 2001; Forscher and Smith, 1988; Kabir *et al.*, 2001; Schaefer *et al.*, 2002). This combination of actin filaments and microtubules, as well as other biopolymers known as intermediate filaments, molecular motors such as myosins and kinesins, and a host of accessory proteins, builds a dynamic cellular cytoskeleton. These cytoskeletal structures give neurons their structural integrity and their ability to generate forces for movement and morphological changes.

Growth cones respond to guidance cues in two- and three-dimensional environments. *In vitro*, it has been shown that nerve growth factor (NGF) is a potent neuronal chemoattractant, even at very low concentrations in three-dimensional gels (Rosoff *et al.*, 2004). Considering a 10-μm-wide growth cone exposed to a 0.1% gradient of 1 nM leads to about a thousand molecules of NGF in the vicinity of the growth cone. The growth cone shows an amazing ability to detect such small number of molecules and to differentiate directions based on even one single molecule (Rosoff *et al.*, 2004), making it a truly impressive natural detector. A statistical analysis of the movement of the growth cone's leading edge has recently revealed that stochastic fluctuations between extension and retraction may explain its ability to follow minute chemical gradients in an extremely chemically noisy environment (Betz *et al.*, 2006).

C. Growth Cone Movement and Guidance

The growth cone is characterized by highly dynamic movements of actin filaments, which continuously incorporate new actin monomer subunits at the leading edge of the lamellipodium. In one scenario, it has been proposed that as the membrane and filaments elastically fluctuate due to thermal energy, actin monomers can slip in between the tips of filaments and the membrane when the fluctuation creates a sufficient space for a 7-nm subunit. The polymerization of monomer subunits in turn exerts a force on the membrane, resulting in forward membrane protrusion in a process generally known as a "thermal ratchet" (Mogilner and Oster, 2003; Theriot, 2000).

Polymerizing actin filaments in the lamellipodium flow away from the leading edge at a fairly conserved rate in an actin and myosin motor-dependent process, which is known as centripetal actin flux or simply actin flow (Jay, 2000; Medeiros *et al.*, 2006). The term "retrograde flow" describes the variable movement of actin filaments rearward with respect to the substrate, as its movement is in opposition to the forward movement of the cell (Jurado *et al.*, 2005; Lin *et al.*, 1995). The migration speed of the growth cone, or a nonneuronal-migrating cell that shows similar processes, is approximately equal to the retrograde flow minus the centripetal actin flux, such that when the retrograde flow is near zero, the cell approaches its maximum speed (Jurado *et al.*, 2005). Similar movements of actin have been found in many types of migrating cells such as fibroblasts, keratocytes, and single-cell organisms such as amoebas, although the molecular details differ between systems (Ponti *et al.*, 2004). In the growth cone, actin retrograde flow also fulfills the additional task of limiting microtubule extension into the peripheral region (Schaefer *et al.*, 2002). Microtubules emanate from the central region around the neurite, and individual filaments extend dynamically throughout the growth cone. It has been shown that microtubule extension is inversely proportional to actin flow since microtubules must extend radially outward against the inward flow of actin (Schaefer *et al.*, 2002; Williamson *et al.*, 1996).

A key difference between the growth cone and other motile systems is the stiff, long axon, or more generally neurite carried in tow behind the growth cone. The neurite is filled with bundles of microtubules, which derive their extreme rigidity from a tubular pipe-like assembled structure of tubulin. Intermediate filaments also contribute to the width of the neurite. Since the growth cone is not free to simply crawl, but must be followed by an neurite, the rate of growth cone extension is limited by the rate of neurite assembly and the rate of material transport to support neurite assembly (Martenson *et al.*, 1993). Manually pulling on growth cones, however, has been shown to increase the rate of neurite extension (Zheng *et al.*, 1991). Eventually, the transport efficiency can drop to a point where growth cone movement becomes significantly diminished.

Although not unique to neurons, filopodia are more prominent in growth cones than in many other cell types in two-dimensional culture. They are believed to serve as extracellular chemical and mechanical sensors, relaying information about the external environment to the cell (Challacombe *et al.*, 1996; Chien *et al.*, 1993). Filopodia are not simply passive sensors. They can generate active pulling forces with a magnitude of ~1 pN per filopodium (Heidemann *et al.*, 1990). Furthermore, microtubules have been observed to target the "focal complexes" (sites of cell adhesion) at filopodia ends precisely (Bershadsky *et al.*, 2006; Gordon-Weeks, 1991; Kaverina *et al.*, 2002), and filopodia movement appears to guide microtubule extension (Schaefer *et al.*, 2002). Since local pharmacological stabilization or depletion of microtubules in the growth cone leads, respectively, to neurite turning toward or away from the drug (Buck and Zheng, 2002), microtubules and filopodia are likely to form a key partnership in axonal turning. However, growth cone guidance may involve additional mechanisms. For example, reducing centripetal actin flux (via myosin Ic inactivation) caused a more significant turn response in growth cones than disabling filopodia (via myosin V inactivation), suggesting that traction forces exerted by filopodia alone are probably not the dominant element in growth cone turning (Diefenbach *et al.*, 2002); however, it is noteworthy that in this study the growth cone somata were detached from the substrate (Wang *et al.*, 2003). Additionally, myosin II inactivation has been shown to decrease retrograde flow rates (Medeiros *et al.*, 2006), but conflictingly also to increase retrograde flow rates in other studies (Brown and Bridgman, 2003). From these results, it is obvious that the critical mechanisms for propelling and turning the growth cone have not yet been clearly identified, which poses a challenge in developing strategies to guide growth cones.

D. Existing Guidance Methods

There exist numerous techniques, natural and synthetic, to guide a neuron to a particular target. In general, complex conditions must be fulfilled in order to obtain the right neuronal response to guidance cues. Researchers have explored various approaches to control neuronal growth, primarily *in vitro* and additionally *in vivo* during nerve regeneration (Chierzi *et al.*, 2005). Even in the 1920s, experiments showed that extracellular electric DC fields could direct nerve outgrowth (Ingvar, 1920). Electric fields reversibly increase neurite growth toward the cathode and reduce extensions in the direction of the anode (Patel and Poo, 1982; Schmidt *et al.*, 1997). Furthermore, they also increase and direct neurite branching (McCaig and Rajinicek, 1991), and the electric fields induced by conducting polymer layers can enhance growth speed (Schmidt *et al.*, 1997).

The search for neuronal guidance techniques *in vitro* has resulted in a variety of methods, many of which use the general phenomenon of contact guidance and surface micropatterning strategies to regulate neuronal growth. For example, surface topography with pits and connecting grooves has been used to control the outgrowth and synapse formation of snail neurons (Merz and Fromherz, 2002). Other micropatterning strategies utilize differential adhesion, whereby a growth cone leads a neurite by sticking preferentially to substrates of stronger adhesiveness (Rudolph *et al.*, 2006; Vogt *et al.*, 2003). Adhesive properties of surfaces have been controlled by polymer patterning (Bohanon *et al.*, 1996; Takezawa *et al.*, 1990; Yamada *et al.*, 1990) or by patterning chemical properties such as hydrophobicity (Dewez *et al.*, 1998) or surface charge (Branch *et al.*, 2000). A more physiological methodology relies on the patterning of cell adhesion proteins (Blawas and Reichert, 1998), peptide sequences derived from these proteins (Herbert *et al.*, 1997; Patel *et al.*, 1998), or extracellular matrix proteins such as laminin and fibronectin. In addition, assays based on patterned surfaces have been used to identify new adhesion factors (Walter *et al.*, 1987a,b). The surface patterns in these studies are typically generated by photolithography or microcontact printing (Clark *et al.*, 1993; Fromherz *et al.*, 1991; Prinz and Fromherz, 2000; Chapter 5 by Spatz and Geiger, and Chapter 19 by Lele *et al.*, this volume).

In vivo, axons find a trail to their targets by exploiting many different guidance cues. A family of conserved chemical cues known as netrins can cause attractive or repulsive responses of growth cones (Livesey, 1999). Once the initial group of axons is in place, other axons can follow the pathfinders' trail by using several cell–cell adhesion molecules (CAMs) to which they selectively adhere (Walsh *et al.*, 1997). Guidance channels have also been used in a clinical context for the regeneration of peripheral nerves (Rivers *et al.*, 2002; Valentini, 1995).

All of the techniques above either directly or indirectly modulate the activity of the cytoskeleton, which as described earlier is the structure responsible for moving the growth cone and changing its shape. In optical guidance, we propose to bypass signaling processes and directly, physically, modify the activity of cytoskeleton to steer the growth cone's movement. The physical principles are similar to those in optical tweezers or the optical stretcher (Chapter 17 by Lincoln *et al.*, this volume), except that in optical guidance the leading edge of the growth cone is the target for the optical forces.

II. Apparatus

A laboratory equipped with optical tweezers will require only a minimal amount of construction to perform optical neuronal guidance. In essence, all that is needed for optical cell guidance is an infrared (IR) laser, a microscope with a high numerical aperture objective [as would be found in a typical optical tweezers setup (Svoboda and Block, 1994)], and a system for ensuring the viability of cells on the microscope. Of course, these three principal components can be incorporated in a myriad of different ways, depending on the researcher's focus, skills, time, and budget. Here, we will present a detailed description of our chosen setup, with some thoughts about other accessible options.

A. Laser Light Sources

An essential part of any experiment with optical forces lies with the laser and optics. In general, a near-IR wavelength of 800 nm is preferable when working with cells due to the low absorption of water at this wavelength (Peterman *et al.*, 2003; Svoboda and Block, 1994). Currently, this wavelength limits the use of high-power, cost-effective diode/fiber options; however, new products are emerging and other near-IR wavelengths may be equally effective (Mohanty *et al.*, 2005). For example, with collaborators we have successfully guided primary embryonic chicken neurons with a ytterbium fiber laser at 1064 nm (Stuhrmann *et al.*, 2006), although heating due to increased water absorption at this wavelength deserves attention, as it is approximately five times higher than using 800-nm laser of comparable power (Svoboda and Block, 1994). Other collaborators have also demonstrated that wavelengths of 780 and 1064 nm optically guide neurons equally well (Stevenson *et al.*, 2006). In addition to potential cell damage, the applied power is also limited by the objective's damage threshold, which is usually between 500 and 1000 mW, with phase-contrast objectives typically occupying the lower part of this range. We have chosen a Coherent 890 Ti:Sapphire (1.6 W at 800 nm), which is pumped by a Coherent Verdi V-10 diode-pumped ring laser (10 W at 532 nm). This Ti:Sapphire has the added convenience that its wavelength can be tuned between 690 and 1100 nm. To maintain a clean Gaussian intensity profile, we use the TEM_{00} mode of the laser. Additionally, we use the laser only in continuous wave (CW) mode, as opposed to mode-locked (pulsed) operation, which would cause significant morphological damage by essentially cooking the cell with high-intensity pulses.

B. Laser Light Control Elements

The laser must then be directed to and optically coupled into the microscope. In Fig. 2, one can see that we first expand the laser to roughly the size of the objective's back aperture (6 mm) using a collimating two-lens system (telescope). We then split the beam (not depicted in figure), to allow the use of a single-laser source in two setups, at a ratio of 60/40 to compensate for the differences in the efficiency between the two paths. In the first setup, the laser enters the acousto-optical deflectors (AODs; AA Opto-Electronic, St. Remy les Chevreuse, France; 6.7-mm aperture, 31-mrad maximum scan angle at 800 nm, 10-μsec random access time, 80% maximum transmission efficiency). An AOD is a solid-state device that is able to change the transmitted laser's deflection and intensity in response to an input radio frequency (RF) signal (AA Opto-Electronic: 91–118 MHz, 2-W max output power, 1-μsec sweep time). The device is aligned such that the Bragg configuration of its internal crystal transmits a first-order diffracted beam whose angle is related to the RF signal's frequency and whose intensity at that angle is controlled by the amplitude of the RF.

The second setup is largely identical to the first, apart from the use of rapidly moving galvo mirrors (Magnet Optical Scanner, VM-1000C, accessible angle of ±45°, 0.8-msec full step access time; GSI Lumonics, Billerica, Massachusetts), instead of AODs as the laser-steering component. Additionally, since the galvo mirrors only control the laser angle, an acoustooptical modulator (AOM; AA Opto-Electronic: AA.MT.110/A1-ir) is included to control the incident power. While the AODs offer a much faster scan rate than with galvos, they are optimized to operate at a specific wavelength ±50 nm, and are less efficient, transmitting only ~80% of the incident light. Both systems allow the implementation of multiple optical traps and arbitrary radiation patterns by quickly rastering between differ-ent points on the sample plane, achieving similar results as those done with holographic optical tweezers (HOTs) (Grier and Roichman, 2006), but due to their much faster scan rate, AODs typically produce more time-continuous radiation patterns. Galvos, however, reflect a broad spectrum of light with highly reflective mirrors that are very efficient, and have a much larger angular range. We find that the AODs suit our purposes better since we have ample incident power and have only used a single wavelength of 800 nm to date, but the preference between these steering devices will vary with researchers and applications.

After passing through the steering device, the laser beam is focused through another telescope to ensure that the beam hits the center of the objective back aperture for any angle of deflection. The laser is coupled into the objective via reflection off a narrow bandwidth dichroic mirror, with >90% transmission from 410 to 660 nm and high reflectivity at other wavelengths (675DCSX, Chroma

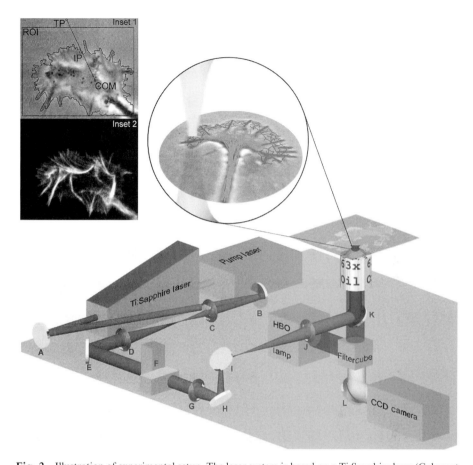

Fig. 2 Illustration of experimental setup. The laser system is based on a Ti:Sapphire laser (Coherent, 890) pumped by a 10-W solid-state laser. The laser beam is repositioned by mirrors A and B and then collimated and expanded to 6-mm diameter in a telescope (C, D). It is deflected by a mirror (E) into a pair of acoustooptical deflectors (F), which control x- and y-deflection as well as the intensity of the beam. The beam is then coupled into the microscope beam path via additional mirrors (H, I), lenses (G, J), and a dichroic mirror (K). The dichroic mirror only reflects infrared light of the laser and thus allows the combination of laser irradiation, fluorescence microscopy, and phase-contrast imaging (recorded with a CCD camera as shown). Circular insert shows a diagram of the focused laser beam positioned with respect to growth cone. Rectangular insert shows automated tracking by the LabVIEW program (top) and the corresponding fluorescence image of GFP actin (bottom). Tracking is computationally done by an edge-detection algorithm within the user-defined region of interest (ROI). The laser beam is automatically targeted just outside the leading edge (target point; TP), and the guidance action presumably takes place near the intersection point (IP), between TP and the center of mass (COM) of the growth cone. (See Plate 30 in the color insert section.)

Technology Corp., Rockingham, Vermont). This mirror is inserted in the nose-piece (Fig. 2, element K) and is aligned to reflect the laser straight through the objective via a homebuilt arrangement with micrometer screws, as shown in

Fig. 2. Since this mirror reflects 800-nm laser light but transmits the rest of the visible spectrum, phase-contrast, fluorescence, or DIC imaging can be conducted simultaneously with optical guidance.

C. Microscope Irradiation and Imaging

Our system is built on a Leica DM IRB inverted microscope; however, most modern microscopes should be adequate, provided there is a video port and some way to couple in the external laser. Due to safety concerns, particularly due to the laser's near-IR emission not being readily visible, we use a CCD camera (Cohu 4910 monochrome CCD camera, 6.4×4.8 mn^2 chip with 768×576 pixels resolution; San Diego, California) as opposed to microscope eyepieces to observe the sample, and record the images to a computer. A modest computer by modern standards (Pentium 4, 1.8 GHz, 1-GB RAM), with an analog-to-digital converter card for image capture (National Instruments IMAQ PCI/PCX-1407 Framegrabber; Austin, Texas), and an analog voltage output card for controlling the AODs (National Instruments PCI-6711 analog voltage output device), is adequate for our applications. Image recording, recognition, and laser control are all performed with LabVIEW programs written by our group (available on request). One of these programs was developed specifically for optical guidance of cells and is detailed by Stuhrmann *et al.* (2005). Our system allows the recording of images with or without showing the laser beam, by using a wide band-pass filter which blocks the laser's IR light to the camera while allowing other light to pass (FM01 Wide Band hot mirror: 90% reflectance 750–1200 nm, Thorlabs GmbH Dachau/ Munich Germany). This filter is moved in or out of the imaging light path with an actuator, which is also computer-controlled through LabVIEW. Even for images including the laser, we use a filter that blocks $\sim$99% of the light to prevent the laser's intensity from dramatically overexposing the image.

D. Cell Culture System

We principally use two cell lines, PC12 and NG108-15, which are both immortalized cells. PC12 cells were originally derived from a transplantable rat pheochromocytoma (tumor of the adrenal gland) (Greene and Tischler, 1976), and behave as not fully differentiated precursors of nerve cells, but reversibly become more "neuron-like" when treated with nerve growth factor (NGF). In the undifferentiated proliferative state, they adhere weakly to plastic dishes and tend to grow in small clusters with a doubling time of $\sim$92 h. They are cultured in a specially formulated medium:

- 85% RPMI-1640 (ATCC 30-2001, containing 2-mM L-glutamine, 10-mM HEPES, 1-mM sodium pyruvate, 4500-mg/liter glucose, 1500-mg/liter sodium bicarbonate)
- 10% horse serum (ATCC 30-2040)

- 5% fetal bovine serum (ATCC 30-2020)
- 100-U/ml penicillin and 100-μg/ml streptomycin (Sigma P0781).

This medium is refreshed every 2 days, and cells are resuspended and subcultured prior to confluency, or about every 5–7 days. To induce the neuronal phenotype, NGF (Sigma N6009) is added at 25–50 ng/ml for at least 24 h.

NG108-15 is an immortalized mouse neuroblastoma cell line, which was developed in 1971 by Hamprecht *et al.* (1985) by fusing mouse neuroblastoma cells with rat glioma cells in the presence of inactivated *Sendai virus*. NG108-15 cells grow neurites spontaneously in culture and are also known to form synapses that are at least functional on the presynaptic side. They require a different medium from PC12 cells:

- 90% DMEM (ATCC 30-2002, containing 4-mM L-glutamine, 110-mg/liter sodium pyruvate, 4500-mg/liter glucose, 1500-mg/liter sodium bicarbonate)
- 10% fetal bovine serum (ATCC 30-2020)
- 100-U/ml penicillin and 100-μg/ml streptomycin (Sigma P0781)
- 10-mM HEPES, diluted from 1-M stock solution.

Unlike PC12 cells, no further chemicals are necessary for neurites to extend. However, 1-mM cyclic AMP (Sigma D0627) is required for presynaptic activities (Chen *et al.*, 2001). In addition, both cell lines seem to show decreased proliferation and increased neurite extension with reduced serum in the medium (total serum ~2.5%).

We have also worked with primary rat and mouse neonatal cortical cells, which are taken directly from freshly euthanized animals. Primary cells are of course the most physiologically relevant cells, especially for neuronal networks. Nevertheless, since neurons with long, well-established neurites are preferred, cell lines are much easier to use on a regular basis than primary cells, which are only recommended for the most advanced stage of study. Primary mouse and rat neonatal cortical cells require a special culture medium:

- Neurobasal NB (Gibco 21103-049)
- 2% B27-supplement (Gibco 17504-044, serum-free supplement)
- 2-mM L-glutamine (Sigma G7513)
- 1% penicillin–streptomycin–neomycin (PSN) antibiotics (Gibco 15640-055).

Cells are usually maintained in an incubator at 37 °C with 5% CO_2 to balance the pH and a humidity close to 100% to prevent evaporation. Without stable cell viability, the described optical guidance experiments are simply impossible. In addition, challenges of cell culture are compounded when cells are transferred to a stage-based culture system.

There are many commercial and homebuilt systems for maintaining mammalian cells on the microscope stage. However, since an oil objective is typically used to create a tightly focused laser trap, the system must heat not only the stage but also the objective as it is in direct thermal contact (through the oil) with the cell substrate. We initially used a rather expensive, commercially available system (Bioptechs FCS2 Focht Live-Cell Chamber and objective heater; Butler, Pennsylvania), which has excellent temperature control and offers an open or closed design. However, pH control and pressure fluctuations from the peristaltic pump were problematic. We have since developed our own stage culture system, which is in our opinion generally easier and more flexible to use.

We modify standard 2-in. tissue culture dishes (Nunc Brand, Denmark), by cutting a 26-mm-diameter hole from the bottom-center, and then gluing a 40-mm-diameter, 170-μm-thick glass coverslip (VWR) from the outside of the dish with silicone adhesive (General Electric, RTV 108) to cover the hole. The substitution of the coverslip for the existing plastic bottom is critical, since the latter is thicker than the oil objective's working distance. Other consumer silicone sealants may also be used for attaching coverslips, although it is preferable to use adhesives with nontoxic solvents like acetic acid, as opposed to methanol. The dishes are ready for use 24–48 h after the silicone has been applied, and are generally stored in 80% ethanol/dH$_2$O to maintain sterility. The assembled dish with a lid can be seen in Fig. 3, left panel.

We also constructed Teflon tops for these dishes in both open (with metal ports for chemical or gas exchange as shown in Fig. 3, left panel) and closed (no ports, for sealed dishes) configurations (dish, right panel). For gas exchange, we connect the open version top to humidified prebottled 5% CO$_2$ in synthetic air (Air Liquide GmbH, Cologne, Germany) to maintain the same atmospheric control afforded by

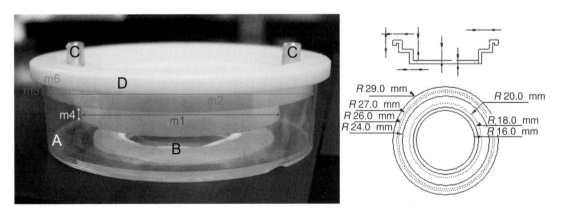

Fig. 3 Design of the experimental dish (left). A 26-mm-diameter hole is cut from the center of a standard 2-in. polystyrene Petri dish (A). The hole is then covered by gluing from the outside a glass coverslip of 40-mm diameter, 170-μm thickness (VWR Germany, Cat No. 631-0177), with silicone (General Electric, RTV 108) (B). Gas exchange is conducted through metal connectors (C), which are 3.5 mm in outer diameter, with a 5.5-mm collar to prevent it from falling into the holes (3.6 mm) where they plug into the Teflon top (D). The dimensions are: m1 = 40.0 mm, m2 = 52.7 mm, m3 = 3.0 mm, and m4 = 4.9 mm. Detailed design of the Teflon top, with the top and cutaway views and dimensions is shown in the right panel.

the incubator. The tops have the same glass coverslip surface as the bottom, providing a window in the center to allow transmitted light illumination.

Before use, the dishes are cleaned again with 80% ethanol and rinsed with sterile phosphate buffered saline (PBS) in a sterile culture hood. Typically, neurons adhere rather poorly to glass, and a surface coating of matrix proteins is often used to promote adhesion. Laminin (Sigma L2020) can be adsorbed at a concentration of 40 μg/ml in PBS, using a volume large enough to cover the glass surface ($\sim$50 μl). After 1–3 h at 37 °C (or room temperature overnight), the surface is gently rinsed twice with sterile PBS, never allowing the adsorbed laminin layer to dry out. Poly-L-lysine (Sigma P8920) may be used in place of laminin as described above by applying 0.5 ml of a 0.1 mg/ml solution to coat 25 cm^2, although it may be allowed to dry out over about 24 h. After rinsing three times with PBS, a laminin layer may be applied as described above, or the cell suspension may be directly added. Cells plated directly onto poly-L-lysine appear less mobile than those on laminin, probably because adhesion to cationic polylysine is electrostatic rather than specific to adhesion receptors. Typically, PC12 cells show substantial extended neurites after about 24–48 h of exposure to NGF at 25–50 ng/ml, whereas NG108-15 cells are often ready for experimentation in as little as 4 h after plating without exposure to NGF.

For short-term experiments of 2 h or less, the dishes are often filled with HEPES-buffered medium (including NGF in the case of PC12 cells), allowed to equilibrate under the appropriate CO_2, and then sealed with the closed top and some parafilm or vacuum grease. Thus, only temperature needs to be controlled, as other conditions do not vary appreciably over this short time period.

Temperature control can be accomplished using a dish heater to keep the bulk medium warm, and (most importantly) an objective heater for oil immersion objectives, which otherwise behave as a heat sink that lowers the temperature of the medium in the observation field. The temperature must not be driven above 37 °C, as cells have very little tolerance for overheating. Underheating is typically far less serious a problem than overheating; while cells at 30 °C may not be optimal, they are unlikely to be damaged as is often seen at temperatures above 40 °C.

The dish heater consists of an aluminum cup, which is fitted to hold the Petri dish snugly, and has the center bottom cut out similarly to the dishes, to allow the passage of light (Fig. 4, right panel). The dish is thermally insulated from the microscope in our case, by having only three small steel-bearing "feet" as the support contacts between heater and stage. Additionally, the cup is held on the stage with a Teflon mount to reduce thermal conduction. The aluminum cup is warmed by two PT100 elements (Conrad Electronics, item 171778-62, Heraeus GmbH) in parallel, one on each side of the cup, and a third PT100 element, which functions as the temperature sensor. The PT100 sensor is connected to a homebuilt electronic controller (detailed schematic available on request), which monitors the temperature relative to the set point and sends a proportionally regulated voltage to heat the dish. One could also connect the PT100 chips to a DC power supply without feedback, and calibrate an appropriate voltage based on dish temperature

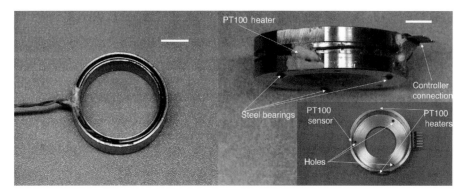

Fig. 4 Photographs of homebuilt heater systems. The left panel depicts the objective heater, which is necessary for use with oil immersion objectives. Thermocoax cable is wound three times within the machined aluminum, glued into place, and then connected via wires to an external power supply. The inner diameter is 29.0 mm and the outer is 36.5 mm, but the dimensions will obviously vary with different objectives. The right panel shows a side view of the Petri dish heater, which has an outer diameter of 60.0 mm and an inner diameter of 54.2 mm. The inner diameter should be small enough to ensure good thermal contact with the Petri dish, but not tight enough to prevent easy removal of the Petri dishes. The PT100 heater and thermometer elements are indicated as are the steel-bearing feet that help to isolate the heater thermally from the stage. The right panel inset shows a view from above of the dish heater; here one can see the PT100 elements, and holes which aid in removing the Petri dish from the heater. Scale bars are 10 mm.

measurements. The voltage output required typically ranges from 0 to 10 V to heat the dish cup to ∼40 °C (the dish cup is heated marginally above 37 °C, as some heat is lost to the environment).

The microscope objective also requires a heater due to the strong thermal contact of oil immersion objectives with the sample. As a heating element, we use thermocoaxial cable (item number 2NcNcAc15, Thermocoax SAS, Cedex, France) wound into a coil and encased in an aluminum cylinder that snugly fits the objective, as shown in the left panel of Fig. 4. Since different objectives vary slightly in diameter, we constructed separate heaters for each objective, keeping all of the objectives constantly warmed so as not to potentially damage internal lens alignments with stresses from frequent heating and cooling. We use a typical DC power supply (Conrad Electronics, Voltcraft PS 303 Pro) to power the objective heater and calibrate the temperature against the voltage with a digital thermometer. However, one could use the same feedback approach as described above for the dish cup heater.

E. Adapting Existing Systems for Optical Guidance

As mentioned earlier, an optical guidance system may be adapted from existing optical tweezers, using a range of near-IR wavelengths with or without the AOD or galvanometer beam-steering optics as described above. A minimal system may consist simply of a focused near-IR laser beam as a static optical trap. Many biological science laboratories will likely already have access to on-stage cell culture systems, confocal laser-scanning microscopes (CLSMs), or optical tweezers, and these systems can be modified to allow optical neuronal guidance experiments without needing to build an entirely new system.

Static optical traps can be used for automated optical guidance in two possible scenarios. With the help of an automated microscope stage, one can move the sample instead of the laser beam. Alternatively, the static trap may be turned into a movable trap by the use of an inexpensive mechanical actuator that controls the x–y position of the telescope lens, at a fraction of the cost compared to using precise, automated microscope stages (Svoboda and Block, 1994; Visscher et al., 1996). Beam and sample positioning represent equally feasible options. While these simple approaches may have a limited range, an accessible area of $\sim 10 \times 10 \ \mu m^2$ is already sufficient for growth cones growing at 1–$2 \ \mu m/min$ over 10 min.

As a further simplification, one can control the position of the laser spot by moving the sample or the laser spot by hand. While computer control certainly eases the burden on the researcher, it is not strictly necessary. In fact, our original experiments (Ehrlicher et al., 2002) were performed with a CLSM by manually adjusting the position of the laser spot every few seconds. The Ti:Sapphire laser, as often found in multiphoton CLSMs, was used in CW mode as opposed to the typical pulsed mode required for multiphoton microscopy, and the CCD images were simply recorded with a VCR.

III. Experiments

We have observed optical guidance in PC12 cells (Ehrlicher et al., 2002), both with the standard cell line (Greene and Tischler, 1976) and a gelsolin-overexpressing version (Furnish et al., 2001). PC12 cells grow in a very regular fashion with well-defined shapes. NG108-15 cells, which appear to be a more robust cell line that tolerates higher deviations from ideal conditions outside incubators, have also been successfully used in optical guidance experiments. Furthermore, for transfections with green fluorescent protein (GFP) constructs, the efficiency with NG108-15 cells is higher than that with PC12 cells using lipofectamin (invitrogen) or nanofectin (PAA).

A. Turns

The general procedure we have used to optically guide cells is as follows. First, one must identify an actively extending cell that displays dynamic filopodia and lamellipodia. This is critical, as optical guidance only presumes to guide the extension of active cells kept under optimal chemical and temperature control. Next, a direction for biased extension is chosen. For establishing the effectiveness of the technique, the chosen guidance direction should be significantly different from the native direction of growth in order to obtain an unambiguous guidance

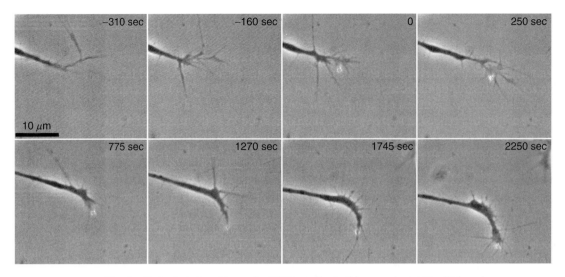

Fig. 5 Optically induced turn of a PC12 neurite. In this example, a 150-mW laser beam, seen as a small white spot, is applied to the lower edge of a small extending PC12 neurite. The neurite subsequently grows toward the laser beam and continues to extend and turn over ~37 min.

response; however, actual direction selections will of course vary with the desired application. The growth is first recorded for ~5–10 min to establish the native growth behavior, after which the laser is introduced at the edge of lamellipodium in the direction of desired biased growth. Optimal results are obtained when approximately a third to a half of the beam spot overlaps the lamellipodium and some filopodia.

Sometimes neurites will stop growing, or even retract, after the experiment has begun; however, it is rare for a neuron to continue to grow along the original direction and away from the beam. Laser powers ranging from ~20 to 200 mW are effective, with most experiments performed in the 100- to 200-mW range. However, there have been no systematic studies on the dependence of the power on guidance effectiveness. Stevenson *et al.* (2006) showed that powers of 9–25 mW using wavelengths of both 780 and 1064 were effective in optically guiding turns.

Figures 5 and 6 show typical examples of optical guidance. In Fig. 5, the laser is positioned at time zero at the right edge of the growth cone. The growth cone asymmetrically extends into the laser and proceeds to grow toward the bottom of the image. In Fig. 6, the laser is introduced on the right side of the growth cone, and again the neurite extends into the focus and grows toward the left side of the image.

B. Accelerated Growth

Both in asymmetric optical guidance and when the laser has been placed directly in front of the extending growth cone, we have often seen a rapid increase in the growth cone's translocation speed. In the left panel of Fig. 7, one can see a direct

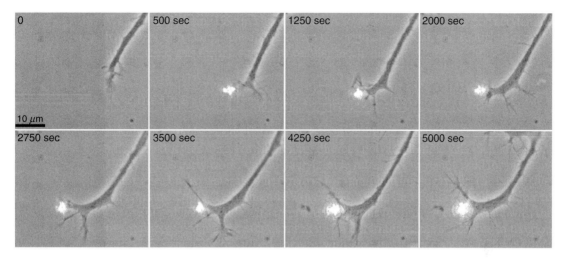

Fig. 6 Optically induced turn of a PC12 neurite. In this example, a 160-mW laser beam is applied to the left edge of a large extending PC12 neurite, causing the neurite to grow toward the laser beam. The neurite extends and turns, pulling the axonal stump behind it, over ~83 min.

comparison between two side-by-side PC12 growth cones, one of which is optically guided. The optically guided one increases in speed from 7 ± 5 μm/h to 37.5 ± 22.3 μm/h.

C. Bifurcations

We have more rarely observed other optically guided phenomena such as splitting of a growth cone or halting extension, as shown in Fig. 7. Splitting or bifurcating of a growth cone seems to occur where it is easier for the growth cone to follow the laser by forming a new branch rather than turning the entire extending neurite, as when the laser is placed further back along the edge of the stump of a neurite, attempting to induce a more radical turn (Fig. 7, center panel). This process may have interesting applications as a method to split neurites. To stop a growth cone, we have placed the laser well within the lamellipodium near the center of the growth cone and observed that local extension ceases, as seen in the right panel of Fig. 7. The extension typically resumes on the removal of the laser beam.

D. Other Observations

Optical guidance is not an indefinitely prolonged effect, and after 10–15 min the growth rate typically decreases, which may be attributable a shortage of cytoskeletal components among other possible causes. In Fig. 8, one observes a sizable extension between panels A ($t = 0$) and B ($t = 500$ sec), followed by a pause until panel E ($t = 1500$ sec), whereupon a wave of external lamellipodium activity has flowed to the growth cone allowed the growth cone to resume its advance.

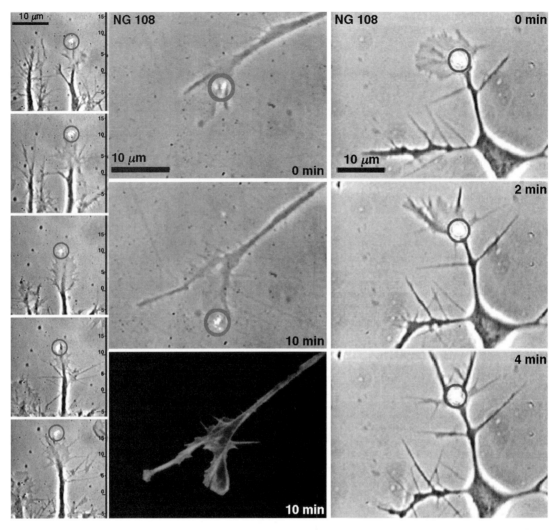

Fig. 7 Enhanced extension, bifurcation, and optically induced halting in growth cones. Position of the laser beam is indicated by red circles. In the left panel, two side by side PC12 growth cones are observed. A 20-mW laser (circled in red) is placed at the leading edge of the growth cone on the right, resulting in an increase in speed from 7 ± 3 μm/h to 37.5 ± 22.5 μm/h. The time interval between frames is 10 min, and the reference marks to the right are in micrometer. In the center panel, an optically induced bifurcation of a growth cone is shown. A growth cone, which is growing toward the lower left, sprouts off an extension to the lower right under the influence of the laser. The last picture displays the distribution of actin filaments by rhodamine–phalloidin staining. Actin filaments have clearly accumulated at the areas of lamellipodia extension. In the right panel, a laser positioned within the lamellipodium causes the growth cone to halt without causing damage, as the growth cone resumed growth after removing the laser. (See Plate 31 in the color insert section.)

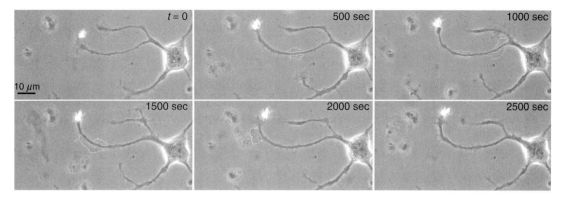

Fig. 8 Time series of an optically guided growth cone in a primary neonatal mouse cortical neuron. This time series shows neurite extension in response to the application of a 90-mW laser beam. In the first period of activity between 0 and 500 sec, the neurite increases in area by ~19.8 μm^2. The translocation pauses after about 600 sec, with little growth in the neurite until ~1840 sec, when a second surge of material was seen flowing up to the growth cone. The resulting temporary increase in translocation velocity leads to an increase in area by ~19.1 μm^2 between 1500 and 2500 sec. Material surges are outlined in white for clarity. The observations suggest that the duration of guidance is likely to be limited by available materials, without necessarily involving a more complicated mechanism of biochemical or genetic "tolerance."

Unguided neurites also tend to grow discontinuously in phases of extension, retraction, and rest, which make unambiguous guidance difficult. We have found that these phases can be stabilized with mood-altering psychoactive drugs such as valproic acid (VPA), lithium, or carbamazepine (CBZ). These drugs have been shown to inhibit the collapse of growth cones and to increase the growth cone area in sensory neurons (Williams *et al.*, 2002). With NG108-15 and PC12 cells, these drugs inhibit proliferation while promoting neurite extension. VPA stood out from the other drugs in its dose-dependent effectiveness, as shown in Fig. 9.

IV. Plausible Mechanisms of Optical Guidance

We reasoned originally that the optical trap may be able to locally bias the diffusion and concentration of actin monomers and oligomers, which may locally increase the rate of actin polymerization and enhance the polymerization-driven extension. While this concept of laser-enhanced actin concentration/polymerization has been detailed previously (Ehrlicher *et al.*, 2002) and remains a possibility, we have also proposed several other possible mechanisms.

A. Filopodial Asymmetries

As discussed in Section I, filopodia are critically important in two- and three-dimensional guidance of neurons. These filopodia control the distribution of exploratory microtubules, which are necessary and sufficient to turn a neurite.

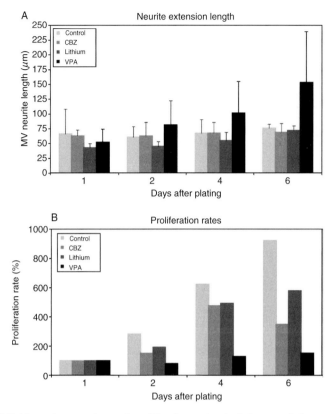

Fig. 9 NG108-15 neurite extension and proliferation under the influence of pharmacological substances. Valproic acid (VPA) at 1 mn, shown in black, causes significantly higher average neurite extension as compared to 100-μM carbamazepine (CBZ), 1-mM lithium, or control (A). VPA also inhibits proliferation of new cells (B). This inverse relationship between extension rate and cell proliferation is not surprising, since cell division and neurite elongation appear to be mutually exclusive events. All observations were done in serum-free DMEM (Sigma D6429) supplemented with 100-U/ml penicillin/100-μg/ml streptomycin and 10-mM HEPES.

The laser may be able to introduce an asymmetry in filopodia distribution around the growth cone, and additionally provide a positive force feedback by holding the filopodia in place.

In considering the potential mechanism for optical guidance, a significant observation is that the method does not appear to work for migrating nonneuronal systems such as keratocytes (unpublished data). These other cells show both centripetal actin flux and retrograde flow but lack the prominent filopodia found in growth cones, implicating filopodia as the primary structures responsible for the detection and/or response to the optical forces.

B. Retrograde Flow

In growth cones, the actin network moves radially inward with respect to the leading edge, as discussed in Section I (Schaefer *et al.*, 2002). In the presence of optical guidance, the applied forces may pull the denser actin structures within the cytosol forward and hinder the retrograde flow, generating an additional traction-like resistance for lamellipodia extension, and/or aiding microtubule extension. Furthermore, the laser might trigger additional connections between the cell and the substrate, thus reducing the rate of retrograde flow and increasing the net rate of extension as explained earlier.

C. Actin Polymerization via Membrane Tweezing

Another possibility is that the laser trap may pull on the membrane, enhancing actin polymerization and protrusion as was described in the thermal ratchet mechanism in the introduction. Pulling the membrane forward with the laser would increase the probability of actin monomers attaching to the extending filament, and thus increase the polymerization rate of actin filaments that push the membrane forward.

Figure 10 shows ~20 nm of movement of a living NG108-15's membrane with 40 mW of applied laser power. This movement indicates that the hypothesis of enhanced actin polymerization due to membrane tweezing is plausible and deserves consideration.

D. Laser-Induced Heating

One possibility is that a high-power focused laser beam can cause localized heating, which might influence reactions in the immediate area and bias growth cone extension. However, this hypothesis does not seem very probable since temperature increases should be very small. There is a strong asymptotic dependence of heating on the distance to the surface of the substrate. As an optical trap comes closer to the surface, the induced heating rapidly declines (Peterman *et al.*, 2003). For 500-nm beads suspended far away (at least 10 μm) from the surface in glycerol, this heating is estimated to be 41.1 ± 0.7 kW, but in water there is far less localized heating due to lower absorbance, as well as a greater thermal conductivity. Thus, one would expect only 7.7 ± 1.2 kW in water under the same conditions. For spherical particles near the glycerol–glass interface, about 10-kW heating is expected. Considering the lower absorption of water, ~1.9 kW is expected at a water–glass interface (Peterman *et al.*, 2003). Given the lower laser powers used in optical guidance (<200 mW and typically around 80 mW), a spherical particle would experience 0.15- to 0.40-kW heating. The heating of a growth cone is expected to be even lower since growth cones are only a few micrometers in thickness with a fairly circular extended shape ~10 μm in diameter. With such a large aspect ratio, the growth cone is completely thermally coupled to the substrate, which would decrease heating to only a fraction of that for a sphere. Therefore, we

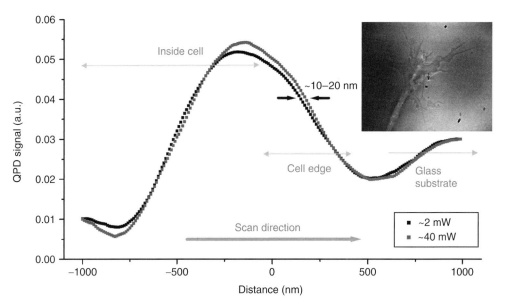

Fig. 10 Effect of membrane tweezing on a living NG108–15 neuron. The laser is scanned with 10-nm steps over a path of 2000 nm perpendicular to the cell's edge (black arrow in inset image indicates end point and the length of the arrow represents the scanned path). Black data points show the scan with an applied laser power of ∼2 mW. Gray data points show the same scan with ∼40-mW laser power. The laser exerts a pulling force on the edge of the cell, resulting in a small but clear difference of ∼20 nm between the curves. This demonstrates that the laser does indeed pull on the membrane.

believe that heating is minimal even from an intense laser beam, and is unlikely to contribute to the guidance effect. Stevenson *et al.* (2006) have also shown identical guidance results from a 780- to 1064-nm laser, suggesting no influence of heating on optical guidance, as the 1064-nm laser should heat the sample more due to higher absorption at that wavelength.

V. Summary

Optical guidance of neurites allows a degree of flexibility in guidance path not available from other methods. In its most minimal form, the apparatus only requires optical tweezers, a microscope, and some method to stabilize the cells for the duration of the experiment. There are numerous future possible applications for this technology. One wishful direction might be in the field of regenerative medicine. It is conceivable that the technique could be used with an optical fiber to guide neurons *in vivo*, possibly to repair connections severed through trauma. Another possibility would be to promote the assembly of specific complex neuronal networks, which would be ideal for preclinical pharmaceutical studies. Finally, a great deal of work has gone into neuron–semiconductor interfaces (Merz and Fromherz, 2002), and the ability to connect the interfaced neurons in a specific fashion would be a considerable advantage.

Acknowledgments

The authors acknowledge generous financial support from the Deutsche Forschungsgemeinschaft (DFG KA 1116/3-2) and Ms. Marianne Duda.

References

Bershadsky, A. D., Ballestrem, C., Carramusa, L., Zilberman, Y., Gilquin, B., Khochbin, S., Alexandrova, A. Y., Verkhovsky, A. B., Shemesh, T., and Kozlov, M. M. (2006). Assembly and mechanosensory function of focal adhesions: Experiments and models. *Eur. J. Cell Biol.* **85,** 165–173.

Betz, T., Lim, K., and Käs, J. (2006). Neuronal growth: A bistable stochastic process. *Phys. Rev. Lett.* **96,** 0981031–0981034.

Blawas, A. S., and Reichert, W. M. (1998). Protein patterning. *Biomaterials* **19,** 595–609.

Bohanon, T., Elender, G., Knoll, W., Koberle, P., Lee, J. S., Offenhausser, A., Ringsdorf, H., Sackmann, E., Simon, J., Tovar, G., and Winnik, F. M. (1996). Neural cell pattern formation on glass and oxidized silicon surfaces modified with poly(N-isopropylacrylamide). *J. Biomater. Sci. Polym. Ed.* **8,** 19–39.

Branch, D. W., Wheeler, B. C., Brewer, G. J., and Leckband, D. E. (2000). Long-term maintenance of patterns of hippocampal pyramidal cells on substrates of polyethylene glycol and microstamped polylysine. *IEEE Trans. Biomed. Eng.* **47,** 290–300.

Brown, M. E., and Bridgman, P. C. (2003). Retrograde flow rate is increased in growth cones from myosin IIB knockout mice. *J. Cell Sci.* **116**(Pt. 6), 1087–1094.

Buck, K. B., and Zheng, J. Q. (2002). Growth cone turning induced by direct local modification of microtubule dynamics. *J. Neurosci.* **22,** 9358–9367.

Challacombe, J. F., Snow, D. M., and Letourneau, P. C. (1996). Actin filament bundles are required for microtubule reorientation during growth cone turning to avoid an inhibitory guidance cue. *J. Cell Sci.* **109,** 2031–2040.

Chen, X. L., Zhong, Z. G., Yokoyama, S., Bark, C., Meister, B., Berggren, P. O., Roder, J., Higashida, H., and Jeromin, A. (2001). Overexpression of rat neuronal calcium sensor-1 in rodent NG108–15 cells enhances synapse formation and transmission. *J. Physiol.* **532,** 649–659.

Chien, C. B., Rosenthal, D. E., Harris, W. A., and Holt, C. E. (1993). Navigational errors made by growth cones without filopodia in the embryonic *Xenopus* brain. *Neuron* **11,** 237–251.

Chierzi, S., Ratto, G. M., Verma, P., and Fawcett, J. W. (2005). The ability of axons to regenerate their growth cones depends on axonal type and age, and is regulated by calcium, camp and ERK. *Eur. J. Neurosci.* **21,** 2051–2062.

Clark, P., Britland, S., and Connoly, P. (1993). Growth cone guidance and neuron morphology on micropatterned laminin surfaces. *J. Cell Sci.* **105,** 203–212.

Dent, E. W., and Kalil, K. (2001). Axon branching requires interactions between dynamic microtubules and actin filaments. *J. Neurosci.* **21,** 9757–9769.

Dewez, J. L., Lhoest, J. B., Detrait, E., Berger, V., Dupont-Gillain, C. C., Vincent, L. M., Schneider, Y. J., Bertrand, P., and Rouxhet, P. G. (1998). Adhesion of mammalian cells to polymer surfaces: From physical chemistry of surfaces to selective adhesion of defined patterns. *Biomaterials* **19,** 1441–1445.

Diefenbach, T. J., Latham, V. M., Yimlamai, D., Liu, C. A., Herman, I. M., and Jay, D. G. (2002). Myosin 1c and myosin IIB serve opposing roles in lamellipodial dynamics of the neuronal growth cone. *J. Cell Biol.* **158,** 1207–1217.

Ehrlicher, A., Betz, T., Stuhrmann, B., Koch, D., Milner, V., Raizen, M. G., and Kas, J. (2002). Guiding neuronal growth with light. *Proc. Natl. Acad. Sci. USA* **99**, 16024–16028.

Forscher, P., and Smith, S. J. (1988). Actions of cytochalasins on the organization of actin filaments and microtubules in the growth cone. *J. Cell Biol.* **107**, 1505–1516.

Franze, K., Grosche, J., Skatchkov, S. N., Schinkinger, S., Schild, D., Uckermann, O., Travis, K., Reichenbach, A., and Guck, J. (2007). Spotlight on glial cells: Living optical fibers in the vertebrate retina. *Proc. Natl. Acad. Sci. USA* (in press).

Fromherz, P., Schaden, H., and Vetter, T. (1991). Guided outgrowth of leech neurons in culture. *Neurosci. Lett.* **129**, 77–80.

Furnish, E. J., Zhou, W., Cunningham, C. C., Kas, J. A., and Schmidt, C. E. (2001). Gelsolin over-expression enhances neurite outgrowth in PC12 cells. *FEBS Lett.* **508**, 282–286.

Gordon-Weeks, P. R. (1991). Evidence for microtubule capture by filopodial actin filaments in growth cones. *Neuroreport* **2**, 573–576.

Greene, L. A., and Tischler, A. S. (1976). Establishment of a noradrenergic clonal line of rat adrenal pheochromocytoma cells which respond to nerve growth factor. *Proc. Natl. Acad. Sci. USA* **73**, 2424–2428.

Grier, D. G., and Roichman, Y. (2006). Holographic optical trapping. *Appl. Opt.* **45**, 880–887.

Hamprecht, B., Glaser, T., Reiser, G., Bayer, E., and Propst, F. (1985). Culture and characteristics of hormone responsive neuroblastoma X glioma hybrid cells. *Methods Enzymol.* **109**, 316–341.

Heidemann, S. R., Lamoureux, P., and Buxbaum, R. E. (1990). Growth cone behavior and production of traction force. *J. Cell Biol.* **111**, 1949–1957.

Herbert, C. B., McLernon, T. L., Hypolite, C. L., Adams, D. N., Pikus, L., Huang, C. C., Fields, G. B., Letourneau, P. C., Distefano, M. D., and Hu, W. S. (1997). Micropatterning gradients and controlling surface densities of photoactivatable biomolecules on self-assembled monolayers of oligo(ethylene glycol) alkanethiolates. *Chem. Biol.* **4**, 731–737.

Ingvar, S. (1920). Reactions of cells the galvanic current in tissue culture. *Proc. Soc. Exp. Biol. Med.* **17**, 198–199.

Jay, D. G. (2000). The clutch hypothesis revisited: Ascribing the roles of actin-associated proteins in filopodial protrusion in the nerve growth cone. *J. Neurobiol.* **44**, 114–125.

Jurado, C., Haserick, J. R., and Lee, J. (2005). Slipping or gripping? Fluorescent speckle microscopy in fish keratocytes reveal two different mechanisms for generating a retrograde flow of actin. *Mol. Biol. Cell* **16**, 507–518.

Kabir, N., Schaefer, A. W., Nakhost, A., Sossin, W. S., and Forscher, P. (2001). Protein kinase C activation promotes microtubule advance in neuronal growth cones by increasing average micro-tubule growth lifetimes. *J. Cell Biol.* **152**, 1033–1044.

Kaverina, I., Krylyshkina, O., and Small, J. V. (2002). Regulation of substrate adhesion dynamics during cell motility. *Intl. J. Biochem. Cell Biol.* **34**, 746–761.

Livesey, F. J. (1999). Netrins and netrins receptors. *Cell Mol. Life Sci.* **56**, 62–68.

Lu, Y. B., Franze, K., Seifert, G., Steinhauser, C., Kirchhoff, F., Wolburg, H., Guck, J., Janmey, P., Wei, E. Q., Kas, J., and Reichenbach, A. (2006). Viscoelastic properties of individual glial cells and neurons in the CNS. *Proc. Natl. Acad. Sci. USA* **103**, 17759–17764.

Martenson, C., Stone, K., Reedy, M., and Sheetz, M. (1993). Fast axonal transport is required for growth cone advance. *Nature* **366**, 66–69.

McCaig, C. D., and Rajinicek, A. M. (1991). Electrical fields, nerve growth and nerve regeneration. *Exp. Physiol.* **76**, 473–494.

Medeiros, N. A., Burnette, D. T., and Forscher, P. (2006). Myosin II functions in actin-bundle turnover in neuronal growth cones. *Nat. Cell Biol.* **8**, 215–216.

Merz, M., and Fromherz, P. (2002). Polyester microstructures for topographical control of outgrowth and synapse formation of snail neurons. *Adv. Mater.* **14**, 141–144.

Mohanty, S. K., Sharma, M., Panicker, M. M., and Gupta, P. K. (2005). Controlled induction, enhancement, and guidance of neuronal growth cones by use of line optical tweezers. *Opt. Lett.* **30**, 2596–2598.

Mogilner, A., and Oster, G. (2003). Force generation by actin polymerization II: The elastic ratchet and tethered filaments. *Biophys. J.* **84,** 1591–1605.

Patel, N., Padera, R., Sanders, G. H., Cannizzaro, S. M., Davies, M. C., Langer, R., Roberts, C. J., Tendler, S. J., Williams, P. M., and Shakesheff, K. M. (1998). Spatially controlled cell engineering on biodegradable polymer surfaces. *FASEB J.* **12,** 1447–1454.

Patel, N., and Poo, M. M. (1982). Orientation of neurite growth by extracellular fields. *J. Neurosci.* **2,** 483–496.

Peterman, E. J., Gittes, F., and Schmidt, C. F. (2003). Laser-induced heating in optical traps. *Biophys. J.* **84**(2 Pt 1), 1308–1316.

Ponti, A., Machacek, M., Gupton, S. L., Waterman-Storer, C. M., and Danuser, G. (2004). Two distinct actin networks drive the protrusion of migrating cells. *Science* **305,** 1782–1786.

Rivers, T. J., Hudson, T. W., and Schmidt, C. E. (2002). Synthesis of a novel, biodegradable electrically conducting polymer for biomedical applications. *Adv. Funct. Mater.* **12,** 33–37.

Rosoff, W. J., Urbach, J. S., Esrick, M. A., McAllister, R. G., Richards, L. J., and Goodhill, G. J. (2004). A new chemotaxis assay shows the extreme sensitivity of axons to molecular gradients. *Nat. Neurosci.* **7,** 678–682.

Rudolph, T., Zimmer, K., and Betz, T. (2006). Microstructuring of UV-transparent functionalized films on glass by excimer laser irradiation. *Mater. Sci. Eng. C* **26**(5–7), 1131–1135.

Schaefer, A. W., Kabir, N., and Forscher, P. (2002). Filopodia and actin arcs guide the assembly and transport of two populations of microtubules with unique dynamic parameters in neuronal growth cones. *J. Cell Biol.* **158,** 139–152.

Schmidt, C. E., Shastri, V. R., Vacanti, J. P., and Langer, R. (1997). Stimulation of neurite outgrowth using an electrically conducting polymers. *Proc. Natl. Acad. Sci. USA* **94,** 8948–8953.

Stevenson, D. J., Lake, T. K., Agate, B., Garces-Chavez, V., Dholakia, K., and Gunn-Moore, F. (2006). Optically guided growth at near infrared wavelengths. *Opt. Express* **14,** 9786–9793.

Stuhrmann, B., Goegler, M., Betz, T., Ehrlicher, A., Koch, D., and Kas, J. (2005). Automated tracking and laser micromanipulation of cells. *Rev. Sci. Instr.* **76,** 0351051–8.

Stuhrmann, B., Jahnke, H.-G., Schmidt, M., Jähn, K., Betz, T., Müller, K., Rothermel, A., Käs, J., and Robitzki, A. A. (2006). A versatile optical manipulation system for inspection, laser processing, and isolation of individual living cells. *Rev. Sci. Instrum.* **77,** 0631161–11.

Svoboda, K., and Block, S. (1994). Biological applications of optical forces. *Annu. Rev. Biophys. Biomol. Struct.* **23,** 247–285.

Takezawa, T., Mori, Y., and Yoshizato, K. (1990). Cell culture on a thermo-responsive polymer surface. *Biotechnology* **8,** 854–856.

Theriot, J. A. (2000). The polymerization motor. *Traffic* **1,** 19–28.

Valentini, R. F. (1995). "The Biomedical Engineering Handbook." CRC, Boca Raton, FL.

Visscher, K., Gross, S. P., and Block, S. M. (1996). Construction of multiple-beam optical traps with nanometer-resolution position sensing. *IEEE J. Select. Topics Quantum Electron.* **2,** 1066–1076.

Vogt, A. K., Lauer, L., Knoll, W., and Offenhausser, A. (2003). Micropatterned substrates for the growth of functional neuronal networks of defined geometry. *Biotechnol. Prog.* **19,** 1562–1568.

Wakatsuki, T., Schwab, B., Thompson, N. C., and Elson, E. L. (2001). Effects of cytochalasin D and latrunculin B on mechanical properties of cells. *J. Cell Sci.* **114,** 1025–1036.

Walsh, F. S., Meiri, K., and Doherty, P. (1997). Cell signaling and CAM-mediated neurite outgrowth. *Soc. Gen. Physiol. Ser.* **52,** 221–226.

Walter, J., Henke-Fahle, S., and Bonhoeffer, F. (1987). Avoidance of posterior tectal membranes by temporal retinal axons. *Development* **101,** 909–913.

Wang, F. S., Liu, C. W., Diefenbach, T. J., and Jay, D. G. (2003). Modeling the role of myosin 1c in neuronal growth cone turning. *Biophys. J.* **85,** 3319–3328.

Williams, R. S., Cheng, L., Mudge, A. W., and Harwood, A. J. (2002). A common mechanism of action for three mood-stabilizing drugs. *Nature* **417,** 292–295.

Williamson, T., Gordon-Weeks, P. R., Schachnet, M., and Taylor, J. (1996). Microtubule reorganization is obligatory for growth cone turning. *Proc. Natl. Acad. Sci. USA* **93**, 15221–15226.

Yamada, N., Okano, T., Sakai, H., Karikusa, F., and Sawasaki, Y. (1990). Thermo-responsive polymeric surfaces: Control of attachment and detachment of cultured cells. *Makromol. Chem. Rapid. Commun.* **11**, 571–576.

Zheng, J., Lamoureux, P., Santiago, V., Dennerl, T., Buxbaum, R. E., and Heidemann, S. R. (1991). Tensile regulation of axonal elongation initiation. *J. Neurosci.* **11**, 1117–1125.

CHAPTER 22

Microtissue Elasticity: Measurements by Atomic Force Microscopy and Its Influence on Cell Differentiation

Adam J. Engler, Florian Rehfeldt, Shamik Sen, and Dennis E. Discher

Biophysical Engineering and Polymers Laboratory
School of Engineering and Applied Science
University of Pennsylvania
Philadelphia, Pennsylvania 19104

Abstract

It is increasingly appreciated that the mechanical properties of the microenvironment around cells exerts a significant influence on cell behavior, but careful consideration of what is the physiologically relevant elasticity for specific cell types is required to produce results that meaningfully recapitulate *in vivo* development. Here we outline methodologies for excising and characterizing the effective microelasticity of tissues; but first we describe and validate an atomic force microscopy (AFM) method as applied to two comparatively simple hydrogel systems.

With tissues and gels sufficiently understood, the latter can be appropriately tuned to mimic the desired tissue microenvironment for a given cell type. The approach is briefly illustrated with lineage commitment of stem cells due to matrix elasticity.

I. Introduction

A principal goal of cell culture is to translate *in vivo* cellular environments to *ex vivo* systems that are accessible for deeper study. Cell cultures are maintained at 37 °C, for example, because many cell processes prove sensitive to small deviations from body temperature, but other physical variables are also likely to be important. In particular, most cell cultures are conducted on rigid glass coverslips and polystyrene dishes, with only recent attention to the fact that cells are isolated from soft tissues that are not nearly as mechanically rigid as plastic and glass. More compliant microenvironments could in principle yield more faithful biological responses.

In most soft tissues—brain, muscle, skin, and so on—a combination of adherent cells and extracellular matrix (ECM) establishes a relatively elastic environment. At the macroscopic scale, elasticity is evident from a solid tissue's ability to recover its shape within seconds after acute deformation, for example poking and pinching, or even after sustained compression such as sitting. Regardless of geometry, the intrinsic resistance of a solid to a stress is measured by the solid's elastic modulus E, which is most simply obtained by applying a force, such as hanging a weight, to a section of tissue or other material and then measuring the relative change in length or strain (Fig. 1) (Chapter 1 by Janmey *et al.*, this volume).

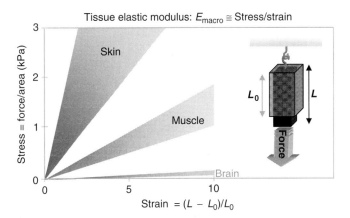

Fig. 1 Tissue elasticity. Typical stress versus strain plot for several soft tissue samples shows the natural variability within the material by the spread of each stress–strain relationship. Such a plot is generated from a tissue extended by a force (per cross-sectional area at right). The advantage of micromechanical measurements over whole tissue techniques illustrated in this chapter is that micromechanical measurements reflect not only tissue differences but also the variation within a given tissue (as indicated at left). Regardless, measurements are typically made on timescales of seconds to minutes and are in SI units of pascal (Pa).

Many tissues and biomaterials exhibit a relatively linear stress versus strain relationship up to small strains of about 10–20%. The slope E of this stress versus strain plot is the elastic modulus and is relatively constant especially in the range of small strains exerted by cells (Saez et al., 2005; Wang et al., 2002), although stiffening (increased E) at higher strains is the norm (Fung, 1994; Storm et al., 2005). On the basis of such measurements or related methods, comparisons of three diverse tissues that contain a number of different cell types and extracellular matrices show that brain tissue with all its neurons and other cell types ($E \sim 0.5$ kPa; Gefen and Margulies, 2004) is clearly softer than unstimulated, striated muscle ($E \sim 10$–20 kPa transverse to the muscle axis; Engler et al., 2004b), and muscle appears to be softer than the combination of epithelial cells, fibroblasts, and other cell types that make up skin ($E \sim 50$ kPa; Diridollou et al., 2000). For comparison, hard plastics and glass have $E \sim 1$ MPa-GPa. Although the mapping of microscale tissue elasticity at a resolution typical for histology is a long-range goal of the methods described in this chapter, Fig. 1 already suggests distinct mechanical *micro*environments for neurons in brain, myotubes in fiber bundles, and epithelial cells and fibroblasts in skin. Translating tissue Es to the types of elastic gel systems useful for cell culture is the focus here, that is $E_{\text{tissue}} \rightarrow E_{\text{gel}}$. Numerous gel systems—particularly polyacrylamide (PA) gels—have tunable elasticity (E_{gel}) that can be adjusted over several orders of magnitude from extremely soft to stiff, mimicking a wide range of tissues by controlling the extent of polymer cross-linking (Chapter 2 by Kandow et al. and Chapter 23 by Johnson et al., this volume).

The underlying mechanisms for cell sensitivity to matrix elasticity are still being established, but several basic principles are clear. First, cells from solid tissues can adhere to a wide range of elastic matrices when appropriate ligands of sufficient density are provided (Engler et al., 2004a; Rajagopalan et al., 2004). Additionally, cells also generally possess an actin-myosin contractile apparatus, which manifests as *stress* fibers in culture as has been known for decades both from light and electron microscopy (Buckley and Porter, 1967) and from fluorescence staining of actin-binding myosin subfragments (Goldman et al., 1979; Schloss et al., 1977). Roles for the actin–myosin complex in cell structure and motility of nonmuscle cells have likewise been appreciated for many years (Pollard et al., 1976), and important ideas of cytoskeletal tension have increasingly focused on filament connections to adhesions and mechanotransduction (Alenghat and Ingber, 2002; Bershadsky et al., 2006). However, feedback of matricellular elasticity on cytoskeletal structure, stress, and remodeling appears to be a newer principle (Pelham and Wang, 1997): the actin–myosin cytoskeleton promotes a physical linkage from the cytoplasm through the membrane to the microenvironment. This linkage is in turn modulated by its contractile pull against the elastic matrix in the microenvironment (Fig. 2A and B), and, with its contractile and mechanotransductive components in series with one another, can be loosely approximated as a linear series of springs as each protein contains an internal compliance or "spring constant."

Myosin generates an intracellular contractile tension, and this force is balanced in steady state through compliant linker and adhesion proteins by strain that

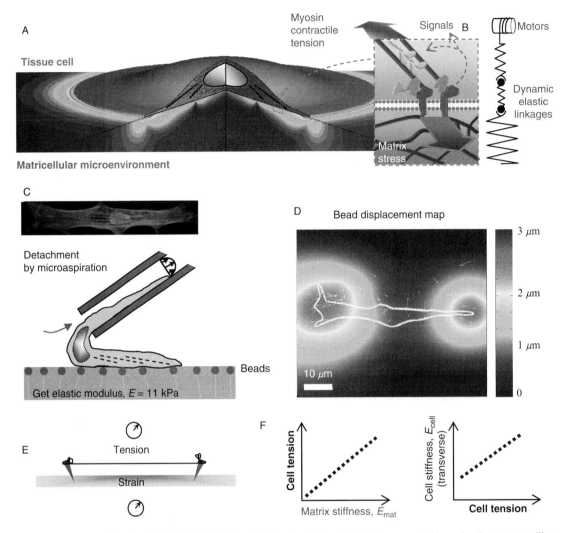

Fig. 2 Matricellular elasticity and cell tension. (A and B) Tension generated by molecular motors pulling on linker and adhesion proteins is mechanically balanced by strains induced in the extracellular environment (illustrated by the color gradient). This balance of elastic linkages is coupled to signaling pathways that can change their signal strength for anchorage and other processes based on a variety of intra- and extracellular cues. (C and D) To quantify the matrix strain under a cell (white outline in D), fluorescent tracer particles embedded in the matrix are imaged while a cell is attached and (during as well as) after the cell is detached—in this case by single-cell aspiration into a micropipette. The differences in the two images yield a displacement map that can be mathematically related to the matrix strain and to the traction forces exerted by the cell. (E and F) Schematically, the strain in the matrix increases with the tension in the cell. It is also found that stiffer matrices induce higher tensions in cells, which suggests cells sense the strain in the matrix or connections to it. The string picture of tension-set elasticity seems suited to stress fibers, and would be expected to yield stiffer cells (when probed transverse to tensed direction) when the cells are more tensed, as found experimentally. (See Plate 32 in the color insert section.)

builds up within the matrix. The balance is dynamic in that myosin motors and other proteins from inside to outside all interact noncovalently and transiently, so that minor force perturbations tend to change protein–protein associations (Chapter 16 by Evans and Kinoshita, this volume) and perhaps also protein conformations, thereby transducing signals for cell shape changes, motility, and so on. Externally imposed forces are already known to influence adhesion growth, protein association, and cell dynamics (Bershadsky *et al.*, 2006) with induction of signals (Riveline *et al.*, 2001). Cell-generated matrix strains have been visualized and quantified by various soft substrate techniques that reveal cell tractions (Dembo and Wang, 1999; Harris *et al.*, 1980; Oliver *et al.*, 1995). Such strains are illustrated here by cell-induced bead displacements that can be quantified after a cell is detached from the elastic gel substrate (Fig. 2C and D). Displacement maps, along with the substrate's elastic modulus (E_{gel}), provide the basic inputs needed to compute the traction forces in a compliant matrix (Lo *et al.*, 2000; Chapter 12 by Lee, this volume). Accurate measurements of E_{gel} are thus required (Chapter 3 by Frey *et al.*, this volume).

Force-dependent dynamics of cytoskeletal–adhesive interactions establish a basis for signaling processes that couple adhesion remodeling to anchorage-dependent viability and other signals, especially signals for differentiation as illustrated at the end of this chapter. Schematically, cell tension induces matrix strain (Fig. 2E), but the cell appears to seek a suitable balance and exerts more tension on a stiffer matrix (Engler *et al.*, 2006; Fig. 2F). Furthermore, as one would expect of a tensed filament system, cell stiffness increases in rough proportion to cell tension (Engler *et al.*, 2006; Wang *et al.*, 2002). Mechanical responses to the matricellular microenvironment include processes on short timescales of hours or less, for example cell spreading and motility, as well as processes on longer time-scales, for example proliferation, cytoskeletal rearrangement, and differentiation. In particular, cell differentiation appears strongly affected and maximally promoted on suitably compliant substrates: neurons branch more on brain-mimetic soft substrates (Flanagan *et al.*, 2002), muscle cells striate to a maximum extent on muscle-mimetic stiff substrates (Engler *et al.*, 2004a), and osteoblasts upregulate their calcification program on suitably compliant gels (Kong *et al.*, 2006). Mesenchymal stem cells (MSCs) differentiate to each of these three cell lineages—nascent neurons, myocytes, and osteoblasts—when cultured in identical media on substrates with tissue-mimetic stiffness (Engler *et al.*, 2006). In addition, breast epithelium tumorigenesis appears more pronounced on stiff substrates that mimic fibrotic tumors (Paszek *et al.*, 2005), and the fibroblast to myofibroblast transition is promoted on gels that mimic fibrotic wound stiffness (Goffin *et al.*, 2006). The latter paper takes an overall approach similar to our work cited above on muscle cells and MSCs with micromechanical characterization of *both* the tissue microenvironments and the culture substrates. The goal of this chapter is to summarize these key methods.

Traditional bulk measurements of macroscopic tissue mechanics have limited sensitivity, and so more appropriate measurement tools should be used to mechanically probe microenvironments of potentially heterogeneous tissues. Amplifying the importance of good microscale measurements are *in vitro* findings that even small microenvironmental variations—as small as 20–25% change in matrix stiffness—can have profound effects on the differentiation processes cited above as well as other cell functions (Lo *et al.*, 2000; Pelham and Wang, 1997; Peyton and Putnam, 2005; Yeung *et al.*, 2005). Here, we describe methods for AFM (atomic force microscopy)-based probing of gels and soft tissues as a type of "histoelastography." The methods are ultimately intended to guide the development of better cell culture methods.

II. AFM in Microelasticity Measurements

A number of variations of the basic AFM design are available from different manufacturers, with prices starting at about half the cost of a high-quality inverted optical microscope. A main requirement that can be readily evaluated in demonstrations by manufacturers is stable and accurate force measurements on soft samples, and we recommend using a systematic range of PA gel samples as described in some detail below. The AFM has long been appreciated as well suited to detailed mechanical assessments at microscale resolution (Binnig *et al.*, 1986), especially in the "force mode" (Weisenhorn *et al.*, 1989) as opposed to the more conventional imaging modes (Chapter 15 by Radmacher, this volume). In force mode, heterogeneous samples are easier to work with if the AFM is mounted on a good optical microscope; many commercial AFMs come integrated with modest magnification lens systems that can prove adequate. Application to soft biological materials in the 1990s (Firtel, 1995; Mahaffy *et al.*, 2000; Radmacher, 1997; Radmacher *et al.*, 1995; Rotsch *et al.*, 1999) provided some of the first submicron scale measurements of the elasticity of cultured cells and reconstituted matrix. However, systematic studies of gel elasticity as well as applications of AFM to freshly isolated tissue samples as summarized in this chapter are.

A. AFM Probing and Analysis

Briefly, the AFM method entails indenting a flexible cantilevered microprobe into a sample while measuring the cantilever tip deflection (Fig. 3A; Chapter 15 by Radmacher, this volume). The base of the cantilever is controllably displaced by a nanometer precision piezoelectric device, and the deflection of the end of the flexible cantilever is measured with similar precision by reflection of a laser beam off of the metallized backside of the cantilever onto a position-sensitive photodetector. The disposable cantilevers are commercially available in a variety of types, with most of them having a micrometer-scale probe at the tip that is pyramidal in shape but this can also be spherical (Bioforce Nanoscience, Ames, Iowa).

AFM light sources also vary in wavelength and intensity and require cantilevers made of materials or coatings that maximize reflected light. Cantilever flexibility proves equivalent to the linear flexibility of a spring, and a main advantage for quantification by AFM is that the spring constant k is precalibrated (to within about 20% accuracy for most manufacturers, we find). For the biological samples discussed here, cantilevers are typically chosen in the range of $k \sim 10\text{–}200$ pN/nm, which means, of course, that forces of only 100–2000 pN will deflect the cantilever by a very detectable 10 nm. For perspective, the low end of this force scale is equivalent to that generated by only about 20 myosin motors at peak force. A more precise calibration of each cantilever's spring constant at the time of use can be done by standard methods generally supplied by the instrument manufacturers. When appropriate, use of multiple tip geometries is preferable as they impart vastly different strain profiles and can lend some insight into the appropriateness of the underlying assumptions. Indeed, estimates of the theoretical displacement field for a sphere-tipped probe show that the maximum average radial strain, u_{rr}, is about 28% at the surface of the sample under the tip when indenting to a depth $\sim R$. This u_{rr} is not excessively large. A sharper tip will generally exert larger strains than a sphere for a given indenting force, but it can also improve the resolution and accuracy of the measurement unless sample porosity exceeds the radius of the pyramidal tip. Sphere-tipped probes distribute the strain over a larger area at the expense of lateral resolution provided by sharper tips.

One principal benefit of AFM over macromechanical testing or even high-resolution imaging techniques, for example electron microscopy, is that AFM is not hampered by gripping, mounting, and fixation problems associated with these other approaches. AFM also allows wet samples, including most cells with their buffered solutions or serum-containing medium, freshly isolated tissues, hydrated gel substrates, and ECM, to be probed, and if desired, even topographically mapped (with account for sample deformation). The disposable cantilever and relatively simple optics of AFM allow living cells and fresh tissues to be maintained in a buffered solution or serum-containing medium during measurement. New problems can arise, however, such as serum deposition on the cantilever surfaces, which can affect reflectivity as well as the spring constant k, sometimes increasing the latter sufficiently to make it important to periodically verify k during the course of a several hour experiment. Nonetheless, the AFM permits real-time assessments of perturbations to biological samples, for example effects of added drugs (Rotsch and Radmacher, 2000), as well as changes in cell mechanical properties or micro-environment over time in culture (Collinsworth et al., 2002; Engler et al., 2004c; Radmacher, 1997). From such measurements, important insight can be obtained regarding physical contributions of the microenvironment or "niche" to cells.

Indentation of a sample (Fig. 3A) from the point of contact (z_0) to a depth δ is determined by subtracting the deflection, d, of the cantilever from the distance driven into the sample by the piezoelectric device, z. The force, f, is calculated simply by Hooke's law as $f = kd$. Force–indentation data (Fig. 3A inset) can

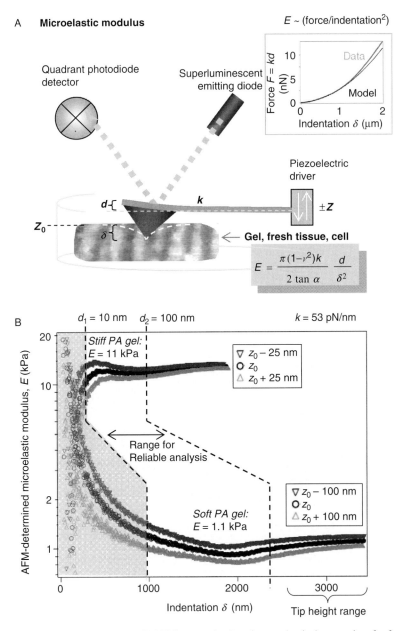

Fig. 3 Atomic force microscopy. (A) AFM can probe the micromechanical properties of soft materials by pressing a tip on a cantilever into the surface and analyzing the resulting deflection, schematically depicted with striated skeletal muscle here. Light from a superintensified diode is reflected off the end of the cantilever onto a segmented photodiode, thus magnifying small tip deflections into a detectable signal. When pressing into the material, tip indentation δ (the difference between tip deflection d and the cantilever position z) can be plotted versus the applied force, yielding a force–indentation plot that can

then be fit with a suitable theoretical model, which depends on tip geometry, to determine the local elastic modulus E (Chapter 3 by Frey *et al.*, this volume). For thin homogeneous samples on sufficiently rigid supports, the sample thickness, h, can also be determined, but for thick samples (relative to the probe tip), use is generally made of classic expressions such as that of Hertz appropriate to a spherical tip (Hertz, 1882) or that of Sneddon for a conical tip (Sneddon, 1965) that approximates a pyramidal tip.

In determining E by AFM, careful attention should be paid to numerous parameters, including indentation velocity (typically 0.1–10 μm/sec), data sampling rate, determination of contact point z_0 (which typically affects E by up to 10%), and also indentation depth which is typically 1–2 μm for thick samples and similar to displacements that cells exert in gels (Fig. 2D). It is also important to note that fitting indentations less than 600–800 nm will add significant uncertainty to the resultant measurements as illustrated in Fig. 3B (Richert *et al.*, 2004). Moreover, since many biological features are thin, for example lamellapodia, secreted matrix, and so on, there is a more appropriate equation for thin samples as expressed in terms of the radius, r, of the spherical probe tip, the Poisson ratio v ($\approx$0.5 for hydrated samples), and the relative film thickness $\chi = h/r$ (Dimitriadis *et al.*, 2002):

$$f = \left[\frac{4Er^{1/2}\delta^{3/2}}{3(1-v^2)}\right]\left\{4\left[1 - \left(\frac{2\alpha_0}{\pi}\right)\chi + \left(\frac{4\alpha_0^2}{\pi^2}\right)\chi^2 + \left(\frac{8}{\pi^3}\right)\left(\alpha_0^3 - \frac{4\pi^2}{15}\right)\chi^3 \right.\right.$$
$$\left.\left. + \left(\frac{16\alpha_0}{\pi^4}\right)\left(\alpha_0^3 - \frac{\beta_0 3\pi^2}{5}\right)\chi^4\right]\right\} \tag{1}$$

where $\alpha_0 = [1.2876 - 1.4789\,v + 1.3442\,v^2]/(1 - v)$ and $\beta_0 = [0.6387 - 1.0277\,v + 1.5164\,v^2]/(1 - v)$. If one ignores the "correction term" in the second brackets in Eq. (1), the classic Hertz model is recovered. Engler *et al.* (2004c) verified the validity of the equation above by casting hydrogels of calibrated height, h, from 70-μm down to several micrometers, as described below.

---

be fit with an appropriate indentation model to obtain the microscopic Young's modulus E (inset). (B) The AFM determined microelastic modulus E of two PA gels (10%/0.1% and 3%/0.1%) as determined pointwise for each indentation δ. For each data point (deflection d, indentation δ), we first used the optimal contact point z_0 as obtained from the optimal model fit in the deflection range 10–100 nm (dashed lines). Subsequent variation of the determined contact point z_0 (circles) by +25 nm (+100 nm for the softer gel) (upward triangles) and −25 nm (−100 nm) (downward triangles) showed only small changes in E that converge at higher indentations. The tip height (of 2.5–3.5 μm) is a maximum for indentation.

B. General Issues in Sample Preparation

Careful preparation of any soft sample is of great importance for reproducible and meaningful AFM experiments. Aging time of artificial matrices, history of probing, and handling of samples, including temperature, buffers, and so on, can all have significant effects. Artificial hydrogel samples are probably the most robust and relatively easy to prepare and measure, but proper attention is required in using fresh reagents and following specific protocols. It is preferred that any sample be attached to an underlying rigid support as shown below for both PA and hyaluronic acid (HA) hydrogels which are covalently bound to glass. Experiments with adherent cells are also relatively straightforward to perform with similar considerations. Samples that are not immobilized on a support, such as floating collagen gels, tissue specimens, or suspended cells, require greater care (or should be avoided entirely), since drift combined with buoyancy of the sample can introduce artifacts that make meaningful indentation difficult if at all possible.

Cells generally have a negative surface charge and will attach nonspecifically to a positively charged rigid support, for example polylysine-coated glass, with a significant tension induced in a cell membrane that spreads on such a surface (Hategan *et al.*, 2003; Sen *et al.*, 2005). Another strategy for cell immobilization involves entrapping cells in microfabricated wells (Rosenbluth *et al.*, 2006), but such techniques are often not sufficient for tissue specimens or larger ECM preparations, which must be attached more strongly to a rigid support to maintain the resting length so as not to induce tensions that would affect the measured elasticity. Attachment of the basal surface of a tissue or ECM can be done with adhesive strips or bonding. It is important to have tissue or ECM samples of a workable size (millimeters to centimeters for most instruments) so that they are small enough to fit on a given AFM stage but large enough to manipulate and attach firmly to the underlying glass support. Additionally, AFM samples should be relatively flat—on the scale of the micron-size probe tip—so that the tip can accurately raster across or indent the surface without complications from large changes in curvature. Furthermore, since tissue specimens are rarely flat and the surface of interest is sometimes inaccessible—the endothelial lining of blood vessels, and so on—sectioned specimens should be dissected further, if possible, to produce a flattened sheet of tissue for probing.

Regardless of attachment, for all types of samples it is important to have a sense of sample thickness. For samples that are too thin, deformation is strongly affected by the underlying rigid support. Although Eq. (1) appears to account for such artificial magnifications, the indentation depth should generally be minimized for thinner matrices as sample thickness is often difficult to estimate or measure with great accuracy, as in cells. It is also important to apply reasonable displacements and forces so that one adequately measures the elasticity. Given a choice, biological samples should therefore be made thick enough (typically $100 \ \mu m$) to assess their true material properties or else made to resemble typical *in vivo* dimensions, for example thin basement membrane.

III. Materials Characterization

A. Artificial Matrices

Compared to biological samples that are detailed in the next section, hydrogels are relatively easy to prepare and measure (Beningo *et al.*, 2002). Polymerization conditions and buffer parameters are nonetheless critically important, and accuracy in formulation and measurement are required to reproducibly obtain results for hydrogels that conform to prescribed mechanical characteristics, giving *E* within the needed 25% accuracy. It is important to not only use buffers appropriate to the matrix but also the same buffer solutions should be used for both matrix storage and the immersion medium for the AFM probe to prevent osmotic swelling or shrinking of the gel during characterization.

Great care should be taken when determining concentrations of monomer and cross-linker as well as other components. The rate of polymerization, for example, has been shown to influence the Poisson ratio, *v*, which quantifies lateral versus elongational strain (Geissler *et al.*, 1988; Hecht *et al.*, 1985; Li *et al.*, 1993), and could account for the formation of microdomains that affect AFM-determined moduli by up to 21% (Hecht *et al.*, 1985). With proper care and precision in preparation, such issues are generally avoided, as shown in the two hydrogel systems described below: PA and HA hydrogels.

1. PA Hydrogels

PA hydrogels are relatively simple in their mechanics, and have been extensively characterized by other traditional techniques, including bulk tensile loading, microindentation, and rheology (Jacot *et al.*, 2006; Pelham and Wang, 1997; Yeung *et al.*, 2005). This makes them excellent samples for assessing and refining the micromeasurement capabilities of an AFM. Moreover, PA matrices have widely tunable mechanical properties, are fairly inert, and are readily polymerized in the presence of free radicals in aqueous solutions. They also exhibit little hysteresis in indentation, that is their indentation and retraction curves are nearly identical, and they generally do not adhere to an AFM probe (or to cells or most materials). In case the cantilever tip does adhere during indentation, it can be translated up several micrometers to disrupt the weak nonspecific adhesion. All of these features and more allow for precise chemical and mechanical manipulations. Moreover, since PA gels are normally polymerized in thick slabs for polyacrylamide gel electrophoresis (PAGE), many protocols are in the literature; the primary extension to the methods described below are (1) covalent attachment of a polymerizing solution of acrylamide to a chemically modified glass coverslip on the "bottom" surface of the gel and (2) covalent attachment of ligand to the top surface of the gel for cell adhesion. Neither of these has a major affect on PA gel elasticity.

a. Gel Preparation

For covalently coupling PA gel samples to the support, glass coverslips are aminosilanized and activated with glutaraldehyde using commercially available reagents from Sigma (St. Louis, Missouri) (Pelham and Wang, 1997). In order to achieve broad control over gel stiffness, the cross-linker N,N'-methylene-bis-acrylamide can be varied from at least 0.03 to 0.3% (w/v in PBS) and the acrylamide (C_3H_5NO) monomer can be varied from 3 to 10% (w/v in PBS) or more. To obtain gels that are ~70- to 100-μm thick (as measured by through-focus microscopy), 25 μl of the mixed monomer/cross-linkers solution is polymerized on a 22-mm coverslip using 1/200 volume of 10% ammonium persulfate (which should be stored frozen or made fresh) and 1/2000 volume of N,N,N',N'-tetramethylethylenediamine. The polymerizing gel is covered with a second coverslip 22 mm in diameter coated with dichlorodimethylsilane to ensure easy detachment and a flat and uniform gel surface once polymerized. For thinner gels, solution volumes can be adjusted appropriately and/or borosilicate beads (1%, w/v) of appropriate size added as spacers (Duke Scientific, North Carolina), with a weight applied to prevent spacer bead aggregation. Once polymerized, gels should be immediately immersed in the buffer, identical to the buffer in which the gel was initially polymerized (PBS) to prevent osmotic swelling, which could wrinkle or detach the film. Amine-containing proteins, such as collagen type 1 (BD Bioscience; Franklin Lakes, New Jersey), are readily cross-linked to the surface of the gel overnight using a photoactivatable cross-linker, Sulfo-SANPAH (Pierce, Rockford, Illinois), or NHS-acrylamide copolymerized into the gel (Pierce, Rockford, Illinois; Chapter 2 by Kandow *et al.*, this volume), and attachment can be confirmed and even calibrated in density by using it in a 9:1 mixture with FITC-conjugated protein (Molecular Probes, Carlsbad, California).

b. Probing PA Hydrogel by AFM

AFM measurements of the elastic modulus should generally be made on a range of gels with various concentrations of both bis-acrylamide cross-linker and acrylamide monomer. In the examples recommended here, samples with monomer and cross-linker concentrations ranging from 3 to 10% and from 0.03 to 3%, respectively, are placed on the AFM stage and submerged in PBS prior to AFM indentation with a blunted pyramid tip. Once the cantilever is lowered to the surface of the gel, indentations are made at a tip velocity of 2 μm/sec over a total vertical distance of 4 μm (for both indentation and retraction). Multiple force–indentation curves are generated at >10 locations across the surface of the gel for an average of 50–60 force curves per matrix condition. The contact point (z_0) for each indentation curve is computed (Chapter 3 by Frey *et al.*, this volume), and the data between the contact point and the end of the indentation is subsequently analyzed and fit to the Hertz sphere or the Sneddon cone model to determine the "microelastic" modulus E. All elasticity data for a given matrix is calculated assuming a Poisson ratio v of 0.4–0.45 as determined by macroscopic tension tests.

The outcome is not dramatically affected by the Poisson ration as variation of the v between 0.3 and 0.5 (Li *et al.*, 1993) affects the modulus by only 10% (Engler *et al.*, 2004a).

It is usually adequate to curve fit models to experimental data (Fig. 3A inset), but it can be useful to assess the accuracy of such fits. Figure 3B shows pointwise calculations of the elastic modulus for a moderately stiff PA gel and for a soft PA gel. Since the tip height is typically a few micrometers, analyses of deep indentations should be avoided. Indeed, the range for reliable analysis is widely recommended to be where the cantilever (of a selected spring constant k) is deflected by 10–100 nm. By extrapolation to zero deflection, this range is used to identify the contact point z_0, but a pointwise analysis allows this point to be shifted (±25 nm for the stiff gel or ±100 nm for the soft gel) so that error propagation from contact point identification can be assessed. The determined modulus proves robust to such perturbations ($<10\%$ effects on E), but it is nonetheless clear that the initial 100–1000 nm of indentation can often deviate (above or below) from the modulus that can be confirmed by other data or other methods.

For simple macroscale measurements, gels that are ~1-mm thick can be cast and tested by both AFM and bulk tensile loading (Engler *et al.*, 2004a; Pelham and Wang, 1997). Stress–strain curves generated from tensile loading of these thick gels of length L are calculated in terms of Cauchy stresses (force per cross-sectional area) and Lagrangian strains in which the latter are calculated with $\varepsilon = (L^2 - L_0^2)/2L_0^2$. The inset to Fig. 4A demonstrates linearity even beyond 50% strain. For a physiological comparison, maximum traction strains generated by 3T3 fibroblasts appear to be less than about 15–25% regardless of substrate modulus (Lo *et al.*, 2000; Saez *et al.*, 2005), which agrees well with the small strain assumptions made in AFM. For macroscopic tension tests performed on ~1-mm matrices also tested via AFM nanoindentation ($n = 50$ points/matrix) (Fig. 4A, $n = 3$ gels), data at each polymer concentration proves to be statistically similar even though the techniques use different approximations and imply that samples that are sufficiently thick and large do not require covalent attachment to substrate for accurate AFM micromeasurements. The collagen-coated matrices probed with AFM also fit the correlation, which indicates little to no mechanical contribution from a thin layer of ligand.

Classic theory of rubber elasticity would predict that the elastic modulus of a polymer gel scales linearly with cross-linker. This linearity of E over a broad range of cross-linker concentration, indeed, holds as shown in Fig. 4B for several independently prepared acrylamide samples (3%, 5%, 10%) as measured by both sphere and pyramid tip AFM geometries, and analyzed with two different AFM analysis algorithms in our laboratory. The plot also shows excellent agreement with published measurements made by others using both AFM (4%) and bulk rheology methods. Deviation at low concentrations of cross-linker (for the 3% gel) is consistent with percolation of bonds and is likely to be a more heterogeneous material that gives larger error bars (as found), and should probably be avoided for

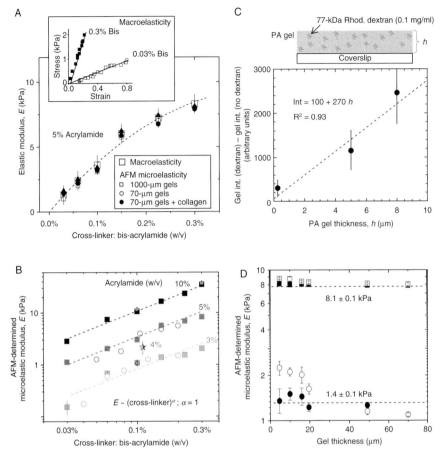

Fig. 4 PA gel elasticity. (A) Elastic moduli were measured for 5% PA hydrogels by both macroscopic tension test ($n = 157$) and nanoindentation with AFM ($n = 50$ per sample). Comparison of the data indicates nearly a 1:1 correlation, with significant overlap for both sample populations regardless of surface functionalization. The inset demonstrates the linearity of elasticity; E is the slope. The macroscopic tension tests suggest a Poisson ratio $v \cong 0.4$–0.45. (B) AFM-determined elastic moduli for various monomer and cross-linker concentrations (both %, w/v), spanning three orders of magnitude (squares). The tight error bars for most points illustrate the precision of the measurement technique. Open circles represent data for 5.5% and 3% PA gels as determined by others using a macrorheology method (Yeung *et al.*, 2005); the open star represents a data point from others for a cross-linked 4% PA gel analyzed by both AFM and rheology (300-mm thick, 30/0.8 acrylamide/bis-acrylamide by weight; Mahaffy *et al.*, 2000). The smaller open diamonds represent three additional data points for PA gels prepared and analyzed independently in our laboratory, and overlaying perfectly with the larger set of data. (C, D) PA gels, when cast as thin films, prove to be thickness dependent in their apparent elasticity as measured by AFM indentation. Five percent acrylamide solutions were mixed with either 0.3% (squares) or 0.03% (circles) bis-acrylamide cross-linker and polymerized with "spacer" beads of varying diameter to set the height. Three such stiff gels were also mixed with fluorescent dextran for simple intensity measurements to confirm thickness control. Closed points use the thin film correction to the Hertz sphere model; open points use the classic Hertz sphere model. The horizontal lines are averages of the thin film corrected results. Significant deviation between the two is seen for matrices less than 20-μm thick.

cell biological studies. Deviations from classic theory also occur when samples become extremely thin, as mentioned above for Eq. (1) and as can be expected for some secreted matrices.

To directly examine thickness effects, identical PA solutions with dissolved fluorescent dextran are cast to various thicknesses or heights h with microbead spacers, and then verified by fluorescence intensity analyses of images (Fig. 4C). Such gels can then be probed by AFM using spherical tips that allows for height corrections. While traditional AFM theory can determine an apparent elasticity, that is the elasticity measured from a composite of the material and underlying support (Fig. 4D, open data points), a correction for the thickness using Eq. (1) yields the expected material properties as found for thicker matrices of the same material (Fig. 4D, closed data points). Without the correction term of Eq. (1), calculations of the elastic modulus for 5-μm-thick gels that are soft ($E = 1.4$ kPa for 5% acrylamide/0.03% bis-acrylamide solution) are found to overestimate the modulus by about 50% while stiffer gels (5%/0.3%) give smaller overestimates.

2. HA Hydrogels

PA hydrogels are mechanically tunable, but they are nonbiological materials (as are glass and polystyrene), and acrylamide monomer must be washed out thoroughly as it is toxic to cells. Use of biologically derived, compatible ECMs can therefore have some advantages provided that the elasticity can be widely tuned. HA, a negatively charged, linear polysaccharide of D-glucuronic acid and D-N-acetylglucosamine, is found abundantly in the ECM and connective tissue. Despite its biocompatibility, native HA gels exhibit poor mechanical properties. While HA can be cross-linked using EDC chemistry (Ladam *et al.*, 2003) before serving as a soft substrate for cell biology experiments, such hydrogel preparations lack long-term stability and tunable elasticity in the required range of 0.1–100 kPa. To meet such mechanical requirements, chemical modification that adds thiol groups to the HA polymer is used (HA–S) (Shu *et al.*, 2002), allowing for HA–HA or HA–polyethylene glycol diacrylate (PEG–DA) cross-linking and mechanically stable hydrogels (Cai *et al.*, 2005; Shu *et al.*, 2006). Such a system is described below to illustrate the utility of the AFM method in characterization of a complex biohydrogel system.

a. Gel Preparation

HA hydrogels are prepared on activated, aminosilanized glass (as described above) using a solution of HA–S in PBS (pH 7.5), mixed with the appropriate amount of PEG–DA, and then polymerized under a hydrophobic glass slide as with the PA gels. After 1 h of gelation, the glass slide is removed. The hydrogel is then rinsed extensively to remove any unbound molecules and stored at 4 °C till further use. Collagen type I can be covalently bound to the HA surface by carbodiimide chemistry (Sehgal and Vijay, 1994), using HEPES at pH 8.5 with 200-mM

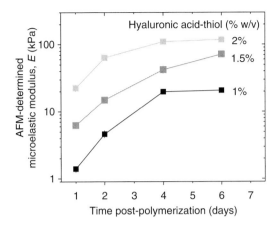

Fig. 5 HA gel elasticity. AFM-determined elastic moduli of PEG–DA cross-linked HA–S hydrogels, polymerized at three different concentrations, show time-dependent increases in their elasticity. Stiffening occurs due to auto cross-linking of the remaining free thiol groups, saturating after 6 days to yield stable gels afterward. By varying HA–S concentration as well as PEG–DA cross-linker concentration, the elastic modulus E can be finely tuned over a wide range.

EDC, 50-mM NHS, and the desired amount of collagen (fluorescent if needed). The reaction is allowed to proceed overnight at 37 °C.

b. Elasticity of HA Gels Probed by AFM

The elasticity E of HA hydrogels is measured by AFM as described above. Figure 5 shows the effective Young's modulus, E, for three different concentrations of HA gels. Progressive formation of disulfide bonds with time under ambient oxidative conditions increases the cross-linking, and the gels consistently appear approximately fivefold stiffer with similar kinetics. Saturation cross-linking is seen for each gel by about 4–6 days. The elasticity of these hydrogels can be stabilized and finely tuned over a wide range, that is 0.1–150 kPa, by varying the HA–S and PEG–DA concentrations. Measurements before and after the collagen type I attachment also show no difference in elasticity within the accuracy of the measurements. Additionally, we find that cells can attach to collagen-coated HA gels, although the emphasis here is on the AFM characterization of a bio-derived hydrogel matrix.

B. Cell-Secreted Matrices

Although cell mechanics has been widely studied by AFM, cell-secreted matrices do not appear to have been studied until recently. The mechanical properties and thickness of matrix can indeed be probed immediately adjacent to adherent cells (Fig. 6). This is important as an intermediate step for comparing the *in vitro* microenvironment with its physiological counterpart, especially for cells that are known to secrete large amounts of ECM, for example fibroblasts and osteoblasts,

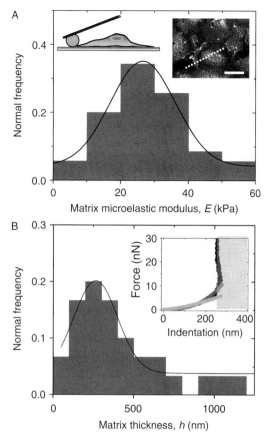

Fig. 6 Mechanical properties of a secreted matrix. (A) After 7 days in culture, hFOB osteoblasts have secreted significant amounts of matrix proteins adjacent to and on top of themselves as represented in the inset with osteocalcin staining. This matrix is indented with a spherical tip as drawn, with the probe rastering across the cell starting at a position far from the cell to ensure that the probe initially indents the bare glass (dotted line). Gaussian fits of the elastic modulus histogram yield a modulus of $E_{matrix} \approx 27 \pm 10$ kPa. The scale bar is 20 μm. (B) Once the probe maximally indents the matrix, the force continues to increase with the AFM's downward motion, without any further indentation as represented by the black data in the inset force–indentation curve. Normal (light gray) and thickness corrected (dark gray) Hertz sphere models help identify the maximal indentation point, the effective modulus from the normal sphere fit and the corrected modulus from the thin film model. Histograms of the effective modulus, when fit again by a Gaussian, give an average thickness of 350 ± 100 nm.

as well as for cells that live near basement membrane matrices, for example epithelium attached to basal lamina.

While spread cells are generally micrometers thick, the matrices that they lay down adjacent to their periphery seem likely to be thinner. To measure the thickness and elasticity of such films, a spherical tip (Fig. 6A inset schematic) is preferred, as it distributes the strain more broadly in the thin film rather than penetrating it as a pyramidal tip will. For the studies here, osteoblasts were plated on collagen-coated glass coverslips and allowed to spread and secrete for 7 days before the AFM measurements were performed. Osteoblasts secrete the matrix molecule osteocalcin, which can be stained by standard immunofluorescence methods and is seen adjacent to cells (Fig. 6A inset image). Under reflected light illumination to image the cell, the AFM was used to probe the adjacent matrix by starting on one side of the cell, 10-μm away from the cell edge, and moving across the cell as traced by the dotted line in the inset. Force–indentation curves distant from the cell resemble those obtained on rigid surfaces and schematically have the shape (_]). However, within 2–3 μm of the cell periphery, force–indentation curves are more rounded (_)) as expected for contact with an elastic matrix; analysis yields a relatively stiff elastic modulus of 27 ± 10 kPa. This level of elasticity was also measured on the opposite side of the cells again for 2–3 μm and proved similar. In analyzing these force–indentation curves, after a given amount of indentation, it became clear that the AFM tip was no longer pressing into a soft material (Fig. 6B inset). This indicates that the probe was maximally indented into the matrix, and a Gaussian fit of this maximum thickness distribution was 350 ± 100 nm (Fig. 6B). Equation (1) was not used because the thin film correction appears most important for much softer matrices (<10 kPa).

C. Passive Tissue Elasticity

As described in Section I, cell adhesion and mechanics as well as differentiation of a number of cell types have an *in vitro* context that motivates improved measurements of the natural microenvironment. While bulk tensile testing may be sufficient to determine "whole tissue" parameters, it cannot adequately assess the microenvironment felt by cells. Here we summarize methods used for measuring the mechanical properties of the *in vivo* microenvironments by applying AFM on tissue sections.

1. Microelastic Modulus of the Arterial Media Layer

Arterial smooth muscle cells (SMCs) are found *in vivo* in the medial layer composed of near-equal parts cells and elastin–collagen matrix (Fig. 7). This layer is located between the inner intima and an outer granular layer. SMCs appear highly sensitive to the elasticity of their microenvironment, showing more stress fibers and focal adhesions on increasingly stiff matrices (Engler *et al.*, 2004a) as do most other cell types *in vitro*. Whole carotid arteries ~2 mm in diameter were excised from 6-month-old pigs, studied within 8 h of isolation, and kept in media typical of whole muscle mechanical testing (Barton-Davis *et al.*, 1998), that is Dulbeco's Modified Eagle Medium (DMEM) supplemented with 10% serum, L-glutamine, and antibiotics. With this type of testing and in such a short time frame post-isolation, rigor does not appear to be an issue. The arteries were sectioned with razor blades under a dissecting microscope into cross-sections

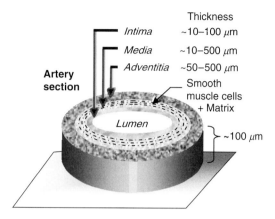

Fig. 7 Schematic of arterial cross-section. Arteries contain three layers of distinct composition and artery-dependent thickness. The *Intima* faces the vessel lumen and is lined with endothelial cells. The *Media* is both elastic and contractile with SMCs typically composing ∼30–60% of the mass (Hermanson, 1996). Collagen is the predominant matrix protein at ∼10–40% by mass, with elastin also being a significant fraction. The outermost, protective layer is the *Adventitia*, which is expected to be far stiffer.

that were ∼100-μm thick, as confirmed by light microscopy. The vessel sections were wet mounted within minutes on cover glass with adhesive double-sided tape in an orientation such that the medial layer was accessible to probing by AFM. These freshly prepared sections were placed in the AFM and probed ∼10 times at each location along the cross-section, using either spherical or pyramidal tips. Curve fits using the respective models agree with the experimental results up to indentations of tip size (e.g., $R \cong 2.5$ μm). The fitted elastic moduli for the medial layer are within a narrow range of $E_{media} = 5\text{--}8$ kPa, and the result appears essentially independent of tip shape, which lends additional confidence to the values reported here. Compared to a typical collagen gel, this level of elasticity is stiffer by an order of magnitude.

2. Skeletal Muscle Stiffness *Ex Vivo*: Normal and Dystrophic

As a microenvironmental probe, AFM appears to be useful in distinguishing small mechanical changes between tissues and within tissues, and might be increasingly used to elucidate changes in tissue stiffness with disease. We illustrate this by measuring the mechanical differences between normal mouse muscle and dystrophic muscle lacking the elastic linker protein dystrophin (Engler *et al.*, 2004b). Humans with Duchenne muscular dystrophy lack this protein, which mechanically couples the contractile apparatus of the cytoskeleton through adhesions to the ECM.

Left and right whole extensor digitorum longus (EDL) muscles were carefully excised from female C57 normal (3, 13, and 21 months old) and *mdx* dystrophic mice (3, 9, and 15 months old), divided into thinner bundles with a razor, wet mounted onto coverslips, and tested within 3 h to prevent the onset of rigor. Tissue section thickness was confirmed by light microscopy to be $\sim$100 μm, and samples were kept in tissue culture media appropriate to muscle as indicated above for arteries. To quantitate the transverse (not longitudinal), passive elasticity of a representative muscle, EDL muscles were probed several hundred times normal to the muscle surface and along the direction relevant to cell attachment to matrix. Data was averaged over multiple locations within each sample ($n = 3$) from normal mice of different ages, where the predominant modulus was $E_{\text{muscle}} = 12 \pm 4$ kPa (Fig. 6, 85% of data, $n = 387$). This agrees with the elasticity of cultured C2C12 cells (up to 1 week): $E_{\text{C2C12}} = 12$–15 kPa (Collinsworth *et al.*, 2002). A secondary peak of 38 ± 8 kPa is present at much lower frequency, and the origin of this peak could be components of matrix such as collagen fibrils.

The dystrophic muscle of the *mdx* mouse is well known to exhibit an increase in fibrosis during muscle degeneration and has many other perturbations that could have mechanical consequences. The EDL muscle from these animals appeared significantly stiffer with a dominant peak in modulus at $E_{mdx} = 18 \pm 6$ kPa (Fig. 8, 55% of data, $n = 143$). Age did not appear to be a factor in either normal or *mdx* samples ($p = 0.42$). However, the microscale measurements of transverse elasticity for *mdx* muscle also proved more heterogeneous than normal muscle: *mdx* probing gave a significant number of indentations with $E_{mdx} > 35$ kPa ($\sim$40%) and also a few with $E_{mdx} < 5$ kPa ($\sim$5%). Further study is needed to understand the molecular basis for this general stiffening and heterogeneity, but significant differences in

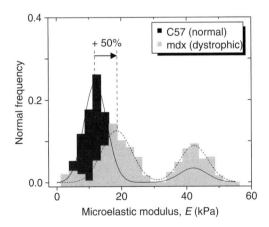

Fig. 8 Muscle elasticity from normal (C57) and dystrophic (*mdx*) mice. Histograms of EDL muscle elasticity shows a dominant peak at 12 ± 4 kPa (dark gray data), corresponding to normal muscle cell elasticity (Collinsworth *et al.*, 2002; Engler *et al.*, 2004b). However, a secondary peak is more prominent in dystrophic *mdx* muscle (light gray data), where fibrosis creates a matrix that is more rigid. This rigid matrix appears to stiffen the normally soft *mdx* cells to create a primary peak that is 50% higher than C57 muscle elasticity.

micromechanics seem clear and reproducible. On the basis of the strong effects of substrate elasticity on muscle cell differentiation in culture, which show optimal muscle striation on hydrogels of modulus $12 \pm 4\,\text{kPa}$, the altered mechanics of diseased muscle is likely to negatively influence differentiation.

IV. Assessing Mechanical Influences on Cells

Hydrogels described above can be particularly useful if they can adequately mimic *in vivo* microenvironments. Various cell types (e.g., myoblasts, fibroblasts, endothelial cells, and so on) have been plated on flexible materials, especially PA matrices (Engler *et al.*, 2004b; Flanagan *et al.*, 2002; Kong *et al.*, 2005; Lo *et al.*, 2000; Paszek *et al.*, 2005; Peyton and Putnam, 2005; Rajagopalan *et al.*, 2004), and provide sufficient background information (Engler *et al.*, 2004a,b; Paszek *et al.*, 2005; Peyton and Putnam, 2005; Yeung *et al.*, 2005), but relationships to cell and tissue physiology have not been made especially clear. Matrix stiffness strongly influences how cells adhere and form focal contacts (Pelham and Wang, 1997), and this in turn influences cell area and cell shape (Engler *et al.*, 2004a; Gaudet *et al.*, 2003). Further assessments of cell response have used a host of standard cell biology methods such as immunofluorescence microscopy, Western blotting, BrdU labeling, or real-time PCR to analyze changes in cell number, cell division, protein expression or mRNA, cytoskeletal, and morphological changes. The results often show correlations with matrix mechanical properties and thus demonstrate how physical cues influence cell function.

Recent experiments with bone marrow-derived MSCs amplify the results found with more committed cell types. Solely by sensing matrix elasticity in the absence of other environmental cues (e.g., additional growth factors), MSCs can be programmed to specify particular lineages versus others (Engler *et al.*, 2006). Naive MSCs plated on collagen-coated PA gels adhere and form adhesions in a stiffness-sensitive manner as do most cell types, but these cells also undergo considerable morphological changes dependent on the elasticity. MSCs also generate increasing amounts of force on increasingly stiff matrices (schematically in Fig. 2E and F), but most important is the fact that the rigidity of a matrix corresponding to specific tissues surprisingly promotes tissue-mimetic differentiation. Figure 9A illustrates this sensitivity in morphology to the elastic modulus of the substrate. Quantification of cell shape in these systems is assessed in terms of branches per millimeter as appropriate to nascent neurons and aspect ratio or spindle factor as appropriate to myoblasts (Fig. 9B and C).

Additional proof of selective differentiation is provided by immunofluorescence for markers in the cells after 1–4 weeks in culture. MSCs are found to express key lineage markers typical of neurons (β3-tubulin), muscle (MyoD), and bone (CBFα1) on matrices that recapitulate the native tissue elasticity of brain, muscle, and nascent bone, respectively, as shown in Fig. 9A. Collectively, these observations show how the elasticity of the underlying matrix alone contributes to cell differentiation.

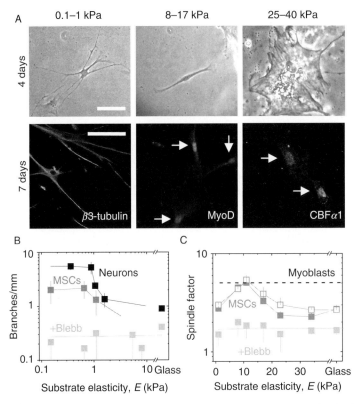

Fig. 9 Influence of matrix elasticity on cell lineage specification. (A) Naive MSCs develop increasingly branched, spindle, or polygonal shapes when grown on gels in the range typical of $\sim E_{\text{brain}}$ (0.1–1 kPa), $\sim E_{\text{muscle}}$ (8–17 kPa), or stiff cross-linked collagen matrices (25–40 kPa), respectively. β3-Tubulin (neurocyte), MyoD (myocyte), and CBFα1 (osteocyte) are selectivity expressed in cells that exhibit the morphology indicative of each lineage. Scale bar is 20 μm. (B, C) Graphs quantify the morphological changes versus stiffness: cell branching per length of primary mouse neurons, MSCs, and blebbistatin-treated MSCs as well as spindle morphology of MSCs, blebbistatin-treated MSCs, and mitomycin-C-treated MSCs (open squares) compared to C2C12 myoblasts (dashed line).

References

Alenghat, F. J., and Ingber, D. E. (2002). Mechanotransduction: All signals point to cytoskeleton, matrix, and integrins. *Sci. STKE* **2002,** PE6.

Barton-Davis, E. R., Shoturma, D. I., Musaro, A., Rosenthal, N., and Sweeney, H. L. (1998). Viral mediated expression of insulin-like growth factor I blocks the aging-related loss of skeletal muscle function. *Proc. Natl. Acad. Sci. USA* **95,** 15603–15607.

Beningo, K. A., Lo, C. M., and Wang, Y. L. (2002). Flexible polyacrylamide substrata for the analysis of mechanical interactions at cell-substratum adhesions. *Methods Cell Biol.* **69,** 325–339.

Bershadsky, A., Kozlov, M., and Geiger, B. (2006). Adhesion-mediated mechanosensitivity: A time to experiment, and a time to theorize. *Curr. Opin. Cell Biol.* **18,** 472–481.

Binnig, G., Quate, C. F., and Gerber, C. (1986). Atomic force microscope. *Phys. Rev. Lett.* **56,** 930–933.

Buckley, I. K., and Porter, K. R. (1967). Cytoplasmic fibrils in living cultured cells. A light and electron microscope study. *Protoplasma* **64**, 349–380.

Cai, S., Liu, Y., Zheng Shu, X., and Prestwich, G. D. (2005). Injectable glycosaminoglycan hydrogels for controlled release of human basic fibroblast growth factor. *Biomaterials* **26**, 6054–6067.

Collinsworth, A. M., Zhang, S., Kraus, W. E., and Truskey, G. A. (2002). Apparent elastic modulus and hysteresis of skeletal muscle cells throughout differentiation. *Am. J. Physiol. Cell Physiol.* **283**, C1219–C1227.

Dembo, M., and Wang, Y.-L. (1999). Stresses at the cell-to-substrate interface during locomotion of fibroblasts. *Biophys. J.* **76**, 2307–2316.

Dimitriadis, E. K., Horkay, F., Maresca, J., Kachar, B., and Chadwick, R. S. (2002). Determination of elastic moduli of thin layers of soft material using the atomic force microscope. *Biophys. J.* **82**, 2798–2810.

Diridollou, S., Patat, F., Gens, F., Vaillant, L., Black, D., Lagarde, J. M., Gall, Y., and Berson, M. (2000). *In vivo* model of the mechanical properties of the human skin under suction. *Skin. Res. Technol.* **6**, 214–221.

Engler, A., Bacakova, L., Newman, C., Hategan, A., Griffin, M., and Discher, D. (2004a). Substrate compliance versus ligand density in cell on gel responses. *Biophys. J.* **86**, 617–628.

Engler, A. J., Griffin, M. A., Sen, S., Bonnemann, C. G., Sweeney, H. L., and Discher, D. E. (2004b). Myotubes differentiate optimally on substrates with tissue-like stiffness: Pathological implications for soft or stiff microenvironments. *J. Cell. Biol.* **166**, 877–887.

Engler, A. J., Richert, L., Wong, J. Y., Picart, C., and Discher, D. E. (2004c). Surface probe measurements of the elasticity of sectioned tissue, thin gels and polyelectrolyte multilayer films: Correlations between substrate stiffness and cell adhesion. *Surf. Sci.* **570**, 142–154.

Engler, A. J., Sen, S., Sweeney, H. L., and Discher, D. E. (2006). Matrix elasticity directs stem cell lineage specification. *Cell* **126**, 677–689.

Firtel, R. A. (1995). Integration of signaling information in controlling cell-fate decisions in dictyostelium. *Genes Dev.* **9**, 1427–1444.

Flanagan, L. A., Ju, Y. E., Marg, B., Osterfield, M., and Janmey, P. A. (2002). Neurite branching on deformable substrates. *Neuroreport* **13**, 2411–2415.

Fung, Y.-C. (1994). A First Course in Continuum Mechanics, 3rd edn. Prentice-Hall, Englewood Cliffs, NJ.

Gaudet, C., Marganski, W. A., Kim, S., Brown, C. T., Gunderia, V., Dembo, M., and Wong, J. Y. (2003). Influence of type I collagen surface density on fibroblast spreading, motility, and contractility. *Biophys. J.* **85**, 3329–3335.

Gefen, A., and Margulies, S. S. (2004). Are *in vivo* and *in situ* brain tissues mechanically similar? *J. Biomech.* **37**, 1339–1352.

Geissler, E., Hecht, A. M., Horkay, F., and Zrinyi, M. (1988). Compressional modulus of swollen polyacrylamide networks. *Macromolecules* **21**, 2594–2599.

Goffin, J. M., Pittet, P., Csucs, G., Lussi, J. W., Meister, J. J., and Hinz, B. (2006). Focal adhesion size controls tension-dependent recruitment of alpha-smooth muscle actin to stress fibers. *J. Cell. Biol.* **172**, 259–268.

Goldman, R. D., Milsted, A., Schloss, J. A., Starger, J., and Yerna, M. J. (1979). Cytoplasmic fibers in mammalian cells: Cytoskeletal and contractile elements. *Annu. Rev. Physiol.* **41**, 703–722.

Harris, A. K., Wild, P., and Stopak, D. (1980). Silicone rubber substrata: A new wrinkle in the study of cell locomotion. *Science* **208**, 177–179.

Hategan, A., Law, R., Kahn, S., and Discher, D. E. (2003). Adhesively-tensed cell membranes: Lysis kinetics and atomic force microscopy probing. *Biophys. J.* **85**, 2746–2759.

Hecht, A. M., Duplessix, R., and Geissler, E. (1985). Structural inhomogeneities in the range 2.5–2500 A in polyacrylamide gels. *Macromolecules* **18**, 2167–2173.

Hermanson, G. (1996). Bioconjuate Techniques. Academic Press, San Diego.

Hertz, H. (1882). Über die Berührung fester elastischer Körper. *J. Reine Angew. Mathematik* **92**, 156–171.

Jacot, J. G., Dianis, S., Schnall, J., and Wong, J. Y. (2006). A simple microindentation technique for mapping the microscale compliance of soft hydrated materials and tissues. *J. Biomed. Mater. Res. A* **79**, 485–494.

Kong, H. J., Boontheekul, T., and Mooney, D. J. (2006). Quantifying the relation between adhesion ligand-receptor bond formation and cell phenotype. *Proc. Natl. Acad. Sci. USA* **103**, 18534–18539.

Kong, H. J., Polte, T. R., Alsberg, E., and Mooney, D. J. (2005). FRET measurements of cell-traction forces and nano-scale clustering of adhesion ligands varied by substrate stiffness. *Proc. Natl. Acad. Sci. USA* **102**, 4300–4305.

Ladam, G., Vonna, L., and Sackmann, E. (2003). Micromechanics of surface-grafted hyaluronic acid gels. *J. Phys. Chem. B* **107**, 8965–8971.

Li, Y., Hu, Z., and Li, C. (1993). New method for measuring Poisson's ratio in polymer gels. *J. Appl. Polymer. Sci.* **50**, 1107–1111.

Lo, C. M., Wang, H. B., Dembo, M., and Wang, Y. L. (2000). Cell movement is guided by the rigidity of the substrate. *Biophys. J.* **79**, 144–152.

Mahaffy, R. E., Shih, C. K., MacKintosh, F. C., and Kas, J. (2000). Scanning probe-based frequency-dependent microrheology of polymer gels and biological cells. *Phys. Rev. Lett.* **85**, 880–883.

Oliver, T., Dembo, M., and Jacobson, K. (1995). Traction forces in locomoting cells. *Cell Motil. Cytoskeleton* **31**, 225–240.

Paszek, M. J., Zahir, N., Johnson, K. R., Lakins, J. N., Rozenberg, G. I., Gefen, A., Reinhart-King, C. A., Margulies, S. S., Dembo, M., Boettiger, D., Hammer, D. A., and Weaver, V. M. (2005). Tensional homeostasis and the malignant phenotype. *Cancer Cell* **8**, 241–254.

Pelham, R. J., and Wang, Y.-L. (1997). Cell locomotion and focal adhesions are regulated by substrate flexibility. *Proc. Natl. Acad. Sci. USA* **94**, 13661–13665.

Peyton, S. R., and Putnam, A. J. (2005). Extracellular matrix rigidity governs smooth muscle cell motility in a biphasic fashion. *J. Cell. Physiol.* **204**, 198–209.

Pollard, T. D., Fujiwara, K., Niederman, R., and Maupin-Szamier, P. (1976). Evidence for the role of cytoplasmic actin and myosin in cellular structure and motility. *In* "Cell Motility" (R. Goldman, T. Pollard, and J. Rosenbaum, eds.), pp. 689–724. Cold Spring Harbor Laboratory, Cold Spring Harbor, New York.

Radmacher, M. (1997). Measuring the elastic properties of biological samples with the AFM. *IEEE Eng. Med. Biol. Mag.* **16**, 47–57.

Radmacher, M., Fritz, M., and Hansma, P. K. (1995). Imaging soft samples with the atomic force microscope: Gelatin in water and propanol. *Biophys. J.* **69**, 264–270.

Rajagopalan, P., Marganski, W. A., Brown, X. Q., and Wong, J. Y. (2004). Direct comparison of the spread area, contractility, and migration of balb/c 3T3 fibroblasts adhered to fibronectin- and RGD-modified substrata. *Biophys. J.* **87**, 2818–2827.

Richert, L., Engler, A. J., Discher, D. E., and Picart, C. (2004). Elasticity of native and cross-linked polyelectrolyte multilayer films. *Biomacromolecules* **5**, 1908–1916.

Riveline, D., Zamir, E., Balaban, N. Q., Schwarz, U. S., Ishizaki, T., Narumiya, S., Kam, Z., Geiger, B., and Bershadsky, A. D. (2001). Focal contacts as mechanosensors: Externally applied local mechanical force induces growth of focal contacts by an mDia1-dependent and ROCK-independent mechanism. *J. Cell. Biol.* **153**, 1175–1186.

Rosenbluth, M. J., Lam, W. A., and Fletcher, D. A. (2006). Force microscopy of nonadherent cells: A comparison of leukemia cell deformability. *Biophys. J.* **90**, 2994–3003.

Rotsch, C., Jacobson, K., and Radmacher, M. (1999). Dimensional and mechanical dynamics of active and stable edges in motile fibroblasts investigated by using atomic force microscopy. *Proc. Natl. Acad. Sci. USA* **96**, 921–926.

Rotsch, C., and Radmacher, M. (2000). Drug-induced changes of cytoskeletal structure and mechanics in fibroblasts: An atomic force microscopy study. *Biophys. J.* **78**, 520–535.

Saez, A., Buguin, A., Silberzan, P., and Ladoux, B. (2005). Is the mechanical activity of epithelial cells controlled by deformations or forces? *Biophys. J.* **89**, L52–L54.

Schloss, J. A., Milsted, A., and Goldman, R. D. (1977). Myosin subfragment binding for the localization of actin-like microfilaments in cultured cells. A light and electron microscope study. *J. Cell. Biol.* **74,** 794–815.

Sehgal, D., and Vijay, I. K. (1994). A method for the high efficiency of water-soluble carbodiimide-mediated amidation. *Anal. Biochem.* **218,** 87–91.

Sen, S., Subramanian, S., and Discher, D. E. (2005). Indentation and adhesive probing of a cell membrane with AFM: Theoretical model and experiments. *Biophys. J.* **89,** 3203–3213.

Shu, X. Z., Ahmad, S., Liu, Y., and Prestwich, G. D. (2006). Synthesis and evaluation of injectable, *in situ* crosslinkable synthetic extracellular matrices for tissue engineering. *J. Biomed. Mater. Res. A* **79,** 902–912.

Shu, X. Z., Liu, Y. C., Luo, Y., Roberts, M. C., and Prestwich, G. D. (2002). Disulfide cross-linked hyaluronan hydrogels. *Biomacromolecules* **3,** 1304–1311.

Sneddon, I. N. (1965). The relation between load and penetration in the axisymmetric boussinesq problem for a punch of arbitrary profile. *Int. J. Eng. Sci.* **3,** 47–57.

Storm, C., Pastore, J. J., MacKintosh, F. C., Lubensky, T. C., and Janmey, P. A. (2005). Nonlinear elasticity in biological gels. *Nature* **435,** 191–194.

Wang, N., Tolic-Norrelykke, I. M., Chen, J., Mijailovich, S. M., Butler, J. P., Fredberg, J. J., and Stamenovic, D. (2002). Cell prestress. I. Stiffness and prestress are closely associated in adherent contractile cells. *Am. J. Physiol. Cell Physiol.* **282,** C606–C616.

Weisenhorn, A. L., Hansma, P. K., Albrecht, T. R., and Quate, C. F. (1989). Forces in atomic force microscopy in air and water. *Appl. Phys. Lett.* **54,** 2651–2653.

Yeung, T., Georges, P. C., Flanagan, L. A., Marg, B., Ortiz, M., Funaki, M., Zahir, N., Ming, W., Weaver, V., and Janmey, P. A. (2005). Effects of substrate stiffness on cell morphology, cytoskeletal structure, and adhesion. *Cell Motil. Cytoskeleton* **60,** 24–34.

CHAPTER 23

Demystifying the Effects of a Three-Dimensional Microenvironment in Tissue Morphogenesis

Kandice R. Johnson,[*] Jennifer L. Leight,[*] and Valerie M. Weaver[†,‡]

[*]Institute for Medicine and Engineering
Department of Bioengineering
University of Pennsylvania
Philadelphia, Pennsylvania 19104

[†]Departments of Surgery and Anatomy
Center for Bioengineering and Tissue Regeneration
University of California, San Francisco, California 94143

[‡]Department of Bioengineering
University of Pennsylvania
Philadelphia, Pennsylvania 19104

Abstract
I. Introduction
II. Rationale
    A. Stromal–Epithelial Interactions
    B. ECM Mechanics and Epithelial Behavior
    C. 3D Organotypic Model Systems
III. Methods
    A. Engineered Cell/Tissue Explants
    B. Isolation of Bulk Proteins
    C. Isolation of Bulk mRNA
    D. Rapid Protein Isolation Techniques
    E. Immunofluorescence
IV. Materials
    A. Engineering Tissue Explants
    B. Isolation of Bulk Proteins
    C. Isolation of Bulk mRNA
    D. Rapid Protein Isolation Techniques
    E. Immunofluorescence

Abstract

Tissue morphogenesis and homeostasis are dependent on a complex dialogue between multiple cell types and chemical and physical cues in the surrounding microenvironment. The emergence of engineered three-dimensional (3D) tissue constructs and the development of tractable methods to recapitulate the native tissue microenvironment *ex vivo* has led to a deeper understanding of tissue-specific behavior. However, much remains unclear about how the microenvironment and aberrations therein directly affect tissue morphogenesis and behavior. Elucidating the role of the microenvironment in directing tissue-specific behavior will aid in the development of surrogate tissues and tractable approaches to diagnose and treat chronic-debilitating diseases such as cancer and atherosclerosis. Toward this goal, 3D organotypic models have been developed to clarify the mechanisms of epithelial morphogenesis and the subsequent maintenance of tissue homeostasis. Here we describe the application of these 3D culture models to illustrate how the microenvironment plays a critical role in regulating mammary tissue function and signaling, and discuss the rationale for applying precisely defined organotypic culture assays to study epithelial cell behavior. Experimental methods are provided to generate and manipulate 3D organotypic cultures to study the effect of matrix stiffness and matrix dimensionality on epithelial tissue morphology and signaling. We end by discussing technical limitations of currently available systems and by presenting opportunities for improvement.

I. Introduction

Tissue development depends on coordinated cycles of transcriptionally regulated cell growth, death, and migration that are controlled by exogenous soluble and physical stimuli and spatially dependent cell–matrix and cell–cell adhesion (Barros *et al.*, 1995; Ingber, 2006; Jacobson *et al.*, 1997; Locascio and Nieto, 2001). Regardless of length scale, understanding the molecular basis of tissue-specific differentiation and homeostasis requires an appreciation of adhesion-dependent cell behavior in the context of a three-dimensional (3D) extracellular microenvironment and a complex adhesion-dependent multicellular tissue. Through genetic and biochemical analysis, we have learned much about cell adhesion, including details of adhesion molecule structure and function, and about how various cell adhesion molecules likely mediate cell–extracellular matrix (ECM) interactions and facilitate cell–cell junctional complexes (Fuchs *et al.*, 1997; Huttenlocher *et al.*, 1998; Springer and Wang, 2004). We have also learned much about how exogenous growth, death, migration, and even mechanical cues activate signaling cascades to influence the fate of individual cells and undifferentiated 2D cell monolayers

(Huttenlocher *et al.*, 1995; McBeath *et al.*, 2004; Stegemann *et al.*, 2005; Thornberry and Lazebnik, 1998; Wang *et al.*, 2001). Yet, all too often, experimental conclusions reached from observations of single cells and simplified 2D monolayer cell sheets do not accurately represent how cells behave within 3D tissues *in vivo* (Green *et al.*, 1999; Sethi *et al.*, 1999). Indeed, developmental models and transgenic animals consistently underscore the importance of studying cell behavior in the correct tissue context. However, live animal experimentation is so inherently complex that systematic assessment of the effect of individual variables, such as cell shape and matrix compliance, on cell behavior is extremely challenging and impractical (Sethi *et al.*, 1999; Wang *et al.*, 2005). At the interface between *in vivo* studies and 2D culture models are the organotypic culture systems that can faithfully recapitulate various aspects of tissue organization and function *ex vivo*. These organotypic 3D models have been employed with varying degrees of success to clarify some of the mechanisms, whereby biological processes such as adhesion-dependent survival (Weaver *et al.*, 2002; Zahir and Weaver, 2004), polarity (O'Brien *et al.*, 2002; Wang *et al.*, 1998), proliferation (Zink *et al.*, 2004), and even epigenetics (Bissell *et al.*, 1999; Zink *et al.*, 2004) regulate cell behavior as well as novel feedback/regulatory mechanisms (Bissell *et al.*, 2002). Organotypic culture models have been effectively applied to study tissue-specific differentiation (Bissell *et al.*, 2002), to understand factors controlling stem cell behavior (Hendrix *et al.*, 2001), and even microenvironmental control of malignant transformation and tumor dormancy (Margulis *et al.*, 2005; Weaver *et al.*, 1997). Through the prudent use of organotypic 3D models, critical disparities between the molecular determinants of cell polarity (reviewed in O'Brien *et al.*, 2002), apoptosis resistance (Weaver *et al.*, 2002), and growth factor responsiveness (Wang *et al.*, 1998) in cells incorporated into a 3D tissue and those propagated as 2D monolayers have been revealed (reviewed in O'Brien *et al.*, 2002). Yet, while 3D organotypic models can faithfully recapitulate some aspects of tissue behavior *ex vivo*, many of the systems routinely used to study tissue-like behaviors employ crudely defined natural ECM molecules that contribute to considerable experimental variance. In addition, many of the approaches used to assemble 3D tissue-like structures in culture operate by simultaneously modifying multiple variables, including restrictions on cell shape, matrix compliance, biochemical cues and metabolites, and even the spatial orientation of the ECM, thereby obscuring definitive experimental conclusions regarding individual experimental parameters (Kleinman *et al.*, 1986; Paszek *et al.*, 2005; Wozniak *et al.*, 2003). Indeed, the engineering of surrogate tissues and the development of tractable approaches to diagnose and treat chronic-debilitating diseases such as cancer and atherosclerosis require both a comprehensive understanding of tissue-specific behavior at the molecular level and highly reproducible systems. Accordingly, considerable effort has been expended toward developing synthetic biomaterials in which individual material properties such as cell shape, matrix presentation (2D vs 3D), ligand density, and elastic modulus can be precisely modulated (Chen *et al.*, 1998; Engler *et al.*, 2004; Tan *et al.*, 2003; Yamada *et al.*, 2003). By applying one of the defined systems in which matrix compliance and ligand density could be rigorously controlled, the critical role of matrix stiffness and integrin adhesions as a key regulator of multicellular mammary epithelial cell (MEC) tissue morphogenesis and malignant transformation has been highlighted (Paszek and Weaver, 2004; Paszek *et al.*, 2005). In this chapter, we discuss the rationale for applying well-defined 3D organotypic culture assays to study adhesion-dependent cell behavior. We describe the use of 3D MEC organotypic cultures to illustrate how matrix compliance plays a critical role in regulating mammary tissue function and signaling. Finally, we outline experimental methods to generate, manipulate, and study the effect of matrix stiffness and matrix dimensionality on epithelial tissue morphology and signaling, and discuss technical limitations of currently available systems and future opportunities for improvement.

II. Rationale

A. Stromal–Epithelial Interactions

During embryogenesis, the epithelium originates from the endoderm and ectoderm and develops into a specialized tissue whose primary functions in the organism are to protect and to control permeation or transport. Unlike skin and esophagus, which are stratified epithelia that provide a critical barrier, the secretory epithelium is composed of a simple layer of epithelial cells lining tubes and ducts, whose principal function is to facilitate secretion and transport of biological materials. *In vivo*, the secretory epithelium abuts on and is surrounded by a stroma, which consists of cellular and noncellular components, including ECM molecules, soluble factors, and various stromal cells such as fibroblasts, adipocytes, and endothelial cells. Directly interacting with the epithelium is the basement membrane (BM), which is a specialized, highly organized ECM composed primarily of laminins 1, 5, and 10, collagen IV, entactin, and heparin sulfate proteoglycans. The BM in turn intersects with the interstitial matrix, which consists of collagens I and III, fibronectin, tenascin, elastins, and various proteoglycans including lumican, biglycan, and decorin (Kleinman *et al.*, 1986). Collectively, the various components of the ECM and stroma provide biochemical (composition) and biophysical (structural modification and organization) cues to the epithelium and operate in concert with soluble factors released from the resident stromal cells to maintain the epithelium's organ-specific function. Perturbations in stromal–epithelial interactions and altered epithelial organization are hallmarks of cancer and many chronic degenerative diseases. Moreover, disrupting tissue organization or altering ECM integrity precipitates disease, and restoring tissue structure or proper ECM interactions normalizes tissue behavior (reviewed in Hagios *et al.*, 1998; Jeffery, 2001). Accordingly, the goal of 3D organotypic culture models is to recreate tissue-specific interactions, organizations, functions, and behavior *ex vivo* through prudent control of the biochemical and biophysical properties of the ECM, in order to understand the role of stromal–epithelial interactions and tissue structures in tissue-specific functions.

Mammary gland organotypic culture models have been used effectively to study the role of stromal–epithelial and ECM interactions in tissue-specific differentiation (Debnath *et al.*, 2003; Petersen *et al.*, 1992; Weaver *et al.*, 1996; Wozniak *et al.*, 2003). Unlike other tissues, the mammary gland undergoes unique developmental cycles in the adult organism and the gland can be readily accessed and manipulated *in vivo* and in culture. Additionally, reasonable quantities of breast tissue can be isolated and propagated *ex vivo* for culture experiments. As such, much of what we know regarding ECM-dependent epithelial differentiation has been derived from organotypic cultures of primary and immortalized MECs. Early studies demonstrated that MECs grown as 2D monolayers on rigid tissue culture substrates or within a physically constrained collagen I gel fail to assemble tissue-like structures (acini) and differentiate [no detectable expression of differentiated proteins such as whey acidic protein (WAP) or β-casein], despite the availability of appropriate growth factors and lactogenic hormones (reviewed in Roskelley *et al.*, 1995). Yet, when the same MECs are grown within unconstrained collagen I gels and allowed to deposit and organize their own endogenous BM, or are embedded within a compliant reconstituted BM (rBM), they are able to assemble multicellular tissue-like structures (acini; reminiscent of terminal ductal lobular units in tissues *in vivo*) and differentiate in response to hormonal cues (expressed β-casein and WAP; reviewed in Roskelley *et al.*, 1995; Fig. 1). Further studies using murine and human MECs have also consistently shown that the composition and spatial context of the ECM profoundly influence the responsiveness of an epithelium to exogenous growth, migration, and death stimuli (Wang *et al.*, 1998, 2005; Weaver *et al.*, 2002). For example, some human luminal epithelial breast tissues *in vivo* express the estrogen receptor (ER) and proliferate in response to hormonal fluctuations in estrogen. When these MECs are isolated and cultured on tissue culture plastic, they spread to form raised ER-negative, 2D cobblestone monolayer colonies that lack estrogenic responsiveness. However, if the isolated MECs are instead grown in the context of a compliant rBM, they retain their ER expression and maintain their estrogenic responsiveness (Novaro *et al.*, 2003). Likewise, undifferentiated MECs grown on tissue culture plastic are highly sensitive to exogenous death cues, whereas their rBM-differentiated counterparts exhibit extremely high resistance to multiple apoptotic stimuli (Weaver *et al.*, 2002). Analogous observations regarding the importance of biochemical and biophysical ECM cues for epithelial morphogenesis and tissue-specific differentiation have also been reported for thyroid, salivary gland, and kidney epithelia studies (Kadoya and Yamashina, 2005; O'Brien *et al.*, 2001; Yap *et al.*, 1995).

B. ECM Mechanics and Epithelial Behavior

Many important discoveries have been made concerning the molecular mechanisms by which the ECM influences epithelial behavior, including the requirement of signaling through laminin-dependent ligation of $\alpha3\beta1$ and $\alpha6\beta4$ integrins. In addition, cooperative ERK-PI3 kinase and RacGTPase-NFκB signaling through

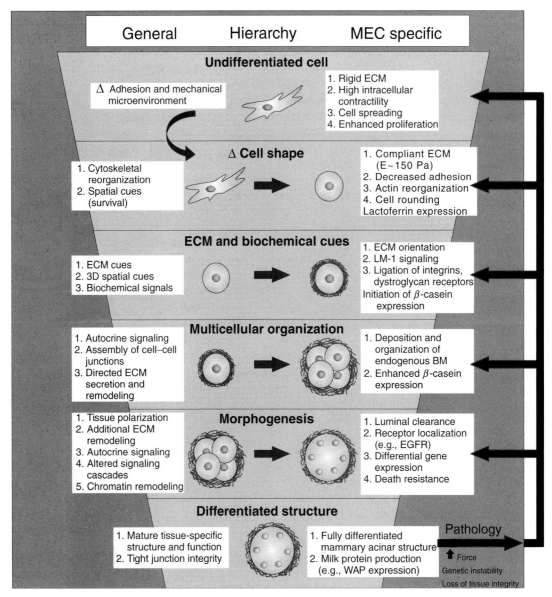

Fig. 1 Biochemical and biophysical cues from the extracellular matrix regulate tissue-specific epithelial differentiation. Illustration depicting ECM regulation of tissue-specific differentiation through a progressively complex hierarchy of adhesion-regulated events functionally linked to changes in cell shape, receptor-initiated biochemical signaling, assembly of multicellular structures, and reciprocal biochemical and physical modification of the ECM microenvironment adjacent to the epithelial tissue. Undifferentiated cell (top): an undifferentiated cell interacting with a highly rigid 2D ECM substratum, such as matrix-coated tissue culture plastic, will adhere rapidly and, if given sufficient ECM ligand, will spread appreciably using multiple adhesion receptors, including integrins, and assemble mature focal

epidermal and insulin growth factor receptors and prolactin-dependent activation of Stat3 have been identified as key biochemical events involved in directing MEC growth, survival, and differentiation (Akhtar and Streuli, 2006; Muschler *et al.*, 1999; Paszek *et al.*, 2005; Zahir *et al.*, 2003). The ECM not only influences epithelial behavior through biochemical signaling but also through the mechanical properties of the microeniverniment.

Early studies with constrained versus released collagen gels revealed the importance of ECM mechanics in directing the cell shape of MECs to promote differentiation (Emerman and Pitelka, 1977). MECs plated on constrained collagen gels or gluteraldehyde-crosslinked rBM fail to differentiate in response to lactogenic stimuli and instead spread to form a 2D cell monolayer despite appropriate integrin-ECM ligation and growth factor signaling (reviewed in Roskelley *et al.*, 1995; Weaver and Bissell, 1999). Furthermore, laminin- and proteoglycan-mediated ligation of dystroglycan (DG) has been strongly implicated as the primary mediator of ECM-directed cell shape fate determination in MECs and as a critical component in establishing a continuous BM (Muschler *et al.*, 1999). The hypothesized mechanism seems to depend only on DG's extracellular domain

---

adhesions. Epithelial cells grown on a rigid 2D matrix proliferate readily to form viable polarized cellular monolayers with adherens and tight junctions as well as prominent focal adhesions. Such cells exhibit robust Rho GTPase activation in response to exogenous stimuli, and require activated PI3 kinase or ERK signaling to survive. Under these conditions, epithelial cells do not assemble 3D tissue-like structures or express differentiated proteins in response to "differentiation cues." Mechanical cues (second tier): an epithelial cell interacting with a highly compliant ECM readily adheres using multiple matrix receptors, including integrins, and assembles small immature focal complexes but fails to spread appreciably. Instead cells interacting with a compliant matrix exhibit profound reorganization of their actin cytoskeleton. MECs grown under these conditions can be induced to express lactoferrin if given the correct exogenous soluble cues. 3D ECM and biochemical cues (third tier): epithelial cells interacting with a highly compliant matrix in three dimensions adhere through multiple adhesion receptors including integrins, syndecans, and DG, and proliferate readily in response to exogenous growth factors. MECs interacting with a highly compliant 3D ECM can be induced to express abundant quantities of the differentiation protein β-casein. Multicellular organization, morphogenesis, and tissue differentiation (fourth to sixth tiers): in response to a 3D compliant ECM, ductal epithelial cells begin to interact with one another and assemble multicellular polarized structures with cell–cell junctions including adherens, scribble, and gap junctions. MECs assembled into multicellular 3D-polarized tissue-like structures begin to deposit and assemble an endogenous basally polarized basement membrane, show enhanced expression of milk protein expression such as β-casein, and exhibit enhanced long-term survival and apoptosis resistance to multiple exogenous stimuli including chemotherapeutics, immune receptor activators, and gamma irradiation. Long term culture of epithelial cells in the context of a 3D compliant ECM permits completion of tissue-like morphogenesis characterized by the assembly of an apically and basally polarized, growth-arrested tissue with a cleared central lumen and spatial restriction of various membrane associated proteins including growth factor receptors. Once a fully polarized and growth-arrested structure has formed, mammary acini can now be induced to express additional milk proteins such as WAP in response to lactogenic hormones. However, in response to an increase in matrix stiffness as occurs following chronic inflammation, injury, or tumorigenesis, or following genetic mutations and oncogene activation, tissue integrity becomes progressively compromised reversing the cell state to a less differentiated condition. In extreme cases, cells can behave analogous to undifferentiated, highly contractile single cells.

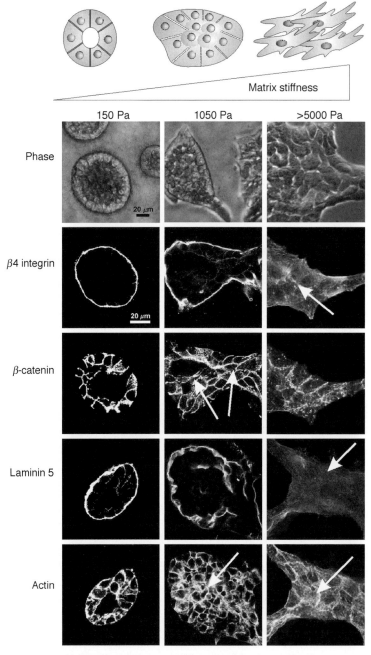

Fig. 2 Matrix stiffness modulates MEC growth and morphogenesis. Phase-contrast and confocal immunofluorescence images of 3D MEC colonies on 3D rBM-crosslinked PA gels of increasing elastic moduli (E = 150–5000 Pa) after 20 days, showing progressively disrupted colony morphology as

and to involve DG binding to laminin, which then polymerizes on the cell surface and onto adjacent DG-expressing cells, ultimately establishing a continuous BM. This process of ligation-driven BM assembly is almost certainly in competition with integrin-based and other adhesive processes. The latter seem more mechanosensitive and might dominate on rigid substrates versus soft substrates.

Although the detailed molecular mechanisms of the mechanosensitivity of MEC differentiation remain to be delineated, recent studies using both nontransformed and transformed human MECs suggest that Rho GTPase-dependent cell contractility regulates adhesion-directed, cell shape-dependent, epithelial tissue-specific functions (Paszek et al., 2005). Transformed human mammary epithelial tumor cells propagated on top of constrained collagen I gels assembled aberrant invasive structures with high Rho and ROCK activity, whereas they could form cell aggregates reminiscent of nontransformed tubules when grown in unconstrained collagen I gels (Keely et al., 1995). In concert with these in vitro observations, transformed mammary tumors were recently shown to exhibit enhanced Rho GTPase activity and exert elevated myosin-dependent cell contractility and aberrant integrin adhesions when compared to nontransformed MECs. Normalizing tumor cell contractility through application of pharmacological inhibitors of Rho, ERK signaling, or myosin could phenotypically revert the malignant phenotype (Paszek et al., 2005). Consistent with a critical role for matrix compliance in epithelial behavior, nontransformed MECs grown within highly compliant rBM gels or nonconstrained and compliant collagen I/rBM gels competently assemble polarized, growth-arrested acini-like structures. However, when grown within constrained collagen I/rBM gels or collagen I/rBM gels of higher concentration and stiffness, they form progressively disrupted, disorganized, and continuously proliferating colonies (Paszek et al., 2005; see also Sections III.A.1 and III.A.2).

Through the application of defined, synthetic, rBM-crosslinked polyacrylamide gels, it was concluded that matrix stiffness and not matrix density or physical presentation constitutes a critical regulator of multicellular epithelial morphogenesis (Paszek et al., 2005; Fig. 2). These studies clearly emphasize the importance of myosin contractility and integrin adhesion maturation as matrix-regulated cell shape and force regulators. They have also identified altered ERK-dependent cell growth and survival, destabilization of cell–cell adhesions, and perturbed matrix assembly, as central mechanisms for further study. Indeed, the proper assembly of an endogenous cell-derived matrix plays a key role in epithelial differentiation, as has been illustrated by the necessity of proper

---

matrix stiffness increases (top). Cell–cell adherens junctions are disrupted and luminal clearance is compromised with even a modest increase in the elastic modulus of the matrix ($E = 1050$ Pa central panel; β-catenin and actin). Basal polarity is perturbed (disorganized $\beta4$ integrin and absence of basally deposited laminin-5) once the matrix stiffness stiffens appreciably ($E > 5000$ Pa; right panel). Scale bar is $20\mu m$. Adapted from Paszek et al. (2005).

laminin–nidogen interactions for mammary tissue differentiation and gene expression in culture (Pujuguet *et al.*, 2000) and for multiple epithelial tissues including the kidney *in vivo* (Willem *et al.*, 2002). Indeed, in lung development, increasing the compliance of the chest wall or decreasing the skeletal muscle fibers that aid in breathing modifies the biophysical properties of the tissue microenvironment by decreasing the applied force to the developing lung, leading to a decrease in lung growth, which further perturbs the tissue ECM and compromises tissue function (reviewed in Liu *et al.*, 1999).

C. 3D Organotypic Model Systems

Key to engineering tissue-specific function is the application of an appropriate ECM in which the biochemical, biophysical, and spatial cues can be defined and controlled. An array of natural ECMs and a growing list of synthetic biomaterials, each with advantages and disadvantages, are available to the experimentalist. Ideally, a comprehensive assessment of what constitutes normal ECM composition, mechanical properties, and organization should be taken into consideration. Unfortunately, our comprehension of these variables has lagged behind, due to the complexity, lack of homogeneity, and anisotropy of biological materials.

rBMs isolated from Engelbreth–Holm–Swarm (EHS) mouse sarcomas have been routinely used to assemble tissue-like structures in culture and have been successfully applied to study mammary, thryoid, salivary gland, lung, and kidney epithelial cell morphogenesis and differentiation, and to distinguish between normal and transformed epithelial cells (Azuma and Sato, 1994; Debnath *et al.*, 2003; Nogawa and Ito, 1995; O'Brien *et al.*, 2001; Petersen *et al.*, 1992; Yap *et al.*, 1995). Similarly, fibrin gels have also been successfully used to assemble 3D normal and transformed tissue-like structures in culture (Alford *et al.*, 1998). However, given that rBM is directly isolated from tissues, the matrix is inherently complex, poorly defined, and subject to complications with lot to lot variability and limitations due to the specific nature of the biochemical and biophysical environment associated with sarcomas. Fibrin gels, while attractive, also suffer from preparation variance. Additionally, fibrin gels are easily proteolyzed by cell-derived MMPs and consequently are not viable for long-term culture experiments. While alternative fibronectin sources that are less proteolytically sensitive have proven useful, these matrices have yet to be routinely applied to epithelial organ culture models.

As an alternative, collagen I gels have been extensively used as a 3D tissue matrix. The application of defined collagen gels to replace the more complex and biologically accurate rBM and fibrin gels has several advantages, including the fact that collagen I is a more biologically defined substrate, is relatively inexpensive to prepare or purchase, and is much more readily available. Because collagen I is the most common protein found in vertebrate animals and is structurally highly conserved, it is generally well tolerated for *in vivo* studies, and multiple cell types readily adhere to this substrate. In addition, the elastic moduli of a collagen I gel can be readily manipulated by varying collagen orientation, fibril crosslinking, concentration, or even biochemical modification or mutation (Christner *et al.*, 2006; Girton *et al.*, 1999; Martin *et al.*, 1996; Roeder *et al.*, 2002), thereby increasing its biological versatility (Elbjeirami *et al.*, 2003; Grinnell, 2003). The magnitude and directional orientation of externally imposed tension can also be easily manipulated with collagen preparations. For example, through the release of collagen gels from the culture vessel, the isometric tension within the gel can be dramatically reduced (Rosenfeldt and Grinnell, 2000). Collagen gels can also be biochemically modified to facilitate epithelial functionality, as for example through the addition of either rBM, purified laminin, or derivatized peptides (Gudjonsson *et al.*, 2002).

Purified, biologically derived materials, such as rBM and collagen I, have an intrinsic amount of biochemical and biophysical variability due to the inherent variability between animals and preparations. This variability leads to inconsistencies between experiments, as well as a high degree of heterogeneity within single gels. Additionally, the dynamic range of elastic moduli that can be reasonably achieved with these systems is limited by biochemical and biophysical constraints of these unique macromolecules. Therefore, although these materials have proven to be useful for clarifying the general influence of matrix on cell and tissue phenotypes, they are not as tractable for defining precise molecular mechanisms mediating mechanotransduction.

To address the issues listed above, especially control over matrix compliance, we and many others use a system first developed by Pelham and Wang (1997) that involves functionalizing synthetic polyacrylamide gels for cell culture (by crosslinking them with precise concentrations of ECM ligands) as 2D model systems for cell spreading, adhesion, and migration. Polyacrylamide gels represent tractable materials to allow studies of molecular pathways and signaling events of cells grown in various mechanical environments. The mechanical properties of these gels, which have been defined using rheology and atomic force microscopy (Engler *et al.*, 2004; Guo *et al.*, 2006; Yeung *et al.*, 2005; Chapter 22 by Engler *et al.*, this volume), can be manipulated by changing the relative concentration of acrylamide and the crosslinker, bis-acrylamide, yielding a system with precisely controlled biochemical and biophysical properties. Polyacrylamide is an exemplary material for studying cell behavior, as it is nonreactive, resistant to nonspecific binding and protein adhesion, and optically clear. The most significant downside to the polyacrylamide gel system is that acrylamide is cytotoxic in its monomeric form, which precludes the extension of its use to 3D cultures in which cells are embedded before polymerization. To overcome this limitation, we have used these polyacrylamide gels to reconstitute 3D conditions by overlaying MECs plated on top of rBM-crosslinked polyacrylamide gels with a blanket layer of rBM. Although the cells undergo normal morphogenesis under these conditions, there are some limitations inherent in this unique technique. Namely, and most importantly, this is neither a true 2D nor a complete 3D system, and the cells behave differently than they do in full 3D cultures (Leight *et al.*, unpublished observations). Although this drawback leads to difficulty in interpretation and definition of these experiments, this system is suitable for approximating the physiological mechanical conditions under which epithelial cells grow and thrive. Alternatively, polyethylene glycol gels combined with bioactive peptides, such as fibronection- and laminin-binding sites, are also attractive biomaterials. However, their 3D organization is significantly different from that found in naturally occurring matrices and *in vivo* in that they typically have a greater matrix density and altered spatial orientation (reviewed in Zhang, 2004). Furthermore, because matrix remodeling is a critical aspect of epithelial morphogenesis, expensive bioactive peptides that can be proteolytically remodeled need to be incorporated into these synthetic biomaterials to permit proper tissue morphogenesis, migration, and to support long-term cell and tissue viability (reviewed in Lutolf and Hubbell, 2005). As an attractive new strategy in the arsenal of synthetic materials, novel matrices that incorporate recombinant natural and synthetic proteins and biopeptides are currently being developed and offer new hope for future applications.

Progress has been made in recapitulating tissue-specific morphology *ex vivo* either for tissue transplantation or for the study of tissue-specific function, but the application of these organotypic model systems to dissect the molecular basis of tissue homeostasis and disease has lagged behind significantly. The failure to exploit current 3D model systems for the study of cell behavior and signaling in the context of a tissue-like microenvironment and structure resides primarily in the lack of appropriate, cost-effective, easy, and reproducible strategies to manipulate, analyze, and assess cell function, signaling, and gene expression in these model systems. We have been studying the effect of cell shape, matrix compliance, adhesion, and dimensionality on cell behavior at the molecular levels, and here we provide a detailed description of the methods we have successfully used to do so.

III. Methods

A. Engineered Cell/Tissue Explants

    1. Natural matrices: rBM

       a. On ice, evenly coat the bottom of a tissue culture plastic dish with ice-cold rBM and incubate the dishes at 37 °C for 10–20 min to permit gel polymerization (see Table I for volume).

       b. Prepare a single-cell suspension of trypsinized/washed cells in log-phase growth, and adjust the final cell concentration to 1×10^6 cells/ml media.

       c. Aliquot cell suspension into individual tubes, adjusting cell number to desired total gel volume [see step (g)].

       d. Centrifuge individual tubes to pellet cells (5 min, $180 \times g$ rcf).

       e. Aspirate the supernatant from the cell pellet, leaving 5% of the media behind.

Table I
Matrix Volume (μl) per Well Size

	Underlay (μl)	Embed (μl)
60 mm	1250	4250
35 mm	450	1400
12-well	200	700
4-well	100	300
24-well	100	300
48-well	50	175
96-well	20	60

    f. Resuspend the cells in the remaining media by vigorously tapping the side of the tube, and place the tube on ice (note: do not vortex).

    g. Add desired volume (Table I) plus an additional 10% of ice-cold rBM, and resuspend cells by gentle pipetting, taking care to avoid bubbles and keep the tube cold.

    h. Transfer the cell/gel solution to the precoated tissue culture chambers, ensuring that the surface is covered with a uniform layer of cell/gel solution. Incubate at 37 °C for 20–30 min to permit gel polymerization.

    i. Gently add complete growth media to the cultures until the gels are fully covered, taking care not to disrupt the gels. Incubate cultures in humidified chambers (37 °C) for desired length of time with media changes every 2–3 days.

2. Natural matrices: collagen I

    a. Prepare collagen/rBM solution following an adapted version of the protocol published by BD (BD Biosciences—Discovery Labware, Catalog No. 354236 Product Specification Sheet).

       i. Place the following on ice: acid-solubilized rat tail collagen I, sterile 10× phosphate-buffered saline (PBS), sterile deionized, distilled water (ddH$_2$O), sterile 1-N sodium hydroxide (NaOH), and an empty sterile tube marked "Final Collagen Solution."

      ii. Calculate the desired volumes required for the experiment. (Note: Prepare 20% extra volume to account for material loss during experimental manipulations. See Table I for suggested total volumes.):

$$\text{Volume of } 10 \times \text{PBS} = \frac{\text{Total final volume}}{10}$$

$$\text{Volume of collagen} = \frac{\text{Total final volume} \times \text{Final collagen concentration}}{\text{Stock collagen concentration}}$$

$$\text{Volume of 1-N NaOH} = \text{Volume of collagen} \times 0\,023$$

$$\text{Volume of ddH}_2\text{O} = \text{Final volume} - \text{Volume of collagen} -$$
$$\text{Volume of } 10\times \text{ PBS} - \text{Volume of 1-N NaOH} - \text{Volume of rBM}$$

iii. To the tube marked "Final Collagen Solution," add the desired volumes of sterile ice-cold $10\times$ PBS, 1-N NaOH, and ddH$_2$O, then mix.

iv. To the tube marked "Final Collagen Solution," add the acid-solubilized collagen [from step (ii)], and mix gently by pipetting several times, taking care to keep the solution ice cold and to minimize air bubble formation. (Note: Do not over mix the gel solution, or the materials properties of the final gel will be altered.)

v. Neutralize the pH to 7.2–7.6 by titrating, drop-wise with 1-N NaOH until the solution turns a slight shade of reddish-purple indicated by the phenol red dye. Mix gently after the addition of each drop.

vi. If desired, add an appropriate amount of rBM to the gel solution and leave on ice until required. The collagen solution can be used immediately or held on ice for 2–3 h.

b. Place the desired tissue culture dish on ice, and coat the bottom of each well with a thin layer of the collagen/rBM gel solution. Incubate the plate at 37 °C for 10–20 min to permit gel polymerization (see Table I for volume).

c. Prepare a single-cell suspension of trypsinized/washed cells in log-phase growth, and adjust the final cell concentration to 1×10^6 cells/ml media.

d. Aliquot cell suspension into individual tubes, adjusting cell number to the desired gel volume [see step (h)].

e. Centrifuge individual tubes to pellet cells (5 min, $180 \times g$ rcf).

f. Aspirate the supernatant from the cell pellet, leaving 5% of the media behind.

g. Resuspend the cells in the remaining media by vigorously tapping the side of the tube and place the tube on ice (note: do not vortex).

h. Add desired volume plus an additional 10% of ice-cold collagen/rBM solution, and resuspend cells by gentle pipetting, taking care to avoid bubbles and to maintain the tube cold.

i. Transfer the cell/gel solution to the precoated tissue culture chambers ensuring that the surface is covered with a uniform layer of cell/gel solution. Incubate at 37 °C for 20–30 min to permit gel polymerization.

j. After polymerization, release the gel from the sides of the dish by running a small sterile spatula around the edge.

   k. Gently add complete growth media to the cultures until the gels are fully covered, taking care not to disrupt the gels. Incubate cultures in humidified chambers (37 °C) for desired length of time with media changes every 2–3 days.

Anticipated results: Because the cells are seeded as single entities, it is possible to monitor the various stages of morphogenesis. Within 24 h all of the cells should be actively dividing, and by 48 h cell doublets should have formed with detectable polarized deposition of laminin-5 ECM protein and E-cadherin and β-catenin localized at cell–cell junctions. By 72–96 h cell proliferation should approach 60–85% [assessed by 5-bromo-2-deoxyuridine (BrdU) incorporation] and basal/apical tissue polarity should be established (determined by basal localization of β4 integrin and basal deposition of laminin-5). Within 10 days, fully embedded MECs within a compliant 3D rBM or collagen/rBM gel should have assembled small, essentially uniform, growth-arrested, polarized acini (Petersen *et al.*, 1992; Weaver *et al.*, 2002). Tissue morphology can be easily assessed by monitoring morphogenesis using immunofluoresence and morphometric assessment markers (Sections III.E.1 and III.E.2; Debnath *et al.*, 2003; Paszek *et al.*, 2005; Weaver *et al.*, 1997). [Note: Cells grown on top of rBM as opposed to those completely embedded tend to form larger and more heterogeneous spheroids and exhibit slightly delayed lumenal clearance and growth arrest dynamics (Leight *et al.*, unpublished observations). In addition, cells embedded within collagen/rBM gels of increasing elastic modulus ($E > 675$ Pa) form larger nonpolarized structures that lack a central lumen (Paszek *et al.*, 2005)].

   3. *Synthetic matrices: functionalized polyacrylamide gels* [Note: For immunofluorescence techniques, small (18–25 mm) coverslips can be used. For total RNA and protein isolation, larger (50 mm) coverslips should be used. The cell number should be optimized for the experiment performed, based on the degree of cell spreading and proliferation anticipated. For example, at least 750,000 cells are needed for total RNA and protein isolation, while significantly less is required for immunofluorescence visualization.]

   a. Flame coverslip quickly and let cool.

   b. Using a cotton swab, evenly and thoroughly coat the coverslip with 0.1-N NaOH. Air dry the coverslips until a filmy coat appears.

   c. Using a p100 pipette tip, spread an even but thin coat of silane onto the surface of the coverslip (refer to Table II for amount). Allow the silane coating to dry (room temperature, 5–10 min), and place the coated coverslips silane-side up in a Petri dish. (Note: Do not incubate longer than 20 min.)

   d. Wash the coverslips thoroughly with ddH$_2$O (minimum 3×; 10 min each), tapping the dish vigorously to remove excess liquid.

   e. Incubate the coverslips in 70% glutaraldehyde (1:140; v:v in PBS) at room temperature for 30 min.

Table II
**Amount of Each Solution to Add During Steps (c), (o), (q), and (s) When
Using the Stated Coverslip Size**

	Silane	Acrylamide solution	rBM solution	Ethanolamine
18-mm circle	20	20	300	300
25-mm circle	30	30	450	450
50-mm circle	100	190	900	900

All values listed are in microliters.

Table III
**Recipes of 1-ml Polyacrylamide Gel Solutions for Given Elastic Modulus
(Yeung *et al.*, 2005)**

Elastic modulus (Pa)	140	400	1050	5000	60,000
Acrylamide (%)	3	3	3	5.5	10
Bis-acrylamide (%)	0.04	0.05	0.1	0.15	0.5
40% Acrylamide (μl)	75	75	75	137.5	250
2% Bis-acrylamide (μl)	20	25	50	75	250
0.5-M HEPES, pH 4.22 (μl)	100	100	100	100	100
TEMED (μl)	0.5	0.5	0.5	0.5	0.5
ddH$_2$O (μl)	648.9	643.9	618.9	541.4	243.9

f. Wash the coverslips thoroughly with ddH$_2$O (minimum 3×; 5–10 min each).

g. Arrange the coverslips face-up to dry.

h. After fully drying, the activated coverslips can be used immediately or stored for several weeks in a dry place. (Note: If the coverslips turn a rust-brown color, they should not be used as this is indicative of excess silane reacting with glutaraldehyde.)

i. In a microcentrifuge tube, mix the solutions required for gel preparation (see Table III for recipe).

j. In another microcentrifuge tube, weigh 5.6 mg of *N*-succinimidyl ester of acrylamidohexanoic acid (N6 crosslinker) per 1 ml of final desired solution. This compound can incorporate into the polyacrylamide gel, rendering it reactive with amine groups of proteins.

k. Add 70 μl of 200 proof ethanol and 80 μl ddH$_2$O (per 1 ml of final solution) to the N6 cross-linker. Briefly, sonicate in a sonicating water bath (average peak power = 45 W), or vortex at highest setting until fully dissolved.

l. Add 844.4 μl of gel solution to the cross-linker/ethanol solution, and vortex.

m. Degas the gel solution using a vacuum flask or chamber for at least 30 min.

n. While the solutions are being degassed, evenly coat an additional set of equivalent-sized coverslips with Rain-x. Allow the Rain-x coating to dry at room temperature for 5–10 min, then gently buff the coverslip using a Kimwipe.

o. Place the activated coverslips face-up on a secured piece of paraffin.

p. Add 10 μl of freshly made 5% ammonium persulphate (w:v in ddH$_2$O) per 1 ml of fully degassed acrylamide solution, mix well, and quickly dispense the desired volume of solution onto each activated coverslip (refer to Table III for volumes). Carefully place the Rain-x treated coverslip on top without trapping air bubbles and allow the gel solution to polymerize at room temperature for 25–60 min. (Note: Gel polymerization is indicated by retraction of the gel from the edge of the coverslip. Do not allow them to set for longer than 60 min, or the gel will dehydrate.)

q. While the gels are polymerizing, prepare a surface amenable to placing the gels on ice by covering the top surface of polystyrene tissue culture dishes with parafilm. (Note: The size is dependent on the size of the coverslips.)

r. Place the parafilm-affixed dishes on ice, and prepare the rBM solution for coating the coverslips. In a prechilled conical tube, prepare a solution of 140-μg/ml rBM, 5-mM EDTA, in 50-mM HEPES buffer, pH 8.0. (Note: Be sure to keep on ice to prevent polymerization of rBM.)

s. After the polyacrylamide gels have fully polymerized, carefully remove the top Rain-x-coated coverslip using a razorblade, taking care not to scratch the gel surface. Rinse each gel with ice-cold ddH$_2$O. If using 18- and 25 mm coverslips, place the coverslips gel-side up on the parafilm-affixed dishes on ice. Immediately dispense the appropriate amount of rBM solution onto the gel (Table III). For 50 mm coverslips, pipette the appropriate amount of rBM solution (Table III) directly onto the paraffin-affixed dishes, and place the rinsed gels face down on top of the solution. Avoid trapping air bubbles under the coverslip. Incubate coverslips on ice for 2 h.

t. While the coverslips are incubating, prepare the ethanolamine solution (1:100 v:v; 50-mM HEPES, pH 8.0) and chill the solution on ice.

u. Following incubation, individually rinse each gel in ice-cold ddH$_2$O and wipe the parafilm-affixed dishes dry. Following the method in step (s), incubate the coverslips with ethanolamine on ice for 30 min to quench the unreacted N6 crosslinker (refer to Table III for volumes).

v. Soak the prepared gels in ice-cold PBS. In a sterile tissue culture hood, move the gels to sterile tissue culture dishes and store for up to 3 days in sterile 2-mM sodium azide/PBS at 4 °C.

w. Prior to cell plating, rinse each gel thoroughly in sterile PBS (minimum 3×), and leave the gels fully immersed in sterile PBS while preparing single-cell suspensions.

x. Prepare a single-cell suspension of trypsinized/washed cells in log-phase growth, and adjust the final cell concentration to 1×10^6 cells/ml media.

y. Aliquot cell suspension into individual tubes, adjusting cell number to desired concentration.

z. Centrifuge individual tubes to pellet cells (5 min, $180 \times g$ rcf).

aa. Aspirate the supernatant from the cell pellet, leaving 5% of the media behind.

bb. Resuspend the cells in the remaining media by vigorously tapping the side of the tube and place the tube on ice (note: do not vortex).

cc. Resuspend the cell suspension in cold growth medium, supplemented with 500-ng/ml fungizone, 50-μg/ml gentamicin sulfate, and 1:100 penicillin/streptomycin (stock concentration: 10,000 units penicillin/ml and 10,000-μg streptomycin/ml).

dd. Pipette the desired cell number onto each gel taking care to ensure an even distribution of cells across the gel surface, and allow the cells to adhere. (Note: The length of time for cells to adhere to the gel surface varies between cell types and needs to be optimized for each experiment.)

ee. To facilitate 3D morphogenesis, after complete cell adhesion (minimum 6–8 h), cover cells in media containing 0.2-mg/ml rBM, and incubate cells for desired number of days. Change the culture media including 0.2-mg/ml rBM, fungizone, gentamicin, and penicillin/streptomycin, the following day and then every other day until termination of experiment.

Anticipated results: After 14 days in culture, MECs plated on top of functionalized polyacrylamide gels with a rBM blanket layer, similar to cells overlaid on rBM gels (Debnath *et al.*, 2003), form larger acini than their counterparts embedded within natural matrices. Analogous to cells grown within a natural matrix or grown on top of rBM with an overlay of rBM, by day 4, cells grown on the rBM PA gels should have acquired detectable cell–cell E-cadherin/β-catenin junctions as well as basal and apical polarity. Thus, by day 4, the cells should be highly proliferative but have acquired basal polarity, detectable by basally localized β4 integrin and deposition of a laminin-5- and collagen IV-rich endogneous BM as well as apically localized cortical actin. Studies have revealed that while the growth rate of MCF10As plated on soft ($E = 140$ Pa) and stiff ($E = 5000$ Pa) rBM-functionalized polyacrylamide gels are similar, cells interacting with a matrix stiffer than 1000 Pa fail to fully growth arrest. We have previously shown that MECs plated with a 3D rBM blanket layer on soft rBM-functionalized polyacrylamide gels undergo normal morphogenesis, while morphogenesis is perturbed in those plated on a stiff gel under the same conditions (Paszek *et al.*, 2005; Fig. 2).

A protocol outlining basic immunofluorescence techniques for each cell culture method is described in the Section III.E, for the visualization of characteristics indicative of morphogenesis. A comparison detailing morphogenetic characteristics of all of the described methods is currently in progress (Leight *et al.*, unpublished observations).

B. Isolation of Bulk Proteins

1. Natural matrices: rBM (Note: This protocol is for isolating proteins from 1-ml rBM gels. Adjust volumes stated if necessary.)

   a. Prepare 25 ml of ice-cold Dulbecco's PBS solution (DPBS) containing 5-mM EDTA (EDTA/DPBS).

   b. Supplement 25% of the EDTA/DPBS solution as prepared above with a cocktail of serine and cysteine protease and tyrosine phosphatase inhibitors. (Note: See Section IV for specific reagents and concentrations.)

   c. Place culture on ice and gently aspirate off the medium.

   d. Add 3 ml of the protease inhibitor/EDTA/DPBS solution prepared in step (b) to the culture dish, and pipette up and down using a p1000 pipette until a uniform suspension is obtained, avoiding the formation of insoluble foam.

   e. Transfer the solubilized rBM gel solution to a 15-ml conical tube on ice.

   f. Repeat steps (d) to (f) once, collecting all of the solubilized rBM into the conical tube.

   g. Angle the conical tubes in a box of ice. Secure the tube and box of ice on a rocker and rock at 4 °C for 45–60 min.

   h. Place 24 nonstick microcentrifuge tubes on ice, and aliquot the cell/rBM solution evenly among the tubes.

   i. Centrifuge tubes at 4 °C for 10 min (3200 × *g* rcf).

   j. Aspirate the supernatant, leaving 5% of the media behind.

   k. Add 500 μl of the EDTA/DPBS solution to one tube, scraping the pellet against the side of the tube to resuspend the pellet. Mix well, and transfer to the next tube. Continue to scrape, mix, and transfer the solution to combine a total of four tubes into one.

   l. Repeat steps (i) to (k) until the original 24 tubes are combined into one.

   m. Centrifuge final tube at 4 °C for 15 min (21,000 × *g* rcf).

   n. Prepare the lysis buffer, supplementing it with a cocktail of serine and cysteine protease and tyrosine phosphatase inhibitors.

   o. Carefully aspirate the supernatant, and resuspend the pellet in 100–300 μl of lysis buffer (see note in Section IV), depending on the size of the pellet and the amount of protein expected.

   p. Incubate on ice for 30 min.

  q. Sonicate the lysis solution on ice with three pulses of 10 sec each, at an output power of 8 W, pausing for 30 sec between each pulse for the sample to cool down.

  r. Centrifuge final tube at 4 °C for 10 min (10,500 × g rcf).

  s. Transfer the supernatant to a clean microcentrifuge tube and fast freeze on dry ice. Store at −80 °C.

2. Natural matrices: collagen I (Note: This protocol is for isolating proteins from 1-ml collagen/rBM gels. Adjust volumes stated if necessary.)

  a. Prepare 10 ml of ice-cold collagen release solution: 2-mg/ml collagenase, 2-mg/ml trypsin, and 5% fetal bovine serum in DMEM:F12. Keep on ice until needed and warm the amount needed just prior to experimentation. (Note: The trypsin should be EDTA-free or cell cadherin junctions will be disrupted.)

  b. Gently aspirate medium from the 3D culture.

  c. Add 2.5 ml of the collagen release solution and 500 μl of full-strength dispase and incubate at 37 °C for 10–15 min. (Note: The dispase should be prewarmed to 37 °C.)

  d. Pipette up and down vigorously using a p1000 pipette to disrupt the gel. Incubate at 37 °C for 10–15 min.

  e. Repeat step (d) until pipetting is easy and the colonies fall freely.

  f. Transfer the solubilized gel solution to a 15-ml conical tube.

  g. Centrifuge tubes for 5 min at 180 × g rcf.

  h. Aspirate the supernatant, leaving 5% of the media behind.

  i. Resuspend the pellet in 5 ml DMEM:F12 with 10% fetal bovine serum. Pellet by centrifugation for 5 min at 180 × g rcf.

  j. Repeat steps (h) and (i) three times to thoroughly wash the pellet.

  k. Aspirate the supernatant, leaving 5% of the media behind.

  l. Resuspend the pellet in 5 ml ice-cold DMEM:F12. Pellet by centrifugation (4 °C, 5 min; 180 rcf).

  m. Repeat steps (k) and (l) three times to thoroughly wash the pellet.

  n. Aspirate the supernatant.

  o. Prepare lysis buffer, supplementing it with a cocktail of serine and cysteine protease and tyrosine phosphatase inhibitors. (Note: See Section IV for specific reagents and concentrations.)

  p. Carefully aspirate the supernatant, and resuspend the pellet in 100–300 μl of lysis buffer, depending on the size of the pellet and the amount of protein expected.

  q. Incubate on ice for 30 min.

  r. Sonicate the lysis solution on ice with three pulses of 10 sec each at an output power of 8 W, pausing for 30 sec between each pulse for the sample to cool down.

s. Centrifuge final tube at 4 °C for 10 min (10,500 × g rcf).

t. Transfer the supernatant to a clean microcentrifuge tube, and fast freeze on dry ice. Store at −80 °C.

3. Synthetic matrices: functionalized polyacrylamide gels (Note: This protocol is for 50-mm polyacrylamide gels.)

a. Prepare the lysis buffer, supplementing it with a cocktail of serine and cysteine protease and tyrosine phosphatase inhibitors. (Note: See Section IV for specific reagents and concentrations.)

b. Aspirate the medium from the gels and rinse with ice-cold DPBS.

c. Aspirate the DPBS from each plate and invert the lids of 60-mm tissue culture dishes onto ice.

d. Remove the coverslip from the plate, and very carefully wipe the edges clean with a cotton swab and/or Kimwipe to remove any cells adhered to the glass or edges of the gel to eliminate cell variability.

e. Pipette 250 μl of lysis buffer into the lid of the tissue culture plate.

f. Place the coverslip gel-side down onto the lysis buffer and incubate on ice for 5 min.

g. With a cell scraper, push the coverslip down and carefully scrape the coverslip against the lid of the culture plate for at least 5 min.

h. Squeeze the excess buffer from underneath the glass and remove the coverslip from the lid.

i. Transfer the solution into a microcentrifuge tube.

j. Incubate on ice for 30 min.

k. Sonicate the lysis solution on ice with three pulses of 10 sec each at an output power of 8 W, pausing for 30 sec between each pulse for the sample to cool down.

l. Centrifuge final tube at 4 °C for 10 min (10,500 × g rcf).

m. Transfer the supernatant to a clean microcentrifuge tube and fast freeze on dry ice. Store at −80 °C.

C. Isolation of Bulk mRNA

1. Natural matrices: rBM

a. Prepare a 3-ml solution/ml of rBM of 4-M guanidine thiocyanate; 25-mM sodium citrate–citric acid, pH 7.0; 0.5% (w:v) N-laurylsarcosine, sodium salt; 100-mM 2-mercaptoethanol. (Note: The 2-mercaptoethanol should be added fresh each time.)

b. Aspirate culture media and add 3 ml guanidine thiocyanate/mercaptoethanol solution as prepared above for each milliliter of rBM.

c. Using a p1000 pipette, pipette the solution up and down to solubilize the rBM.

   d. Transfer the solubilized solution to an RNase-free polypropylene tube.

   e. Add 1/10 volume {note: for steps (e) through (g), 1 volume refers to the sum of the residual culture volume [from step (b)] and the guanidine thiocyanate/mercaptoethanol volume} of DEPC-treated 2-M acetic acid–sodium acetate, pH 4.0, and mix thoroughly by vortexing at the highest setting. (Note: Be sure to obtain a completely homogeneous solution before proceeding to the next step.)

   f. In a fume hood, add 1 volume of 0.1-M citrate, pH 4.3 buffered/saturated phenol, and mix thoroughly by vortexing at the highest setting. (Note: Protective safety attire should be worn. Be sure to obtain a completely homogeneous solution before proceeding to the next step.)

   g. In a fume hood, add 2/10 volume of 49:1 (v:v) chloroform:isoamyl alcohol. Incubate at room temperature for 5–10 min. Then, briskly shake the tube by hand 8–10 times. Do not vortex here. (Note: A milky solution should form.)

   h. Place the tube on ice for at least 15 min. The emulsion should break into two phases at this point, with a clear aqueous layer forming on top of a milky organic layer.

   i. Centrifuge at 4 °C for 30 min (3200 × g rcf) to clarify the upper aqueous phase.

   j. Transfer the upper aqueous phase to a clean RNase-free tube, taking care to avoid disturbing the interface, and add equal volume (the volume of the aqueous phase) of isopropanol, prechilled to −20 °C. Mix well and incubate at −20 °C for 2 h to overnight to precipitate the RNA.

   k. Pellet RNA by centrifugation at 4 °C for 30 min (21,000 × g rcf).

   l. Carefully aspirate the supernatant, being careful not to disturb the loosely adherent pellet.

   m. Add 500 μl of very cold (−20 °C) 75% (v:v) ethanol in DEPC-treated ddH$_2$O, and immediately pellet the total RNA by centrifugation at 4 °C for 15 min (21,000 × g rcf).

   n. Repeat steps (l) and (m) six times to thoroughly wash the pellet and incubate the last wash in −20 °C 75% ethanol overnight.

   o. Pellet total RNA by centrifugation at 4 °C for 15 min (21,000 × g rcf), and carefully aspirate the supernatant.

   p. Air dry the pellet at room temperature until the pellet appears glassy (usually 10–15 min, depending on the volume of residual ethanol).

   q. Dissolve the RNA pellet in DEPC-treated ddH$_2$O. The volume depends on the expected amount of total RNA. (Note: When starting with 250,000 preformed spheroids, the expected RNA yield usually ranges between 10 and 15 μg.) Incubate on ice for 30 min to 1 h to thoroughly solubilize RNA.

   r. Store at −80 °C until required.

    2. Natural matrices: collagen I

      a. Aspirate culture medium. Use a flamed RNase-free razor blade to cut the collagen gel into small pieces for easier RNA extraction, taking care to remove residual medium released prior to gel solubilization.

      b. Add 3 ml of the prepared guanidine thiocyanate/mercaptoethanol solution (as described in C.1) per 1 ml of collagen, and allow to solubilize for 5 min at room temperature on a rocking platform.

      c. Follow steps (C.1.e) to (C.1.r) to extract the total RNA.

    3. Synthetic matrices: functionalized polyacrylamide gels

      a. Prepare a solution of 4-M guanidine thiocyanate; 25-mM sodium citrate–citric acid, pH 7.0; 0.5% (w:v) N-laurylsarcosine, sodium salt; 100-mM 2-mercaptoethanol. (Note: The 2-mercaptoethanol should be added fresh each time.)

      b. Invert the lids of 60-mm tissue culture dishes on the bench top. Add 600 μl of the prepared guanidine thiocyanate/mercaptoethanol solution for each 50 mm polyacrylamide gel to each lid.

      c. Remove the coverslip from the culture dish, and very carefully wipe the edges clean with a cotton swab and/or Kimwipe to remove any cells adhered to the glass or edges of the gel to eliminate cell variability.

      d. Place the coverslip gel-side down onto the guanidine thiocyanate/mercaptoethanol solution, and incubate on ice for 5 min.

      e. With a cell scraper, push the coverslip down and carefully scrape the coverslip against the lid of the culture plate for at least 5 min.

      f. Squeeze excess buffer from underneath the glass and remove the coverslip from the lid.

      g. Transfer the solution to an RNase-free polypropylene tube and follow steps outlined in (C.1.e) to (C.1.r) to extract the total RNA.

D. Rapid Protein Isolation Techniques

    1. Synthetic matrices: functionalized polyacrylamide gels (Note: This protocol is for isolating Rac-GTP. For rapid isolation of other proteins, the assay will be similar, but may need to be modified and/or optimized. At least 600 μg of total protein is needed. Adjust the number and size of gels to obtain enough protein.)

      a. Prepare glutathione-sepharose beads for glutathione-S-transferase-tagged p21-binding domain of Pak1 (GST-PBD) binding.

        i. Centrifuge 1 ml of 50% glutathione-sepharose slurry at 4 °C for 30 sec (21,000 $\times$ g rcf).

        ii. Aspirate the supernatant and add 500 μl MLB (Section IV. D.I). Pellet by centrifugation at 4 °C for 30 sec (21,000 $\times$ g rcf).

      iii. Repeat step (ii) three times to wash the beads.

      iv. Aspirate the supernatant and resuspend the sepharose beads in an equal volume of MLB (500 μl) to produce a 50% slurry.

      v. Incubate the 20- to 30-μl GST-PBD with 20- to 30-μl sepharose slurry (4 °C; 20 min).

b. Supplement MLB with a cocktail of serine and cysteine protease and tyrosine phosphatase inhibitors. (Note: See Section IV for specific reagents and concentrations.)

c. Aspirate the medium from the gels and rinse with ice cold DPBS.

d. Aspirate the DPBS from each plate and invert the lids of 60-mm tissue culture plates onto ice.

e. Remove the coverslip from the plate, and very carefully wipe the edges clean with a cotton swab and/or Kimwipe to remove any cells adhered to the glass or edges of the gel to eliminate cell variability.

f. Pipette 350- to 400-μl MLB into the lid of the tissue culture plate.

g. Place the coverslip gel-side down onto the lysis buffer and incubate on ice for 5 min.

h. With a cell scraper, push the coverslip down and carefully scrape the coverslip against the lid of the culture plate for at least 5 min.

i. Squeeze the excess buffer from underneath the glass and remove the coverslip from the lid.

j. Pipette the MLB solution into a microcentrifuge tube, combining like samples.

k. Centrifuge the tubes at 4 °C for 5 min (10,500 × g rcf).

l. Transfer at least 800 μl of lysate into the tube containing GST-PBD. Leave at least 50 μl of lysate for determining total Rac separately.

m. Gently rock the solution at 4 °C for 60 min.

n. Collect the GST-PBD-Rac mixture by centrifugation at 4 °C for 30 sec (21,000 × g rcf).

o. Carefully aspirate the supernatant and resuspend the beads in 500 μl of ice-cold MLB.

p. Centrifuge at 4 °C for 30 sec (21,000 × g rcf).

q. Repeat steps (o) and (p) three times.

r. Resuspend the beads in 15 μl of sample loading buffer for electrophoresis and vortex briefly. Heat the samples to 95 °C for 10 min. (Note: Visually confirm that the sample loading buffer penetrates the beads before heating.)

s. Centrifuge the solution at room temperature for 30 sec (21,000 × g rcf).

t. Fast freeze the samples in a dry ice/ethanol bath and store at −80 °C.

E. Immunofluorescence

1. Natural matrices: rBM

    a. Aspirate cell culture medium, and wash in DPBS (containing Ca^{2+} and Mg^{2+}), if cultures were grown in serum.

    b. Add equal-volume neutralized collagen solution [see step (A.2.a) for directions, omitting ddH_2O] and mix thoroughly. (Note: Collagen is added to strengthen the matrix and permits easier cryosectioning.)

    c. Incubate the gels at 37 °C for 30 min to polymerize.

    d. If desired, sections can be triton-extracted prior to fixation to facilitate cytoskeletal and nuclear visualization.

        i. Prepare ice-cold cytoskeletal extraction buffer (see Section IV for details) containing Triton X-100 (0.005%; v:v) and 5-mM EGTA, supplemented with protease and phosphatase inhibitors.

        ii. Add equal-volume extraction buffer and incubate at room temperature for 30 min.

    e. Fix with 2% paraformaldehyde (pH 7.4) at 4 °C overnight.

    f. Rinse cultures with PBS/glycine at 4 °C (minimum 3×, 5 min each).

    g. Incubate cultures with 18% sucrose-PBS/glycine at 4 °C for 3 h.

    h. Incubate cultures with 30% sucrose-PBS/glycine at 4 °C for 3 h.

    i. Rinse cultures with PBS/glycine at 4 °C for 5 min.

    j. Add OCT Tissue Tek compound for cryosection, and rapidly freeze on a bed of dry ice and ethanol or in liquid nitrogen. Store culture blocks at −80 °C until required.

    k. Prepare activated gelatin-coated microscope slides:

        i. Autoclave (121 °C, 30 min) 0.5 g gelatin in 25 ml ddH_2O and cool to room temperature.

        ii. Add 0.05 g chromium potassium sulfate dissolved in 75 ml of ddH2O to precooled gelatin solution.

        iii. Store at 4 °C until required.

    l. Prepare microscope slides for tissue culture sections:

        i. To minimize antibody solution requirement, generate a hydrophobic incubation ring on a microscope slide. This can be done using a hydrophobic (wax) pen to draw small rings slightly larger than the tissue samples to be stained. Additionally, this can be achieved by melting paraffin at 95 °C around the circumference of a microcentrifuge tube or the lid of a 15 ml conical tube lid, and then gently, but firmly placing the tube on the slide. Incubate the slide on a heating block at 58 °C for 1 min. Carefully, remove the tube/lid from the slide while the slide is on the heating block, and let the paraffin solidify at room temperature.

     ii. Evenly coat the interior of the paraffin ring with activated 0.5% gelatin, and air dry at room temperature overnight.

m. Using a cryostat, cut frozen sections of 3D tissue blocks (5–20 μm), and transfer sections to the gelatin-coated paraffin ring. Store the slides at −80 °C until required.

n. For immunostaining, remove sections from freezer, thaw, and air dry at room temperature for 5–20 min.

o. Rehydrate sections in IF buffer at room temperature for 20 min.

p. Incubate sections in blocking buffer at room temperature for 1 h or at 4 °C overnight in a humidified chamber.

q. Incubate sections in primary antibody solution at room temperature for 1–2 h or at 4 °C overnight in a humidified chamber.

r. Wash sections in IF buffer at room temperature, minimum three times for 15 min each.

s. Incubate in secondary antibody in IF buffer at room temperature for 45 min. (Note: If the secondary antibody is fluorescent, keep slides under foil.)

t. Wash sections in IF buffer at room temperature, minimum three times for 10 min each.

u. To visualize nuclei, counterstain with 1-μg/ml DAPI in PBS/glycine at room temperature for 5 min.

v. Rinse each gel in PBS/glycine at room temperature, minimum three times for 5 min each.

w. Aspirate residual liquid from the gels and mount sections with mounting media. Leave under foil to dry at room temperature for 15–30 min. Secure with nail polish when dry.

x. Store at −20 °C until visualization.

2. Natural matrices: collagen I

    a. Follow the steps outlined in Section III.E.1 , omitting steps (b) and (c).

3. Synthetic matrices: functionalized polyacrylamide gels

    a. Aspirate cell culture medium and rinse cells grown in serum with DPBS (containing Ca^{2+} and Mg^{2+}).

    b. Fix with 2% paraformaldehyde, pH 7.4 at room temperature for 30 min.

    c. Rinse with IF buffer at room temperature, minimum three times for 5 min each.

    d. With fine tip forceps, remove the coverslip from the cell culture dish, and place the coverslips gel-side up on a secured piece of paraffin.

    e. Follow protocol outlined in steps (Section III.E.1.p) to (Section III. E.1.v).

    f. Aspirate residual liquid from the gels and mount gels onto microscope slides with mounting media. Leave under foil to air dry at room temperature for 15–30 min. Secure with nail polish when dry.

    g. Store at $-20\,^{\circ}$C until visualization.

======== **IV. Materials**

A. Engineering Tissue Explants

1. Natural matrices: rBM

    a. Wet ice

    b. rBM, BD Biosciences BD Matrigel™ [Note: As there is inherent lot-to-lot variability, each lot should be tested prior to use. When deciding which lots to choose, we prefer lots that have endotoxin levels less than 2 units/ml and protein concentrations ranging from 9 to 12 mg/ml. Additionally, we have found that MECs behave similarly in Matrigel and Growth Factor Reduced Matrigel (both from BD), although this should be tested for each cell line. Each lot should be tested for compatibility with the various cell lines by looking for changes in morphology in 3D cultures and ensuring low background nucleic acid and IgG levels that would interfere with RNA isolation and immunofluorescence procedures. It should be noted also that the rheological properties of commercially available rBM preparations can also vary from $E = 50$ to 200 Pa (unpublished observations).]

    c. Cell culture supplies (medium, trypsin, trypsin-inhibiting agent)

2. Natural matrices: collagen I/rBM

    a. Wet ice

    b. Acid-solubilized rat tail collagen I [Note: Although this protocol is designed for acid-solubilized rat tail collagen I, we have previously used acid-solubilized bovine dermal collagen I (ICN Biomedicals Catalog No. 152394) and acid-solubilized rat tail collagen I (BD Labware Catalog No. 354236) to embed MECs in 3D ECM gels. As there is inherent lot variability, each lot should be tested prior to use to determine the appropriate concentration and gelling time for each system. Additionally, the elastic modulus of the gels can be manipulated by varying the concentration of collagen I ($E = 20$–1800 Pa; Leight *et al.*, unpublished observations). We have found that the elastic modulus varies between lots, so each lot should be tested prior to experimentation (unpublished observation).]

    c. $10\times$ DPBS containing 1.33-g/ml Ca^{2+} and 1.0-g/ml Mg^{2+}, supplemented with phenol red.

    d. Deionized, distilled water (ddH$_2$O)

    e. 1-N sodium hydroxide (NaOH)

    f. rBM

    g. Cell culture supplies (medium, trypsin, trypsin-inhibiting agent)

    h. Small sterile spatula

3. Synthetic matrices: functionalized polyacrylamide gels (Note: When possible, presterilized materials should be used.)

    a. Bunsen burner

    b. Coverslips (No. 1 thickness, hydrolytic class 1 borosilicate coverslips, Fisher; see note in Section III regarding the size of the coverslip required.)

    c. Cotton swabs

    d. 0.1-N NaOH

    e. 3-Aminopropyltrimethoxysilane, 97%, Sigma Aldrich

    f. Glutaraldehyde, 70%, Sigma Aldrich

    g. PBS

    h. ddH$_2$O

    i. 40% Acrylamide

    j. 2% Bis-acrylamide

    k. 0.5-M *N*-(2-hydroxyethyl)-piperazine-*N'*-2-ethanesulfonic acid (HEPES buffer), pH 4.22

    l. *N,N,N',N'*-Tetramethylethylenediamine (TEMED)

    m. *N*-Succinimidyl ester of acrylamidohexanoic acid (N6 cross-linker) [Note: The N6 cross-linker can be synthesized following the protocol outlined in Pless *et al.* (1983).]

    n. 200-Proof ethanol

    o. Rain-X™ (available at an automobile parts store).

    p. Parafilm™

    q. Ammonium persulfate (APS)

    r. 50-mM HEPES buffer, pH 8.0

    s. 0.5-M ethylenediaminetetraacetic acid (EDTA), pH 8.0

    t. rBM

    u. Ethanolamine, Sigma Aldrich

    v. Sterile PBS

    w. Sodium azide

    x. Cell culture supplies (medium, trypsin, trypsin-inhibiting agent)

    y. Cell culture antibiotics and antimycotics [fungizone (amphotericin B, Sigma Aldrich), gentamicin sulfate (Gibco™, penicillin G/streptomycin sulfate)]

B. Isolation of Bulk Proteins

1. Natural matrices and synthetic matrices
   a. DPBS (for rBM and functionalized polyacrylamide gels only)
   b. 0.5-M EDTA, pH 8.0 (for rBM and functionalized polyacrylamide gels only)
   c. Collagenase, Roche Applied Sciences (for collagen/rBM gels only)
   d. Trypsin (for collagen/rBM gels only)
   e. Dispase, BD Biosciences (for collagen/rBM gels only)
   f. Fetal bovine serum (for collagen/rBM gels only)
   g. DMEM:F12 (for collagen/rBM gels only)
   h. Wet ice
   i. Serine and cysteine protease and tyrosine phosphatase inhibitor cocktail: 2-μg/ml aprotinin (Roche Applied Sciences), 1-μg/ml leupeptin (Sigma Aldrich), 1-μg/ml E-64 (Sigma Aldrich), 50-mM sodium fluoride, 10-μg/ml pepstatin A (Sigma Aldrich), 0.5-mM benzamidine (Sigma Aldrich), 1-mM sodium orthovanadate, 1-mM Pefabloc SC (Roche Applied Sciences). [Note: Activate 125-mM sodium orthovanadate with 100-mM hydrogen peroxide at room temperature for 20 min just prior to use (Zhang *et al.*, 2005). Add activated sodium orthovanadate and Pefabloc SC to solution just prior to use.]
   j. Appropriate lysis buffer (Note: The choice of lysis buffer is dependent on the nature of the protein to be studied and should be optimized for each experiment. Radioimmunoprecipitation assay (RIPA) and Laemmli buffers are common choices. RIPA buffer enables suitable extraction of cytoplasmic proteins, while Laemmli buffer is effective for membrane-bound and nuclear proteins.)
      i. RIPA buffer: 50-mM Tris–HCl pH 8.0, 50-mM sodium chloride, 0.5% (w:v) sodium deoxycholate, 1% IGEPAL® CA-630 (Sigma Aldrich), 0.1% (w:v) SDS
      ii. Laemmli buffer: 33.3-mM Tris–HCl pH 8.0, 50-mM EDTA, 2% (w:v) SDS
   k. Cell scraper, Sigma Aldrich
   l. Dry ice

C. Isolation of Bulk mRNA

1. Natural matrices and synthetic matrices
   a. Diethyl pyrocarbonate (DEPC)-treated ddH$_2$O (Note: Prepare by adding 1-ml DEPC (Sigma Aldrich) to 500-ml ddH$_2$O in a glass bottle. Mix vigorously, and incubate overnight in a fume hood, leaving the cap slightly loose. Autoclave for 45 min at 121 °C, 20 psig, and store at room temperature.)

b. 4-M guanidine thiocyanate; 25-mM sodium citrate–citric acid, pH 7.0; 0.5% (w:v) *N*-laurylsarcosine (ICN Biomedicals), sodium salt; in DEPC-treated ddH$_2$O

c. 2-Mercaptoethanol

d. DEPC-treated 2-M acetic acid–sodium acetate, pH 4.0 (Note: Prepare by adding 4.022-ml glacial acetic acid to 31-ml ddH$_2$O. Adjust the pH to 4.0 with 2-M sodium acetate. Add 2 μl/ml total volume of DEPC. Mix well and incubate at room temperature overnight. Autoclave for 15 min at 121 °C, 18 psig, and store at room temperature.)

e. 0.1-M citrate, pH 4.3 buffered/saturated phenol, Sigma Aldrich

f. 49:1 (v:v) Chloroform:isoamyl alcohol

g. Wet ice

h. Isopropanol

i. 75% Ethanol in DEPC-treated ddH$_2$O

D. Rapid Protein Isolation Techniques

1. Synthetic matrices: functionalized polyacrylamide gels

   a. MLB: 25-mM HEPES, pH 7.5; 150-mM sodium chloride; 1% Igepal CA-630; 10-mM MgCl$_2$; 1-mM EDTA; 10% glycerol

   b. Serine and cysteine protease and tyrosine phosphatase inhibitor cocktail: 2-μg/ml aprotinin (Roche Applied Sciences), 1-μg/ml leupeptin (Sigma Aldrich), 1-μg/ml E-64 (Sigma Aldrich), 50-mM sodium fluoride, 10-μg/ml pepstatin A (Sigma Aldrich), 0.5-mM benzamidine (Sigma Aldrich), 1-mM sodium orthovanadate, 1-mM Pefabloc SC (Roche Applied Sciences) [Note: Activate 125-mM sodium orthovanadate with 100-mM hydrogen peroxide at room temperature for 20 min just prior to use (Zhang *et al.*, 2005). Add activated sodium orthovanadate and Pefabloc SC to solution just prior to use.]

   c. Sample loading buffer: 0.25-M Tris, pH 6.8; 50% (v:v) glycerol; 2% (w:v) sodium dodecyl sulfate (SDS); 50-μl/ml 2-mercaptoethanol in ddH$_2$O, supplemented with bromophenol blue

   d. Wet ice

   e. Ice-cold DPBS

   f. Glutathione-sepharose slurry, Sigma Aldrich

E. Immunofluorescence

1. Natural matrices and synthetic matrices

   a. DPBS (containing 1.33-g/ml Ca^{2+} and 1.0-g/ml Mg^{2+})

   b. Wet ice

   c. Acid-solubilized rat tail collagen I

  d. 10× DPBS with Ca^{2+} and Mg^{2+}, supplemented with phenol red

  e. ddH_2O

  f. 1-N NaOH

  g. 1.5× Cytoskeletal extraction buffer: 150-mM sodium chloride; 450-mM sucrose; 15-mM piperazine-1,4-bis(2-ethanesulfonic acid) (PIPES) buffer, pH 6.7; 5-mM magnesium chloride in ddH_2O

  h. Triton X-100

  i. Serine and cysteine protease and tyrosine phosphatase inhibitor cocktail: 2-μg/ml aprotinin, 1-μg/ml leupeptin, 1-μg/ml E-64, 50-mM sodium fluoride, 10-μg/ml pepstatin, 0.5-mM benzamidine, 1-mM sodium orthovanadate, 1-mM Pefabloc SC (Note: Activate sodium orthovanadate with hydrogen peroxide just prior to use. Add activated sodium orthovanadate and Pefabloc SC to solution just prior to use.)

  j. 2% (w:v) Paraformaldehyde, pH 7.4

  k. PBS/glycine (1× PBS, supplemented with 75-mg/ml glycine)

  l. Dry ice

  m. 0.5% (w:v) Porcine gelatin (Sigma Aldrich) in ddH_2O, supplemented with 0.5-mg/ml chromium potassium sulfate

  n. Microscope slides

  o. Paraffin

  p. IF buffer: 7.6-mg/ml sodium chloride, 1.9-mg/ml sodium phosphate, 0.4-mg/ml potassium phosphate monobasic, 0.5-mg/ml sodium azide, 1-mg/ml bovine serum albumin, 0.2% (v:v) Triton X-100, 0.05% (v:v) Tween 20 in ddH_2O

  q. Blocking buffer: 10% normal goat serum, 0.13-mg/ml appropriate Fab fragments, in IF buffer

  r. Primary and secondary antibodies

  s. Aluminum foil

  t. 4′,6-Diamidino-2-phenylindole (DAPI), Sigma Aldrich

  u. Mounting medium (e.g., Vectashield Mounting Medium®, Vector)

  v. Nail polish

V. Discussion

Epithelial tissues are highly complex, organized 3D structures that evolve incrementally during development to generate these specialized functional tissues through spatially and temporally controlled stromal–epithelial interactions. The tissue microenvironment of the epithelium is composed of multiple stromal cell types, and these cellular components, together with the epithelium, are embedded within a proteinaceous ECM. It is the combination of cellular and ECM interactions, operating through controlled biochemical and physical cues, that ultimately regulates epithelial cell fate and function. The goal of an epithelial experimentalist is to recreate at least some of the intricate relationships that exist between the various cell types and the ECM *in vivo*, but in a simple format in culture, such that the recreated system is more amenable to molecular studies without severely compromising the epithelial cell's normal tissue behavior. The idea is that experimental observations made using such contrived but simplified systems will ultimately be distilled into the critical information that is necessary to systematically engineer surrogate tissues for replacement therapy or to develop tractable treatments to prevent and cure various diseases. Toward this lofty goal, considerable research has been successfully directed at determining how each individual cell variable and microenvironmental component influences epithelial cell behavior (Mostov *et al.*, 2005; Paszek *et al.*, 2005; Petersen *et al.*, 1992).

Despite the efforts, our understanding of what controls the epithelial cells' behavior within the complex 3D tissue-like structure and how combinations of microenvironmental cues might cooperate to influence epithelial function remains rudimentary at best. Moreover, although we and others have been successful in generating functional data using these "crude" systems, it remains difficult to isolate specific responses to allow the identification of the precise molecular mechanisms linked to the generation of a given tissue phenotype. For example, MECs grown within a 3D rBM simultaneously and acutely change their shape, matrix adhesion, cell contractility, and signaling, as well as growth factor and apoptosis responsiveness, as compared to MECs interacting with a 2D rigid substrate (Debnath *et al.*, 2003; Wang *et al.*, 1998; Weaver *et al.*, 2002). To address this difficulty, tractable culture models that reproducibly reconstruct individual aspects of tissue organization and function and that encompass controllable homotypic and heterotypic cell–cell interactions and ECM cues are needed. Preferably, these newly engineered culture systems will be amenable to precise biochemical and physical manipulation and will be sufficiently robust and versatile for routine experimentation. Additionally, they should be inexpensive and lend themselves to easy and reproducible manipulation.

Conventional organotypic systems are often expensive, labor intensive to generate, and suffer from experimental inconsistencies. It is now feasible to synthesize biocompatible matrices to study the effect of individual parameters, such as matrix binding and ECM orientation, receptor expression and activity, cell shape, matrix compliance, and even ECM dimensionality through a combination of nonreactive hydrogels with cell-adhesive sites. Recombinant synthetic proteins have also been used to promote specific adhesion and to foster cell-specific degradation and remodeling of the matrix by incorporating proteolytically degradable peptide sequences. We and others have applied similar strategies to successfully study the phenotypic behavior of individual cells in response to various physical, architectural, and biochemical cues including issues pertaining to the regulation of cell survival (Buckley *et al.*, 1999; Capello *et al.*, 2006; Chen *et al.*, 1997; Friedland *et al.*, unpublished observations), migration (Gobin and West, 2002; Wong *et al.*, 2003), stem cell fate (McBeath *et al.*, 2004), differentiation (Bokhari *et al.*, 2005; Mauck *et al.*, 2006), and growth regulation (Bokhari *et al.*, 2005; Georges *et al.*, 2006; Paszek *et al.*, 2005). However, such specialized systems have limited application for studying cell behavior in multicellular structures and 3D tissues and have only sparingly been applied to the study of heterotypic cell–cell interactions (Georges *et al.*, 2006). Moreover, many of the currently available synthetic biomaterials exhibit incompatible material properties such as high stiffness, elevated matrix density, and random matrix presentation that render them less than suitable for the study of epithelial tissue morphogenesis (reviewed in Zhang, 2004). To address these concerns, newer generations of biomaterials are currently being developed, including highly compliant synthetic matrices generated using combinations of polyethylene glycol and methylcellulose conjugated with various bioactive peptides and MMP-cleavable proteins (Leach JB, personal communication), polyethylene glycol gels with functionalized recombinant proteins (Rizzi and Hubbell, 2005), electrically spun collagen gels with precisely controlled orientations (Matthews *et al.*, 2002), and synthetic gels with gradients of ECM compliance that recreate durotactic-directed cell migration during development, wound closure, and tumor metastasis (Lo *et al.*, 2000; Wong *et al.*, 2003; Zaari *et al.*, 2004). The application of these novel materials together with the availability of pluripotent and tissue-specific stem cells provide encouragement that we are at least moving closer to our idealized model systems, to begin to elucidate the mechanisms regulating multicellular epithelial tissue-specific structure and function.

In addition to these important considerations, it is recognized that tissues develop progressively and evolve through reciprocal and dynamic dialogues between the cellular and stromal components and tissue milieu, and this temporal relationship must also provide the mature tissue with physiological advantages that need to be identified and assessed. For example, although bioengineers have been able to successfully reconstruct blood vessels that are phenotypically and functionally identical to differentiated arteries *in vivo*, the engineered vessels rapidly fatigue when transplanted into a host *in vivo*. One must also consider that our ultimate goal should be the engineering of complex 3D microenvironments that are amenable to dynamic physical and biochemical modification. When seeded with pluripotent and tissue-specific stem cells, they should allow systematic development *ex vivo* of viable, live tissues to be used for routine and faithful experimentation and for various clinical applications. Clearly, we have our work cut out for us.

Acknowledgments

We thank J. C. Friedland and J. N. Lakins for their contributions. This work was supported by NIH grants CA078731 and BRP HL6438801A1 (to V.M.W.) and T32HL00795404 (to K.R.J.), DOD grants W81XWH-05-1-330 and DAMD17-01-1-0367 (to V.M.W.), and a NSF graduate fellowship (to J.L.L.).

References

Akhtar, N., and Streuli, C. H. (2006). Rac1 links integrin-mediated adhesion to the control of lactational differentiation in mammary epithelia. *J. Cell Biol.* **5,** 781–793.

Alford, D., Baeckstrom, D., Geyp, M., Pitha, P., and Taylor-Papadimitriou, J. (1998). Integrin-matrix interactions affect the form of the structures developing from human mammary epithelial cells in collagen or fibrin gels. *J. Cell Sci.* **111**(Pt. 4), 521–532.

Azuma, M., and Sato, M. (1994). Morphogenesis of normal human salivary gland cells *in vitro*. *Histol. Histopathol.* **4,** 781–790.

Barros, E. J., Santos, O. F., Matsumoto, K., Nakamura, T., and Nigam, S. K. (1995). Differential tubulogenic and branching morphogenetic activities of growth factors: Implications for epithelial tissue development. *Proc. Natl. Acad. Sci. USA* **10,** 4412–4416.

Bissell, M. J., Radisky, D. C., Rizki, A., Weaver, V. M., and Petersen, O. W. (2002). The organizing principle: Microenvironmental influences in the normal and malignant breast. *Differentiation* **9–10,** 537–546.

Bissell, M. J., Weaver, V. M., Lelievre, S. A., Wang, F., Petersen, O. W., and Schmeichel, K. L. (1999). Tissue structure, nuclear organization, and gene expression in normal and malignant breast. *Cancer Res.* **59**(7 Suppl.), 1757–1763s; discussion 1763s–1764s.

Bokhari, M. A., Akay, G., Zhang, S., and Birch, M. A. (2005). The enhancement of osteoblast growth and differentiation *in vitro* on a peptide hydrogel-polyHIPE polymer hybrid material. *Biomaterials* **25,** 5198–5208.

Buckley, S., Driscoll, B., Barsky, L., Weinberg, K., Anderson, K., and Warburton, D. (1999). ERK activation protects against DNA damage and apoptosis in hyperoxic rat AEC2. *Am. J. Physiol.* **1**(Pt. 1), L159–L166.

Capello, A., Krenning, E. P., Bernard, B. F., Breeman, W. A., Erion, J. L., and de Jong, M. (2006). Anticancer activity of targeted proapoptotic peptides. *J. Nucl. Med.* **1,** 122–129.

Chen, C. S., Mrksich, M., Huang, S., Whitesides, G. M., and Ingber, D. E. (1997). Geometric control of cell life and death. *Science* **5317,** 1425–1428.

Chen, C. S., Mrksich, M., Huang, S., Whitesides, G. M., and Ingber, D. E. (1998). Micropatterned surfaces for control of cell shape, position, and function. *Biotechnol. Prog.* **3,** 356–363.

Christner, P. J., Gentiletti, J., Peters, J., Ball, S. T., Yamauchi, M., Atsawasuwan, P., Beason, D. P., Soslowsky, L. J., and Birk, D. E. (2006). Collagen dysregulation in the dermis of the Sagg/+mouse: A loose skin model. *J. Invest. Dermatol.* **3,** 595–602.

Debnath, J., Muthuswamy, S. K., and Brugge, J. S. (2003). Morphogenesis and oncogenesis of MCF-10A mammary epithelial acini grown in three-dimensional basement membrane cultures. *Methods* **3,** 256–268.

Elbjeirami, W. M., Yonter, E. O., Starcher, B. C., and West, J. L. (2003). Enhancing mechanical properties of tissue-engineered constructs via lysyl oxidase crosslinking activity. *J. Biomed. Mater. Res. A.* **3,** 513–521.

Emerman, J. T., and Pitelka, D. R. (1977). Maintenance and induction of morphological differentiation in dissociated mammary epithelium on floating collagen membranes. *In Vitro* **13,** 316–328.

Engler, A., Bacakova, L., Newman, C., Hategan, A., Griffin, M., and Discher, D. (2004). Substrate compliance versus ligand density in cell on gel responses. *Biophys. J.* **1**(Pt. 1), 617–628.

Fuchs, E., Dowling, J., Segre, J., Lo, S. H., and Yu, Q. C. (1997). Integrators of epidermal growth and differentiation: Distinct functions for beta 1 and beta 4 integrins. *Curr. Opin. Genet. Dev.* **5,** 672–682.

Georges, P. C., Miller, W. J., Meaney, D. F., Sawyer, E. S., and Janmey, P. A. (2006). Matrices with compliance comparable to that of brain tissue select neuronal over glial growth in mixed cortical cultures. *Biophys. J.* **8,** 3012–3018.

Girton, T. S., Oegema, T. R., and Tranquillo, R. T. (1999). Exploiting glycation to stiffen and strengthen tissue equivalents for tissue engineering. *J. Biomed. Mater. Res.* **1,** 87–92.

Gobin, A. S., and West, J. L. (2002). Cell migration through defined, synthetic ECM analogs. *FASEB J.* **7,** 751–753.

Green, S. K., Frankel, A., and Kerbel, R. S. (1999). Adhesion-dependent multicellular drug resistance. *Anticancer Drug Des.* **2**, 153–168.

Grinnell, F. (2003). Fibroblast biology in three-dimensional collagen matrices. *Trends Cell Biol.* **5**, 264–269.

Gudjonsson, T., Ronnov-Jessen, L., Villadsen, R., Rank, F., Bissell, M. J., and Petersen, O. W. (2002). Normal and tumor-derived myoepithelial cells differ in their ability to interact with luminal breast epithelial cells for polarity and basement membrane deposition. *J. Cell Sci.* **115**(Pt. 1), 39–50.

Guo, W. H., Frey, M. T., Burnham, N. A., and Wang, Y. L. (2006). Substrate rigidity regulates the formation and maintenance of tissues. *Biophys. J.* **6**, 2213–2220.

Hagios, C., Lochter, A., and Bissell, M. J. (1998). Tissue architecture: The ultimate regulator of epithelial function? *Philos. Trans. R. Soc. Lond., B. Biol. Sci.* **1370**, 857–870.

Hendrix, M. J., Seftor, E. A., Meltzer, P. S., Gardner, L. M., Hess, A. R., Kirschmann, D. A., Schatteman, G. C., and Seftor, R. E. (2001). Expression and functional significance of VE-cadherin in aggressive human melanoma cells: Role in vasculogenic mimicry. *Proc. Natl. Acad. Sci. USA* **14**, 8018–8023.

Huttenlocher, A., Lakonishok, M., Kinder, M., Wu, S., Truong, T., Knudsen, K. A., and Horwitz, A. F. (1998). Integrin and cadherin synergy regulates contact inhibition of migration and motile activity. *J. Cell Biol.* **2**, 515–526.

Huttenlocher, A., Sandborg, R. R., and Horwitz, A. F. (1995). Adhesion in cell migration. *Curr. Opin. Cell Biol.* **5**, 697–706.

Ingber, D. E. (2006). Mechanical control of tissue morphogenesis during embryological development. *Int. J. Dev. Biol.* **2–3**, 255–266.

Jacobson, M. D., Weil, M., and Raff, M. C. (1997). Programmed cell death in animal development. *Cell* **3**, 347–354.

Jeffery, P. K. (2001). Remodeling in asthma and chronic obstructive lung disease. *Am. J. Respir. Crit. Care Med.* **10**(Pt. 2), S28–S38.

Kadoya, Y., and Yamashina, S. (2005). Salivary gland morphogenesis and basement membranes. *Anat. Sci. Int.* **2**, 71–79.

Keely, P. J., Fong, A. M., Zutter, M. M., and Santoro, S. A. (1995). Alteration of collagen-dependent adhesion, motility, and morphogenesis by the expression of antisense alpha 2 integrin mRNA in mammary cells. *J. Cell Sci.* **108**(Pt. 2), 595–607.

Kleinman, H. K., McGarvey, M. L., Hassell, J. R., Star, V. L., Cannon, F. B., Laurie, G. W., and Martin, G. R. (1986). Basement membrane complexes with biological activity. *Biochemistry* **2**, 312–318.

Liu, M., Tanswell, A. K., and Post, M. (1999). Mechanical force-induced signal transduction in lung cells. *Am. J. Physiol.* **4**(Pt. 1), L667–L683.

Lo, C. M., Wang, H. B., Dembo, M., and Wang, Y. L. (2000). Cell movement is guided by the rigidity of the substrate. *Biophys. J.* **1**, 144–152.

Locascio, A., and Nieto, M. A. (2001). Cell movements during vertebrate development: Integrated tissue behaviour versus individual cell migration. *Curr. Opin. Genet. Dev.* **4**, 464–469.

Lutolf, M. P., and Hubbell, J. A. (2005). Synthetic biomaterials as instructive extracellular microenvironments for morphogenesis in tissue engineering. *Nat. Biotechnol.* **1**, 47–55.

Margulis, A., Zhang, W., Alt-Holland, A., Crawford, H. C., Fusenig, N. E., and Garlick, J. A. (2005). E-cadherin suppression accelerates squamous cell carcinoma progression in three-dimensional, human tissue constructs. *Cancer Res.* **5**, 1783–1791.

Martin, R. B., Lau, S. T., Mathews, P. V., Gibson, V. A., and Stover, S. M. (1996). Collagen fiber organization is related to mechanical properties and remodeling in equine bone. A comparison of two methods. *J. Biomech.* **29**(12), 1515–1521.

Matthews, J. A., Wnek, G. E., Simpson, D. G., and Bowlin, G. L. (2002). Electrospinning of collagen nanofibers. *Biomacromolecules* **2**, 232–238.

Mauck, R. L., Yuan, X., and Tuan, R. S. (2006). Chondrogenic differentiation and functional maturation of bovine mesenchymal stem cells in long-term agarose culture. *Osteoarthr. Cartil.* **2**, 179–189.

McBeath, R., Pirone, D. M., Nelson, C. M., Bhadriraju, K., and Chen, C. S. (2004). Cell shape, cytoskeletal tension, and RhoA regulate stem cell lineage commitment. *Dev. Cell* **4**, 483–495.

Mostov, K., Brakeman, P., Datta, A., Gassama, A., Katz, L., Kim, M., Leroy, P., Levin, M., Liu, K., Martin, F., O'Brien, L. E., Verges, M., *et al.* (2005). Formation of multicellular epithelial structures. *Novartis Found. Symp.* **269**, 193–200; discussion 200–205, 223–230.

Muschler, J., Lochter, A., Roskelley, C. D., Yurchenco, P., and Bissell, M. J. (1999). Division of labor among the alpha6beta4 integrin, beta1 integrins, and an E3 laminin receptor to signal morphogenesis and beta-casein expression in mammary epithelial cells. *Mol. Biol. Cell* **9**, 2817–2828.

Nogawa, H., and Ito, T. (1995). Branching morphogenesis of embryonic mouse lung epithelium in mesenchyme-free culture. *Development* **4**, 1015–1022.

Novaro, V., Roskelley, C. D., and Bissell, M. J. (2003). Collagen-IV and laminin-1 regulate estrogen receptor alpha expression and function in mouse mammary epithelial cells. *J. Cell. Sci.* **116**(Pt. 14), 2975–2986.

O'Brien, L. E., Jou, T. S., Pollack, A. L., Zhang, Q., Hansen, S. H., Yurchenco, P., and Mostov, K. E. (2001). Rac1 orientates epithelial apical polarity through effects on basolateral laminin assembly. *Nat. Cell Biol.* **9**, 831–838.

O'Brien, L. E., Zegers, M. M., and Mostov, K. E. (2002). Opinion: Building epithelial architecture: Insights from three-dimensional culture models. *Nat. Rev. Mol. Cell Biol.* **7**, 531–537.

Paszek, M. J., and Weaver, V. M. (2004). The tension mounts: Mechanics meets morphogenesis and malignancy. *J. Mammary Gland Biol. Neoplasia* **4**, 325–342.

Paszek, M. J., Zahir, N., Johnson, K. R., Lakins, J. N., Rozenberg, G. I., Gefen, A., Reinhart-King, C. A., Margulies, S. S., Dembo, M., Boettiger, D., Hammer, D. A., and Weaver, V. M. (2005). Tensional homeostasis and the malignant phenotype. *Cancer Cell.* **3**, 241–254.

Pelham, R. J., Jr., and Wang, Y. (1997). Cell locomotion and focal adhesions are regulated by substrate flexibility. *Proc. Natl. Acad. Sci. USA* **25**, 13661–13665.

Petersen, O. W., Ronnov-Jessen, L., Howlett, A. R., and Bissell, M. J. (1992). Interaction with basement membrane serves to rapidly distinguish growth and differentiation pattern of normal and malignant human breast epithelial cells. *Proc. Natl. Acad. Sci. USA* **19**, 9064–9068.

Pless, D. D., Lee, Y. C., Roseman, S., and Schnaar, R. (1983). Specific cell adhesion to immobilized glycoproteins demonstrated using new reagents for protein and glycoprotein immobilization. *J. Biol. Chem.* **258**, 2340–2349.

Pujuguet, P., Simian, M., Liaw, J., Timpl, R., Werb, Z., and Bissell, M. J. (2000). Nidogen-1 regulates laminin-1-dependent mammary-specific gene expression. *J. Cell. Sci.* **113**(Pt. 5), 849–858.

Rizzi, S. C., and Hubbell, J. A. (2005). Recombinant protein-co-PEG networks as cell-adhesive and proteolytically degradable hydrogel matrixes. Part I: Development and physicochemical characteristics. *Biomacromolecules* **3**, 1226–1238.

Roeder, B. A., Kokini, K., Sturgis, J. E., Robinson, J. P., and Voytik-Harbin, S. L. (2002). Tensile mechanical properties of three-dimensional type I collagen extracellular matrices with varied microstructure. *J. Biomech. Eng.* **2**, 214–222.

Rosenfeldt, H., and Grinnell, F. (2000). Fibroblast quiescence and the disruption of ERK signaling in mechanically unloaded collagen matrices. *J. Biol. Chem.* **5**, 3088–3092.

Roskelley, C. D., Srebrow, A., and Bissell, M. J. (1995). A hierarchy of ECM-mediated signalling regulates tissue-specific gene expression. *Curr. Opin. Cell Biol.* **5**, 736–747.

Sethi, T., Rintoul, R. C., Moore, S. M., MacKinnon, A. C., Salter, D., Choo, C., Chilvers, E. R., Dransfield, I., Donnelly, S. C., Strieter, R., and Haslett, C. (1999). Extracellular matrix proteins protect small cell lung cancer cells against apoptosis: A mechanism for small cell lung cancer growth and drug resistance *in vivo*. *Nat. Med.* **6**, 662–668.

Springer, T. A., and Wang, J. H. (2004). The three-dimensional structure of integrins and their ligands, and conformational regulation of cell adhesion. *Adv. Protein Chem.* **68**, 29–63.

Stegemann, J. P., Hong, H., and Nerem, R. M. (2005). Mechanical, biochemical, and extracellular matrix effects on vascular smooth muscle cell phenotype. *J. Appl. Physiol.* **6**, 2321–2327.

Tan, J. L., Tien, J., Pirone, D. M., Gray, D. S., Bhadriraju, K., and Chen, C. S. (2003). Cells lying on a bed of microneedles: An approach to isolate mechanical force. *Proc. Natl. Acad. Sci. USA* **4,** 1484–1489.

Thornberry, N. A., and Lazebnik, Y. (1998). Caspases: Enemies within. *Science* **5381,** 1312–1316.

Wang, F., Weaver, V. M., Petersen, O. W., Larabell, C. A., Dedhar, S., Briand, P., Lupu, R., and Bissell, M. J. (1998). Reciprocal interactions between beta1-integrin and epidermal growth factor receptor in three-dimensional basement membrane breast cultures: A different perspective in epithelial biology. *Proc. Natl. Acad. Sci. USA* **25,** 14821–14826.

Wang, H. B., Dembo, M., Hanks, S. K., and Wang, Y. (2001). Focal adhesion kinase is involved in mechanosensing during fibroblast migration. *Proc. Natl. Acad. Sci. USA* **20,** 11295–11300.

Wang, W., Goswami, S., Sahai, E., Wyckoff, J. B., Segall, J. E., and Condeelis, J. S. (2005). Tumor cells caught in the act of invading: Their strategy for enhanced cell motility. *Trends Cell Biol.* **3,** 138–145.

Weaver, V. M., and Bissell, M. J. (1999). Functional culture models to study mechanisms governing apoptosis in normal and malignant mammary epithelial cells. *J. Mammary Gland Biol. Neoplasia* **2,** 193–201.

Weaver, V. M., Fischer, A. H., Peterson, O. W., and Bissell, M. J. (1996). The importance of the microenvironment in breast cancer progression: Recapitulation of mammary tumorigenesis using a unique human mammary epithelial cell model and a three-dimensional culture assay. *Biochem. Cell Biol.* **6,** 833–851.

Weaver, V. M., Lelievre, S., Lakins, J. N., Chrenek, M. A., Jones, J. C., Giancotti, F., Werb, Z., and Bissell, M. J. (2002). Beta4 integrin-dependent formation of polarized three-dimensional architecture confers resistance to apoptosis in normal and malignant mammary epithelium. *Cancer Cell* **3,** 205–216.

Weaver, V. M., Petersen, O. W., Wang, F., Larabell, C. A., Briand, P., Damsky, C., and Bissell, M. J. (1997). Reversion of the malignant phenotype of human breast cells in three-dimensional culture and *in vivo* by integrin blocking antibodies. *J. Cell Biol.* **1,** 231–245.

Willem, M., Miosge, N., Halfter, W., Smyth, N., Jannetti, I., Burghart, E., Timpl, R., and Mayer, U. (2002). Specific ablation of the nidogen-binding site in the laminin gamma1 chain interferes with kidney and lung development. *Development* **11,** 2711–2722.

Wong, J. Y., Velasco, A., Rajagopalan, P., and Pham, Q. (2003). Directed movement of vascular smooth muscle cells on gradient-compliant hydrogels. *Langmuir* **19,** 1908–1913.

Wozniak, M. A., Desai, R., Solski, P. A., Der, C. J., and Keely, P. J. (2003). ROCK-generated contractility regulates breast epithelial cell differentiation in response to the physical properties of a three-dimensional collagen matrix. *J. Cell Biol.* **3,** 583–595.

Yamada, K. M., Pankov, R., and Cukierman, E. (2003). Dimensions and dynamics in integrin function. *Braz. J. Med. Biol. Res.* **8,** 959–966.

Yap, A. S., Stevenson, B. R., Keast, J. R., and Manley, S. W. (1995). Cadherin-mediated adhesion and apical membrane assembly define distinct steps during thyroid epithelial polarization and lumen formation. *Endocrinology* **10,** 4672–4680.

Yeung, T., Georges, P. C., Flanagan, L. A., Marg, B., Ortiz, M., Funaki, M., Zahir, N., Ming, W., Weaver, V., and Janmey, P. A. (2005). Effects of substrate stiffness on cell morphology, cytoskeletal structure, and adhesion. *Cell Motil. Cytoskeleton* **1,** 24–34.

Zaari, N., Rajagopalan, P., Kim, S. K., Engler, A. J., and Wong, J. Y. (2004). Photopolymerization in microfluidic gradient generators: Microscale control of substrate compliance to manipulate cell response. *Adv. Mat.* **23–24,** 2133–2137.

Zahir, N., Lakins, J. N., Russell, A., Ming, W., Chatterjee, C., Rozenberg, G. I., Marinkovich, M. P., and Weaver, V. M. (2003). Autocrine laminin-5 ligates alpha6beta4 integrin and activates RAC and NFkappaB to mediate anchorage-independent survival of mammary tumors. *J. Cell Biol.* **6,** 1397–1407.

Zahir, N., and Weaver, V. M. (2004). Death in the third dimension: Apoptosis regulation and tissue architecture. *Curr. Opin. Genet. Dev.* **1,** 71–80.

Zhang, S. (2004). Beyond the Petri dish. *Nat. Biotechnol.* **2,** 151–152.

Zhang, X., Huang, J., and McNaughton, P. A. (2005). NGF rapidly increases membrane expression of TRPV1 heat-gated ion channels. *EMBO J.* **24,** 4211–4223.

Zink, D., Fischer, A. H., and Nickerson, J. A. (2004). Nuclear structure in cancer cells. *Nat. Rev. Cancer* **9,** 677–687.

INDEX

VOLUMES IN SERIES

Founding Series Editor
DAVID M. PRESCOTT

Volume 1 (1964)
Methods in Cell Physiology
Edited by David M. Prescott

Volume 2 (1966)
Methods in Cell Physiology
Edited by David M. Prescott

Volume 3 (1968)
Methods in Cell Physiology
Edited by David M. Prescott

Volume 4 (1970)
Methods in Cell Physiology
Edited by David M. Prescott

Volume 5 (1972)
Methods in Cell Physiology
Edited by David M. Prescott

Volume 6 (1973)
Methods in Cell Physiology
Edited by David M. Prescott

Volume 7 (1973)
Methods in Cell Biology
Edited by David M. Prescott

Volume 8 (1974)
Methods in Cell Biology
Edited by David M. Prescott

Volume 9 (1975)
Methods in Cell Biology
Edited by David M. Prescott

Volume 10 (1975)
Methods in Cell Biology
Edited by David M. Prescott

Volume 11 (1975)
Yeast Cells
Edited by David M. Prescott

Volume 12 (1975)
Yeast Cells
Edited by David M. Prescott

Volume 13 (1976)
Methods in Cell Biology
Edited by David M. Prescott

Volume 14 (1976)
Methods in Cell Biology
Edited by David M. Prescott

Volume 15 (1977)
Methods in Cell Biology
Edited by David M. Prescott

Volume 16 (1977)
Chromatin and Chromosomal Protein Research I
Edited by Gary Stein, Janet Stein, and Lewis J. Kleinsmith

Volume 17 (1978)
Chromatin and Chromosomal Protein Research II
Edited by Gary Stein, Janet Stein, and Lewis J. Kleinsmith

Volume 18 (1978)
Chromatin and Chromosomal Protein Research III
Edited by Gary Stein, Janet Stein, and Lewis J. Kleinsmith

Volume 19 (1978)
Chromatin and Chromosomal Protein Research IV
Edited by Gary Stein, Janet Stein, and Lewis J. Kleinsmith

Volume 20 (1978)
Methods in Cell Biology
Edited by David M. Prescott

Advisory Board Chairman
KEITH R. PORTER

Volume 21A (1980)
Normal Human Tissue and Cell Culture, Part A: Respiratory, Cardiovascular, and Integumentary Systems
Edited by Curtis C. Harris, Benjamin F. Trump, and Gary D. Stoner

Volume 21B (1980)
Normal Human Tissue and Cell Culture, Part B: Endocrine, Urogenital, and Gastrointestinal Systems
Edited by Curtis C. Harris, Benjamin F. Trump, and Gray D. Stoner

Volume 22 (1981)
Three-Dimensional Ultrastructure in Biology
Edited by James N. Turner

Volume 23 (1981)
Basic Mechanisms of Cellular Secretion
Edited by Arthur R. Hand and Constance Oliver

Volume 24 (1982)
The Cytoskeleton, Part A: Cytoskeletal Proteins, Isolation and Characterization
Edited by Leslie Wilson

Volume 25 (1982)
The Cytoskeleton, Part B: Biological Systems and *In Vitro* Models
Edited by Leslie Wilson

Volume 26 (1982)
Prenatal Diagnosis: Cell Biological Approaches
Edited by Samuel A. Latt and Gretchen J. Darlington

Series Editor
LESLIE WILSON

Volume 27 (1986)
Echinoderm Gametes and Embryos
Edited by Thomas E. Schroeder

Volume 28 (1987)
***Dictyostelium discoideum:* Molecular Approaches to Cell Biology**
Edited by James A. Spudich

Volume 29 (1989)
Fluorescence Microscopy of Living Cells in Culture, Part A: Fluorescent Analogs, Labeling Cells, and Basic Microscopy
Edited by Yu-Li Wang and D. Lansing Taylor

Volume 30 (1989)
Fluorescence Microscopy of Living Cells in Culture, Part B: Quantitative Fluorescence Microscopy—Imaging and Spectroscopy
Edited by D. Lansing Taylor and Yu-Li Wang

Volume 31 (1989)
Vesicular Transport, Part A
Edited by Alan M. Tartakoff

Volume 32 (1989)
Vesicular Transport, Part B
Edited by Alan M. Tartakoff

Volume 33 (1990)
Flow Cytometry
Edited by Zbigniew Darzynkiewicz and Harry A. Crissman

Volume 34 (1991)
Vectorial Transport of Proteins into and across Membranes
Edited by Alan M. Tartakoff

Selected from Volumes 31, 32, and 34 (1991)
Laboratory Methods for Vesicular and Vectorial Transport
Edited by Alan M. Tartakoff

Volume 35 (1991)
Functional Organization of the Nucleus: A Laboratory Guide
Edited by Barbara A. Hamkalo and Sarah C. R. Elgin

Volume 36 (1991)
***Xenopus laevis:* Practical Uses in Cell and Molecular Biology**
Edited by Brian K. Kay and H. Benjamin Peng

Series Editors
LESLIE WILSON AND PAUL MATSUDAIRA

Volume 37 (1993)
Antibodies in Cell Biology
Edited by David J. Asai

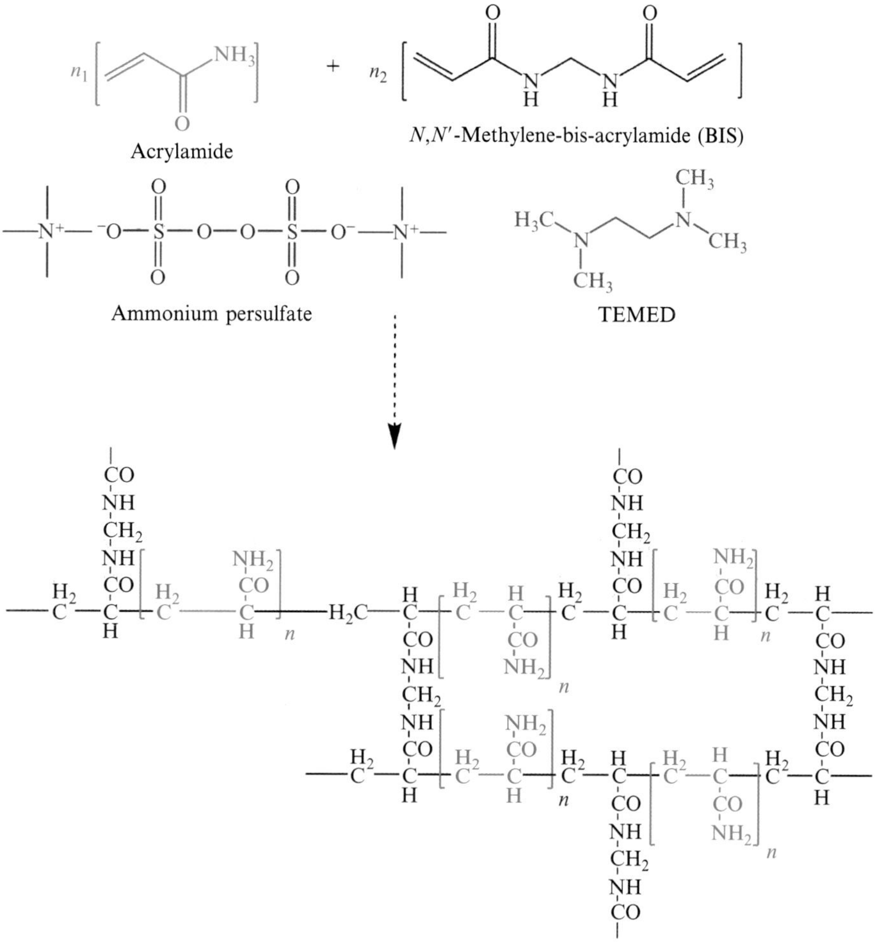

Plate 1 (Figure 2.1)

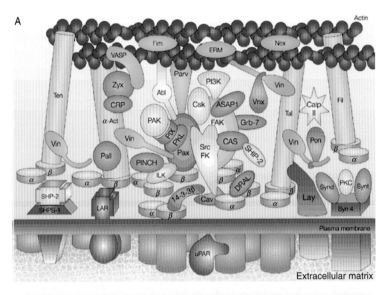

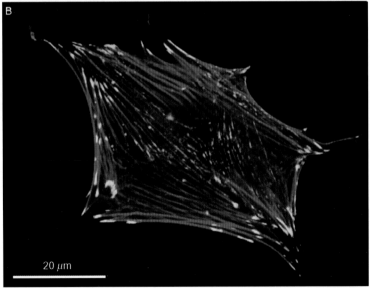

Plate 2 (Figure 5.2)

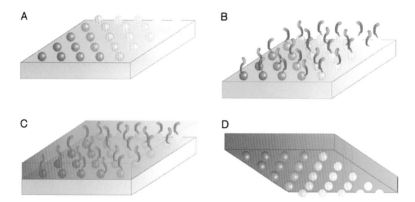

Plate 3 (Figure 5.6)

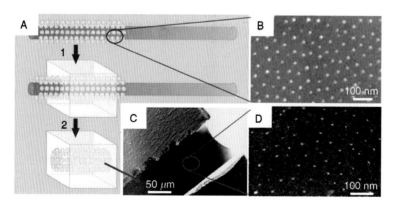

Plate 4 (Figure 5.8)

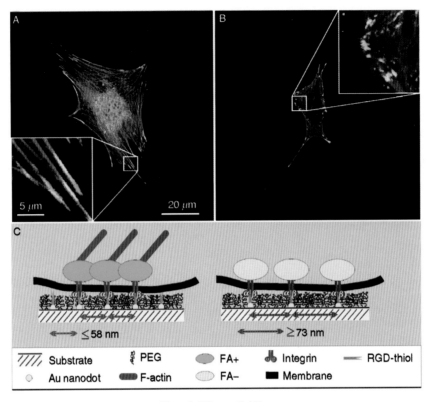

Plate 5 (Figure 5.12)

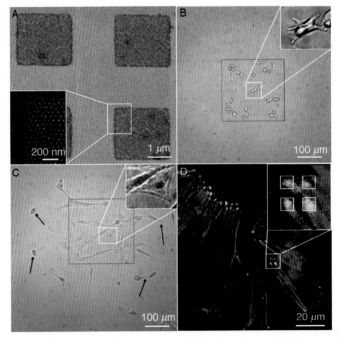

Plate 6 (Figure 5.13)

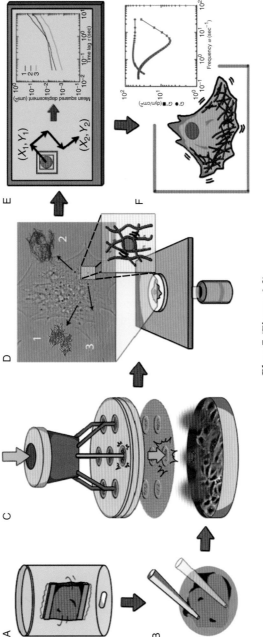

Plate 7 (Figure 6.2)

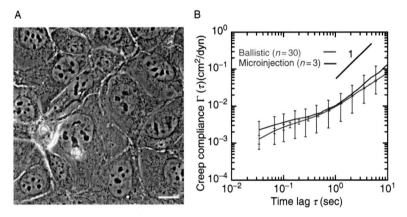

Plate 8 (Figure 6.3)

A

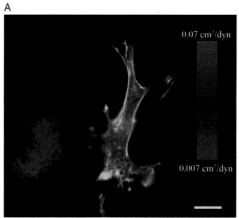

0.07 cm²/dyn

0.007 cm³/dyn

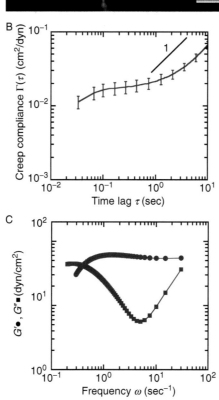

Plate 9 (Figure 6.4)

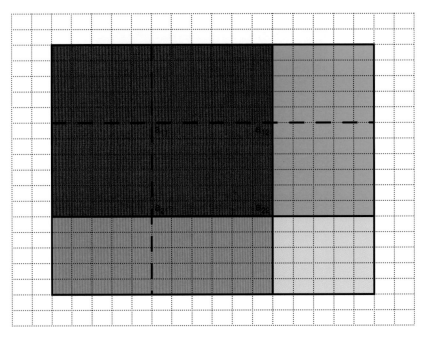

Plate 10 (Figure 8.2)

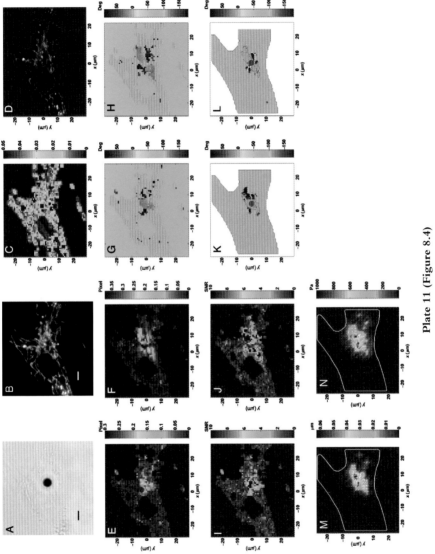

Plate 11 (Figure 8.4)

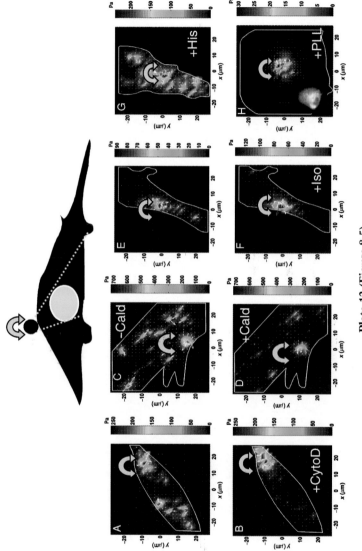

Plate 12 (Figure 8.5)

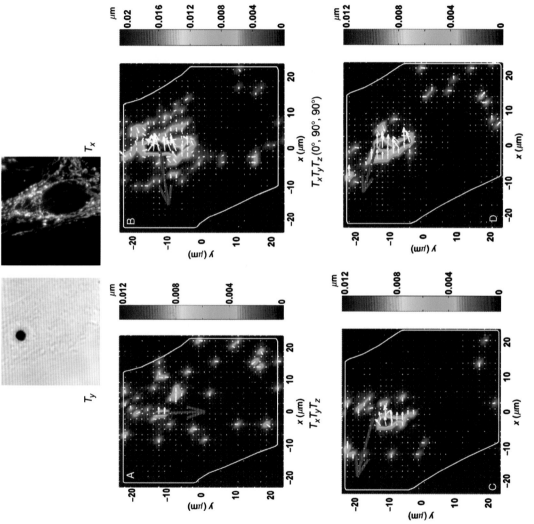

Plate 13 (Figure 8.8)

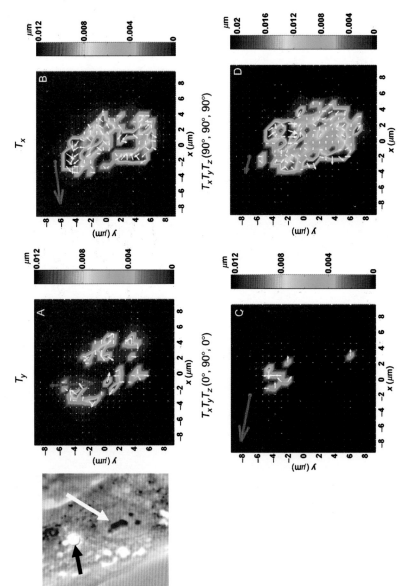

Plate 14 (Figure 8.9)

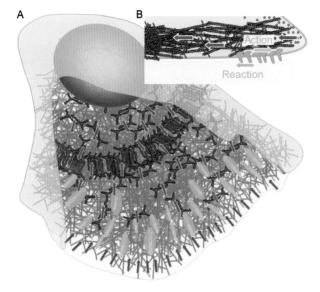

Plate 15 (Figure 9.1)

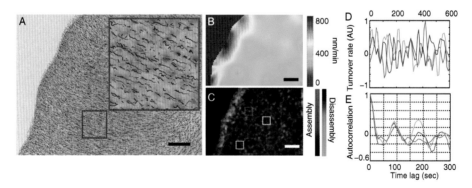

Plate 16 (Figure 9.4)

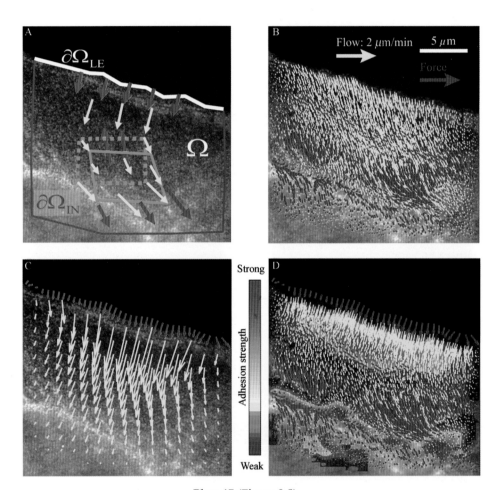

Plate 17 (Figure 9.5)

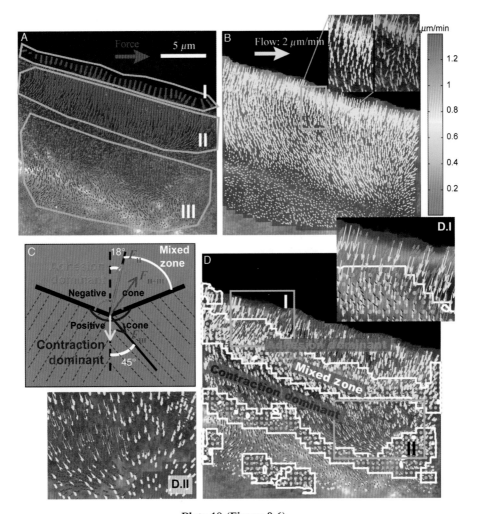

Plate 18 (Figure 9.6)

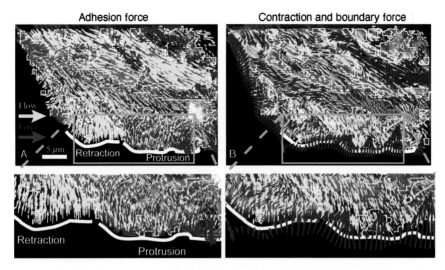

Plate 19 (Figure 9.7)

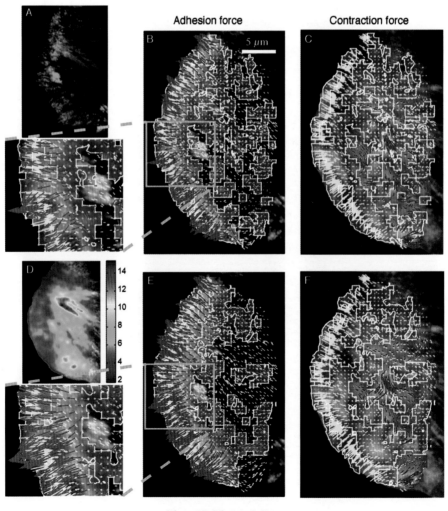

Plate 20 (Figure 9.8)

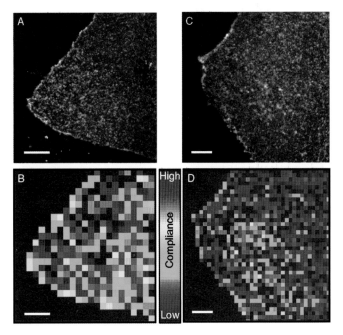

Plate 21 (Figure 9.10)

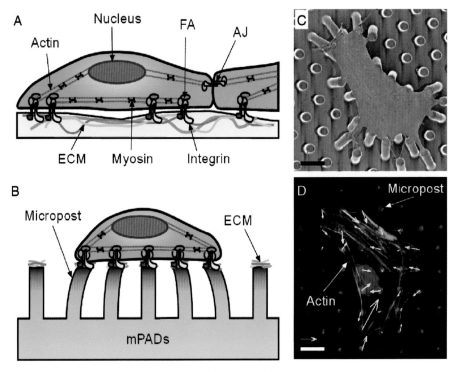

Plate 22 (Figure 13.1)

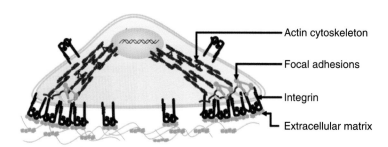

- Actin cytoskeleton
- Focal adhesions
- Integrin
- Extracellular matrix

Plate 23 (Figure 14.1)

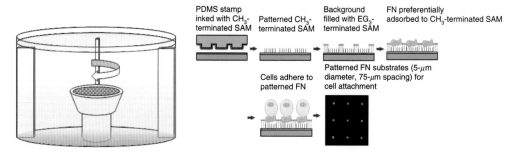

PDMS stamp inked with CH$_3$-terminated SAM

Patterned CH$_3$-terminated SAM

Background filled with EG$_3$-terminated SAM

FN preferentially adsorbed to CH$_3$-terminated SAM

Patterned FN substrates (5-μm diameter, 75-μm spacing) for cell attachment

Cells adhere to patterned FN

Plate 24 (Figure 14.2)

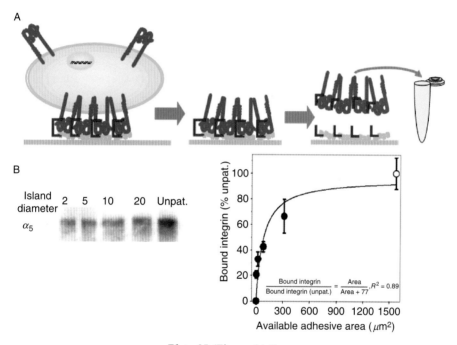

Plate 25 (Figure 14.5)

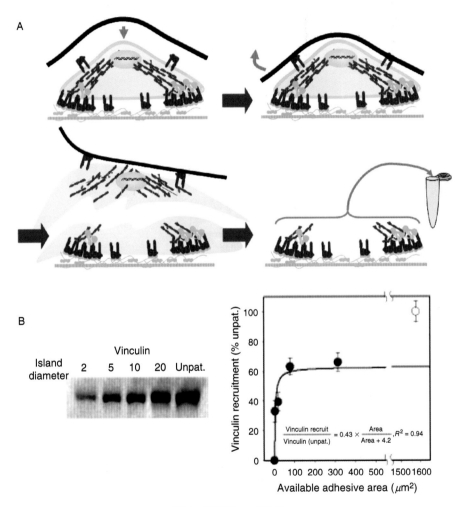

A

B

Island diameter   Vinculin

2   5   10   20   Unpat.

$$\frac{\text{Vinculin recruit}}{\text{Vinculin (unpat.)}} = 0.43 \times \frac{\text{Area}}{\text{Area} + 4.2}, R^2 = 0.94$$

Vinculin recruitment (% unpat.)

Available adhesive area (μm^2)

Plate 26 (Figure 14.6)

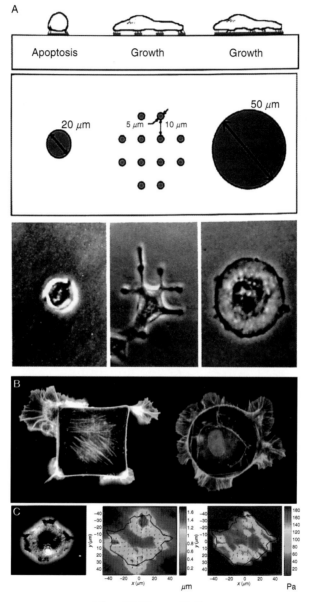

Plate 27 (Figure 19.3)

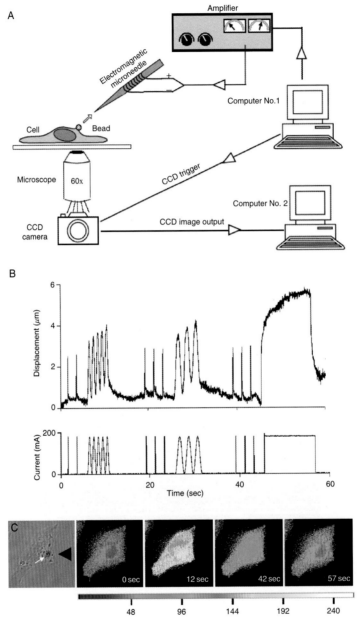

Plate 28 (Figure 19.7)

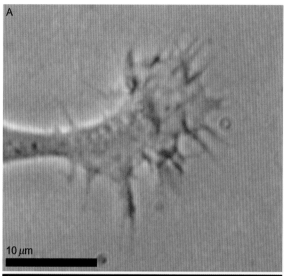

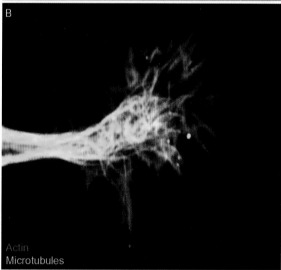

Plate 29 (Figure 21.1)

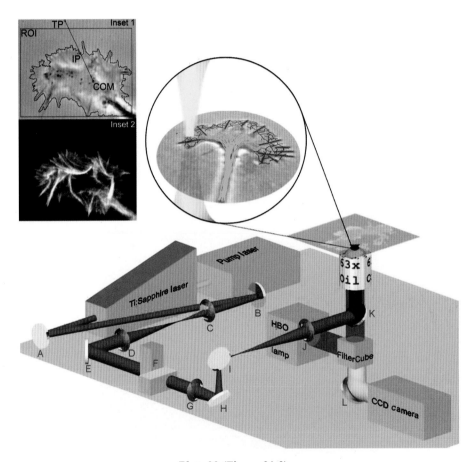

Plate 30 (Figure 21.2)

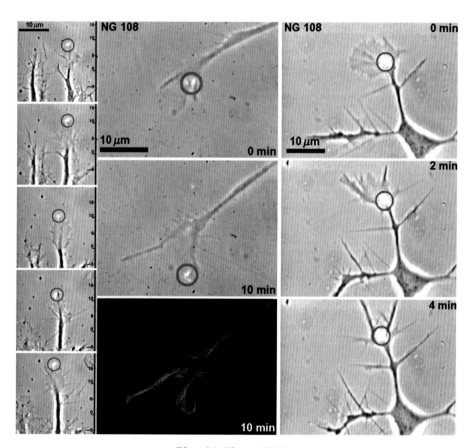

Plate 31 (Figure 21.7)

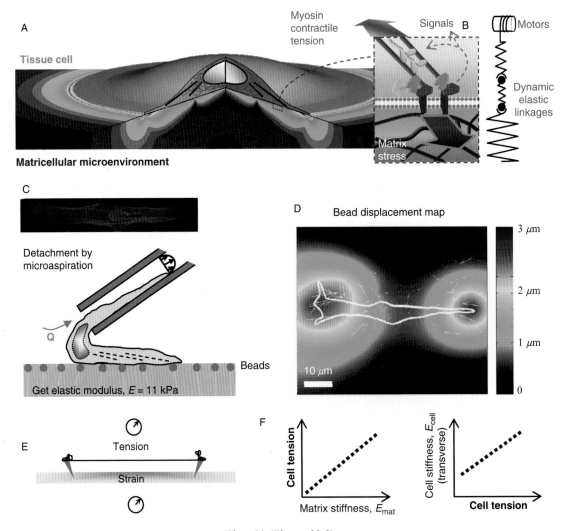

A

Tissue cell

Matricellular microenvironment

Myosin contractile tension

Signals B

Motors

Dynamic elastic linkages

Matrix stress

C

Detachment by microaspiration

Q

Beads

Get elastic modulus, E = 11 kPa

D

Bead displacement map

3 μm

2 μm

1 μm

0

10 μm

E

Tension

Strain

F

Cell tension

Matrix stiffness, E_{mat}

Cell stiffness, E_{cell} (transverse)

Cell tension

Plate 32 (Figure 22.2)